P9-EDF-140

WITHDRAWN
UTSA LIBRARIES

IEEE Press Series on Digital & Mobile Communication

s Digital and Mobile Communication Series is written for research and
ngineers and graduate students in communication engineering. The
reless and personal communication fields receive special emphasis. Books
es, graduate texts and the latest monographs about theory and practice.

John B. Anderson, *Series Editor*
Ericsson Professor of Digital Communication
Lund University, Sweden

Books in the IEEE Press Series on Digital & Mobile Communication

B. Anderson, *Digital Transmission Engineering, Second Edition*

Johannesson and Kamil Sh. Zigangirov, *Fundamentals of Convolutional Coding*

Pandya, *Mobile and Personal Communication Systems and Services*

os Hanzo, P. J. Cherriman, and J. Streit, *Video Compression & Communications over Wireless Channels: Second to Third Generation Systems and Beyond*

jos Hanzo, F. Clare, A. Somerville and Jason P. Woodard, *Voice Compression and Communications: Principles and Applications for Fixed and Wireless Channels*

ansoor Shafi, Shigeaki Ogose and Takeshi Hattori (Editors), *Wireless Communications in the 21st Century*

aj Pandya, *Introduction to WLLs: Application and Development for Fixed or Broadband Services*

Christian Schlegel and Lance Perez, *Trellis and Turbo Coding*

Kamil Zigangirov, *Theory of Code Divison Multiple Access Communication*

DIGITAL TRANSMIS
ENGINEERING

The IEEE Pres
development e
burgeoning w
are of two typ

DIGITAL TRANSMISSION ENGINEERING

Second Edition

JOHN B. ANDERSON
Lund University

John B. Anderson, *Series Editor*

IEEE PRESS

A JOHN WILEY & SONS, INC., PUBLICATION

IEEE Press
445 Hoes Lane
Piscataway, NJ 08854

For general information on our other products and services please contact our Customer Care Department within the U.S. at 877-762-2974, outside the U.S. at 317-572-3993 or fax 317-572-4002.

Wiley also publishes its books in a variety of electronic formats. Some content that appears in print, however, may not be available in electronic format. For information about Wiley products, visit our web site at www.wiley.com.

Library of Congress Cataloging-in-Publication Data is available.

ISBN-13 978-0-471-67910-3
ISBN-10 0-471-46964-9

Printed in the United States of America.

10 9 8 7 6 5 4 3 2 1

CONTENTS

PREFACE TO THE FIRST EDITION

This book is about the transmission of digital data through time and space, as it is practiced in the sophisticated systems of today. Since the ideas behind today's designs can be subtle, theory and concepts play a major role, but engineering practices and judgments are an essential part of the story as well and need equal prominence. This book seeks to combine a clear conceptual grounding with an engineering attitude.

Digital transmission is not an elementary subject. The author remembers learning both digital communication theory and computer programming in the mid-1960s, and being told that these were hard subjects that could not be taught but rather must simply be absorbed. Both were new at the time. Since then we have learned how to teach them, fortunately. Programming turned out to be rather easy, and as it turned out, children could learn it and even get into mischief with it. Digital transmission, on the other hand, is still an advanced discipline. After 30 or more years, it is accessible to the same persons it was in the 1960s, namely, students and practicing engineers with at least 4–5 years' training.

Especially in the United States, where undergraduate education ends a year earlier than in Europe, there has been a trend to bring digital communication down into the undergraduate years. No doubt this is possible for some of the simpler ideas, but the details of a CD player or a cellular phone are too subtle and complex to be grasped by an engineer without significant training. Reed–Solomon codes and maximum-likelihood sequence estimation are not part of ordinary experience. The aim of this book cannot be to dilute these subjects, but rather to explain them in the simplest and most efficient way.

There is a natural evolution in any new technical subject, as it moves away from the half-understood discoveries of a few researchers and inventors. If the subject is important enough, it will move from journal articles and patents to research monographs and early textbooks. These passages are difficult and time-consuming. Discoveries do not come easily, and finding any pedagogical explanation at all, let alone a good one, is a stiff challenge. Textbooks in the engineering of digital communication, as opposed to pioneering monographs, began to appear in the 1975–1985 period. We are now well past the early textbook stage. Digital communication is mature and contains relatively few mysteries. It is time to think about what are the best explanations of the subject, and hopefully some of this has found its way into this book. Our material on baseband pulses and carrier modulation has been completely recast in order to achieve a simpler exposition and introduce a stronger engineering flavor. Two critical parts of a real communication link are the synchronizers and the physical nature of the channel. These are half or more of any complete design and they each have an extensive chapter here, for the first time in a book at this level. Our treatment of equalization is a complete reworking of existing material.

The material in Chapters 1–5 is the basis of the first-year graduate introduction to digital communication taught for some years by the author. It covers pulse and carrier modulation in detail, interference and distortion, synchronization up to the network level, and the engineering details of channels, antennas, and propagation. Additional sections on error control coding, mobile channels, and some advanced techniques make up Chapters 6 and 7 and Section 5.4. These have been included for those who have the need and the space for them.

The prerequisites for this book are good courses in probability and transform and system theory. A course in analog communication is helpful. We have taken pains to avoid significant use of stochastic processes, although that subject provides many valuable insights and is a prerequisite for further study in fields such as detection theory.

We have sought to avoid the encyclopedic approach of some books, which cover more details in many more pages but are barely legible to the student. These books have their place, but it is not at the first-year level.[1] We seek to explain a limited number of topics fully in a book of moderate size. Many important but smaller topics are left out, and we mean no disrespect to the scientists and engineers who devised them. In the references cited in the text we follow a priority that has meant injustice to some. Our first priority has been to refer to the most readable sources of further information for the student. Establishing the history of an idea takes second priority and recent developments take third.

Signal processing based on software algorithms plays an ever-growing role in communication systems, and all engineers need a computing engine for their work. For these reasons, short programs based on MATLAB are incorporated in many places in the text. A number of important homework exercises are based on these simple programs. Student and instructor alike are warned that gaining experience with software tools and the allied signal processing ideas takes significant time. There is a major difference between enjoying a classroom demonstration of software and using it oneself. The author's experience is that this perfecting of the tools adds about 15% (two weeks) to the course.

The pedagogical ideas behind this book and its choice of contents evolved over 15 years teaching the subject. It is inconceivable to the author that a good class text could arise without many years of encounter with a great many students. Most lines in this book trace back to a class question, a discussion afterward, or perhaps simply

[1]This is the final-year level in the European diploma engineer curriculum.

a pained expression. Thus the dedication of this book: To these students, who made it possible. We hope their pain may now be a little less.

It is a pleasure to acknowledge some special organizations and individuals. First and foremost, this book would not have been possible without the support of two foundations: The Humboldt Foundation of Germany, who awarded the author a Humboldt Research Prize Re-Invitation in 1995, and the Science and Engineering Research Council (TFR) of Sweden, who appointed the author National Visiting Chair in Information Technology during 1996–1997. Awards of this type for the support of scholarly work and books are an old tradition which has somewhat fallen by the wayside, and the author wishes especially to recognize the generosity and ideals of these two foundations.

The original LaTeX manuscript was ably typeset by Priscilla Magilligan and Lena Månsson. Their cheerful fortitude and many suggestions were of great value. We would like to recognize the support and encouragement of Dudley Kay, John Griffin, and Savoula Amanatidis at IEEE Press. A large number of colleagues helped through their readings, counsel, and willingness to debate at all hours. Particular thanks are due to Rolf Johannesson and Per Ståhl at Lund University, Joachim Hagenauer and Robert Schweikert at Deutsche Luft und Raumfahrt and Technische Universität München, and Gary Saulnier at Rensselaer Polytechnic Institute. Asad Islam of R.P.I. helped with the Solution Book. The support of industrial colleagues is particularly important to a book like this one. These especially include Sandeep Chennakeshu and Kumar Balachandran at L. M. Ericsson, Krishna Balachandran at Lucent Technologies, and Richard Frey at General Electric Research and Development. Finally, the author must acknowledge the continuous support of the other author in the family, his wife Janet. When the work here gets a little dreary, the reader can turn with profit to one of her books.

JOHN B. ANDERSON

Lund, Sweden
June 1997

PREFACE TO THE SECOND EDITION

The years have fled by, digital communication has evolved, and it is time for a revised edition. The guiding concepts of the first edition focused on core topics and effective combining of theory and practice. These met with enthusiastic acceptance, and we have therefore sought to strengthen them in the second edition. There are new programs and examples, and a few topics that seem to stray too far are reduced. Veterans of the first edition will notice that significant additions are a section on spread spectrum and a more complete and pedagogical treatment of mobile radio; these reflect the increasing importance over the last seven years of cellular and mobile networks. There is also more treatment of fiber transmission, and the end-of-chapter problems have been doubled. While these additions are the most visible ones, almost every sentence in the text has been modified in some way. Hopefully, the unclear spots and unfortunate sentences are much reduced in number, and all the errors in the first edition are repaired.

We have benefited over seven years from a great many discussions with colleagues and instructors who adopted the book. It is a pleasure to mention especially Gary Saulnier, James Leblanc, and Göran Lindell, as well as the reviewers of the new edition. We must mention also our hardworking editors Tony VenGraitis and John Griffin, as well as the production staff at Wiley. Finally, there are all the students, as many now in Sweden and the rest of Europe as in the United States and Canada. They continue to argue, inquire, and find weak spots in the book. As before, the dedication of the book is to them.

John B. Anderson

Lund, Sweden
September 2004

LIST OF PROGRAMS

LIST OF MAJOR EXAMPLES

Worked examples are included with each program.

DIGITAL TRANSMISSION ENGINEERING

CHAPTER 1

INTRODUCTION TO DIGITAL TRANSMISSION

What is digital communication? How did it come about? Why did it come about? How can we engineer and anyalze systems? These questions are the motivation for this book, and the last question is its principal topic.

Modern telecommunication is the confluence of three great trends over the last two centuries: First, the invention of electromagnetic signaling technology in the form of the telegraph, the telephone, and the radio; second, the development of mathematical theories that made these inventions practical and efficient; and finally microcircuitry, the "chip," which made these inventions small, fast, reliable, and very cheap.

To these must be added an intangible that has no explanation and is perhaps simply an absolute: the need of people to communicate with each other. Time and again this urge has financed new communication service, whether it be telegraphing across London in the 1860s or cellular telephones in the 1990s. An axiom of investing once said that American Telephone and Telegraph was always a good investment: When times were good, all stocks go up; when times are bad, people call each other and complain. The AT&T company, for a hundred years a giant, has largely disappeared since 1985 and has been replaced by others. Why this happened is an interesting question but not one for this book. What is important to us now is that only a few fields—transport and health care come to mind—attract the investment and eager public support that communication does. This fact is the economic basis for digital communication.

Although this is a book about engineering analysis, there is room in this first chapter for some history and for the fundamental question of why analog has given way to digital communication. Without these, no one would have been interested in the engineering. The plan of the book concludes the chapter.

Digital Transmission Engineering, Second Edition. By John B. Anderson

1.1 SOME HISTORY AND SOME THEMES

Some major events in the long and fascinating history of electrical communication are listed in Table 1.1. A listing of dates can be misleading, because trends and inventions are most often born in confusion and in many places at once. For this reason, many dates are approximate, and it is especially difficult to define recent innovations. Some innovations, such as the writing of what we now call software algorithms, evolved only slowly over time. A particularly readable study of this history, its less tangible side, and its human consequences has been published by I. Lebow [1]. A briefer introduction appears in Ref. [4].

A reason to study history is to identify trends, which lead to the structure of the present day. We will look now at Table 1.1 and see what it says about trends in digital communication. Parts of the following will seem obvious, but we discuss them to avoid a greater danger, which is that we miss the subtleties among the commonplace.

Communication as we know it today began in the nineteenth century. It is interesting that except for light fibers, the electromagnetic transmission technologies that we use today, namely radio and wireline telegraph and telephone, all appeared in the nineteenth century. So also did the scientific understanding of electricity and electromagnetism, without which the inventions would have made little sense. Radio found immediate use on board ships, but at first it was not at all the channelized medium that we use today. A major contribution to the *Titanic* disaster of 1912 was that all radio operators used the same spectrum space. Since the *Titanic* had the strongest transmitter and constantly used it, other nearby operators were overwhelmed and shut down their radio sets. Consequently, the early distress signals from the sinking ship went unheard by nearby ships.

Here arises a theme that plays a major role in the book—namely, the fact that radio transmission uses a channel and a *carrier*. It is often said that only by conversion to electromagnetic form can an audio signal be made strong enough for transmission. This may have been true in 1912, but kilowatt audio "transmitters," in the form of car stereos, are all too common today. The critical point is rather that electromagnetic transmission can be translated in spectrum, by shifting its carrier, to whatever frequency works best. For communication with submarines, the physics of salt water imply that the best frequencies lie near the audio range. But efficient antennas here are hundreds of kilometers long, and so any radio service subject to ordinary economics uses a much higher frequency. Transmissions that need to follow the Earth for a few hundred kilometers need medium-wave frequencies (these terms are defined in Table 1.2); for a few thousand kilometers they need shortwave frequencies, and space communication must take place in the micro- and millimeter-wave spectrum. All this derives from various physical laws, and translation by carrier is what makes radio economic. A second important reason for carrier transmission is that many users can share the same spectrum if each is translated by a different carrier; this was dramatized by the *Titanic*.

Considerable technology had to be invented before the idea of narrowband channelized radio, which we take for granted today, became practical. New transmitters

TABLE 1.1 Important Milestones in History of Digital Communication

Year	Event
ca. 1820	Oersted shows electric currents create magnetic fields
1830–1840	Henry discovers induction; Faraday and others show changing magnetic fields produce electric fields
1834–1842	Various telegraphs demonstrated
1844	Morse commercial telegraph, Baltimore to Washington
1864	Maxwell publishes his theory of electromagnetism
1866	First permanent transatlantic telegraph
1860–1876	Various telephone demonstrations by Bell and others
1878	First telephone exchange installed by Bell, Hamilton, Canada
1887	Experiments by Hertz verify Maxwell
1894–1898	Marconi and others demonstrate radio over significant distances
1901	First transatlantic radio message by Marconi, United Kingdom to Canada
1904–1906	Fleming announces diode tube; DeForest announces triode
1907	Fessenden transmits speech 320 km
ca. 1918	Armstrong devises superheterodyne receiver
1920	First modern radio broadcast by KDKA, Pittsburgh, PA
ca. 1925	Mechanical TV system demonstrations by Baird, London
1928	Gaussian thermal noise papers of Johnson and Nyquist
1929	Zworykin demonstrates electronic TV system
ca. 1933	Armstrong devises FM
1936	Commercial TV broadcasting by British Broadcasting Company, London
ca. 1940	First use of radar
ca. 1945	Matched filter devised, for radar
1945–1950	Early computers constructed; proofs of sampling theorem appear; signal space theory applied to communication
1948	Transistor demonstrated by Brattain, Bardeen, and Shockley, United States; Shannon publishes his theory of information
1950–1955	Beginnings of computer software; beginnings of microwave long-haul transmission
1953	First transatlantic telephone cable
ca. 1958	Matched filter applied to communication; first chips demonstrated
ca. 1960	Error-correcting codes begin rapid development
ca. 1960	Laser announced, United States.
ca. 1965	Communication satellites using active transponders; long-distance communication to space probes begins
1967	Forney proposes the trellis; Viterbi proposes his algorithm
1970	Low-loss optical fibers demonstrated
1970–1975	Microprocessors appear; large-scale integrated circuits appear; speech and image digitization begins rapid development
ca. 1976	Bandwidth-efficient coded modulations begin to appear; digital telephone trunks first installed
1979	Images received from Jupiter
ca. 1980	Digital optical fiber telephone trunks begin to be installed
1985–1990	Cellular mobile telephones become widespread in Europe
ca. 1990	Use of the Internet accelerates
1992	First digital mobile telephone system, GSM, begins in Europe

Note: Most dates are approximate.

TABLE 1.2 Frequency Bands in Radio Spectrum

Frequency Band	Band Name	Comments
< 100 kHz	Extra low frequency (ELF)	Submarine communication
100–500 kHz	Low frequency (LF)	Follows Earth surface
500–3000 kHz	Medium wave (MW)	AM broadcasting; follows Earth with loss
3–30 MHz	High frequency (HF)[a]	Reflected by ionosphere
30–300 MHz	Very high frequency (VHF)	TV and FM broadcasting
300-1000 MHz	Ultrahigh frequency (UHF)	Mobile radio
1–10 GHz	Microwave	Wideband links, Earth and space
10–100 GHz	Millimeter wave	Space links
> 200,000 GHz	Infrared	Optical fiber links

[a]Also called shortwave.

were developed that radiated a true amplitude-modulated (AM) signal. Receivers that could effectively reject all but one narrow frequency range came with Armstrong's superheterodyne concept in the 1920s.

Another invention by Armstrong in the 1930s, frequency modulation (FM), was the first inkling of another major theme in electromagnetic communication, the *bandwidth and energy trade-off.* Armstrong found that FM had a better signal-to-noise ratio at the same transmitter power than AM, apparently in proportion to its bandwidth expansion. Armstrong found it hard to convince his colleagues of this fact, but today we know that the same trade-off appears in many other places. Pulse-code-modulated (PCM) digitization of speech, for example, essentially freezes its signal-to-noise ratio, no matter how many retransmissions the signal passes through, but PCM also increases the transmission bandwidth.

An even more radical idea was to appear in Shannon's 1948 article, "A mathematical theory of communication" [5]. Shannon showed that error could not only be traded for bandwidth but could in principle be reduced to zero by the use of coded signals. In a parallel development during the 1940s and 1950s, Kotelnikov, Shannon, Wiener, and others developed a new theory of optimal communication, a theory that worked best with symbolic, or "digital," transmission. This theory was mathematical, not electromagnetic, and its ideas were rooted in the twentieth century.

These new digital ideas involved all sorts of complex processing and coding, and were it not for another evolving trend, they would have remained for the most part a curiosity. This trend was the idea of written processing algorithms on the one hand and ever cheaper hardware to run them with on the other. An algorithm is a step-by-step procedure to attain an end. Some algorithms like long division are ancient, but a major leap in the algorithm concept took place in the mid-twentieth century. With the ideas of von Neumann and the perfecting of computer languages and the stored program computer, algorithms took a mighty jump upward in complexity. What has evolved today is a technology in which various functions in a system can all be standard processor chips, each taking its function from the algorithm loaded into it.

It seems likely that the algorithm concept will continue to evolve beyond the sequential programs that dominate today, but the mid-century concept has been more than enough to implement the new theory of communication.

Everyone knows the story of the large-scale integrated circuit, one of the major technology drivers of the last 40 years. But it is worth reiterating how revolutionary the chip really is. A single vacuum tube active circuit element in the 1950s cost about five late-1990s U.S. dollars. Transistor technology soon reduced this cost manifold and at the same time made the element more efficient and reliable. But photolithography and successive waves of miniaturization came, until the cost of this single device had dropped a millionfold and more. Further drops are still to come. It is interesting to imagine what would happen if another part of the economy, say the cost of energy or of cars, were to drop this much. The upheaval would be hard to imagine and hard to plan for. Just such a revolution is in progress in communication, because it is based on key commodities, processing and transmission, whose price has collapsed.

As a small example, take the detector, a part of every receiver. With Fleming's 1904 diode, a great advance in its time, an AM detector was a vacuum tube that cost at least $100. By the 1950s, the detector was a vastly more reliable and cheaper solid-state diode. By the 1970s, this detector and considerably more sophisticated ones were a part of a larger integrated circuit chip that performed several functions. Today the detector and most of the radio can be a low-cost chip on which runs a stored program, whose identity as a "radio" can be changed at will.

What has happened here is (a) a move away from communication as the study of physics and devices and (b) a move toward algorithms, information as symbols, and complex processing. Just as energy and bandwidth were found to trade off, now we see a three way trade-off among energy, bandwidth, and *processing complexity*. An increase in any one means the other two can be reduced, more or less, for the same performance. A PCM digitizer, for example, can be replaced by a more complex digitizer, which puts out fewer bits, which consume less transmission bandwidth. For another example, take Shannon's concept of coded communication. Originally a mathematical theory, it now has a practical meaning: Transmission error may be driven down by more complex coding, instead of more transmission energy. With modern coded modulation, coding complexity can even be exchanged for bandwidth. *The cheapest of energy, bandwidth, and processing is now processor complexity*. Digital communication is so full of concepts, processes, algorithms, and complexity because this is the route to lower cost.

Although the processor revolution dominates our story, it should not be forgotten that many analog components have seen large cost declines. Perhaps the most significant of these is the optical fiber and its codevice, the laser. The much lower per-bit transmission cost of fibers is driving us to a two-choice world, in which fixed channels of any length are fibers and mobile channels are radio. Aside from fiber technology, radio-frequency (RF) components have been miniaturized and reduced in cost, a prime example being the analog RF chips in cellular telephones. Still, digital processor technology is steadily working its way toward the front of radios and eliminating more and more of the RF technology.

While the book takes its structure from these evolving trends, we should admit in closing that predicting the future and characterizing the present are hazardous, even pointless exercises. One needs to learn with caution and prepare for the unexpected. Historians of technology tell us that a major innovation is not felt completely for 50 years, and, if so, it is chastening to look at Table 1.1 and realize how much of it is not yet 50 years old.

1.2 WHY DIGITAL?

Why indeed is communication more and more digitally based? When the stereo shop says its products are digital quality, is this an advertising slogan, or is digital really better? Let us look at the slogans and the reality and see why this revolution is really occurring. There are many solid reasons, and here they are in rough order of importance.

1. *Cheap Hardware.* First and foremost, digital hardware has become very cheap, as we have just stressed. This makes all the other advantages cheap to buy.
2. *New Services.* We live in an age of e-mail, airline booking systems, computer modems, and electronic banking. Whereas voice and images are originally analog and may be transmitted either way, all these newer services are fundamentally digital and must be transmitted symbolically. Whether or not there is a new thirst for data by the human psyche is debatable, but it is clear that widely distributed enterprises and the "virtual workplace" and telephones that work "anytime, anywhere" are key to the way many of us choose to live. These new ways of life are more difficult in an analog world.
3. *Control of Quality; Error Control.* Here the digital story becomes more subtle. For high-quality music and television, digital format is not necessarily better per se. It is easily demonstrated that any reasonable frequency response, signal-to-noise ratio, and dynamic range may be achieved by analog recording and broadcasting. Where the digital format has a major advantage is in systems that are a chain of many tranmission links. These systems appear in many places, and two important ones are worth summarizing here. First, long-distance voice channels are usually chains of many repeaters. In former times, the links were microwave and nowadays they are more likely fiber-optic, but in either case a transcontinental channel can require 50–100 links. We will see in the chapters that follow that with analog links the noises add, while with digital links the symbol error probabilities add. What this means is that a given signal-to-noise ratio over 100 analog links requires a ratio in each link that is $10 \log_{10} 100$, or 20 dB, better; we will see that an error probability p over 100 digital links needs only 1–2 dB more energy in each link. The better telephone quality we enjoy over fiber lines is not so much a

matter of fibers but rather digital transmission by whatever means when there are repeaters. A second, perhaps more subtle, example of a repeater chain is music recording. Now the chain is a microphone, amplifiers, mixing, more mixing, conversion to a medium, storage, conversion back, amplification, speakers. While this is not a sequence of similar links, there is still a sequence of troublesome points where distortion can enter. Again, digital signal handling has a major advantage. Once a quality is agreed upon and set in the initial analog-to-digital conversion, the digital chain can be designed to carry the music essentially error free to the speaker.

4. *Compatibility and Flexibility.* Once signals are digitized, it is possible at least in principle to think about transmitting them all by the same shared medium. Network functions such as switching and multiplexing are much easier. New network topologies and modes of multiple access become possible. Control and servicing information can be combined with revenue-producing signals. All sorts of features, from telephone voice mail to disc-player music search, become economic.
5. *Cost of Transmission.* In certain channels, the cost of digital transmission is less. We have already seen this in the channel with repeaters. Another case where digital transmission wins is a channel with low power, such as the deep-space channel. Since AM schemes cannot have better signal-to-noise ratios than the underlying channel, a bandwidth-expanding modulation such as FM is necessary at the least. This is the energy-bandwidth trade-off. Another alternative is digital transmission, combined with coding: This is trading energy for both complexity and bandwidth. Communication theory shows, in fact, that the most effective use of a weak channel is binary modulation combined with coding. Yet another case where digital modulation can win is a channel with a lot of nearby interferers, such as happens in cellular radio.
6. *Message Security.* With the growth of mobile telephony and electronic information banks, message security has attracted more attention. Older mobile systems are analog FM, which offers little security. Analog encryption is fundamentally difficult, but the encryption of symbols is just as fundamentally not difficult.

1.3 CONTENTS OF THE BOOK

Figure 1.1 shows the overall plan of a digital communication link, and this is the plan of the book as well. To make the plan a little more concrete, the rough parameters of some common data sources are given in Table 1.3. These sources play a role in the examples of the succeeding chapters. Some additional orientation can be obtained from the spectral bands in Table 1.2.

The core of a transmission link is of course its transmitter and receiver. It is convenient to break that subject down into baseband signaling, which has no carrier,

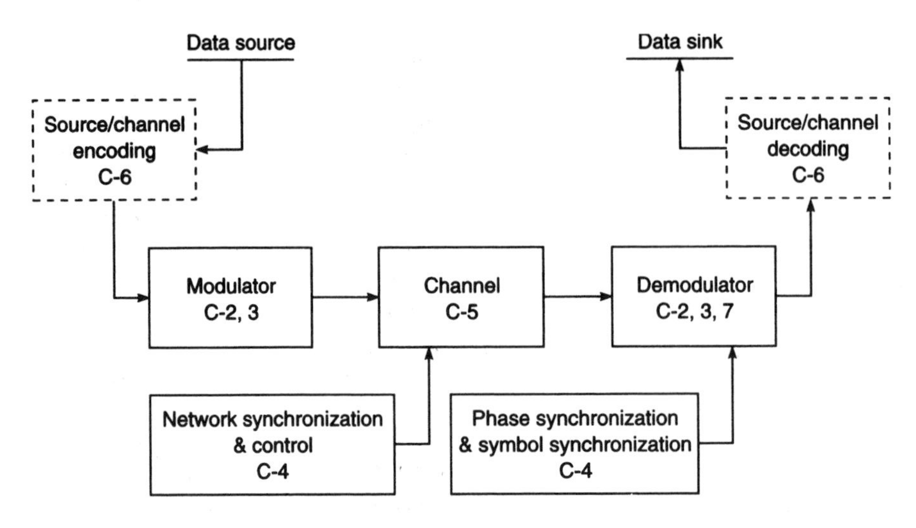

Figure 1.1 Plan of a communication link (and plan of the book).

and bandpass (or RF) signaling, which works with a carrier. The baseband case is presented in Chapter 2, and here we encounter the ideas of pulse forming, modulation by linear superposition of pulses, modulation spectrum, and modulation probability of error. The last is derived via an elegant and general method called *signal space theory*, a major achievement of communication theory. Chapter 3 is the extension to carrier transmission. The ideas of pulses, spectrum, and error probability extend easily to the RF case. Another thrust of the chapter is the parts that are unique to carrier transmission: nonlinear modulations like frequency-shift keying and distortions that occur in RF signaling.

Signals must pass through a channel medium. Some of the many channel types are simple wires, coaxial cables, free space, the terrestrial surface and atmosphere, and optical fibers. Each has its own character and distortions, and these are collect-

TABLE 1.3 Parameters of Some Representative Digital Signal Sources

Signal Source	Analog Bandwidth	Bits	Comments
Packet	—	50–1000 b	Not real time; retransmission allowed
Voice, simple	4kHz	64kb/s	PCM telephone
Voice, complex coding	4kHz	4–6kb/s	Latest cellular speech coding
Phone modem	3.4 kHz	2.4–56 kb/s	Quadrature amplitude modulation (QAM)
Phone trunk	100 kHz	1.5 Mb/s	Lowest level line concentrator; 24 one-way lines
Video, simple	4.5 MHz	40 Mb/s	PCM video
Video, complex coding	4.5 MHz	< 1 Mb/s	Advanced coding; real time

ed in Chapter 5. The chapter also provides the opportunity to study a particularly important and challenging medium, the fading mobile channel. Some further material appears in Chapter 7.

Probably the most underrated area in transmission design is synchronization. Little time is spent in most books on this subject, yet synchronization of various types probably consumes half of transmission design effort. In reality, there are several layers to the synchronization of a system: An RF system needs receiver carrier synchronization, all systems need to identify the symbol boundaries, and most data transmissions have a frame structure that needs identifying. As well, the entire network needs to stay in synchronization. All of this is in Chapter 4. In addition, the chapter is introduces some of the physical side of networks.

Chapters 2–5 are the core of the book. There are a great many allied topics to pursue, and space allows only two of these. Error-correction coding and an introduction to information theory form Chapter 6. Some advanced techniques for distorted and fading channels make up Chapter 7.

Communication engineering goes much further, and indeed it is a constant challenge for the working engineer to keep up with developments. Here are some allied fields. Digital signal processing and software engineering play an important role in modern system design. The areas of speech and of image processing are fields in their own right. A rapidly evolving field is communication devices and electronics and RF engineering; unfortunately, it is hard to find courses—and consequently engineers—in these fields. For those with a theoretical bent, communication theory, detection theory, and information and coding theory are fascinating subjects. Further topics usually taught as separate subjects are antennas, cryptography, and communication networking.

The prerequisites for this book are good courses in probability, transforms, linear system theory, and some introduction to analog communication. Stochastic process theory has been kept to a minimum in order to make the book more accessible, but this and other advanced disciplines are sometimes referred to in the text as a guide for the more advanced reader.

Some useful notation for the rest of the book as well as some further insight into communication electronics are given in Fig. 1.2. This figure defines the parts of a standard transmitter and *superheterodyne receiver*. The last, devised by E. Armstrong before 1920, is the basis of almost all receivers and is one of the great inventions of electrical engineering. The receiver begins with a roughly tuned amplifier, the low noise amplifier (LNA), which is designed for moderate gain at the lowest possible noise contribution (for more, see Section 5.2); the LNA sets the noise level in the receiver. Next comes a mixer, whose purpose is to move all signals from their location in the RF spectrum to a single spectrum location, called the intermediate frequency (IF). The IF amplifier and bandpass filters illustrated in Fig. 1.2 achieve two purposes. They hugely amplify the signal to a level suitable for the detection circuit, and they strongly reject all neighboring signals, thus setting the narrowbandpass characteristic of the receiver (a property called selectivity). The genius in this arrangement is subtle to appreciate because it lies buried in the lore of circuit design: The goals of low noise, tunability, gain, and selectivity in circuit design all

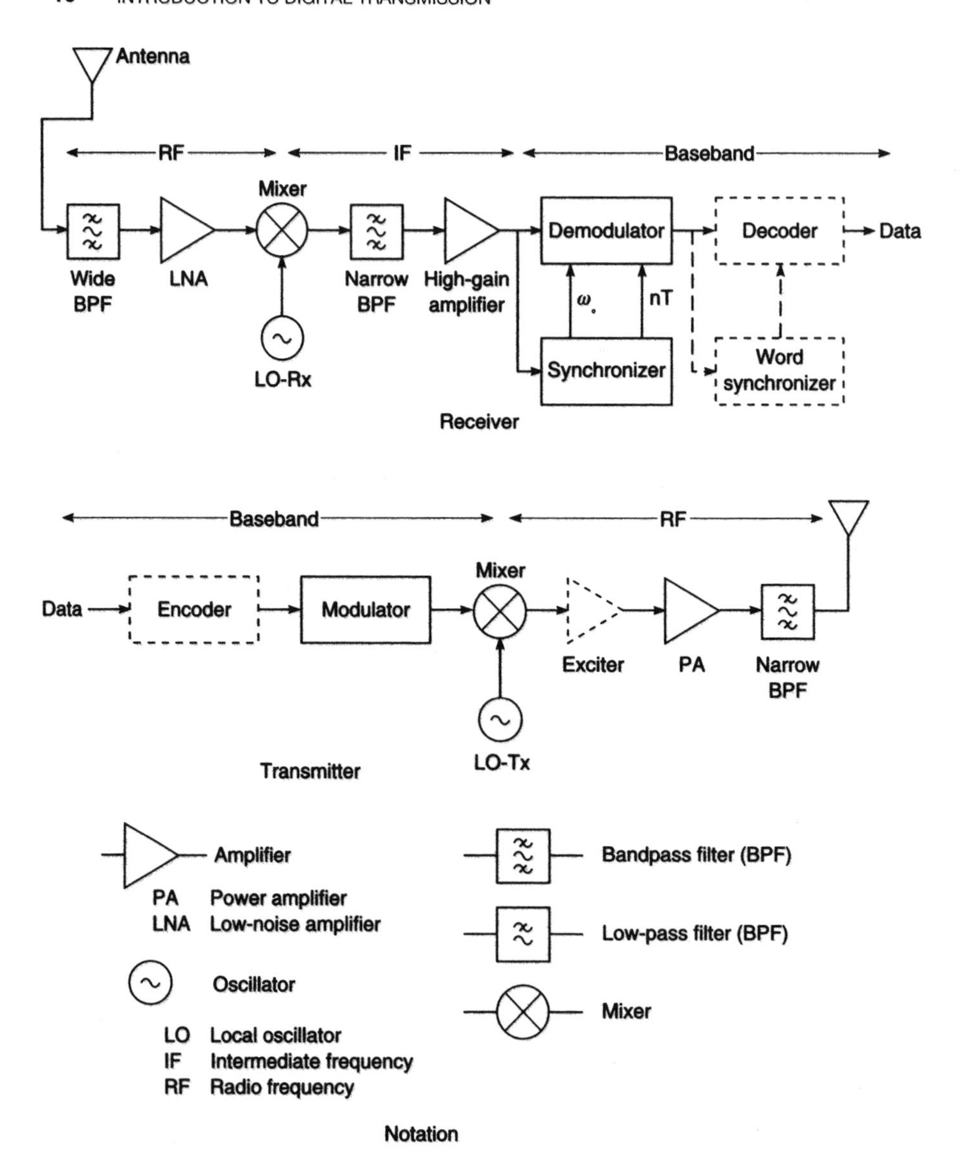

Figure 1.2 Simplified transmitter and superheterodyne receiver, showing notation used in practice and throughout this book. Not all parts are present in all systems.

mildly *contradict*. Somehow, they need to be accomplished in separate, single-purpose circuits. Armstrong was one of the first to appreciate this fact about circuitry and to propose a solution.

The top of Fig. 1.2 shows the superheterodyne circuit and the remaining detector, synchronizer, and decoder subsystems, and it also defines some common names and symbols in communication circuitry. These will be used on and off throughout the book. The electronic design of tuned amplifiers and such blocks as mixers is explained in other books—for example, Refs. [2] and [3].

The reminder of the figure shows a simplified transmitter design and some relevant acronyms and notation. Not all transmitters follow this chain: Many lack an exciter and of course many feed fibers or wires rather than antennas.

1.4 THE COMPUTER PROGRAMS

A major aid, both for learning and for practice of engineering, is some sort of computing engine, especially one with a vector capability. Perhaps the most common such software tool is MATLAB®,[1] but there are several others. For concreteness, MATLAB is used in this book. As the book progresses, short routines are given for the special calculations that arise in each subject. These are called Programs. The Programs and routines for basic operations such as convolution and the fast Fourier transform (FFT) also play a role in many of the homework problems.

Software tools are related to the pervasive shift in communication engineering toward software-based signal processing and away from hardware circuitry. Filtering, for example, is increasingly time-domain convolution instead of analog hardware. The Programs and related homeworks help keep a focus on this shift.

Rather than include the Programs on a disk at the back of the book, we have kept them short and simply written them out. This brevity is made possible by the vector nature of MATLAB. The fact is that disks-in-the-back seldom work the first time: The reader's platform is wrong, the tool version is wrong, the disk is incompatible, and so on. Murphy's Law is against us here. We believe it is easier for the reader to copy out short programs as need be, making the necessary modifications. Inexpensive student editions of MATLAB are widely available.

Experience shows that significant time must be set aside to learn software tools. Not only are there the usual problems with the computer system, but many of us need practice with such underlying techniques as sampling, convolution, and numerical analysis. The problem is not limited to the readers of this book but affects all communication engineers. Time needs to be set aside for these techniques, in addition to that needed for studying digital communication.

In closing, we offer some words of encouragement. Digital communication is a subtle subject, but it contains only a few methods and relatively few central lessons, which with a little care and patience can be recognized and mastered. This book has been designed to make that process easier. Here, as in engineering in general, cleverness is not so important as resourcefulness and dedication to the job. The book is not encyclopedic and it leaves out many smaller topics, so that the larger lessons can be stressed more often. Digital communication also has its engineering lore and its experience aspect, and we have included some of this along the way. The reward

[1]MATLAB is a trademark of The MathWorks, Inc. Further information is available from The MathWorks, Inc., 3 Apple Hill Drive, Natick, MA 01760-2098 USA, or www.mathworks.com. The simplest available versions of such tools are recommended for this book.

for studying this book is entry into a fascinating subject that is welcome and thriving everywhere in the world.

REFERENCES

1. I. Lebow, *Information Highways and Byways,* IEEE Press, New York, 1995.
2. K. K. Clarke and D. T. Hess, *Communication Circuits: Analysis and Design,* Addison-Wesley, Reading, MA, 1971.
3. H. L. Krauss, C. W. Bostian, and F. H. Raab, *Solid State Radio Engineering,* Wiley, New York, 1980.
4. J. B. Anderson and R. Johannesson, *Understanding Information Transmission,* IEEE Press, New York, 2005.
5. C. E. Shannon, A mathematical theory of communication, *Bell Syst. Tech. J.,* 27, 379–423; 623–656.

CHAPTER 2

BASEBAND PULSE TRANSMISSION

2.1 INTRODUCTION

In this and the next chapter, we discuss how to transmit digital data symbols. Symbols, of course, are weightless and intangible, and they need to be converted into material form before they can be sent or stored. Very often this is done by associating the symbols with *pulses* in some way, and a sequence of these pulses are added up to form a pulse *train*, which carries the full message. If the pulse train itself, or a similar original waveform, is transmitted, the communication system is said to be *baseband* and the signal is a baseband signal. Baseband signaling is the subject of this chapter. If the pulse train is translated in frequency to a new band, the system is a *carrier* system. Most, but not all, communication systems employ carriers, because the new frequency range offers some important advantage. We study these systems in Chapter 3. We begin with baseband transmission because carrier systems build directly on baseband systems.

The term *linear modulation* is applied to signals that are made up from a linear superposition of pulses, whether baseband or carrier. What makes a good pulse? Usually, the two most important criteria are low probability of error and narrow bandwidth. The error probability of a pulse train is quite a deep subject, especially when the pulse is complicated, and we will take this up in Sections 2.5–2.6 with the study of *signal space theory*. If a pulse is complicated, it is probably because its bandwidth is narrow. Narrowband signals carry more data symbols in a given bandwidth and are therefore more efficient. According to Fourier theory, the product of pulse bandwidth and time spread is approximately constant; consequently, bandwidth may only be reduced by dispersing the pulse over a longer time. Even for signals with only modest efficiency, pulses are long enough so that they overlap and interfere with each other. This complicates detection and (in Chapter 4) synchronization.

Digital Transmission Engineering, Second Edition. By John B. Anderson

As the pulse bandwidth narrows, time dispersion widens and a good detector becomes harder to build. Pulses with particular sets of properties are needed in order to simplify the detector. These properties are the subject of Section 2.2. While working through the mathematical details, we need to keep in mind the underlying object, which is to work out the compromises that exist among three basic variables: transmission bandwidth, error probability, and the cost of the detector.

2.2 PROPERTIES OF PULSE TRAINS

We begin by thinking about pulses that last only as long as the time between data symbols and that do not overlap with succeeding pulses. The shape and relations between pulses viewed at this level are called their *format*. Next we let the pulses overlap and explore the conditions under which the amplitudes of individual pulses may still be observed from samples of the entire pulse train; these are called *Nyquist* pulses. Finally, we study *orthogonal* pulses. These overlap more seriously, but in a way that makes all but one pulse in the train invisible to a properly designed detector. These two pulse classes introduce two basic detectors. The first, called the *sampling receiver*, works with Nyquist pulses by observing the data symbol values at the right moments in the pulse train. The second observes the entire signal and depends on the orthogonality property to separate out the data symbols. This detector is called the *linear receiver*.

2.2.1 Pulse Formatting

Digital data appear as a sequence of symbols in time. We think of a piece of transmission time as devoted to each symbol. This is the *symbol time* T_s or, when no confusion will result, simply T. The reciprocal $1/T_s$ is called the *baud rate* of the transmission; one baud equals one symbol per second.[1] One must be careful not to confuse T_s with the time devoted to a data *bit*, which will be denoted T_b. Both times are important, and each has its own use. The term T_s is the physical time devoted to a transmission symbol, and it defines the organization and timing of the transmitter and receiver circuits. The symbol itself is defined for the convenience of the transmission system; it may be two-valued, in which case the symbol is a bit and $T_s = T_b$, but more often it takes four or more values, because such a symbol may be more efficient to transmit. On the other hand, we generally measure data—and it usually arrives for transmission—in bits. It is easier, too, to measure the costs of transmission in a per-bit way: bandwidth in hertz per data bit, signal energy in joules per data bit, and equipment costs in dollars or euros per data bit. The actual transmission system might work with four-level symbols, in which case $T_s = 2T_b$, since each symbol carries two data bits. We will constantly encounter such conversions between transmission symbols and data bits.

Now suppose a sequence of transmission symbols $a_0, a_1, \ldots$ modulates a basic pulse $v(t)$ and superpose linearly to form the pulse train

[1]This Système International (SI) unit is named after the French telegraph pioneer, J. M. E. Baudot.

$$s(t) = \sum_{n=0}^{N} a_n v(t - nT_s) \tag{2.2-1}$$

This pulse train constitutes linear modulation. Pulses appear at the baud rate of the modulation. For example, if $T_s = 1/2400$ and the symbols are ±1, the modulation is said to run at 2400 baud; its bit rate is also 2400 b/s. If the symbols take values in the set {–3, –1, 1, 3}, the modulation still runs at 2400 baud, but its bit rate is 4800 b/s. Such a four-level scheme is called a *quaternary* modulation. An eight-level scheme is *octal*, and a general M-level scheme is called M-ary.

The set of values is the symbol *alphabet*. In the information processing world, we give the arbitrary names 0 and 1 to the values of the binary information-bearing symbol. These could have as well been A and B, or aleph and beth, or any other two names. In digital communication, the values not only are symbolic, but also take on electrical and mathematical meanings, and it is more convenient if they are taken as +1 and –1. Similarly, an M-ary symbol is taken as $\pm(M-1), \pm(M-3), \ldots, \pm 1$, that is, as antipodal, equally spaced odd integers. In this case, T_b is $T_s/\log_2 M$. This is the convention we will follow.

If $v(t)$ is a simple width-T_s square pulse and the transmission symbols are ±1, then Eq. (2.2-1) for a typical symbol sequence is the pulse train in Fig. 2.1a. This $v(t)$ has various names, including simply square pulse, but its most common technical name is the NRZ pulse, an acronym for nonreturn-to-zero. NRZ refers to the fact that the pulse train does not return to zero during part of the symbol interval. Figure 2.1b shows the same pulse train with an RZ, or return-to-zero, $v(t)$. Here, NRZ and RZ are examples of a pulse format.

The motivation for an RZ format is that it provides a pulse transition halfway through each interval, which makes synchronization to the symbol interval framework much easier. A disadvantage, compared to the NRZ format, is that the signal spectrum is wider, as we will see in Section 2.3. The decision between the two thus depends partly on whether some extra bandwidth is at hand. Both RZ and NRZ are *polar* formats, meaning in the binary case that the ±1 transmission symbols create corresponding positive and negative excursions in the pulse train. In a unipolar format, all the excursions are positive, meaning in the NRZ case that the +1 symbol is sent as a positive square pulse and the –1 symbol is sent as a zero value. The NRZ example appears in Fig. 2.1c; this format is also called on–off signaling. A true polar format has zero average value (for equiprobable symbol values), while a unipolar format does not.

The term format is often applied in a more general sense to describe how successive symbols change the symbol–pulse relationship. Especially in this context, a format is sometimes called a *line code*. Examples of the so-called bipolar line code, or format,[2] are shown in Fig. 2.1d (for bipolar NRZ) and Fig. 2.1e (for bipolar RZ). In both cases, the pulses representing successive +1 symbols are inverted, while the –1 symbol is always represented by a zero value; +1, +1, +1, +1, . . . would be sent as a sequence of pulses with alternating sign. Line codings such as the bipolar one

[2]Some engineers take bipolar to mean "polar," as we have used it. Formats take a bewildering array of alternate names. We have adopted the most common names.

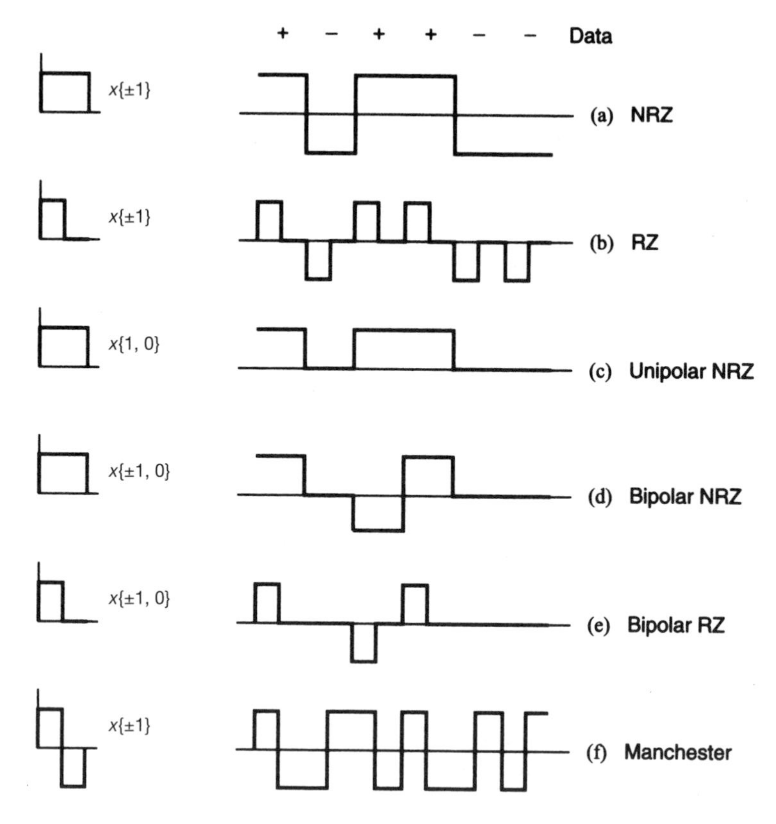

Figure 2.1 Examples of binary pulse formats based on square pulse. Data are +1, –1, +1, +1, –1, –1 in all cases. Basic pulse $v(t)$ and transmission symbol alphabet appear at left.

are used to ensure that there are sufficiently many transitions in the pulse train, without which synchronization to the interval framework will be difficult. The bipolar format breaks up runs of +1s; more complex line codings have been devised to break up runs of both –1s and +1s and to attend to other needs. Gibson [1, p. 250ff] provides some further explanation.

A format such as these last two shows the interplay of several concepts. There is a basic pulse, $v(t)$, which in this case is the square one. The basic pulse may be returned to zero early, as in the RZ case, or even omitted, as in the unipolar case. Finally, the basic pulse may be inverted or deleted, depending on past history. All of these operations should be distinguished from pulse *shaping*, which we will take up shortly; with shaping, the idea is to smooth the basic pulse.

Example 2.2-1: The T1 Carrier System. Data transmission is often organized into *hierarchies*, in which a number of data streams are grouped into a few large ones, which are grouped into even fewer larger ones, and so on. This is discussed in detail in Section 4.8. A very common first level in hierarchies in Japan and North America is the T1 system. As a generic term, T1 has come to mean any system that carries 1.544 Mb/s. In some contexts, T1 retains its original, specific meaning, which is that it is a baseband bipolar format system that carries twenty-four 64-kb/s streams over a wire cable (see, e.g., Ref. 2, p. 315ff). The bipolar format is the one in Fig. 2.1d but with shaped pulses, as will be discussed in Example 2.2-3; it is chosen to make the pulse synchronization easy and also because a DC null is desired in the spectrum. The 24 streams, which can be telephone calls or data, are multiplexed into a 1.536-Mb/s stream. To this are added 8-kb/s "housekeeping" bits that help manage the system. The total stream runs at 1.544 Mb/s. The actual rates of the 24 streams must agree within a certain tolerance.

One last format in common use is the *Manchester* line code, which is demonstrated in Fig. 2.1f. Here a +1 transmission symbol leads to a positive then a negative square half-pulse, while a –1 symbol leads to the reverse, a negative followed by a positive. The Manchester pulse guarantees a signal transition for the synchronizer in the middle of each symbol interval, without a return to zero (a return would increase the error probability). There is also a spectral null at DC, which is an advantage in baseband systems that carry DC power along the same line that carries the data-bearing signal. A disadvantage is that the Manchester format has twice the bandwidth for the same data rate. A common application of Manchester is the Ethernet system.

2.2.2 Overlapping Pulses: The Nyquist Criterion

Although the simple, square NRZ pulse is an obvious way to send data, its short duration and its discontinuities cause it to have a very wide bandwidth. Two ways to reduce the bandwidth of any pulse are to round off its corners and transitions and to lengthen the duration of the pulse. Overlapping pulses interfere with each other. However, some kinds of interference still allow an effective detector to be built. The idea is to employ these special pulses and to gain their bandwidth advantage, yet not lose detector performance.

The first class of such pulses obeys a zero-crossing criterion called the *Nyquist pulse criterion*. For convenience, let the basic pulse $v(t)$ be centered at time zero.

Definition 2.2-1. A pulse $v(t)$ satisfies the Nyquist pulse criterion if it passes through 0 at $t = nT$, $n = \pm 1, \pm 2, \ldots$, but not at $t = 0$.

For short, we will call $v(t)$ a *Nyquist pulse*. Some examples appear in Fig. 2.2, all normalized to have unit energy. A look at the pulse train expression (2.2-1) shows that Nyquist pulses are ideal for forming such signals: A sample at time nT directly gives the value of the a_n, the nth transmission symbol. The detector that works by simply taking this sample, as portrayed in Fig. 2.3, is called the *sampling receiver*.

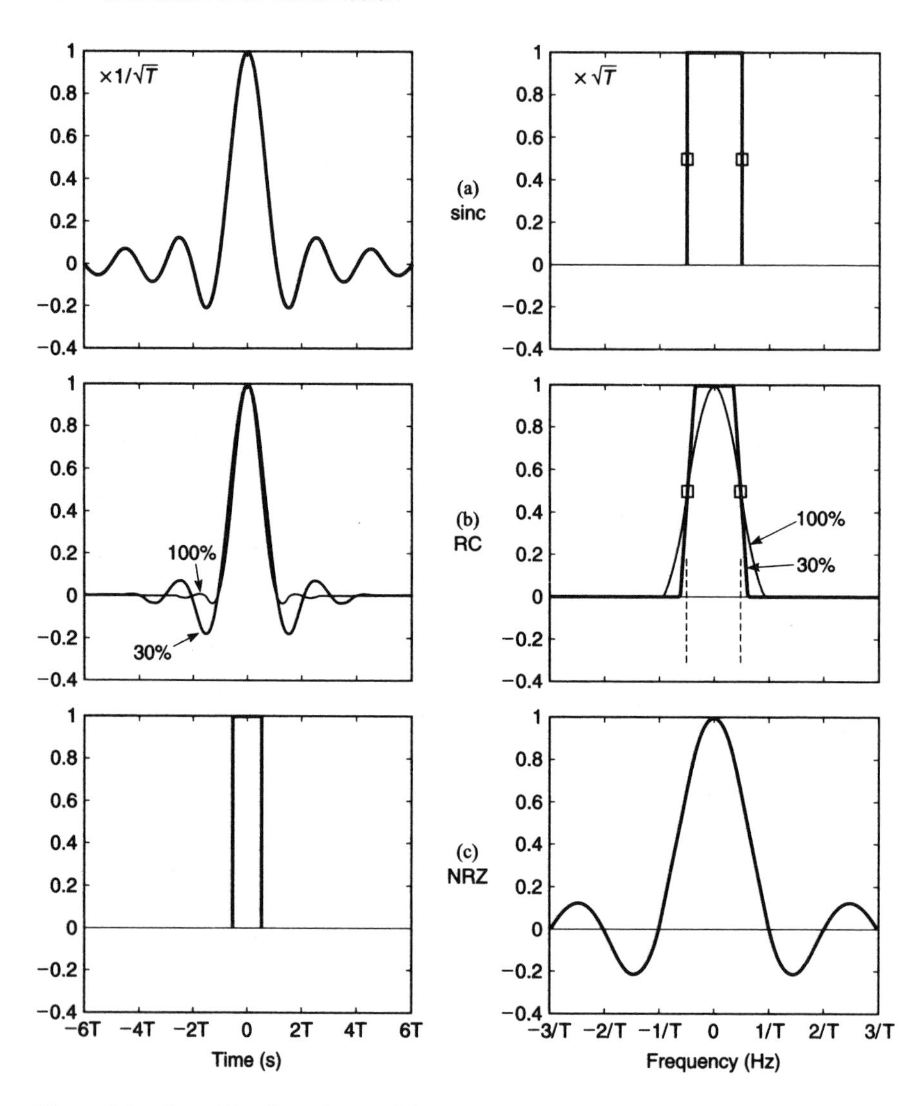

Figure 2.2 Some Nyquist pulses and their Fourier transforms: **(a)** sinc, **(b)** raised cosine, **(c)** NRZ. Note symmetries about the blocks in (a) and (b). To maintain constant energy as T changes, scale pulses as shown.

The top part of the figure is a linear modulator that generates the standard pulse train. Noise may be added in transmission. In the detector, the signal gets sampled at each time nT, and if there is no noise, this sample is already the transmission symbol a_n. Otherwise, the sample is compared to the noise-free sample values in a threshold comparator, and the closest value is selected as the estimate $\hat{a}_n$ for the symbol. For concreteness, the figure shows the binary case, in which the comparator decides between ± 1 by comparing its input to the threshold 0.

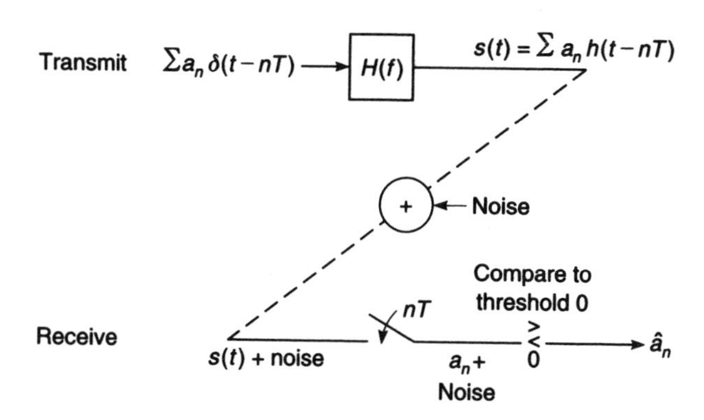

Figure 2.3 Sampling receiver with linear modulator. Receiver consists of a sampler and a comparator (binary case assumed for clarity).

The sampling receiver requires accurate synchronization to the instants nT, and as we will see in Section 2.6, its error probability can be poor in a channel with some noise, but in a noise-free channel with good synchronization one would be foolish not to use it.

One can directly construct time functions that satisfy Definition 2.2-1, but H. Nyquist [3] in 1924 discovered a characterization in terms of the Fourier transform of $v(t)$ that has proven much more useful. To generate a Nyquist pulse, one need only construct a filter with such a transfer function. The result has become known as Nyquist's first criterion.[3] His result has evolved considerably since 1924, and we give now a modern version due to Gibby and Smith [5].

Theorem 2.2-1 (*Nyquist's First Criterion*). $v(t)$ satisfies the Nyquist Pulse Criterion (Definition 2.2-1) if and only if

$$\sum_{n=-\infty}^{\infty} V\left(f - \frac{n}{T}\right) = K_0 \tag{2.2-2}$$

where $V(f)$ is the Fourier transform of $v(t)$ and K_0 is a real constant.

Proof. By using the Poisson sum formula of Fourier analysis, the left-hand side of (2.2-2) may be written as

$$\sum_{n=-\infty}^{\infty} V\left(f - \frac{n}{T}\right) = T \sum_{m=-\infty}^{\infty} v(mT) e^{-j2\pi mTf} \tag{2.2-3}$$

[3]Nyquist [3] also gave criteria for pulses to have a transition halfway through the symbol interval and to have a certain weight; these later were sometimes called the Second and Third Criteria. Nyquist himself did not number his results.

Since the time samples $v(mT)$ of a Nyquist pulse are all zero except for $v(0)$, Eq. (2.2-3) becomes

$$\sum_{n=-\infty}^{\infty} V\left(f - \frac{n}{T}\right) = Tv(0)$$

if and only if $v(t)$ is Nyquist. Here, $Tv(0)$ is the real constant K_0.

It is worth remarking that for a real pulse, $V(f)$ will in general be complex, so that criterion (2.2-2) breaks down into two criteria:

$$\sum_{n=-\infty}^{\infty} \text{Re } V\left(f - \frac{n}{T}\right) = K_0$$
$$\sum_{n=-\infty}^{\infty} \text{Im } V\left(f - \frac{n}{T}\right) = 0 \qquad (2.2\text{-}4)$$

There is a simple interpretation of the criterion in Eq. (2.2-2), which is clear in the first two examples in Fig. 2.2. If the roll-off of $V(f)$ is symmetric about the square blocks in the figure, about the points $[\pm 1/2T, \sqrt{T}v(0)/2]$, then the sum of all the frequency shifts in (2.2-2) will indeed be a constant. Figure 2.4 shows how the summing up works for Fig. 2.2b. In practice, one can approximate the symmetric roll-off with a relatively simple filter and thus generate a Nyquist pulse. The symmetry condition was in fact the criterion originally proposed by Nyquist, and it is easy to see that it is not a necessary condition by looking at the NRZ pulse spectrum in Fig. 2.2c. The NRZ pulse is obviously Nyquist, yet the transform $V(f)$ fails the symmetry condition. Still, it can be shown that $V(f) = v(0)\sqrt{T}\,\text{sinc}(fT)$ satisfies (2.2-2).[4]

In later work, Nyquist [4] noted that there seemed to be a lower limit to the bandwidth of a Nyquist pulse, namely, $1/2T$ Hz. Transmission is certainly possible below that bandwidth, but the pleasing zero-crossings are no longer present and the sampling receiver no longer works. In later years, formal methods of proof of this fact developed and the result came to be called the sampling theorem. A significant milestone in this evolution was Shannon's proof [14]. Here is a version attuned to modulation theory, which also specifies what pulse achieves the narrowest bandwidth.

Theorem 2.2-2 (*The Modulation Sampling Theorem*). The narrowest bandwidth of any Nyquist pulse [i.e., a pulse satisfying Definition 2.2-1 or Eq. (2.2-2)] is $1/2T$ hertz; this pulse is $v(t) = \text{sinc}(t/T)$.

Proof. A proof of the full sampling theorem appears in any standard undergraduate text. A proof of this limited version follows directly from the criterion in Theorem

[4]Here and throughout, $\text{sinc}(x) = (1/\pi x)\sin(\pi x)$.

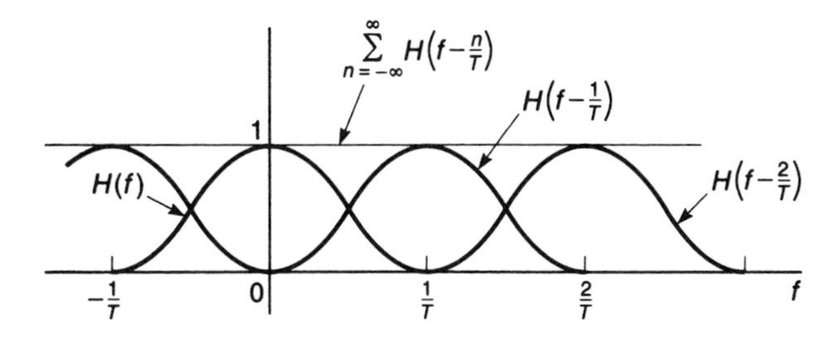

Figure 2.4 Illustration of spectra with roll-off symmetry summing to a constant.

2.2-1: If the bandwidth of $V(f)$ were truly narrower than the range $(-1/2T,\ 1/2T)$, then the sum (2.2-2) could not possibly be constant for all f. The $\text{sinc}(t/T)$ pulse and its transform appear in Fig. 2.2a.

We turn now to some standard examples of Nyquist pulses.

Example 2.2-2: Pulse sinc(t/T). This is the narrowest bandwidth Nyquist pulse. An example of a pulse train, made with the same symbols as in Fig. 2.1, appears in Fig. 2.5a. Figure 2.6a shows how the individual $\text{sinc}(t/T)$ pulses add to form the train. The pulse train has the same bandwidth as the pulse, $1/2T$ Hz, as will be shown in Section 2.3. There is some tendency to look down on the $\text{sinc}(t/T)$ pulse and think of it as an exotic outer limit to communication. It is true that well-behaved filters with a spectral discontinuity are difficult to realize and that the $\text{sinc}(t/T)$ pulse is noncausal. Even so, a practical filter can come close to $V(f)$, and by truncating the left and right tails of $\text{sinc}(t/T)$, an acceptable causal substitute with finite support may be obtained. A more serious difficulty appears when the sampling instants $nT + \varepsilon$ are off by a consistent small offset ε. It can be shown that with a sufficiently long unlucky data sequence (e.g., $+1, -1, +1, -1, \ldots$), the sampled value ceases to resemble a transmission symbol and diverges. The heart of the problem is the observed sum

$$\sum a_n \frac{\sin[(\pi/T)(nT+\varepsilon)]}{(\pi/T)(nT+\varepsilon)} = \sum (-1)^n a_n \frac{\sin(\pi\varepsilon/T)}{\pi n + \varepsilon\pi/T}$$

which need not converge. If accurate sample timing is available, this problem disappears.

Example 2.2-3: Raised-Cosine (RC) Pulse. By allowing a little extra bandwidth, much of the difficulty with the $\text{sinc}(t/T)$ pulse is removed. A smooth spectral roll-off means that a four- to six-pole filter can generate an accurate pulse; time-domain tails will fall off more rapidly so that truncation down to a causal approximation is easier and timing is less critical. The most common example in practice is the *RC pulse*, defined in frequency by

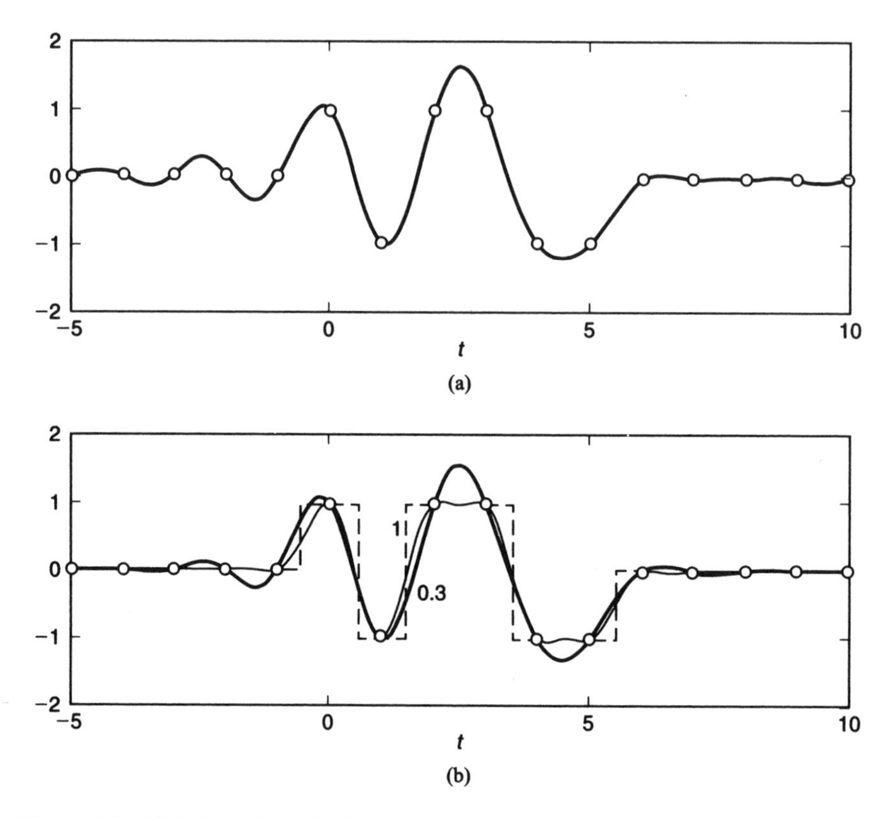

Figure 2.5 **(a)** Pulse train made from superposition of sinc pulses, driven by the data +1, −1, +1, +1, −1, −1; **(b)** pulse trains from the same data, made from 30% and 100% (lighter curve) RC pulses. Note samples at nT (circles), which are data plus leading and trailing zeros. The NRZ pulse train is shown for reference in (b).

$$V(f) = \begin{cases} 1, & 0 \le |f| \le \dfrac{1-\alpha}{2T} \\ \cos^2\left[\dfrac{\pi T}{2\alpha}\left(|f| - \dfrac{1-\alpha}{2T}\right)\right], & \dfrac{1-\alpha}{2T} < |f| < \dfrac{1+\alpha}{2T} \\ 0, & \text{elsewhere} \end{cases} \tag{2.2-5}$$

and in the time domain by

$$v(t) = \frac{\sin(\pi t/T)}{\pi t/T} \frac{\cos(\alpha \pi t/T)}{1 - 4\alpha^2 t^2/T^2} \tag{2.2-6}$$

(Both these expressions have unit peak value.) The parameter α is called the excess bandwidth factor, since the bandwidth of this pulse is $(1 + \alpha)/2T$, while the narrow-

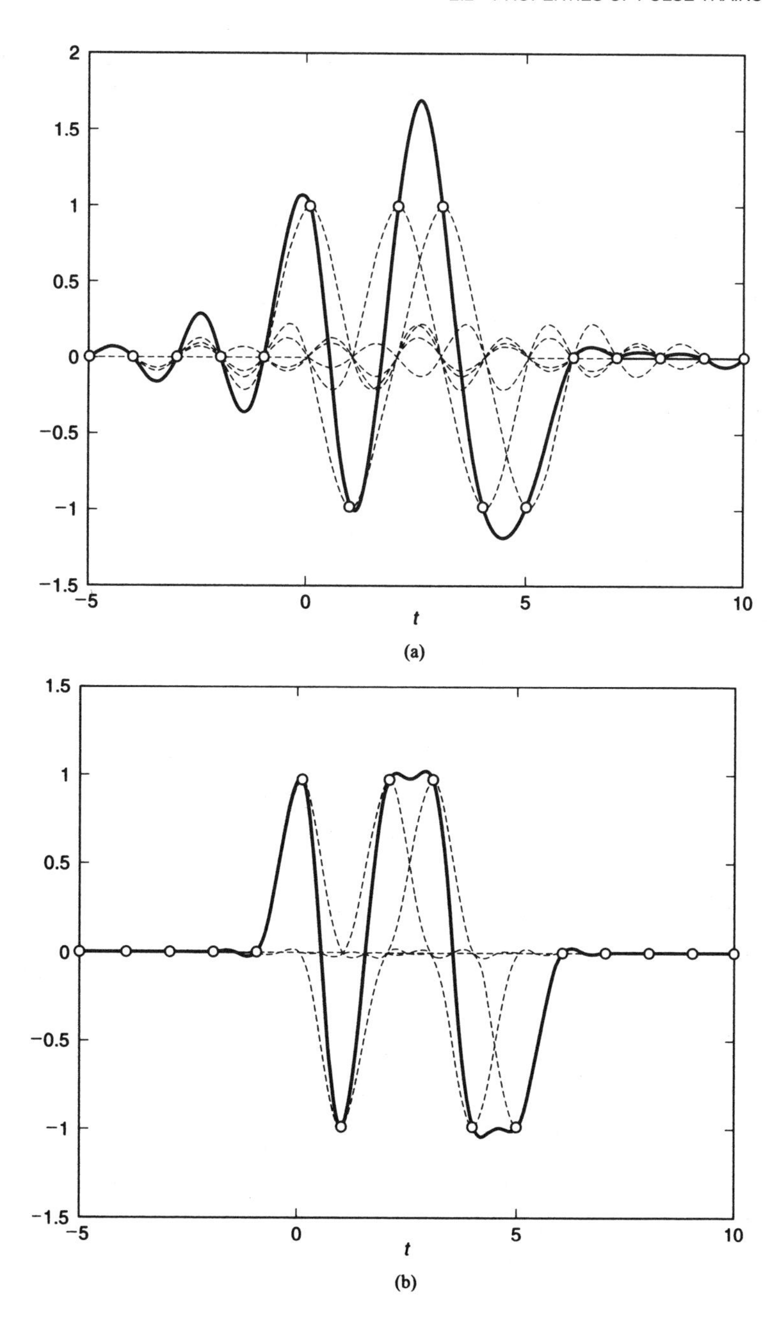

Figure 2.6 **(a)** Individual sinc pulses adding to form pulse train of Fig. 2.5a; **(b)** 100% RC pulses adding to form pulse train. Note much larger tails and excursions in narrow band pulse of (a); tails may not be truncated without widening the bandwidth. Data are the same as in Figs. 2.1 and 2.5.

est possible bandwidth is $1/2T$. Figure 2.2b plots the cases $\alpha = 0.3$ and $\alpha = 1$, which in percentage terms represent 30% and 100% excess bandwidth; the 30% pulse is commonly used in practice. The effect of the extra RC pulse bandwidth is to reduce the excursions in the pulse train and to greatly reduce the tails of the pulse. It is easy to see from (2.2-6) that the tails of an RC pulse decay asymptotically in time as $1/t^3$; the effect of a larger α is to bring this in nearer to the time origin. Figure 2.5b plots 30% and 100% pulse trains together with an NRZ train, and it is clear that the extra bandwidth brings the RC trains much closer to an NRZ. Figure 2.6b shows the orderly way that the pulses in the 100% case add to form the pulse train. For a practical application, we can revisit the $T1$ carrier system of Example 2.2-1. The $T1$ system is a binary baseband system with $T = (1.544 \times 10^6)^{-1} = 0.647$ μs, whose bipolar format doubles the spectrum of its basic pulse form. If the basic pulse is RC with 30% excess bandwidth, the pulse bandwidth is $\frac{1}{2}(1.544)2(1.33) = 2.01$ MHz; with 100% excess bandwidth, this becomes 3.09 MHz. The least possible bandwidth for Nyquist pulses is 1.544 MHz.

Since the RC pulse so commonly occurs, we will now give a short program to compute it.

Program 2.2-1: The RC Pulse. The MATLAB function that follows generates a standard unit peak value RC pulse at the time points given by the vector `t`. The parameter `a` is the excess bandwidth factor, which is denoted α in (2.2-5) and (2.2-6) above. The symbol time T, the variable `tau`, is 1, unless modified in the program. The pulse values are in the vector `y`. The program follows.

```
function y = rcpuls(a,t)
tau = 1;                              %Set symbol time
t = t+.0000001;                       %Insert offset to prevent NANs
tpi = pi/tau; atpi = tpi*a; at = 4*a^2/tau^2;
y = sin(tpi*t).*cos(atpi*t)./(tpi*t.*(1-at*t.^2));
```

As an example, set `t = [-6:.1:6]`. Then `rcpuls(.3,t)` generates the $\alpha = 0.3$ time-domain pulse in Fig. 2.2.

Occasionally, the RC idea is applied in the time domain, meaning that the pulse time response has the RC shape; with an early enough transition, the pulse may even be Nyquist. The time RC shape means the pulse has everywhere a first derivative, which reduces the outer parts of the pulse spectrum. In this book and in the engineering world, RC by itself generally means a spectral raised cosine.

Example 2.2-4: NRZ Pulse as a Nyquist Pulse. The NRZ pulse is trivially a Nyquist pulse. The penalty for such simplicity is a very wide spectrum, as Fig. 2.2 shows (much of the spectrum is truncated there). The spectrum rolls off only as $1/f$, a 6-dB/octave roll-off, which is not sufficient in systems where several signals occupy adjacent bandwidths. For example, the spectral side peaks near $1.5/T$, $2.5/T$, and $3.5/T$ reach 21%, 13%, and 9% of the spectrum amplitude; by comparison, even the 100% excess bandwidth RC pulse has zero spectral energy outside $1/T$. A mathematical name for the width-1 NRZ pulse is

$$\mathrm{rect}(t) = \begin{cases} 1, & -\frac{1}{2} \le u \le \frac{1}{2} \\ 0, & \text{otherwise} \end{cases} \tag{2.2-7}$$

The unit-energy width-T NRZ pulse is $(1/\sqrt{T})\mathrm{rect}(t/T)$.

2.2.3 Orthogonal Pulses

At this point, we have designed a class of pulse trains with relatively narrow bandwidth, whose underlying symbols are easy to extract. The problem, which will develop as the chapter progresses, is that a simple, economic detector can have a poor error probability. The solution to this problem is to make the pulse orthogonal.

Definition 2.2-2. A pulse $v(t)$ is *orthogonal under T-shifts* (or orthogonal for short) if

$$\int_{-\infty}^{\infty} v(t)v(t - nT)\, dt = 0, \qquad n = \pm 1, \pm 2, \ldots$$

where T is the symbol interval.

An orthogonal pulse is uncorrelated with itself shifted by any integer multiple of T. Consequently, a correlation of the whole pulse train with $v(t - nT)$ gives the symbol a_n:

$$\int_{-\infty}^{\infty} \left[\sum_m a_m v(t - mT) \right] v(t - nT)\, dt = a_n \int_{-\infty}^{\infty} v^2(t - nT)\, dt \tag{2.2-8}$$

If $v(t)$ has unit energy, the right-hand side is directly a_n. Furthermore, the correlation in (2.2-8) may be realized by simple linear filtering. To see this, let $s(t)$ represent the pulse train and rewrite (2.2-8) as

$$\begin{aligned} \int s(t)v(t - nT)\, dt &= \int s(t)v[-(nT - t)]dt \\ &= s(t) * v(-t)\Big|_{nT} = a_n \end{aligned}$$

In words, the desired correlation is the value at time nT of the convolution of the pulse train with $v(-t)$. We can implement this by applying the train to a filter with transfer function $V^*(f)$ and sampling the output at time nT.

A detector to do this is sketched in Fig. 2.7. The top half is a linear modulator that generates the standard pulse train; noise may be added in transmission. In the detector, the signal passes through the filter $H^*(f)$ and gets sampled at nT. If there is no noise and $h(t)$ is orthogonal, this sample is already the transmission symbol a_n. Otherwise, the sample is compared to the noise-free symbol values in a threshold comparator, just as in the sampling receiver. This detector circuit has come to be called the *linear receiver*. To be sure, parts of the circuit are not linear, but the name stems from the fact that the core of the signal processing is a free-running linear filter.

It is clear that the linear receiver is simple and practical, but it is not yet clear why we should take the trouble to synthesize the orthogonal pulse and abandon the sampling receiver. The answer lies in the error probability, which must await the analysis in Sections 2.5–2.6. For now, we can give the end result, which is easy to state: For orthogonal pulses and Gaussian noise, *no detector has a lower error probability than the linear receiver.*

Is there a spectral criterion for orthogonal pulses, as there is for Nyquist pulses? To see that there might be, consider that the total impulse response of Fig. 2.7 is $v(t) * v(-t)$, in the absence of noise, which is equivalent to a correlation operation. The correlator output is to be zero at mT, when $m \neq n$; that is, the correlator output should look like a Nyquist pulse. The next theorem states this as a formal criterion.

Theorem 2.2-3 (*Orthogonal Pulse Criterion*). The function $v(t)$ is orthogonal (satisfies Definition 2.2-2) with respect to T if and only if

$$\sum_{n=-\infty}^{\infty} \left| V\left(f - \frac{n}{T}\right) \right|^2 = K_0 \tag{2.2-9}$$

where $V(f)$ is the transform of $v(t)$ and K_0 is a real constant.

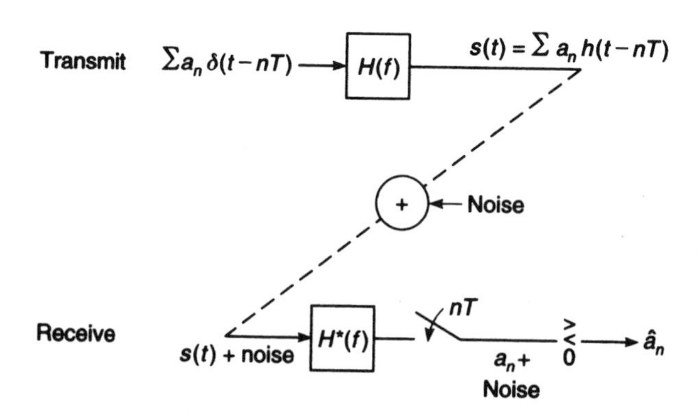

Figure 2.7 Linear receiver with a linear modulator. A second filter $H^*(f)$ is added to sampling receiver.

Proof. The autocorrelation of $v(t)$ is given by

$$R(\tau) = \int v(t)\, v(t-\tau)\, dt$$

The Fourier transform of this is

$$\begin{aligned}\mathcal{F}\{R(\tau)\} &= \iint v(t)v(t-\tau)\, dt\, e^{-j2\pi f\tau} d\tau \\ &= \iint v(t)\, e^{-j2\pi ft} v(t-\tau)\, e^{-j2\pi f(\tau-t)}\, dt\, d\tau\end{aligned}$$

Making the variable change $u = t - \tau$, we get

$$\begin{aligned}\mathcal{F}\{R(\tau)\} &= \int v(t)\, e^{-j2\pi ft}\, dt \int v(u) e^{j2\pi fu}\, du \\ &= V(f)\, V(-f) = V(f)\, V^*(f) = |V(f)|^2\end{aligned} \tag{2.2-10}$$

We can now use the Poisson sum formula on (2.2-10), just as in Theorem 2.2-1, and get that

$$\sum_{n=-\infty}^{\infty} \left| V\left(f - \frac{n}{T}\right)\right|^2 = T \sum_{m=-\infty}^{\infty} R(mT) e^{-j2\pi mTf}$$

The right-hand side is the constant $TR(0)$, if and only if $v(t)$ is orthogonal. This proves the theorem.

An important special case occurs when the pulse $v(t)$ is symmetric. Then the transform $V(f)$ is real and the condition (2.2-9) applies simply to $V^2(f)$ rather than to $|V(f)|^2$. Just as with the Nyquist Pulse Criterion, pulses whose $|V(f)|^2$ has a symmetrical roll-off about the square block, as in Fig. 2.2, satisfy (2.2-9) and are therefore orthogonal. This time, however, the symmetry must apply to $|V(f)|^2$, not $V(f)$.

To conclude, we review some standard examples. It is easy to see that the sinc(t/T) pulse in Fig. 2.2a is orthogonal, since the shifts of its square-block transform trivially satisfy criterion (2.2-9). It is not so easy to work with the transform of the NRZ pulse (Fig. 2.2c), but now the orthogonality is trivial to prove just by looking at the NRZ pulse itself. The RC pulse leads to a more interesting example.

Example 2.2-5: Root RC Pulse. Probably the most commonly used pulse in sophisticated systems is the *root raised-cosine* pulse.[5] The pulse is symmetric in time, so that we can work with $V^2(f)$. In a root RC pulse, $V^2(f)$ satisfies the spectrum

[5]Systems appear in practice that use RC pulses, where the design seems to intend root RC pulses. The notions of Nyquist pulse and orthogonal pulse are easily confused, and so also are RC and root RC pulses. One must beware.

shown in Fig. 2.2b, or the formulas given in (2.2-5). The pulse transform itself does not! It is the square root of the raised-cosine shape. The result is a pulse that satisfies the orthogonality constraint and has the same excess bandwidth parameter α as the RC pulse. An example pulse train appears in Fig. 2.8. It can be seen that the root RC train differs little from the RC train, despite the added feature of orthogonality; the chief difference is that it lacks the zero-crossing property. It has been shown [6] that the unit energy time-domain formula for the pulse is

$$v(t) = \begin{cases} \dfrac{1}{\sqrt{T}} \dfrac{\sin[\pi(1-\alpha)t/T] + (4\alpha t/T)\cos[\pi(1+\alpha)t/T]}{(\pi t/T)[1-(4\alpha t/T)^2]}, & t \neq 0, \neq \pm\dfrac{T}{4\alpha} \\ \dfrac{1}{\sqrt{T}}\left[1-\alpha+\dfrac{4\alpha}{\pi}\right], & t = 0 \\ \dfrac{\alpha}{\sqrt{2T}}\left[\left(1+\dfrac{2}{\pi}\right)\sin\left(\dfrac{\pi}{4\alpha}\right)+\left(1-\dfrac{2}{\pi}\right)\cos\left(\dfrac{\pi}{4\alpha}\right)\right], & t = \pm\dfrac{T}{4\alpha} \end{cases} \quad (2.2\text{-}11)$$

Since the root RC pulse is the sophisticated orthogonal pulse that finds the widest use, it is important to have a program to compute it.

Program 2.2-2: The Root RC Pulse. The next MATLAB function generates a standard unit-energy root RC pulse at time points given by the vector `t`. As in Program 2.2-1, `a` is the excess bandwidth factor, denoted α in (2.2-11); the symbol time T, the variable `tau`, is 1 unless modified in the program. The pulse values appear in vector `y`.

```
function y = rtrcpuls(a,t)
tau = 1;                              %Set symbol time
t = t+.0000001;                       %Insert offset to prevent NANs
tpi = pi/tau; amtpi = tpi*(1-a); aptpi = tpi*(1 + a);
ac = 4*a/tau; at = 16*a^2/tau^2;
y = (sin(amtpi*t) + (ac*t).*cos(aptpi*t))./(tpi*t.*(1-at*t.^2));
y = y/sqrt(tau);                      %Unit energy
```

As an example let `t` = [-6:-1:6]. Then `rtrcpuls(.3,t)` generates the $\alpha = 0.3$ pulse that may be compared to the RC pulse in Fig. 2.2. Note that the root RC pulse lacks the zero-crossing property. The example pulse here was used to construct the train in Fig. 2.8.

The reader can construct other pulse generators in this form. There is a need for one more program, namely a program to form a whole pulse train from a given pulse shape and set of transmission symbols.

Program 2.2-3: Pulse Train. This MATLAB function produces a pulse train `y`, given a transmission pulse sequence in `data` and a function subroutine `fun`, that gen-

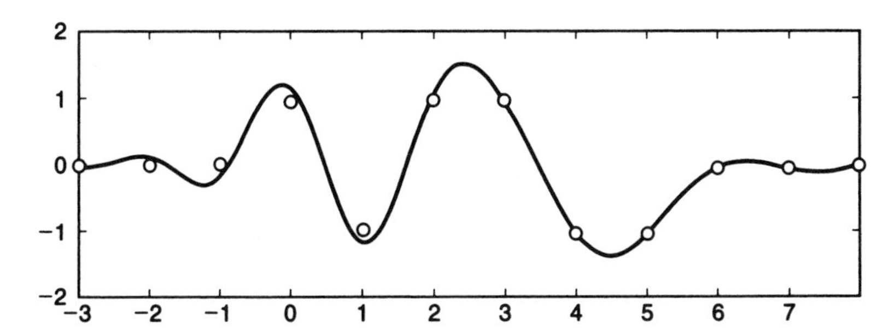

Figure 2.8 Root RC pulse train, made up from the data +1, –1, +1, +1, –1, –1. Samples at nT do not correspond to data. Compare Fig. 2.5b, 30% case.

erates the basic pulse. Additional inputs are `n`, the number of points to be evaluated per symbol interval; `width`, the total width of the basic pulse support (in symbol intervals; must be multiple of $1/n$); and `a`, a pulse parameter. The program assumes the symbol time `tau` is 1; otherwise change the first line.

```
function y = pulsetr(fun,a,n,width,data)
tau = 1;                                  %Change symbol time here
           %Compute working variables
int = tau/n; len = length(data); tmp = tau*width/2;
t = -tmp:int:tmp; lt = length(t);         %t is basic pulse points
num = n*(width+len-1) + 1;                %Total output width
           %Superpose pulses one by one, left to right
y = zeros(1,num); x = feval(fun,a,t);
for k = 1:len,
   tmp = n*(k - 1);
   y = y + [zeros(1,tmp) x*data(k) zeros(1,num-lt-tmp)];
end
```

As an example, set `data = [1 -1 1 1 -1 -1]` and take the pulse to be `rtrcpuls(a,t)` in Program 2.2-2. Set `a = .3`, `n = 10`, and `width = 6`. Then `y = pulsetr('rtrcpuls',.3,10,6,data)` yields Fig. 2.8.

2.3 PULSE TRAIN SPECTRA

Measurement of digital modulation spectra is important because many communications services, such as cellular telephone and broadcasting, work in very crowded radio bands. In any case, transmission bands are strictly regulated by government authorities. These rules generally require that radiation outside a fixed bandwidth lie below a certain fraction of the total transmission power, the object being to avoid interference to neighboring channels. Even when the channel is a coaxial cable or a pair of wires that are completely under our control and free of regulation, calculating the bandwidth is important because it will tell us how much data can flow through the channel.

The object of this section is to calculate the spectrum of a pulse train. Virtually all baseband modulations can be viewed as pulse trains, and so also can the majority of carrier modulations. The end result is simplicity itself: The spectrum of these linear modulations is essentially the spectrum of the individual pulse. By this means, we can find the spectra of most modulations.

We will start with a summary of the general modulator spectrum problem, which is quite a difficult one, and then specialize to the pulse train calculation. The easy extension from the baseband case to the carrier case is put off until Chapter 3.

2.3.1 General Spectrum Problem

Consider a full transmitted signal $s(t)$. It can be carrier or baseband, a pulse train, or otherwise; it may be coded, and the data source itself may contain pattern or correlation. The signal $s(t)$ has Fourier transform $S(f)$, and all of these factors affect $S(f)$. The general modulation problem is to find the average of the square magnitude of $S(f)$, $\mathcal{E}[|S(f)|^2]$, where the expectation is over the transmission symbols. This function is called the *average power spectral density*, abbreviated (despite the "average") as PSD.

It is easy to confuse the PSD with the power spectral density of a random process. The formal definition of the latter is

$$S_{xx}(f) \triangleq \int R_{xx}(\tau)e^{-j2\pi f\tau}\,d\tau \qquad (2.3\text{-}1)$$

in which $R_{xx}(\tau)$ is the normalized autocorrelation of a stationary random process $x(t)$. A modulated signal is seldom stationary, since it has at the least an internal structure that repeats every symbol interval[6]; in any case, (2.3-1) is the Fourier transform of an autocorrelation, not of a signal.

If the signal $s(t)$ contains correlated transmission symbols, or it is not a pulse train, taking the expectation in the PSD definition can be difficult. Such a correlation across L M-ary symbol intervals could mean transforming M^L signals and taking the average, and for many signals of interest, M^L will be very large. The solution to this computational problem that has evolved over the years is to find the PSD for a special version of $s(t)$ that has been randomized in time. Specifically, $s'(t) = s(t + \Delta)$ is formed, in which Δ is a random variable, uniformly distributed in the interval $[0, T)$. One then hopes that $s'(t)$ has the same properties, insofar as engineering is concerned, as $s(t)$. The motivation for adopting the randomized $s'(t)$ is that $s'(t)$ is a stationary random process, and its autocorrelation is

$$R(\tau) = \mathcal{E}[s(t + \Delta + \tau)\, s^*(t + \Delta)] \qquad \text{for any } t$$

in which the expectation is over both the transmission symbols and the offset Δ, and its random process PSD in (2.3-1) is now well-defined. Powerful, relatively simple

[6]A process that is stationary except for a cyclic repetition in its autocorrelation is called *cyclostationary*. A separate random process theory that is sometimes used in the study of modulated signals has developed for these.

methods exist to calculate a true random process PSD, even when the transmission symbols are correlated (see, e.g., Ref. 7) or the modulation is not linear (see Chapter 4 of Ref. 8).

We leave this general case to advanced texts and turn to signals that can somehow be viewed as a superposition of pulses.

2.3.2 Linear Modulation Spectra

Consider now the case of a pulse train modulation $s(t) = \Sigma a_n v(t - nT)$. Actually, we will generalize the signal to the full extent possible, in order to make the coming theorem more useful later, and make the following two formal assumptions:

$$s(t) = \sum_{n=1}^{N} a_n w_n(t) \qquad \text{with } |W_n(f)| = |W_1(f)|, \text{ all } n \tag{2.3-2a}$$

$$a_n \text{ IID} \qquad \text{and} \qquad \mathcal{E}[a_n] = 0, \text{ all } n \tag{2.3-2b}$$

where $W_n(f)$ is the Fourier transform of $w_n(t)$ and IID means independent and identically distributed.

The pulse train clearly satisfies (2.3-2a), since

$$|\mathcal{F}\{v(t - nT)\}| = |V(f)e^{-j2\pi fnT}| = |V(f)| = |\mathcal{F}\{v(t)\}]$$

Condition (2.3-2b) excludes coding before the modulation in general and correlation in the a_n caused by correlation in the original data source.

Theorem 2.3-1: (*Linear Modulation Spectrum Theorem*). Suppose a signal $s(t)$ satisfies (2.3-2). Then its average energy spectrum is

$$\mathcal{E}[|S(f)|^2] = N\mathcal{E}[|a_1|^2]|W_1(f)|^2 \tag{2.3-3}$$

that is, it has the same shape as the spectrum of one pulse alone and N times the energy. The PSD is (2.3-3) times $1/NT$.

Proof. The transform of $s(t)$ is just

$$S(f) = \sum_{i=1}^{N} a_i W_i(f)$$

which has energy spectrum

$$\begin{aligned} |S(f)|^2 &= \left[\sum_{i=1}^{N} a_i W_i(f)\right]\left[\sum_{k=1}^{N} a_k W_k(f)\right]^* \\ &= \sum_{i=1}^{N}\sum_{k=1}^{N} a_i a_k^* W_i(f) W_k^*(f) \end{aligned} \tag{2.3-4}$$

Taking the expectation of this, we get

$$\mathcal{E}[|S(f)|^2] = \sum_{i=1}^{N} \mathcal{E}[a_i a_i^* W_i(f) W_i^*(f)] = \sum_{i=1}^{N} \mathcal{E}[|a_i|^2]|W_i(f)|^2$$

since condition (2.3-2b) implies that the off-diagonal terms are zero in the double sum (2.3-4). Condition (2.3-2a) implies that this line is in fact (2.3-3).

The condition $\mathcal{E}[a_n] = 0$ is included in the theorem because almost all modulations satisfy it, or are assumed to. A nonzero expectation leads to a more complicated theorem and a skewed spectrum. It is worth pointing out that no random time offset was needed to obtain this PSD, nor is the PSD necessarily the power spectral density of a random process.

Since the PSD of a linear modulation is just the power spectrum of one of its component pulses, we have already seen examples of pulse train spectra in the pulse spectra of Fig. 2.2. Generally, modulation spectra consist of a strong central portion, called the *main lobe*, and a number of lesser peaks off to the side called *side lobes*. The sinc(fT) spectrum in Fig. 2.2c (which needs to be squared to give a PSD) shows a good example of these features. A good rule of thumb is that the main lobe carries the content of the modulation, and as long as it survives the channel intact, the receiver can detect the transmission without added difficulty. The chief significance of the side lobes is that they interfere with neighboring transmissions in the band. Pulse design seeks to reduce side lobes, and if they are not so reduced, they often need to be filtered away before or during passage through the channel.

The pulse spectra in Figs. 2.2a and 2.2b lack side lobes, but both of these spectra are strictly band-limited and are therefore not physically realizable. Realizable pulses always have spectral side lobes. These may be reduced by adding to the order of the pulse generation filter or, if pulse replicas are stored as time functions, by adding to their length.

Linear carrier modulation, too, has these spectral features. We will look at these spectra in Chapter 3.

2.3.3 Spectral and Power Out-of-Band Plots

A common way to portray spectra in communication engineering is by plotting $S(f)$ in dB—that is, $20 \log_{10}|S(f)|$—with the zero point in the scale taken as the peak of the main lobe. As well, the frequency scale is often taken as fT, frequency normalized to the symbol period, with the dimensions Hertz-second/symbol.[7]

Figure 2.9 shows the spectra of NRZ signaling with these scalings, together with RZ and Manchester spectra from the examples below. The scalings make it much easier to see the side-lobe structure. The nulls, here and elsewhere, tend to relate in

[7]Sometimes T is T_b, the data bit time, and then fT_b in Hz-s/b measures the bandwidth consumption per data bit. These SI units become cycles/symbol or cycles/bit if one takes hertz as cycles per second; hertz per symbol is used colloquially but is incorrect.

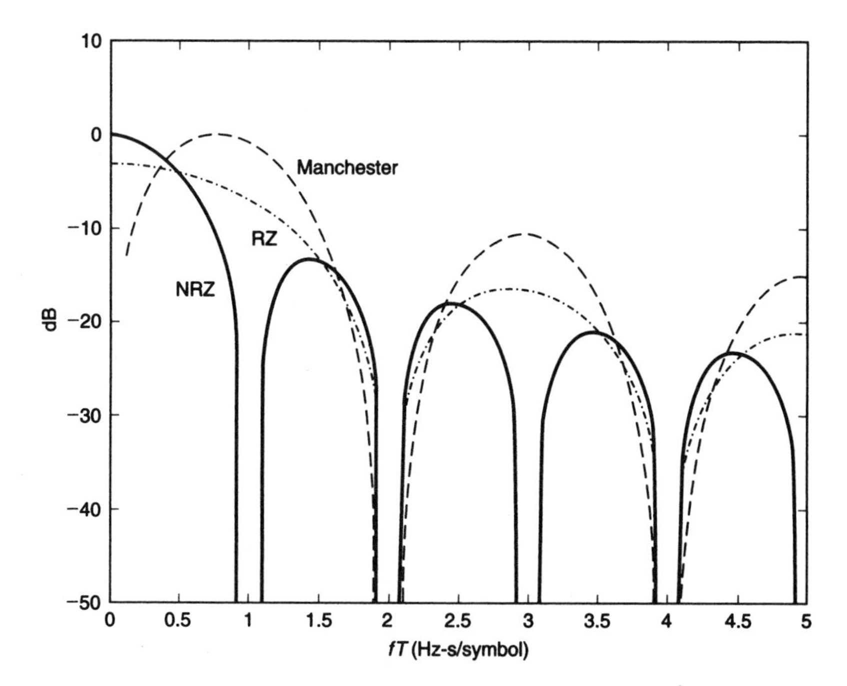

Figure 2.9 Manchester, RZ, and NRZ power spectral densities. Decibels versus frequency normalized to data bit time. Pulses have unit energy.

a simple way to fT. Side-lobe peaks as much as 60 dB below the main lobe are important in mobile communication, and the decibel scale makes these as well as their rate of decay visible.

Another common portrayal of spectra is the *power out-of-band* (POB) plot. The idea of this is to plot at frequency f the total energy outside f, rather than the density at f. Mathematically,

$$P_{OB}(f) \triangleq \frac{\int_f^{\infty} |S(f)|^2 \, df}{\int_0^{\infty} |S(f)|^2 \, df}, \qquad f > 0 \tag{2.3-5}$$

The POB is used to calculate other spectral measures. When $P_{OB}(f) = 0.01$, f is the frequency inside which lies 99% of the signal power, a quantity called the 99% power bandwidth. Typically, when a signal passes through a channel of this width, its detection is unaffected by band limitation, and so 99% bandwidth is often quoted. Analogously, one can define 99.9% and 99.99% power bandwidths; these are important as measures of interference to adjoining channels.

Figure 2.10 is a POB plot for the modulations in Fig. 2.9, with the same decibel and fT scalings. It is easy to see from the different power bandwidths how the channel might affect the modulations and vice versa. The 99% power bandwidth of

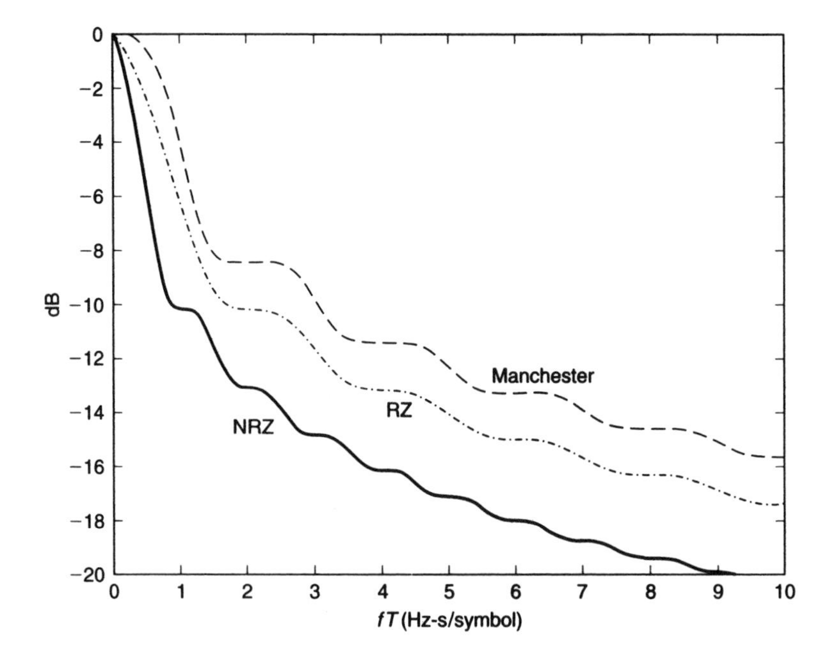

Figure 2.10 Power-out-of-band plots for Manchester, RZ, and NRZ pulses; decibels versus frequency normalized to symbol time. The 99% power bandwidth of NRZ pulse is around

NRZ, the bandwidth where the NRZ curve crosses $10 \log_{10} .01 = -20$ dB, is about 9.5 times $1/T$, counting positive frequencies. Manchester and RZ signaling are twice this. By comparison, the root RC pulse 99% bandwidth is about $0.6/T$! This shows that NRZ signaling is impractical in a band-limited environment.

We will conclude the section with calculation of some basic spectra. These appear in Figs. 2.9 and 2.10.

Example 2.3-1: Binary RZ Format. This polar format replaces the ±1 square NRZ pulse with a half-length square followed by a return to zero (see Fig. 2.1). Consequently, the unit-energy NRZ pulses $v(t) = (\sqrt{1/T})\text{rect}(t/T)$ is replaced by

$$v(t) = \sqrt{\frac{2}{T}}\,\text{rect}\left(\frac{2t}{T} + \frac{1}{2}\right) \qquad \text{(RZ pulse)} \qquad (2.3\text{-}6)$$

Taking the Fourier transform gives

$$V(f) = \sqrt{\frac{T}{2}}\,\text{sinc}\left(\frac{1}{2}fT\right)e^{j2\pi fT/4} \qquad \text{(RZ spectrum)} \qquad (2.3\text{-}7)$$

As a power spectrum, this is simply the NRZ spectrum expanded twofold in frequency and renormalized to unit energy.

Example 2.3-2: Binary Manchester Format. This format replaces the NRZ pulse with a positive half-pulse followed by a negative half-pulse, as in Fig. 2.1. As a unit-energy pulse centered at the origin, this is

$$v(t) = \sqrt{\frac{1}{T}}\left[\text{rect}\left(\frac{2t}{T} + \frac{1}{2}\right) - \text{rect}\left(\frac{2t}{T} - \frac{1}{2}\right)\right] \qquad \text{(Manchester pulse)} \qquad (2.3\text{-}8)$$

The transform is

$$V(f) = 2j\sqrt{T}\,\text{sinc}(\tfrac{1}{2}fT)\sin(\tfrac{1}{2}\pi fT) \qquad \text{(Manchester spectrum)} \qquad (2.3\text{-}9)$$

The power spectrum is close to a double-width RZ spectrum but has an additional null at $f = 0$.

2.4 TRANSMISSION IMPAIRMENTS

Distortion during transmission is a fact of life. One source of distortion to a pulse train is random noise, and we will take up this in the next section. A host of non-noise distortions, or “impairments” as they are called, can affect signals. Channel filters are a necessity because they make a signal bandwidth efficient and reduce interference to neighboring transmissions, but filters also distort signals. So also do nonlinearities in signal amplifiers. Another kind of distortion occurs when the timing of the sampler in Figs. 2.3 or 2.7 is faulty, so that the signal there is sampled too early or too late. We will study some of these impairments with the aid of a convenient tool called the *eye pattern*. Another set of impairments affect carrier-modulated signals; we will take up these in Chapter 3.

2.4.1 Intersymbol Interference

Much of what happens to pulse trains can be characterized as *intersymbol interference*, or ISI for short. Loosely defined, ISI is the effect of a pulse on the detection in other symbol intervals. Figure 2.11 shows the standard 30% excess-bandwidth RC pulse of Fig. 2.2 together with a delayed version of the same pulse and a low-pass filtered version. The original pulse is Nyquist. Consequently, in the sampling receiver (Fig. 2.3) with accurate sampling times nT, the pulse is zero at all sample times other than its own and does not affect the other sampler outputs. In the linear receiver (Fig. 2.7), with root RC pulses, the response at the receiver filter output to each root RC pulse is the RC pulse, and here, again, with accurate sample timing each transmitted pulse affects only one sampler output.

The second pulse in Fig. 2.11 is delayed by $0.2T$, so that the sampling occurs $0.2T$ too early. Now the pulse contributes at other sampling times, in the amounts shown in the figure. This is ISI stemming from sampling time error. The third

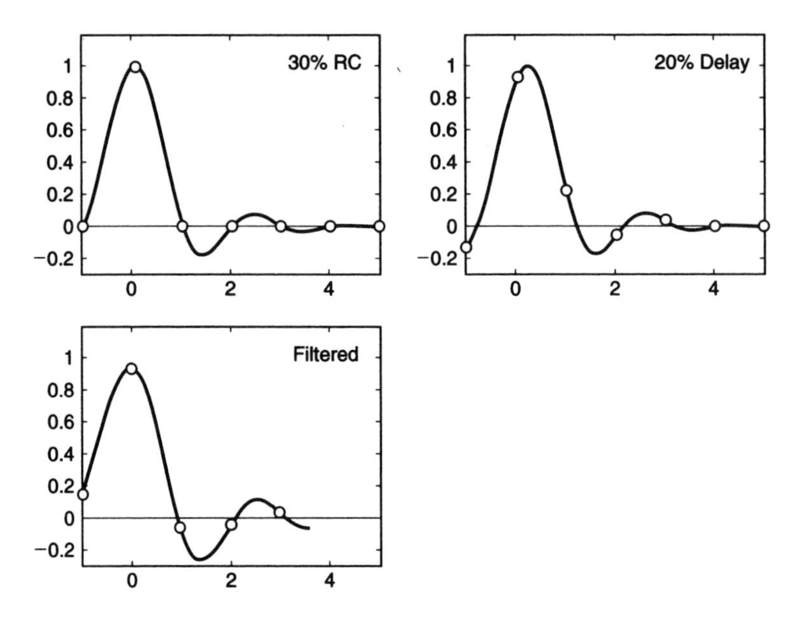

Figure 2.11 Intersymbol interference with 30% RC pulse: Response versus time for original pulse, pulse delayed 20%, and pulse filtered by six-pole Butterworth filter with cutoff $0.5/T$ hertz. Left part of each pulse is suppressed for clarity. Filter delay of $1.4T$ is subtracted from bottom pulse time scale.

pulse is the RC pulse filtered by a six-pole Butterworth filter with cutoff frequency at $0.5/T$ Hz (details about this filter are in Example 2.4-1). The filter has a delay of about $1.4T$ seconds; but even with this taken out, significant contributions remain at the other sampling times. This ISI stems from the time dispersion of the filter.

2.4.2 Eye Patterns

The effect of all these impairments is easily seen from a diagram called an *eye pattern.* There is an easy way to explain the construction of an eye diagram, which is also a way to measure one in the field. Imagine that the baseband waveform is connected to an oscilloscope whose time base is triggered by the receiver sampler timing once each T seconds; feed a long sequence of random data to the transmitter. The result will be a superposition of the possible T-second movements in the waveform. Figure 2.12a shows what happens with a binary NRZ pulse train, and Fig. 2.13a shows an RC train. Both NRZ and RC pulses are Nyquist; consequently, both eye patterns pass through ± 1 at the sample times, which are at the center of the time scale. The NRZ case is simply a square block, but the RC pulse train passes through ± 1 in many different ways, in a pattern that is typical of practical modulations. The pattern forms a kind of eye, from which the name of the display comes. The eyes in Figs. 2.12a and 2.13a are said to be fully open.

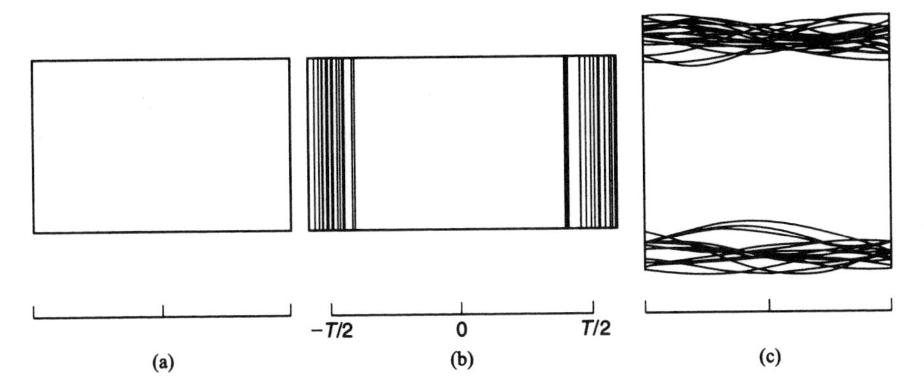

Figure 2.12 Some NRZ pulse train eye patters. **(a)** Undistorted pulse train; **(b)** 7% rms Gaussian timing jitter, 20 random transmission symbols; **(c)** Gaussian additive noise, single-sided bandwidth $1/T$ hertz, and signal-to-noise ratio 19 dB, with random symbols. Plot axes are pulse train amplitude versus symbol time; sampling is nominally at time 0.

The effect of impairments is to partially close the eye some. As long as the eye is open at least a little in the middle at the sampling time, it is clear that the sampling and linear receivers will detect correctly. If all the space is filled with signal transitions, then some of the +1s are being driven to −1s and vice versa, and detection can be erroneous.

Figure 2.12b shows the effect of sampling time error on the NRZ pulse eye pattern, and Fig. 2.13b shows the effect on the RC pulse pattern. In both cases the error is fixed during each symbol interval and varies from interval to interval according to a zero-mean Gaussian distribution whose standard deviation is 7% of T.[8] The timing error narrows the root-mean-square (rms) RC eye opening by about 8%. The sampling receiver looking at this pulse train will thus continue to make correct decisions, as will a linear receiver that views the waveform at the output of its receiver filter. But there is less resistance now to other narrowings, such as those caused by noise. Timing error does not reduce the NRZ eye at all. Here we can see that the narrower bandwidth of the RC train has been taken in trade for less resistance to timing error.

Figures 2.12c and 2.13c show the effect of Gaussian channel noise on the two eye patterns. Now both NRZ and RC eyes are narrowed the same amount in the center of the eye. The position of each signal trajectory at the sample time is a Gaussian random variable whose σ in these pictures is about 12% of the no-noise eye opening. If the transmission is impaired by both timing error and noise, the variances of the effects in plots (b) and (c) will add. The rms sum of the 8% and 12% in the figures would be about 14% total narrowing in an rms sense.

The narrowing of an eye pattern is sometimes expressed as a loss in signal-to-noise ratio (SNR), since a narrowing from whatever source is equivalent to scal-

[8]Sample timing error is called *jitter*; it will be discussed further in Chapter 4.

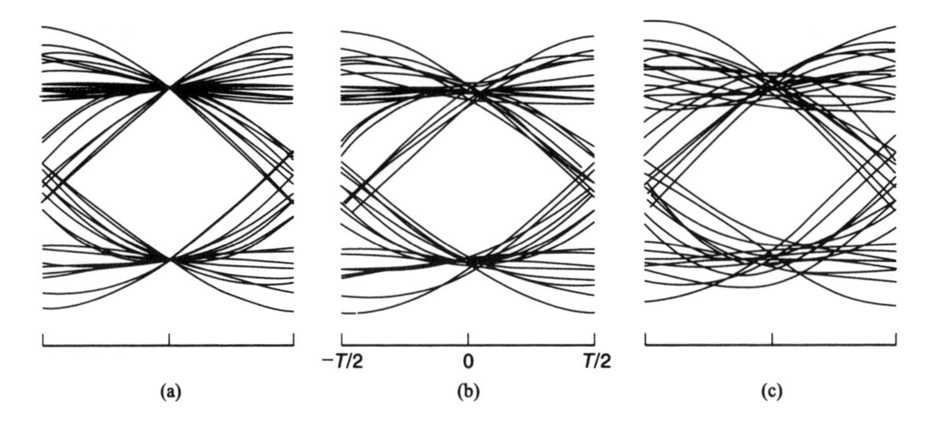

Figure 2.13 Some 30% RC pulse eye patterns. Plots **(a)–(c)** as in Fig. 2.12.

ing back the signal amplitude while the noise stays fixed. An 8% timing error narrowing, for example, would be reckoned as an SNR loss of $20 \log_{10}(0.92) = -0.72$ dB.

An eye pattern may also be plotted for nonbinary transmission symbols. A four-level eye appears in Fig. 2.14; 80 random symbols are chosen from the set $\{-3, -1, 1, 3\}$, and the pulse is 30% RC. There are somewhat more ± 1 than ± 3 pulses in this example.

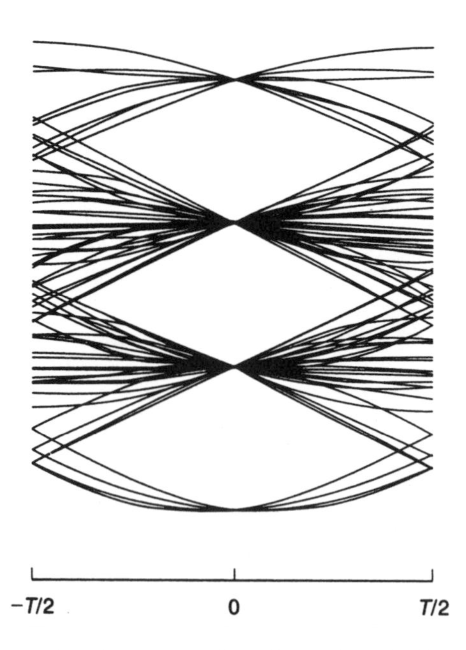

Figure 2.14 Four-level eye pattern from 80 symbols; basic pulse is 30% RC.

2.4.3 Channel Filter Effects

The next two examples illustrate the effect of filtering on pulse trains. Both the pulses and the filters are typical of those encountered in practice. The eye pattern closures in these examples would add to those we have already discussed. Figure 2.15 shows the system setup. For the sampling receiver, the channel filter G follows the Nyquist pulse-forming filter V; for the linear receiver, $\sqrt{V}$ forms orthogonal pulses that are filtered by G in the channel and $\sqrt{V^*}$ in the receiver. What reaches the sampler has been filtered by VG in either case.

Example 2.4-1: Filtering of Narrowband Pulses. This example looks at an RC pulse train (root RC in the linear receiver case) with 30% excess bandwidth that is filtered by a six-pole Butterworth low-pass filter. The frequency response of a Butterworth filter is given by

$$|H(jf)| = \left[\frac{1}{1+(f/f_0)^{2N}}\right]^{1/2} \tag{2.4-1}$$

in which N is the filter order and f_0 is the 3-dB cutoff frequency. A six-pole filter cuts off rapidly after this frequency. Three cutoffs will be employed, namely, f_0 values of $1.0/T$, $0.5/T$, and $0.4/T$ hertz; the spectra of these three filters are plotted in Fig. 2.16. Recall that the magnitude spectrum of the pulse train falls to half at $0.5/T$ and to zero at $0.65/T$ hertz. The impulse responses of the three appear in Fig. 2.17, from which it is clear that the filters disperse and delay the pulse train to some degree.[9] Figures 2.18a–c show the filtered eye pattern from 40 random data for cutoffs 1.0, 0.5, and 0.4, respectively. At cutoff 1.0 in plot (a), the filter bandwidth exceeds the pulse train bandwidth, and there is no measurable effect in the eye pattern. A narrowing by about 20% occurs with cutoff 0.5 in (b), which rapidly increases with cutoff 0.4 in (c); a cutoff of 0.35 will close the eye completely. Since there is a delay in the filters, the eye plotting must be similarly delayed in order to center the eyes. Plots (a), (b), and (c) are delayed by $0.6T$, $1.4T$, and $1.7T$ s, respectively, which correspond roughly to the delays in the peaks in Fig. 2.17. The lessons here are that successful detection can take place even when the filter cutoff lies well within the pulse bandwidth, although there will be some loss in noise resistance[10]; however, the detector must take into account the filter delay.

Example 2.4-2: Filtering of Wideband Pulses. This example looks at the same filters as the previous one, but the RC (or root RC) pulse train has 100% excess bandwidth. Now the pulse train spectrum reaches out to $1/T$ Hz, which is far beyond the two narrower cutoff frequencies. In fact, half the root RC pulse train bandwidth

[9]The filters in these examples are actually discrete-time equivalents, achieved with the bilinear transformation and sample rate $10/T$. This is typical practice.

[10]Some of the loss may be recouped by using a more advanced receiver than the linear receiver.

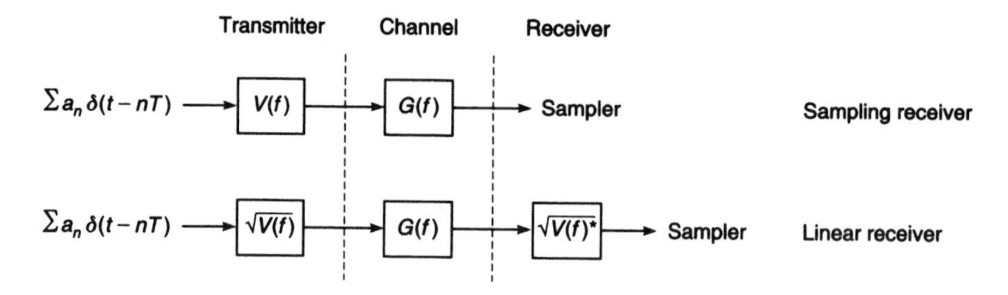

Figure 2.15 Channel filtering models for Examples 2.4-1 and 2.4-2. Eye patterns measured prior to samplers.

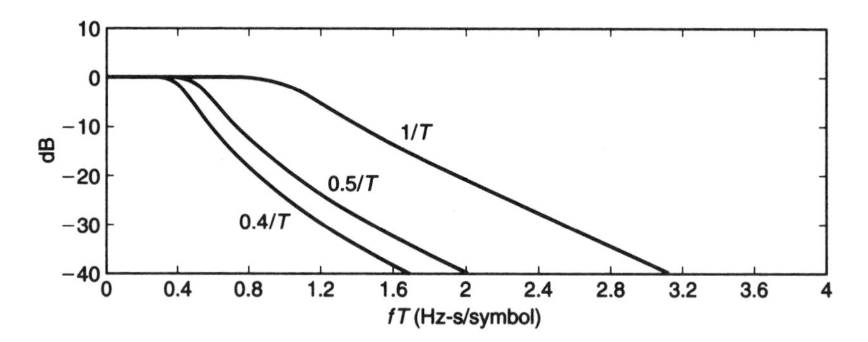

Figure 2.16 Three six-pole Butterworth magnitude spectra for filters in Examples 2.4-1 and 2.4-2. Cutoff frequencies as shown. Axes in decibels versus normalized frequency fT.

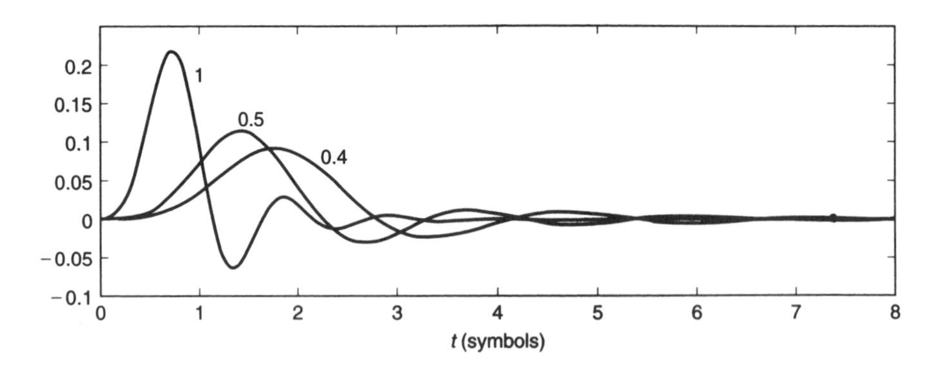

Figure 2.17 Impulse responses of three filters in Fig. 2.16.

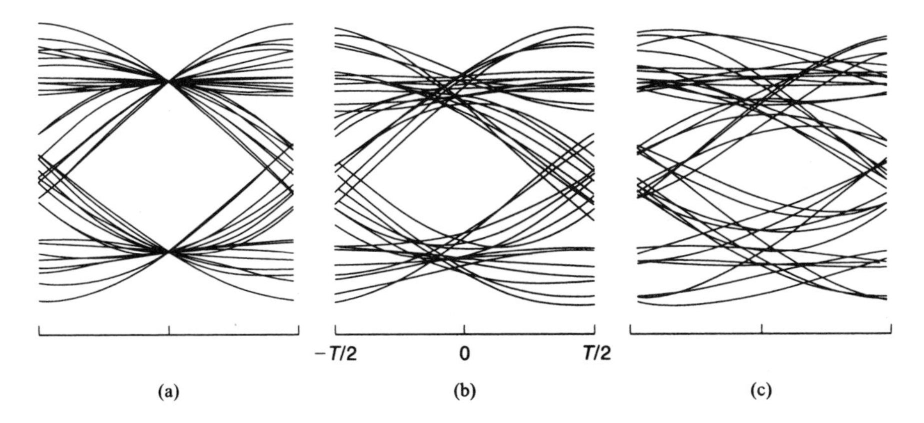

Figure 2.18 Filtered 30% RC pulse train eye patterns: **(a)** $1/T$ hertz filter bandwidth; **(b)** $0.5/T$ hertz; **(c)** $0.4/T$ hertz. Forty random binary symbols. Filters as in Fig. 2.16; delays of $0.6T$, $1.4T$, and $1.7T$, respectively, subtracted from pulse train time scales. Filter of bandwidth $0.35/T$ will completely close the eye.

lies outside bandwidth $0.5/T$ hertz (cf. the spectrum in Fig. 2.4). Figures 2.19a–c show the filtered eye patterns for cutoffs $1.0/T$, $0.5/T$, and $0.4/T$. The patterns are delayed as they were in Example 2.4-1. For cutoff 1.0, the pattern in (a) shows a little closure at the widest opening; it shows also that the wideband RC pulse train has only a few different transitions. At cutoff 0.5 in (b), the eye narrows by about the same as it did with the 30% pulses in Fig. 2.17b, although there is a different character to the transitions; a similar conclusion holds for plot (c). It is interesting to compare the actual filtered pulse trains for some of these cases. Figure 2.20 compares the 30% and 100% cases with and without filtering at cutoff $0.4/T$ hertz. For

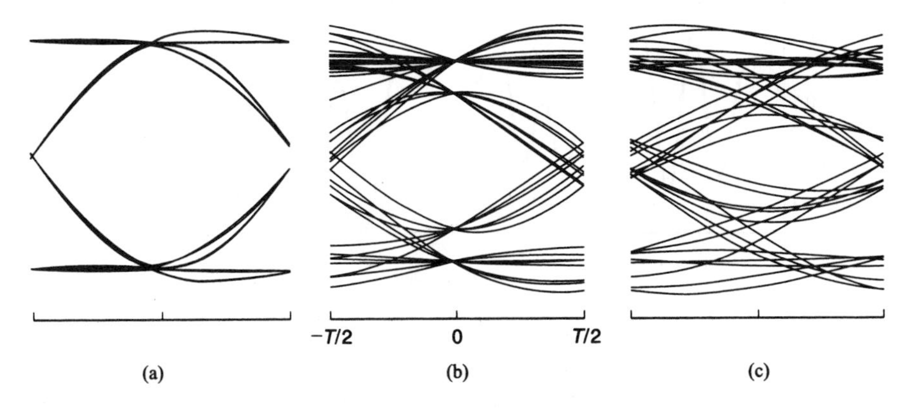

Figure 2.19 Filtered 100% RC pulse train eye patterns: **(a)** $1/T$ hertz filter bandwidth; **(b)** $0.5/T$ hertz; **(c)** $0.4/T$ hertz. All other data as in Fig. 2.18.

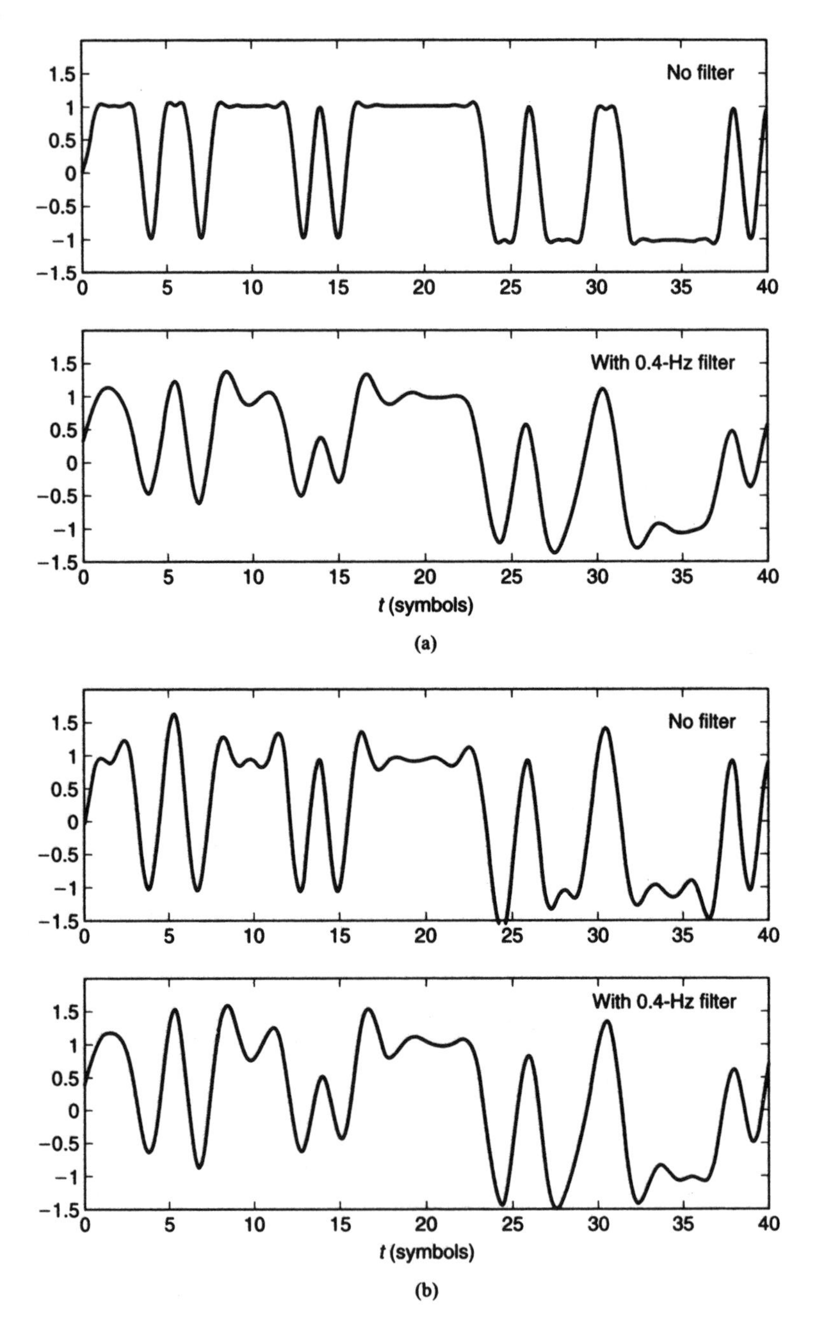

Figure 2.20 Actual pulse trains for Figs. 2.18 and 2.19: **(a)** 100% RC pulse train with and without $0.4/T$ filtering; **(b)** 30% RC pulse train with and without same filtering. There is little difference in the trains after filtering.

pulses that are already reduced to 30% excess bandwidth, the filtering has little effect. When it is applied to a 100% pulse, the end result is similar to the 30% pulse train. The lesson here is that filtering can be applied before or after transmission, with a similar gross effect on the signal. But the Butterworth filter destroys the zero-crossing property of the RC train and thus reduces the noise resistance of the linear receiver.

2.5 SIGNAL SPACE ANALYSIS

The major remaining task in this chapter is to calculate the error probability of a digital detector. Our goal in this section is to design a detector that decides which symbol is the most likely cause of a signal that the detector observes. The overall train of thought in this design has come to be called *signal space* analysis. Its origin, and indeed much of its modern form, traces back to the 1947 Ph.D. thesis of V.A. Kotelnikov [10]. Its passage to an everyday engineering tool stems from the classic 1965 text by Wozencraft and Jacobs [11], which is still available and remains the premier reference for those in need of 100+ pages on the subject. The main features of signal space analysis are the reduction of continuous-time signals to vectors, which enables the error probability calculation, and the portrayal of signals as points in an ordinary Euclidean geometry. Parts of the analysis are subtle and mathematical, but others—such as viewing signals as geometric constellations—are simple and intuitive and have found universal acceptance, even among those not mathematically inclined.

2.5.1 Maximum-Likelihood and MAP Receivers

A schematic diagram of a communication system is shown in Fig. 2.21. The figure also establishes notation for what is to come. The nth symbol in a sequence has appeared at the transmitter; it takes on one of the M values $m_1, m_2, \ldots, m_M$, and the transmitter converts it to one of the signals $s_1(t), s_2(t), \ldots, s_M(t)$. In the channel,

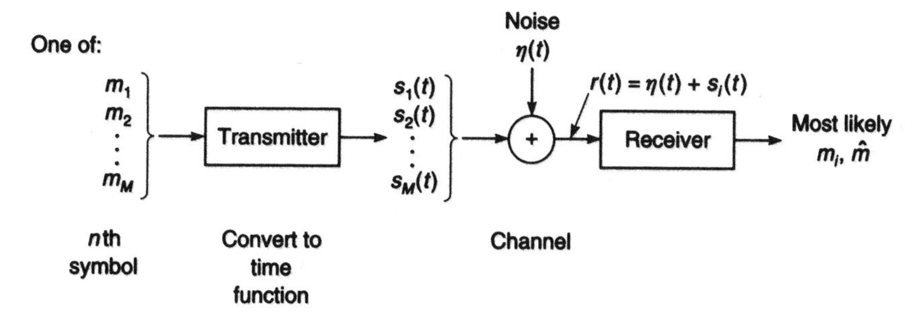

Figure 2.21 Setup for transmission of M messages.

noise $\eta(t)$ is added to the transmitted signal $s_i(t)$ to form the received signal $r(t) = s_i(t) + \eta(t)$. The receiver decides which signal was most likely sent: It selects the most likely of the set of signals $\{s_i(t)\}$ from the information that it has available. Its decision is the symbol $\hat{m}$.[11]

What information does the receiver have? To make sense of this question, we need to take a probabilistic view of the communication system. Obviously, the noise is random. We will assume also that the source is random. Many real sources of data take on no consistent symbol distribution, and some, such as an idling data source, even put out long repeating streams. Some are effectively random, but the symbols in different positions have different distributions. An example is ordinary 8-bit PCM; here the initial sign bit is equiprobable but correlated from PCM word to PCM word, the higher-order bits are skewed toward one symbol outcome, and the lower-order bits are nearly IID. What is to be done about this? We do not wish to redesign the detector at each change in the nature of the data. A practical compromise often followed is to model the data as random and uncorrelated, with some long-term distribution. Or, one can take a truly distant view and simply take the data as IID random.

We can organize all this information into two groups, the received signal $r(t)$, which is an observation, and the suppositions and guesses about the source, which are called the *a priori information*. To find out which transmitted signal is most likely given the information, the detector must calculate the conditional probabilities $P[s_1(t)|r(t)], P[s_2(t)|r(t)], \ldots, P[s_M(t)|r(t)]$ and find the largest one. From the definition of conditional probability, the ith of these probabilities is given by

$$P[s_i(t)|r(t)] = \frac{P[s_i(t) \text{ sent and } r(t) \text{ received}]}{P[r(t) \text{ received}]} \tag{2.5-1}$$

The numerator here may be written as $P[s_i \text{ sent}|r(t) \text{ received}] \cdot P[r(t) \text{ received}]$ or, equivalently, as $P[r(t) \text{ received}|s_i \text{ sent}] \cdot P[s_i \text{ sent}]$ (this is a statement of Bayes' rule). Consequently, (2.5-1) may be written as

$$\frac{P[r(t) \text{ received}|s_i(t) \text{ sent}]P[s_i(t) \text{ sent}]}{P[r(t) \text{ received}]} \tag{2.5-2}$$

The i that maximizes (2.5-2) indicates the signal most likely transmitted.

Some simplifications to (2.5-2) can be made. First, the probability $P[r(t)|s_i(t)]$ is identical to the probability that the noise equals the difference between $r(t)$ and $s_i(t)$, $P[\eta(t) = r(t) - s_i(t)]$, since the channel noise is additive. Second, $P[r(t)]$, whatever it might be, does not depend on i, and it can be struck out of (2.5-2) during the maximizing over i. What remains after these two steps is

$$\text{Find } i \text{ that achieves} \qquad \max_i P[\eta(t) = r(t) - s_i(t)]\, P[s_i(t)] \tag{2.5-3}$$

[11]Here and throughout, the hat symbol over a variable indicates an estimate of the variable.

A detector that performs (2.5-3) is called a *maximum a posteriori* (MAP) receiver. The meaning of MAP is that the observation $r(t)$ and prior knowledge about the data source help in the decision a posteriori, that is, after the observation.

Although the MAP receiver is becoming more common, it has not found wide use in digital communication because members of the set $\{P[s_i(t)]\}$—the prior probabilities in the language of probability—are often unknown or hard to define. In this case the best that can be said is simply that the data symbols are IID. Then $P[s_i(t)] = 1/M$ for all i, and the probability in (2.5-3) becomes $(1/M)P[\eta(t) = r(t) - s_i(t)]$. As before, we can remove any factor that is constant during the maximizing over i. This leaves

$$\text{Find } i \text{ that achieves} \qquad \max_i P[\eta(t) = r(t) - s_i(t)] \tag{2.5-4}$$

A detector that performs (2.5-4) is called a *maximum likelihood* (ML) receiver.[12] In finding $\hat{m}$, it considers only the channel noise probability distribution.

The ML receiver is the basis of most detectors in digital communication. This simple form is often very easy to construct, and even when it is not, it is the inspiration behind simplified versions. When the symbols really are IID, the ML receiver is also MAP; otherwise, it is often the best we can do.

One subtlety in the detector setup needs clarifying. In a simple view of Fig. 2.21, the m_i are directly data symbols, in the binary case, they are just ± 1s. In a more sophisticated view, the m_i are collections of symbols—that is, whole *messages*. For example, in PCM it might make more sense to take a whole 8-bit PCM word as a "symbol" and to capture the complexity of the PCM word structure in the probabilities of the 256 possible symbol outcomes. One might even go further and think of groups of PCM words as a symbol and thus capture in the symbol distribution the correlation between PCM words. The result will be better detection because the data source is better modeled. The point here is that *Eqs. (2.5-1)–(2.5-4) are unchanged*. In the spirit of Section 2.2, $s_i(t)$ could be an entire pulse train.

Before continuing, we reiterate the assumptions made so far:

- The channel noise is additive and a random variable; it may be correlated with itself but not with the transmitted signal.
- The data source is a random variable, possibly correlated.
- The source puts out M different well-defined messages; M may be large.

2.5.2 Definition of Signal Space

Unfortunately, it is often true that the probabilities in (2.5-3) and (2.5-4) cannot be evaluated directly because they contain a fatal flaw: There is no way to consistently assign probabilities to the outcomes of the random process $\eta(t)$. If the noise process takes on discrete realizations, we might assign probabilities to each one, but for most cases, including the Gaussian case, the time function $\eta(t)$ takes on an uncount-

[12]The name comes from estimation theory where the form (2.5-4) is called an ML estimator; sometimes, as when we have a secret hunch about the data, the name conflicts with common sense.

able infinity of outcomes and it is impossible to assign a nonzero probability to any one of them. For a precise rendering of the problem, we defer to a stochastic processes text. For now, we can simply write down the way out of the difficulty: *Construct an orthogonal basis for the outcome functions and then work with their vector space components.*

The orthogonal basis will not be different in principle from, for example, the Fourier basis that underlies the Fourier transform, but we need a basis that is better adapted to the problem at hand. The transmitted signals $s_i(t)$, the noise $\eta(t)$, and the received signal $r(t)$ will all be expressed as weighted sums of the basis functions.

To begin, assume that a set of J orthogonal basis functions $\{\phi_1(t), \ldots, \phi_J(t)\}$ has been obtained; we will look at how to do this shortly. Express the ith transmitted signal $s_i(t)$ as the vector[13]

$$\boldsymbol{s}_i = (s_{i1}, s_{i2}, \ldots, s_{iJ}), \qquad i = 1, \ldots, M \tag{2.5-5}$$

The jth component here is the inner product $\int_{\mathcal{T}} s_i(t)\phi_j^*(t)\, dt$ of the signal $s_i(t)$ with the basis function $\phi_j(t)$, and $\mathcal{T}$ is the interval over which the signals are active. Each of the M signals is exactly given by

$$s_i(t) = \sum_{j=1}^{J} s_{ij}\phi_j(t), \qquad i = 1, \ldots, M \tag{2.5-6}$$

In the same way we express the noise $\eta(t)$ as

$$\boldsymbol{\eta} = (\eta_1, \ldots, \eta_J, \eta_{J+1}, \ldots) \tag{2.5-7}$$

The noise is usually not confined to the J dimensions represented by the signal basis functions, and so we have shown some extra dimensions in (2.5-7). But it will turn out that these dimensions play no role, and consequently we can be obscure about what the corresponding basis functions might be. The received signal $r(t)$ is a signal plus the same noise, and so its full representation is

$$\boldsymbol{r} = (r_1, \ldots, r_J, r_{J+1}, \ldots) \tag{2.5-8}$$

Once again, we will not need to know the basis functions beyond the Jth.

Figure 2.22 repeats Fig. 2.21, but this time in terms of the vector components. Because of the additive noise, each received signal component r_j satisfies $r_j = \eta_j + s_{ij}$. The process just described comprises *forming the signal space* for the given set of transmitted signals $\{s_i(t)\}$.

We are now in a position to write the MAP and ML receivers in terms of well-defined probabilities. The expression is

$$\text{Find } i \text{ that achieves} \qquad \max_i P[\boldsymbol{\eta} = \boldsymbol{r} - \boldsymbol{s}_i]P[s_i(t)] \tag{2.5-9}$$

[13]A boldface letter ordinarily denotes a row vector; a column vector is denoted by the transpose—for example, $\boldsymbol{s}'$. $\boldsymbol{s}_i$ denotes the ith of a set of vectors, s_j is the jth component of $\boldsymbol{s}$, and s_{ij} is the jth component of vector $\boldsymbol{s}_i$.

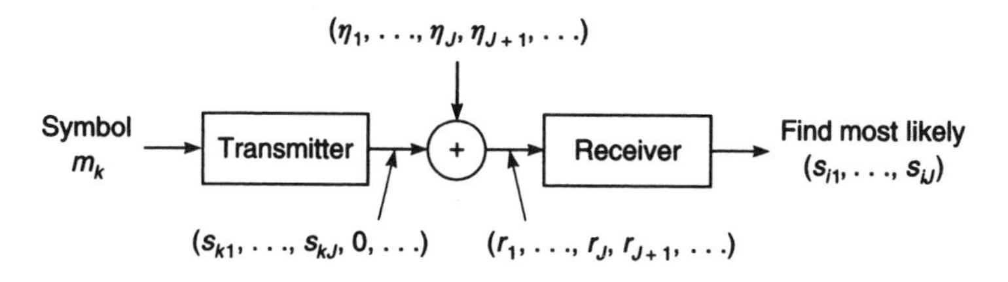

Figure 2.22 Vector version of Fig. 2.21.

for MAP receivers and

$$\text{Find } i \text{ that achieves} \qquad \max_i P[\boldsymbol{\eta} = \boldsymbol{r} - \boldsymbol{s}_i] \tag{2.5-10}$$

for ML receivers. The meaning of the probabilistic events is that all the vector components on either side of the equal sign must match. Because the components are scalars and not time functions, random process theory tells us that the probabilities may be well-defined.

Before going further, we need to review how to find an acceptable orthonormal basis for the transmitted signals. As an exercise in random process theory, the process consists of two parts, finding a basis for $s_i(t)$, $\eta(t)$, and $r(t)$ and showing that the probabilities of all events of interest may be found from the vector component probabilities. The Karhunen–Loève expansion performs the job and creates a complete, possibly infinite-dimensional orthonormal basis for all the signals, including the noise. Furthermore, for white Gaussian noise any complete orthonormal basis is acceptable for the noise if its first J functions are complete and orthonormal for the signals taken alone.[14] The consequence is that we need only find a basis for the set $\{s_i(t)\}$.

An orthonormal basis for a target set of functions is found by the *Gram–Schmidt procedure*, which we will now outline. The version here is for real functions, but it easily extends to the complex case.

Step 1: Construction of $\phi_1(t)$. The first basis function may be any of the $s_i(t)$; taking $s_1(t)$, form

$$\phi_1(t) = \frac{1}{\sqrt{E_1}} s_1(t) \tag{2.5-11}$$

where

$$E_1 = \int_{\mathcal{T}} |s_1(t)|^2 \, dt$$

The effect of the $\sqrt{E_1}$ factor is to normalize $\phi_1(t)$ to unit energy. (The interval $\mathcal{T}$, as before, is the interval over which the set of signals is active.)

[14]The details of the general case appear in, e.g., van Trees [12].

Step 2: Construction of $\phi_2(t)$. To find the second basis function, first define the auxiliary function

$$\theta_2(t) = s_2(t) - (s_2, \phi_1)\ \phi_1(t) \tag{2.5-12}$$

in which

$$(s_2, \phi_1) \triangleq \int_{\mathcal{T}} s_2(t)\phi_1^*(t)\, dt$$

is the inner product between s_2 and ϕ_1. The effect of Eq. (2.5-12) is to remove any part of s_2 that lies along ϕ_1, leaving a remainder that is orthogonal to ϕ_1. The second basis function is then

$$\phi_2(t) = \frac{1}{\sqrt{E_2}}\ \theta_2(t) \tag{2.5-13}$$

where

$$E_2 = \int_{\mathcal{T}} |\theta_2(t)|^2\, dt$$

Again, $\sqrt{E_2}$ normalizes ϕ_2.

Step 3: Construction of the jth Basis Function. If $\phi_1(t), \ldots, \phi_{j-1}(t)$ are already found, $\phi_j(t)$ is found by first forming the auxiliary function

$$\theta_j(t) = s_j(t) - \sum_{k=1}^{j-1} (s_j, \phi_k)\phi_k(t) \tag{2.5-14}$$

It is easy to see that $\theta_j(t)$ is orthogonal to every previous basis function $\phi_\ell(t)$, $\ell < j$:

$$\begin{aligned}(\theta_j, \phi_\ell) &= (s_j, \phi_\ell) - \sum_{k=1}^{j-1} (s_j, \phi_k)\ (\phi_k, \phi_\ell)\\ &= (s_j, \phi_\ell) - (s_j, \phi_\ell)[0 + \cdots 0 + 1 + 0 \cdots]\\ &= 0\end{aligned}$$

The second line follows because all the previous basis functions are mutually orthogonal. The new basis function is $\theta_j(t)$ normalized, which is

$$\phi_j(t) = \frac{1}{\sqrt{E_j}}\ \theta_j(t) \tag{2.5-15}$$

where

$$E_j = \int_{\mathcal{T}} |\theta_j(t)|^2\, dt$$

The Gram–Schmidt process produces at most one new orthonormal function for each function in the target signal set. If it happens that the next target function can be expressed as a linear combination of earlier ones, then $\theta_j(t)$ in (2.5-14) is zero and no new basis function is produced; the vector that represents the next function is linearly dependent on the previous vectors.

All these ideas are illustrated in the following example with simple pulses. There would be little point in applying the process to the set of T-shifts of an orthogonal pulse, since this set is already orthogonal and need only be normalized to make up an acceptable basis for itself. The same set translated by any t_o is also its own orthogonal basis. We will apply the Gram–Schmidt process to carrier signals in Chapter 3.

Example 2.5-1: Gram–Schmidt Process with Simple Pulses. The three waveforms in Fig. 2.23 are to be expressed in terms of an orthonormal basis. In step 1, $s_1(t)$ needs only to be normalized by $\sqrt{2}$ to produce $\phi_1(t) = (1/\sqrt{2})\, s_1(t)$. In step 2,

$$\theta_2(t) = s_2(t) + \tfrac{1}{2}\, s_1(t)$$

and

$$\phi_2(t) = \frac{2}{\sqrt{6}}\left[s_2(t) + \frac{1}{2}\, s_1(t)\right]$$

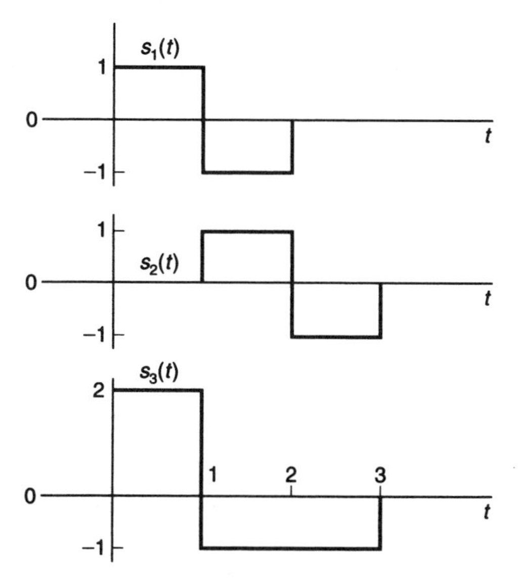

Figure 2.23 Three signal waveforms for Example 2.5-1.

However, an attempt to find $\theta_3(t)$ leads to $\theta_3(t) = 0$, since

$$s_3(t) = \frac{3}{\sqrt{2}}\phi_1(t) + \frac{3}{\sqrt{6}}\phi_2(t)$$

is linearly dependent on s_1 and s_2. A two-dimensional space expresses all three functions. The vectors that represent the functions are

$$\mathbf{s}_1 = (\sqrt{2}, 0), \qquad \mathbf{s}_2 = \left(\frac{-1}{\sqrt{2}}, \frac{3}{\sqrt{6}}\right), \qquad \mathbf{s}_3 = \left(\frac{3}{\sqrt{2}}, \frac{3}{\sqrt{6}}\right)$$

Component s_{ij} is $(s_i(t), \phi_j(t))$. These vectors are plotted in Fig. 2.24.

2.5.3 Signal Space for White Gaussian Noise

The most commonly occurring communications channel is one with additive white Gaussian noise, universally abbreviated as AWGN. Roughly speaking, channels fall into three types: non-Gaussian noise channels, channels in which the noise stems from other users, and AWGN channels, including those where the signal strength fades. In the AWGN case, the noise usually stems from the receiver itself, specifically from thermal noise in the first amplifier stage; its level is precisely known, and it is unquestionably white. This is a lucky break indeed, because AWGN simplifies signal space analysis tremendously.

However ubiquitous it may seem, white Gaussian noise is still an abstraction and its use in analysis can lead to odd conclusions. Its random process power spectral density is a constant for all frequency, which means that an observation of it through an unbounded bandwidth will have unbounded power. Can it even be safely measured? Do receiver front ends generate infinite power? The answer to these puzzles in the practical world is that AWGN in fact never manifests itself in an infinite way. A measuring instrument or a receiver always has finite bandwidth and is

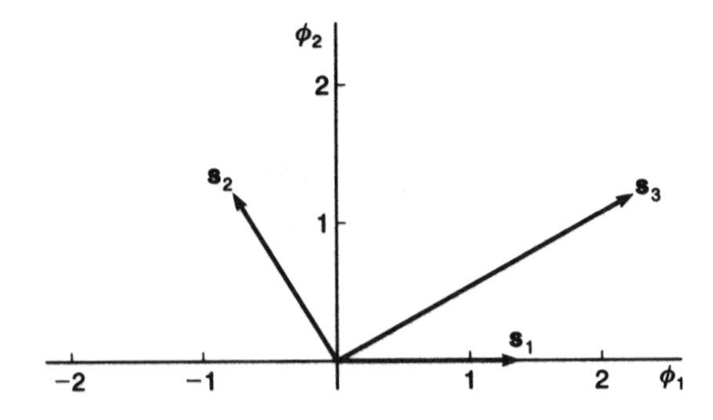

Figure 2.24 Signal space vectors for signals in Example 2.5-1. Note that three vectors span two dimensions.

thus safe from destruction. As discussed in Section 5.2.2, thermal noise is perfectly white in the radio spectrum but not so beyond 100 GHz or so. With the AWGN idea, we pretend that the input to our signal processing is true white noise, and we get away with it because the processors have finite bandwidth.

Our aim now is the evaluation of (2.5-9)–(2.5-10) when $\eta(t)$ is AWGN. Specifically, we need to evaluate $P[\boldsymbol{\eta} = \boldsymbol{r} - \boldsymbol{s}_i]$, the probability that the vector of random variables $\boldsymbol{\eta}$ takes the value $\boldsymbol{r} - \boldsymbol{s}_i$. Each component is given by $\eta_j = (\eta(t), \phi_j(t)) \triangleq \int \eta(t)\phi_j^*(t)dt$. The critical facts appear in the next theorem.

Theorem 2.5-1. If $\eta(t)$ is a white Gaussian random process with power spectral density $\frac{1}{2}N_0$,[15] then the random variables $\{\eta_j\}$, the projections of the outcome $\eta(t)$ onto any set of orthonormal basis functions, are IID Gaussian variables with mean 0 and variance $\frac{1}{2}N_0$. That is,

$$\mathcal{E}[\eta_j] = 0, \qquad \text{all } j$$

$$\text{cov}(\eta_j, \eta_k) = \begin{cases} 0, & j \neq k \\ \frac{1}{2}N_0, & j = k \end{cases}$$

Proof. The results follow from standard methods of random process theory. We will give a broad outline, and those in need of more detail can consult a random process text.

Since the variables η_j are obtained from a Gaussian random process by a linear operation, namely integration, they themselves are Gaussian random variables. It remains only to find their mean and covariance, since these alone specify such a variable. For the mean,

$$\mathcal{E}[\eta_j] \triangleq \mathcal{E}\left[\int_{\mathcal{T}} \eta(t)\phi_j^*(t)\,dt\right] = \int_{\mathcal{T}} \mathcal{E}[\eta(t)]\phi_j^*(t)\,dt = 0$$

because $\eta(t)$ is a zero-mean process. The order of expectation and integration may be reversed here because both are linear operations. For the covariance, we have

$$\begin{aligned} \text{cov}(\eta_j, \eta_k) &\triangleq \mathcal{E}\left[\int_{\mathcal{T}} \eta(t)\phi_j^*(t)\,dt \int_{\mathcal{T}} \eta(u)\phi_k(u)\,du\right] \\ &= \iint \mathcal{E}[\eta(t)\eta(u)]\phi_j^*(t)\phi_k(u)\,dt\,du \end{aligned}$$

where we have again reversed operations. The inner expectation is the autocorrelation of the white Gaussian process $\eta(t)$, which is $(N_0/2)\delta(t-u)$. Carrying out the integration of the impulse gives

[15] A factor of 2 must appear somewhere in the signal space analysis, and by tradition it is placed here.

$$\iint \frac{1}{2} N_0 \delta(t-u) \phi_j^*(t) \phi_k(u)\, dt\, du = \begin{cases} \frac{1}{2}N_0, & j = k \\ 0, & j \neq k \end{cases}$$

Thus the Gaussian variables η_j are uncorrelated, and consequently they must be independent.

Now to the MAP and ML receivers. Because the η_j are independent, the maximization of $P[\boldsymbol{\eta} = \boldsymbol{r} - \boldsymbol{s}_i]$ may be written as a product,[16]

$$\max_i P[\boldsymbol{\eta} = \boldsymbol{r} - \boldsymbol{s}_i] = \max_i \prod_j P[\eta_j = r_j - s_{ij}]$$

which we can break into two products,

$$\max_i \prod_{j=1}^{J} P[\eta_j = r_j - s_{ij}] \prod_{j>J} P[\eta_j = r_j] \tag{2.5-16}$$

The second group of factors uses the fact that the signal vectors contain no components beyond the Jth: It does not depend on the signal vectors. Once again we find an independent factor in the receiver derivation, and once again it may be dropped. This step is so crucial to an understanding of white-noise detection that it has been given a special name:

Theorem 2.5-2 *(The Theorem of Irrelevance).* Only noise in the dimensions of the signals affects detection.

The proof is simply the fact that the receiver optimization may ignore the right-hand factors in (2.5-16). The implications of the theorem can be deep, but a simple illustration of it is tuning in an ordinary analog radio station by tuning out all the others. These exist in dimensions of the radio spectrum that are irrelevant to the desired station, and so the ideal receiver completely ignores them. White noise has the property that it projects a Gaussian variable onto every signal space dimension, no matter how we select them. There is no escape. But we are saved, sufficiently, by the fact that the second moment of the variable is always just $\frac{1}{2}N_0$ and that all the multitude of components may be ignored except for a few that align with the signals. For additional discussion of the Theorem of Irrelevance, beyond the AWGN case, consult Wozencraft and Jacobs [11].

Now we drop the second group in (2.5-16) and substitute the standard Gaussian density function for each remaining factor. The result is

$$\max_i (\pi N_0)^{-J/2} \exp\left[-\sum_{j=1}^{J} (r_j - s_{ij})^2 / N_0 \right]$$

[16]Technically, the notation here assumes discrete variables. For a Gaussian $\eta(t)$, η_j is a continuous variable and $P[\eta_j = a]$ needs to be replaced by $dP[\eta_j = a]$, by which is meant the probability assigned to a differential interval around a. This is $f(a)da$, where $f(a)$ is the density function at a for the variable η_j. During the maximization the differential da acts as a constant and may be dropped, so that the optimization is over density function expressions.

Yet, again, much of the verbiage does not affect the maximization and can be dropped. The factor $(\pi N_0)^{-J/2}$ may be dropped. We may take natural logs of what remains, since log is a monotone function and does not change which outcome achieves the max. This removes the exp. Finally, dropping the minus after the exp replaces the max with a min. The end result is just

$$\min_i \sum_{j=1}^{J} (r_j - s_{ij})^2/N_0 \tag{2.5-17}$$

The sum in (2.5-17) without N_0 is the *ordinary Euclidean distance* between the two vectors $\boldsymbol{r}$ and $\boldsymbol{s}_i$.

We can now give the final form of the ML and MAP receivers, in terms of the Euclidean norm

$$||\boldsymbol{r} - \boldsymbol{s}_i||^2 = \sum_{j=1}^{J} (r_j - s_{ij})^2$$

between two vectors:

$$\text{Find } i \text{ that achieves} \qquad \min_i ||\boldsymbol{r} - \boldsymbol{s}_i||^2 - N_0 \ln P[\boldsymbol{s}_i] \tag{2.5-18}$$

for the MAP receiver and

$$\text{Find } i \text{ that achieves} \qquad \min_i ||\boldsymbol{r} - \boldsymbol{s}_i||^2 \tag{2.5-19}$$

for the ML receiver. Both receivers *seek the signal closest to the received* **r** in signal space; the MAP receiver weights the signals by their probabilities.

2.5.4 Decision Regions and Signal Constellations

It has been a long trek to arrive at (2.5-18), and it is useful to reconnoitre the discoveries so far. By expressing signals as vectors in signal space, we can compute their probabilities given the observations. This leads to the ML and MAP receivers. For white Gaussian noise, the receivers simply measure Euclidean distance.

Since AWGN receivers and the physical world about us share the same Euclidean geometry, there is much intuition in the signal space view, so much so that communication engineers of all types habitually think about signals geometrically. The urge carries over to cases that are not white and not Gaussian, where, fortunately, the intuition still more or less holds. One of the first to espouse the geometric point of view was Shannon [14], whose exposition is still informative to read.

A plot of signal points is called a *signal constellation*. Figure 2.25 shows a three-point constellation that will serve to illustrate the Euclidean nature of receivers (2.5-18)–(2.5-19). The three signals $\boldsymbol{s}_1$, $\boldsymbol{s}_2$, and $\boldsymbol{s}_3$ in this example all have the same energy, a fact that is immediately clear from the geometry because the points lie on a circle centered at the origin. The ML receiver seeks the signal closest to a received vector $\boldsymbol{r}$. The job is performed by the following geometric procedure:

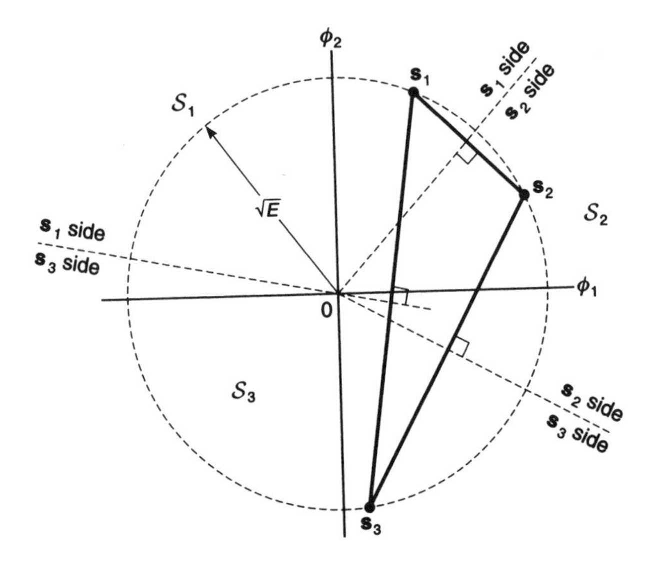

Figure 2.25 Example of decision region construction for three equal-energy signals.

- Draw the perpendicular bisector of the line that joins s_1 and s_2; this divides the points that are closer to s_1 from those that are closer to s_2.
- Repeat for the pair s_2 and s_3 and the pair s_1 and s_3.
- The bisectors form the boundaries of three *decision regions* $\mathcal{S}_1$, $\mathcal{S}_2$, and $\mathcal{S}_3$, where $\mathcal{S}_i$ is the set of received vectors $\boldsymbol{r}$ that are closest to the transmitted signal s_i.

For a detector to reach a maximum likelihood decision, it needs only to find out which decision region contains $\boldsymbol{r}$.

Some other facts are clear from the figure. The decision regions cover the whole space. The region boundaries terminate at the origin (this is true in general only for equal-energy signals). The signals most easily mistaken for each other are s_1 and s_2, because these lie closest and the shortest noise vector $\boldsymbol{\eta}$ will throw $s_1 + \boldsymbol{\eta}$ or $s_2 + \boldsymbol{\eta}$ into the other decision region. The distance $||s_1 - s_2||$ between the closest signal pair is called the *minimum distance* of the constellation.

Several basic types of signal constellations appear in Fig. 2.26. These represent the signals most commonly used in communication. All constellations but (e) have the same bit energy E_b, which may be verified as follows. If all M signals in a set are equiprobable, the average energy in a transmission symbol (i.e., signal) is

$$E_s = \frac{1}{M}\sum_{i=1}^{M} ||s_i||^2 \tag{2.5-20}$$

This gives E for (a) and (b), but $2E$ for (c) and (d). From the discussion in Section 2.2.1, the data bit energy is $E_b = E_s/\log_2 M$. Thus for all four cases E_b is E.

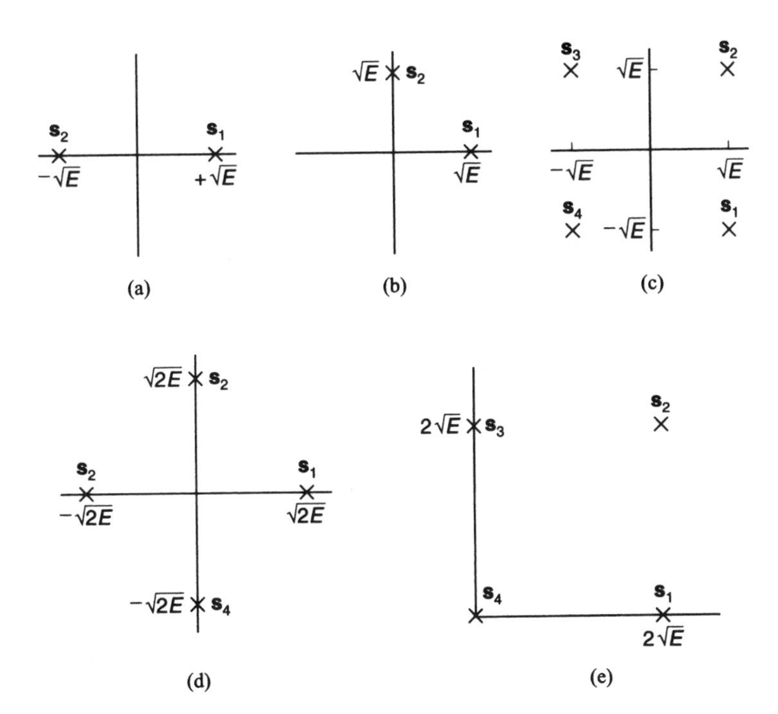

Figure 2.26 Five simple signal constellations in common use: (a)–(d) have the same bit energy E_b; **(a)** antipodal, $D_{\min} = 2\sqrt{E}$; **(b)** orthogonal, $D_{\min} = \sqrt{2E}$; **(c)** square $D_{\min} = 2\sqrt{E}$; **(d)** diamond, $D_{\min} = 2\sqrt{E}$; **(e)** set (c) translated, $D_{\min} = 2\sqrt{E}$.

The vectors $(\sqrt{E}, 0)$ and $(-\sqrt{E}, 0)$ in Fig. 2.26a form a very common case called *antipodal signaling*, so named because the vectors are equal and opposite to one another. If $v(t)$ is any orthogonal pulse (see Definition 2.2-2), then $v(t)$ and $-v(t)$ are antipodal signals. All of the energy in the signals appears in the first signal space dimension; in terms of the basis functions, the signals are $\sqrt{E}\ \phi_1(t)$ and $-\sqrt{E}\ \phi_1(t)$. The vectors in Fig. 2.26b, which are $(\sqrt{E}, 0)$ and $(0, \sqrt{E})$, form a set called *orthogonal* signaling, because the inner product of the two is zero. In many dimensions, an orthogonal signal set is one with one vector in each dimension. Both E_s and E_b are E in (a) and (b), but it is interesting to observe that the minimum distance is a factor $\sqrt{2}$ smaller in (b). This makes (b) a weaker signal constellation, as we will develop further in the next section.

Figure 2.26c is composed of two sets of antipodal signals that are summed up to form the vectors $(\sqrt{E}, \sqrt{E})$, $(-\sqrt{E}, \sqrt{E})$, $(\sqrt{E}, -\sqrt{E})$, and $(-\sqrt{E}, -\sqrt{E})$. These vectors could be formed from two orthogonal pulses, say, $v(t)$ and the time shift $v(t - T)$. Figure 2.26d is a 90-degree rotation of (c). Like (b), each signal has energy in one signal space dimension only, but the minimum distance is now the higher value in (a). Rotations of a constellation leave both the minimum distance and the signal energies unchanged. A consequence, in Section 2.6, is that the detector error probability is unchanged.

Figure 2.26e is a translation of (c), diagonally upward by $\sqrt{2E}$. The average energy of the set is now $4E$, twice that in (c), even though the minimum distance has not changed. The consequence is that twice the energy is required in this signal set to produce the same error probability. We can get a better feel for what is happening here by letting the translation be z and computing the average energy for the new set:

$$E_z = \frac{1}{M}\sum_{i=1}^{M} \|s_i - z\|^2$$

Some calculus proves that E_z is minimized when

$$z_0 = \frac{1}{M}\sum_{i=1}^{M} s_i \qquad (2.5\text{-}21)$$

In geometry z_0 is called the *centroid* of the M vectors.

The constellation $\{s_i - z_o\}$ represents the minimum energy translation of the signals $\{s_i\}$. Since all the translations have the same intersignal distances, and therefore the same error probability, the centroid-translated set is obviously a desirable one. Constellations (a), (c), and (d) are already minimum energy, since their centroids are the zero vector and they need no further translation.

2.6 MAXIMUM LIKELIHOOD RECEIVERS

We focus now on maximum-likelihood detectors for the AWGN channel, first on their error probability and then on their implementation. The error probability is, of course, a fundamental measure of detector performance, and the probabilities for ML detection of the simple constellations in Fig. 2.26 form the building blocks for the error calculation in more complex cases. On the implementation front, there are some pleasant surprises. It turns out that the ML receiver operation of finding the closest signal to the received $r(t)$ can be performed by a simple linear filtering of r. With the right pulse design, these filters are straightforward to build.

2.6.1 ML Receiver Error Calculations

First to the error probability. In the practical world, it goes without saying that receivers are suboptimal, whether by cost of an optimal design or because the signals are distorted by nonlinearities and filters in the channel. In addition, the noise may not be white Gaussian, which means that the find-the-nearest-signal receiver in these sections is not ML. Still, the error performance of an ML receiver in AWGN is the engineer's benchmark, against which practical designs are compared. The essence of that calculation is contained in the case of just two transmitted signals, and this is where we begin. Much further information about Gaussian noise and how it arises is given in Section 5.2.

The Two-Signal Error Probability. Suppose a signal constellation has just two signals separated by Euclidean distance D, as in Fig. 2.27. White Gaussian noise has the peculiar property that its projection on any set of orthogonal axes is an independent zero-mean Gaussian variate with variance $\frac{1}{2}N_0$. This is an odd property indeed, but rather than question it too hard just now, we will let it work for us. Let us adopt a new basis for the signal space, which is a rotation of the old one. A rotation will not affect the signal energies, and it will not change the reality of the physical time signals. The rotation will be such that the first axis runs parallel to the vector connecting $\mathbf{s}_i$ and $\mathbf{s}_k$; the other axes do not matter. The projection of the AWGN onto the new axis is the usual IID Gaussian variate η_1, and it is independent of the projections onto all the other axes.

If $\mathbf{s}_i$ is transmitted, an ML receiver decides $\mathbf{s}_k$ if the received signal $\mathbf{r} = \mathbf{s}_i + \boldsymbol{\eta}$ lies beyond the perpendicular bisector of the line $\mathbf{s}_k - \mathbf{s}_i$. But this event occurs if and only if η_1, an independent Gaussian, exceeds $\frac{1}{2}D$, where D is the distance $\|\mathbf{s}_i - \mathbf{s}_k\|$. The event probability is the *two-signal error probability* defined by

$$p_2(k|i) = \int_{D/2}^{\infty} \frac{1}{\sqrt{\pi N_0}} \exp\left(-\frac{\eta_1^2}{N_0}\right) d\eta_1 \tag{2.6-1}$$

which is the probability that η_1 exceeds $D/2$. We get such a simple form because of the mathematical oddity that is white noise.

Equation (2.6-1) is a standard Gaussian integral. By custom, communication engineers write it in terms of the *Q function*

$$Q(x) \triangleq \frac{1}{\sqrt{2\pi}} \int_x^{\infty} \exp\left(-\frac{1}{2}u^2\right) du, \qquad x \geq 0 \tag{2.6-2}$$

which is the integral of the tail of the unit-variance zero-mean Gaussian density function. The Q function and the related error function erf(·) are widely tabulat-

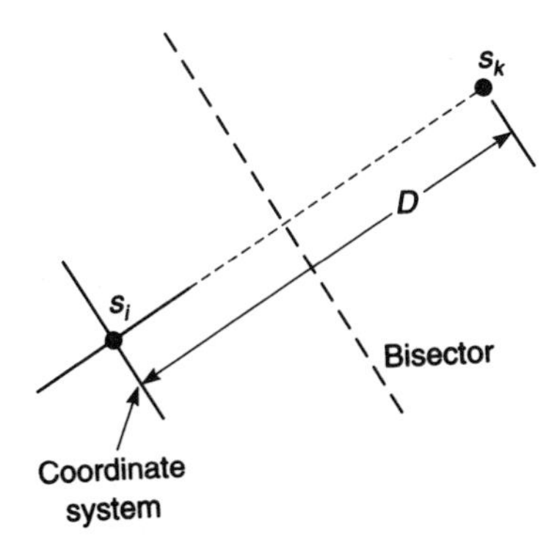

Figure 2.27 Axes for calculating two-signal error probability.

TABLE 2.1 Often Used Values of $Q(\sqrt{2E_b/N_0})$ Expressed as a Function of E_b/N_0

E_b/N_0 (dB)	x	$Q(x)$
–0.86	1.28	10^{-1}
4	2.24	1.25×10^{-2}
4.33	2.33	10^{-2}
5	2.51	5.95×10^{-3}
6	2.82	2.39×10^{-3}
6.79	3.09	10^{-3}
7	3.17	7.73×10^{-4}
8	3.55	1.91×10^{-4}
8.40	3.72	10^{-4}
9	3.99	3.36×10^{-5}
9.59	4.27	10^{-5}
10	4.47	3.87×10^{-6}
10.53	4.75	10^{-6}
11	5.02	2.61×10^{-7}
12	5.63	9.01×10^{-9}

ed.[17] These tables can be difficult to apply, however, and Table 2.1 gives a tabulation especially adapted to communication. A very useful bound to $Q(x)$ is

$$Q(x) \leq \tfrac{1}{2} \exp(-\tfrac{1}{2} x^2), \qquad x \geq 0 \tag{2.6-3}$$

Its utility stems from the fact that it has the same exponential behavior as $Q(x)$; that is, the unit-Gaussian tail integral from x onward tends to zero as $\exp(-x^2/2)$. The bound is compared to $Q(x)$ in Fig. 2.28.

If Eq. (2.6-1) is written as a Q function, the result is

$$p_2(k|i) = Q\left(\frac{D}{\sqrt{2N_0}}\right) \tag{2.6-4}$$

where we have used the variable change $\eta_1 = u\sqrt{N_0/2}$. We can call this the *standard Gaussian detection problem*: What is the probability of confusing two values separated by D when the additive noise variate is zero mean with variance $\frac{1}{2}N_0$?

Now turn to constellations shown in Figs. 2.26a and 2.26b. In these, $E = E_s = E_b$: The average symbol and data bit energies both equal E. Writing the distance D in terms of E_b, we get for cases (a) and (b) two fundamental probabilities in modulation theory. For the antipodal case (a), $D = 2\sqrt{E_b}$ and there are just two signals, $\boldsymbol{s}_1$ and $\boldsymbol{s}_2$, so that

[17] The complementary error function is defined as $\operatorname{erfc}(z) = (2/\sqrt{\pi}) \int_z^\infty \exp(-u^2)\, du$, all z; $\operatorname{erf}(z) = 1 - \operatorname{erfc}(z)$. This function is widely used in the sciences. The Q function is given by $Q(x) = 1/2 \operatorname{erfc}(x/\sqrt{2})$. Here, $Q(\cdot)$ and $\operatorname{erfc}(\cdot)$ are easily confused; some older books use $\operatorname{erfc}(\cdot)$ and a few even define $\operatorname{erfc}(\cdot)$ to be $Q(\cdot)$! Check definitions carefully before using an unfamiliar book.

$$p_2(1|2) = Q\left(\sqrt{2\frac{E_b}{N_0}}\right) \qquad \text{(antipodal signals)} \qquad (2.6\text{-}5)$$

For the orthogonal case (b), $D = \sqrt{2E_b}$ and

$$p_2(1|2) = Q\left(\sqrt{\frac{E_b}{N_0}}\right) \qquad \text{(orthogonal signals)} \qquad (2.6\text{-}6)$$

If there were more than two orthogonal signals, (2.6-6) would apply to any pair of them. Strictly speaking, (2.6-5)–(2.6-6) are the conditional probabilities of deciding s_1 given that s_2 was sent. But $p_2(1|2) = p_2(2|1)$ by symmetry, so that the overall probability of error in these binary cases is

$$p_e = p_2(1|2)P[s_2] + p_2(2|1)P[s_1] = p_2(1|2) \qquad (2.6\text{-}7)$$

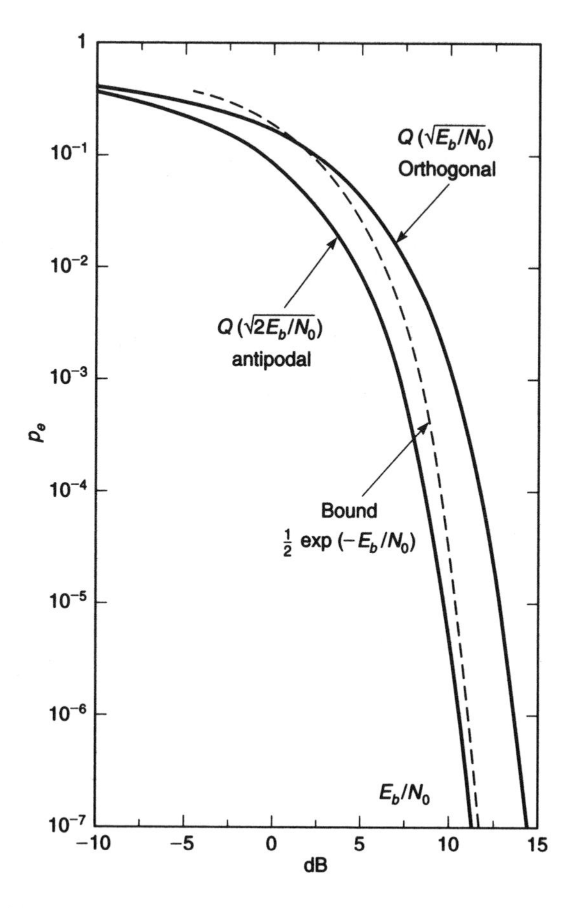

Figure 2.28 Binary orthogonal and antipodal modulation error probabilities with bound (2.6-3) for comparison.

Here, then, is the error probability for two common binary signaling schemes. In terms of orthogonal baseband pulses, (2.6-5) is the case where the binary symbol is carried by the antipodal pulses $v(t)$ and $-v(t)$. Equation (2.6-6) would apply if the symbol were carried by, for example, $v(t)$ and $v(t-T)$. The assumption here is that when $v(t)$ is sent, pulse $v(t-T)$ is shut off and replaced by zero, and vice versa when $v(t-T)$ is sent. (Effectively, it is binary data sent in a unipolar format.) From the error expressions we can draw an important conclusion: *For the same error probability, orthogonal signals need twice the energy per data bit compared to antipodal signals.* Another binary scheme with the poor (2.6-6) performance uses $v(t)$ for one symbol and zero for the other. This is called *on–off signaling*.

The two probabilities are plotted in Fig. 2.28.

Many-Signal Error Probabilities. We turn now to constellations with more than two signals. The two-signal probability plays the major role here because it turns out that a good, tight estimate to the multisignal probability is provided by the two-signal probability for the closest pair of signals. We start with a common case in practice, namely, the four-point constellation in Fig. 2.26c.

The complete four-signal error probability is

$$p_e = \sum_{i=1}^{4} P[\text{Erroneous decision}|\boldsymbol{s}_i \text{ sent}]\, P[\boldsymbol{s}_i]$$

but under the ML scenario and AWGN, the $P[\boldsymbol{s}_i]$ are all $\frac{1}{4}$ and all the conditional probabilities are equal by symmetry. Consequently,

$$\begin{aligned} p_e &= P[\text{Erroneous decision}|\boldsymbol{s}_i \text{ sent}] \\ &= P[\boldsymbol{r} \text{ not in } \mathcal{S}_1|\boldsymbol{s}_1 \text{ sent}] \end{aligned} \tag{2.6-8}$$

Figure 2.29a shows the situation when the first signal is sent. The four decision regions divide up the signal space, and if $\boldsymbol{r}$ lands in $\mathcal{S}_2$, $\mathcal{S}_3$, or $\mathcal{S}_4$, the ML receiver makes an error. Thus

$$p_e = P[\boldsymbol{r} \text{ in } \mathcal{S}_2 \cup \mathcal{S}_3 \cup \mathcal{S}_4|\boldsymbol{s}_1 \text{ sent}] \tag{2.6-9}$$

which is portrayed in Fig. 2.29a. It is possible to integrate the two-dimensional Gaussian density over an area like this, but it is not easy, and an effective substitute is to replace (2.6-9) with

$$p_e \leq \sum_{i=2,3,4} P[\boldsymbol{r} \text{ in } \mathcal{S}_i|\boldsymbol{s}_1 \text{ sent}] \tag{2.6-10}$$

which is an application of the union bound of probability theory. Then we overbound each term in the sum with a two-signal error probability. Figure 2.29b shows the regions of integration needed to compute these probabilities: The $\boldsymbol{s}_2|\boldsymbol{s}_1$ calculation integrates over the vertical-striped area, the $\boldsymbol{s}_3|\boldsymbol{s}_1$ over the diagonal area, and the

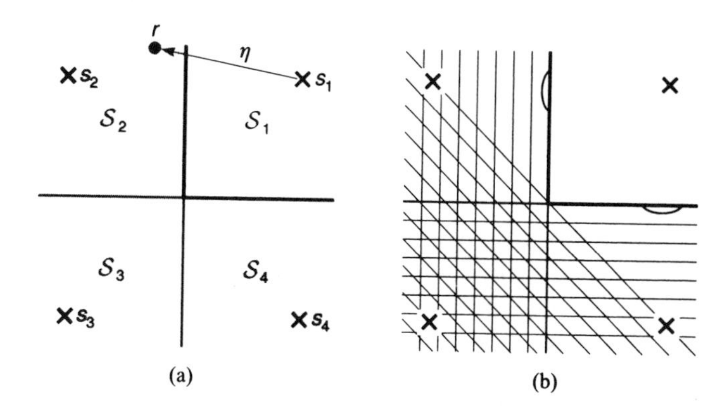

Figure 2.29 **(a)** Incorrect decision region for four-signal rectangular constellation, given that s_1 is sent; **(b)** integration regions for three terms in the union bound. Only small regions nearest s_1 count significantly.

$\boldsymbol{s}_4|\boldsymbol{s}_1$ over the horizontal area. The important point to realize here is that each calculation is taken in isolation and is a two-signal probability. We can add the two-signal calculations to produce an overbound to (2.6-10), which is

$$p_e \leq \sum_{i=2,3,4} p_2(i|1) \tag{2.6-11}$$

From the integration regions in Fig. 2.29b it can be seen that some probabilities are counted twice and some even three times; this is the effect of the overbounds. Nonetheless, (2.6-11) is in general a tight bound. The reason is that the two-dimensional joint Gaussian density of $\boldsymbol{\eta}$ about $\boldsymbol{s}_1$ diminishes rapidly, especially when E_b/N_0 is large, and only the areas closest to $\boldsymbol{s}_1$ count much in the integration. It can be seen that these are counted only once.

Substituting the Q function in (2.6-11), we get (2.6-12). We use the fact that $E_b = \frac{1}{2}E_s = E$ for the constellation shown in Fig. 2.26c:

$$\begin{aligned} p_e &\leq Q\left(\frac{D}{\sqrt{2N_0}}\right)\Bigg|_{D=2\sqrt{E}} + Q\left(\frac{D}{\sqrt{2N_0}}\right)\Bigg|_{D=2\sqrt{2E}} + Q\left(\frac{D}{\sqrt{2N_0}}\right)\Bigg|_{D=2\sqrt{E}} \\ &= Q\left(\sqrt{\frac{2E}{N_0}}\right) + Q\left(2\sqrt{\frac{E}{N_0}}\right) + Q\left(\sqrt{\frac{2E}{N_0}}\right) \\ &= 2Q\left(\sqrt{\frac{2E_b}{N_0}}\right) + Q\left(\sqrt{\frac{4E_b}{N_0}}\right) \end{aligned} \tag{2.6-12}$$

With the bound (2.6-3) one can estimate the relative size of these two terms. For example, at $E_b/N_0 = 8$ dB, where the error rates in Fig. 2.28 begin to be useful, the second term in (2.6-12) is a factor 10^{-3} smaller than the first. It can be shown (see the

Problems) that the true error probability of the four-point constellation shown in Fig. 2.26c is

$$p_e = 2Q\left(\sqrt{\frac{2E_b}{N_0}}\right) - Q^2\left(\sqrt{\frac{2E_b}{N_0}}\right) \tag{2.6-13}$$

The dominant terms in (2.6-12) or (2.6-13) come from the signals $\boldsymbol{s}_2$ and $\boldsymbol{s}_4$ nearest to $\boldsymbol{s}_1$, and this illustrates a general principle about our error calculation: *The nearest neighbors to the transmitted signal dominate the error probability.*

The four-point case generalizes easily to a bound for an arbitrary signal constellation. The principles are the same: Two-signal overbounds lead to a tight estimate, which is dominated by the nearest neighbors in the constellation. Since the signals may not be symmetrical, we begin with the full expression

$$p_e = \sum_{i=1}^{M} P[\text{Erroneous decision}|\boldsymbol{s}_i \text{ sent}]\, P[\boldsymbol{s}_i] = \frac{1}{M}\sum_{i=1}^{M} P[\text{Error}|\boldsymbol{s}_i]$$

The $1/M$ factor stems from the ML-receiver assumption that all signals are equally likely to be sent. By the union bound, each $P[\text{Error}|\boldsymbol{s}_i]$ is overbounded by a sum of two-signal probabilities, one for each remaining signal in the constellation; that is,

$$P[\text{Error}|\boldsymbol{s}_i] = \sum_{k\neq i} P[\boldsymbol{r} \text{ in } S_k|\boldsymbol{s}_i] \leq \sum_{k\neq i} p_2(k|i)$$

$$= \sum_{k\neq i} Q\left(\frac{\|\boldsymbol{s}_k - \boldsymbol{s}_i\|}{\sqrt{2N_0}}\right)$$

Combining everything, we get

$$p_e \leq \frac{1}{M}\sum_{i=1}^{M}\sum_{k\neq i} Q\left(\frac{\|\boldsymbol{s}_k - \boldsymbol{s}_i\|}{\sqrt{2N_0}}\right) \tag{2.6-14}$$

A close look at (2.6-14) with bound (2.6-3) reveals that it contains $M(M-1)$ exponentials that decay rapidly as the distance argument $\|\boldsymbol{s}_k - \boldsymbol{s}_i\|$ grows. The term with the least distance will dominate the sum. There may be several signal pairs lying at this least distance, and the double sum counts each pair twice. The upshot is that a tight estimate to (2.6-14) is given by

$$p_e \leq \frac{2K}{M}\, Q\left(\frac{D_{\min}}{\sqrt{2N_0}}\right) \tag{2.6-15}$$

in which K is the number of distinct signal pairs that lie at the least distance $D_{\min}$.

The distances in Fig. 2.26 and in Eq. (2.6-15) contain within them the effect of the signal energy, and this turns out to be an inconvenience in making calculations and designing systems. One can always improve distance by increasing energy;

what is of much greater interest is improving distance by rearranging the constellation points without changing their average energy. The result will be better error probability at the same cost. In order to factor out the effect of energy, it is customary to work with the *normalized minimum distance*

$$d_{\min} = \frac{D_{\min}}{\sqrt{2E_b}} \tag{2.6-16}$$

The factor of 2 is by convention; another nearly universal convention is that lowercase d, whether a minimum distance or otherwise, denotes a normalized distance, while uppercase D means $\|\mathbf{s}_k - \mathbf{s}_i\|$, the true distance between two signals in signal space.[18] Normalizing by the data bit energy as opposed to the symbol energy E_s serves to remove the effect of the transmission symbol choice. Direct substitution of (2.6-16) into (2.6-15) gives

$$p_e \leq \frac{2K}{M} Q\left(\sqrt{\frac{d_{\min}^2 E_b}{N_0}}\right) \tag{2.6-17}$$

a form that is directly in terms of E_b/N_0 and three parameters that depend only on the *arrangement* of the constellation, namely K, M, and $d_{\min}$. It is clear from this form that $d_{\min}^2$ and E_b/N_0 directly trade off in the error calculation.

For the four-point constellations of Figs. 2.26c and 2.26d, $d_{\min} = 2\sqrt{E}/\sqrt{2E_b} = \sqrt{2}$. Also, $M = 4$, and there are four distinct pairs of points at the minimum distance, so that (2.6-17) becomes the first term of (2.6-12). For Fig. 2.26e, $K = M = 4$ again, but (2.6-16) gives $d_{\min} = 2\sqrt{E}/\sqrt{4E} = 1$, which leads to the estimate $2Q(\sqrt{E_b/N_0})$. This is much worse than the $2Q(\sqrt{2E_b/N_0})$ of case (c) or (d) because E_b/N_0 must be twice as high for the same error probability; the reason is the translation in set (e).

The results in this section are used continually in the following pages. Let us summarize the more important ones.

- $d_{\min}^2 = 2$ for binary antipodal signaling; $p_e = Q(\sqrt{2E_b/N_0})$.
- $d_{\min}^2 = 1$ for M-ary orthogonal signaling; a tight bound for p_e is $(M - 1) Q\sqrt{(E_b/N_0)}$.
- $d_{\min}^2 = 2$ for a four-point constellation symmetrical about the origin [Fig. 2.26c or 2.26d], and a tight bound for p_e is $2Q(\sqrt{2E_b/N_0})$.
- The general error estimate is $(2K/M)\, Q(\sqrt{d_{\min}^2\, E_b/N_0})$ [Eq. (2.6-17)].

2.6.2 ML Receiver Implementations

We need to investigate another startling fact about optimal receivers under white Gaussian noise, which is that they can be constructed out of correlators or ordinary

[18]Note especially that the term $\|\mathbf{r}_k - \mathbf{s}_i\|^2$ the the MAP receiver (2.5-18) includes the effect of E. The balance between E in this term and the density N_0 in the second term determines how much the probabilities $P[\mathbf{s}_i]$ will be weighted in the receiver.

linear filters. Signal space theory leads directly to these conclusions. Not only can receivers be built out of these familiar elements, but the filters for common modulations are often very simple.

The Correlator Receiver. The receiver derivation begins with Parseval's identity, which for our vector signal space says that

$$\|\boldsymbol{r} - \mathbf{s}_i\|^2 = \int_{\mathcal{T}} [r(t) - s_i(t)]^2 \, dt \tag{2.6-18}$$

According to Eq. (2.5-19), the ML receiver works by minimizing the left-hand side of this identity. But we can just as well minimize the right-hand side; after expanding the integral, this operation becomes

$$\min_i \left\{ \int r(t)^2 \, dt + \int s_i(t)^2 \, dt - 2\int r(t)s_i(t) \, dt \right\} \tag{2.6-19}$$

The first term here is a constant during the minimization and can therefore be dropped. The second term is the *energy* of the ith signal. Let ξ_i denote this energy; conceivably, all M of the ξ_i could be precomputed before the receiver begins to operate. If all the signals have the same energy, a common case, then the second term is constant over i and can also be dropped. What remains of (2.6-19) is then equivalent to just

$$\text{Find } i \text{ that achieves} \qquad \max_i \int r(t)s_i(t) \, dt \tag{2.6-20}$$

If the signals are not equal-energy, (2.6-19) can be written as

$$\text{Find } i \text{ that achieves} \qquad \max_i \int r(t)s_i(t) \, dt - \tfrac{1}{2}\xi_i \tag{2.6-21}$$

The heart of these expressions is a correlation integral. It evaluates the similarity between the received signal $r(t)$ and each possible transmitted signal, an entirely reasonable idea for a receiver, which we have just shown is the optimal design, the one that minimizes the error probability in the AWGN channel case. This kind of detector is called a *correlator receiver*. Figure 2.30 sketches the receiver circuit for the general case with unequal energies.

When the receiver processing is realized in discrete time, the correlator receiver is often the more natural implementation of the ML idea. We will therefore take time now for an example of how such receivers work. With analog processing, the filter-based ML receiver that follows after is probably best. In discrete time, filtering is discrete convolution, and correlation is just the same computation in a different order.

Example 2.6-1: Discrete-Time ML Detection. Let all transmitted signals be equal energy and suppose that the received signal $r(t)$ and a target transmitted signal

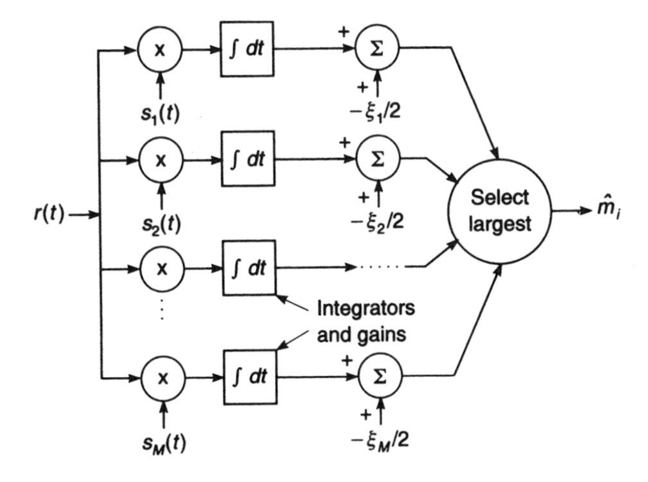

Figure 2.30 Correlator receiver.

$s_i(t)$ are available as MATLAB vectors `rr` and `ss`. There are `fs` of these samples per symbol interval of length `tau`. Take the energy of `ss` to be 1 and of `rr` without noise to be E (this means, for example, that `tau*sum(ss.^2)/fs` is 1). Then the correlation integral in (2.6-20) is given by

```
cor = tau*sum(rr.*ss)/fs
```

In the absence of noise the value here will lie in the interval $[-\sqrt{E}, \sqrt{E}]$, and the correlation receiver seeks the target `ss` with the most positive such value. An important special case occurs when the transmitted signal is a train of orthogonal pulses in the style of Section 2.2.3. Then we can find the ML estimate of transmission symbol a_n directly by evaluating the correlation (2.2-8). Now `ss` is samples of a pulse $a_n v(t - nT)$ that has unit expected energy, and `rr` is the samples from N symbol times of the whole signal $\Sigma a_n v(t - nT)$ plus noise. We can take as illustration standard root RC pulse transmission. Let a single unit-energy pulse centered on a 10-interval width be given by `ss = rtrcpuls(.3,tt)` (from Program 2.2-2), in which `tt` is the 101-sample time vector `[-5:.1:5]`; here the excess bandwidth is 0.3, `fs` is 10 samples per symbol, and the symbol time is 1 second. The pulse width 10 is the parameter `width` in the Pulse Train Program 2.2-3. The full transmission `rr` in the same time frame depends on the 10 symbols before and after the target one. We can set these symbols $a_{n-10}, \ldots, a_n, \ldots, a_{n+10}$ equal to `dat = [1 1 1 1 1 1 1 1 1 1 x -1 -1 -1 -1 -1 -1 -1 -1 -1 -1]`, with $x = a_n = \pm 1$ the target symbol. Then execute

```
rr = pulsetr('rtrcpuls',.3,fs,10,dat);
rr = rr(101:201);
cor = tau*sum(rr.*ss)/fs
```

The first line produces a noise-free received signal with energy 21 (20.93 with the precision here). The second line is needed to match the lengths of `rr` and `ss`, since `ss` has support on the interval [–5, 5], whereas `rr` has `1+3*fs*width` (= 301) points spanning time [–15, 15]. When x is set to 1, `cor` is 0.9998. With higher precision and wider pulse support, this value tends to 1, indicating unit energy in each of the pulses and $a_n = 1$.

The Matched Filter Receiver. With a little more work, the multiplying and integrating in Fig. 2.30 can be reduced to a linear filtering. The key is seeing that the correlation integral is really a convolution of $r(t)$ with a certain filter impulse response; that is, if $h(\tau) = s_i(-\tau)$, we obtain

$$\int_{\mathcal{T}} r(\tau)s_i(\tau)\, d\tau = \int r(\tau)h(t-\tau)\, d\tau\bigg|_{t=0} = r * h\bigg|_{t=0}$$

This is the output at time 0 of a filter $H(f)$ whose input is $r(\tau)$, where $H(f)$ is the transform of $h(\tau)$. From elementary transform properties, $H(f) = S_i^*(f)$.

Actually, the operation is a little more complicated than this, because $h(\tau)$ needs to be the response of a causal, realizable filter. The repairs are sketched in Fig. 2.31, which shows the case where $s_i(\tau)$ is a long Nyquist or other pulse from Section 2.2. The origin of the time axis is taken as the instant assigned to the present transmission symbol. First the signal is truncated at some time NT chosen so that not too

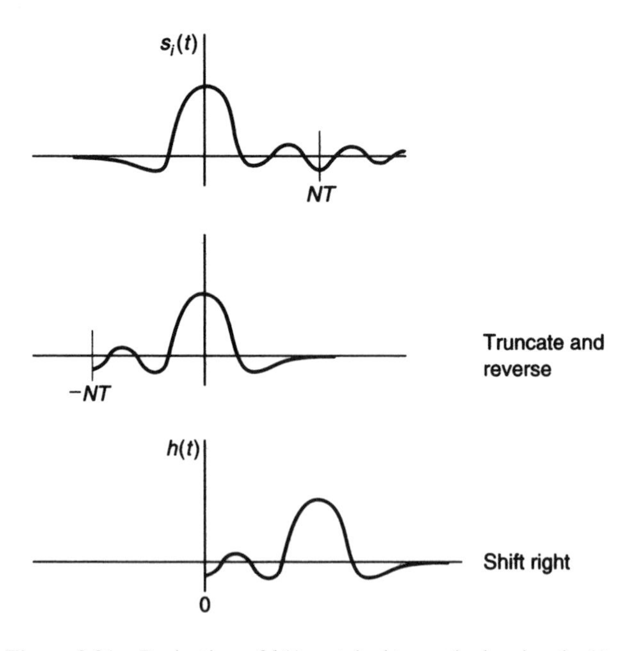

Figure 2.31 Derivation of $h(t)$ matched to particular signal $s_i(t)$.

much distortion is created in the correlation. Next, the signal is reversed to create $s_i(-\tau)$. Finally, the signal is shifted right by NT, to create $h(\tau) = s_i(NT - \tau)$. The desired correlation is now the convolution with this h, sampled at time NT; that is,

$$\int r(\tau)s_i(\tau)\,d\tau = \int r(\tau)h(t-\tau)\,d\tau\Big|_{t=NT} = \int r(\tau)s_i(NT+\tau-t)\,d\tau\Big|_{t=NT}$$

This is the output at time NT of $H(f)$ filtering $r(\tau)$, where $H(f)$ is the transform of the new $h(\tau)$.

A receiver with filters that work in this way is a *matched filter* receiver. One that mimics the operation of the correlator receiver in Fig. 2.30 is shown in Fig. 2.32. It consists of the bank of filters matched to each of the M transmitted signals, samplers that close at time NT, an offset to account as in (2.6-21) for unequal signal energies, and a block that selects the largest of the bank of outputs.

The term *matched filter* traces back to the 1940s, when such filters were developed for radar receivers. It was found that the one-shot radar returns were detected with the highest SNR when the receiver filter was matched to the pulse waveform. Some of this pioneering work is recounted in the classic text by Davenport and Root [15]. The application of matched filtering to digital communication, in which a stream of signals reach the receiver, began in earnest around 1960 with the publication of an *IRE Transactions* special issue and, particularly, with the article by G. Turin (see Ref. 16).

In general, the signals $s_i(t)$—and therefore the filter impulse responses $h_i(t)$—can be long and complicated and can depend on many transmission symbols. In the $s_i(t)$ is contained the response of the channel itself, which could extend over several symbol intervals. In the worst case, a transmission of L M-ary symbols would lead to a bank of L^M filters in the receiver in Fig. 2.32.

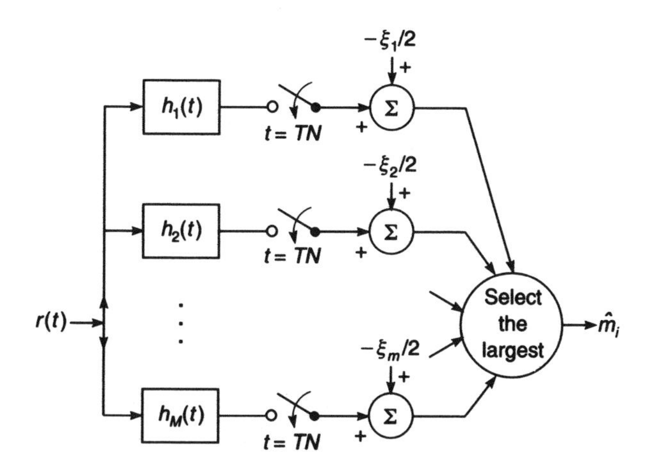

Figure 2.32 Matched filter receiver.

When the transmission that arrives is a superposition of orthogonal pulses, however, *only a single matched filter suffices*. It is matched to a single pulse but detects the whole train. The matched filter receiver becomes the linear receiver that was introduced in Section 2.2.3. To see why, imagine that the transmission consists of a single binary orthogonal pulse, $a_0v(t)$, with $a_0 = \pm 1$. Then the bank in Fig. 2.32 becomes just two filters, $+V^*(f)$ and $-V^*(f)$; these plus the "Select the Largest" block can be replaced by one filter, $V^*(f)$ in Fig. 2.7, plus the comparator block. With M-ary symbols, one filter plus comparison to $M - 1$ thresholds will do the job. In both figures the filters need to be realized in a causal way and the samplers delayed accordingly. To finish the argument, replace the single pulse with the pulse train $s(t) = \Sigma a_n v(t - nT)$. As a consequence of Eq. (2.2-8), the same matched filter can be used to detect each pulse in the sum, even as the others pass through it, so long as the samplers in both figures close precisely at T-second intervals. Thus pulse orthogonality reduces the bank-of-matched-filters receiver to a single filter. The argument is easily extended to a nonbinary symbol alphabet. The simple linear receiver is consequently a maximum likelihood receiver for the a_n.

With all the mathematics of signal space theory, it is easy to get the impression that matched filtering is difficult and mathematical, but as we have just seen, this is often not the case. It is also worth pointing out that the matched filter is in essence just a low-pass filter. If no filter is placed between the white-noise channel and the receiver sampler, the noise component of the sample will have infinite variance. This is because the variance of a sample of density $\frac{1}{2}N_0$ white noise filtered by $H(f)$ satisfies, in general,

$$\sigma^2 = \int_{-\infty}^{\infty} \frac{1}{2} |H(f)|^2 N_0 \, df \tag{2.6-22}$$

On the other hand, a very narrow receiver filter will destroy the transmitted signal. Both extremes destroy the detection, and a compromise is called for. Clearly, a good receiver filter has approximately the bandwidth of the signal expected, since a larger bandwidth lets in unnecessary noise while a smaller one begins to damage the signal. The optimal compromise is the matched filter. If the fine structure of the signal is not known or is subject to change, an acceptable replacement is simply a low-pass filter whose cutoff frequency roughly matches the signal cutoff.

The implementation of individual matched filters is not difficult. At high data rates, more than 20 Mb/s or so, the implementation is necessarily analog, either a bandpass filter centered at the IF frequency or a low pass at baseband. A four- to six-pole filter design is usually sufficient; if the pulses are symmetrical and orthogonal, the receiver filter is a copy of the pulse-forming filter. For IF-based filtering, surface acoustic wave devices are an interesting alternative since they are small and an entire bank of matched filters can be realized on one crystal substrate.

At low rates, matched filtering is simply discrete-time convolution of the input sample stream with the filter impulse response. The sampling rate need not be high (5–10 times the symbol rate is more than enough), and the small amplitudes of the impulse response can be ignored. As a rule, approximations in the transmit filter are risky, since they can dramatically increase the spectral side lobes of the transmission, but in the receive filter they cause little harm.

Matched Filter Properties. We conclude with some properties that concern for the most part what happens when a matched filter is not matched or is replaced with a simple low-pass filter.

An important property is that matched filters maximize the SNR at their output. This is why they were originally introduced to radar receivers.

Theorem 2.6-1. Suppose a received signal $r(t) = s(t) + \eta(t)$, in which $\eta(t)$ is AWGN with density $\frac{1}{2}N_0$, passes through a receive filter $H(f)$. Define the output SNR at the sample time t_0 to be $|g(t_0)|^2/\sigma_h^2$, in which σ_h is given by (2.6-22) and $g = h * s$. Then the SNR is maximized if and only if $H(f) = S^*(f)e^{-j2\pi f t_0}$, or a constant multiple thereof.

Proof. The proof is an application of the Schwarz inequality. The variance of a sample of the noise alone filtered by $H(f)$ is $\sigma_h^2 = (N_0/2)\int |H(f)|^2\, df$. The square magnitude at time t_0 of $s(t)$ alone filtered by $H(f)$ is

$$|g(t_0)|^2 = \left| \int_{-\infty}^{\infty} S(f)H(f)e^{j2\pi f t_0}\, df \right|^2 \tag{2.6-23}$$

in which $S(f)$ is the Fourier transform of $s(t)$. We wish to maximize the ratio $|g(t_0)|^2/\sigma_h^2$ over the choice of $H(f)$. We can just as well maximize $(N_0/2)|g(t_0)|^2/\sigma_h^2 E$, in which E is the energy in $s(t)$, since $N_0/2E$ is constant during the maximization. For E we have

$$E = \int_{-\infty}^{\infty} s^2(t)\, dt = \int_{-\infty}^{\infty} |S(f)|^2\, df$$

by Parseval's theorem. The new object of the maximization is

$$\frac{(N_0/2)|g(t_0)|^2}{\sigma_h^2 E} = \frac{\left| \int H(f)S(f)e^{j2\pi f t_0}\, df \right|^2}{\int |H(f)|^2\, df \int |S(f)|^2\, df} \tag{2.6-24}$$

By the Schwarz inequality,

$$\left| \int u(f)w(f)\, df \right|^2 \le \int |u(f)|^2\, df \int |w(f)|^2\, df \tag{2.6-25}$$

for any two complex functions $u(f)$ and $w(f)$, with equality if and only if $u(f) = Kw^*(f)$, K a constant (integrals are over the real line). We can apply this to Eq.

(2.6-24), with $u(f) = H(f)$ and $w(f) = S(f)e^{j2\pi f t_0}$. We see that (2.6-24) is maximized if and only if $H(f) = S^*(f)\, e^{-j2\pi f t_0}$.

A modification of this theorem exists which shows how to maximize the SNR when the noise is Gaussian but not white (see Refs. 12 or 15).

An important special case occurs when the filtered noise variates in Theorem 2.6-1 are *independent.* Since they are samples of a filtered Gaussian process, they are always Gaussian, but they are also independent when the matched filter impulse response is *orthogonal.* The consequence is that the continuous-time AWGN model of the channel can be replaced by a discrete-time model having IID additive Gaussians, as shown in Fig. 2.33. Simulations and analysis of new communication systems are much easier with the new model.[19] The formal statement of this noise independence is the next theorem.

Theorem 2.6-2. Suppose T-shifts of $h(t)$, the impulse response of $H(f)$, form an orthogonal set in the sense of Definition 2.2-2. Let white Gaussian noise $\eta(t)$ with spectral density $\frac{1}{2}N_0$ be applied to a filter $H^*(f)$ that is matched to $H(f)$. Let w_k be the outputs at any set of times $t_0 + kT$, k an integer. If $h(t)$ has unit energy, the w_k are IID Gaussian variates with zero mean and variance $\frac{1}{2}N_0$.

Proof. The theorem is actually a restatement of Theorem 2.5-1 and follows from the proof given there. The filter outputs are given by

$$w_k = \int \eta(\tau)h(-(t_0 + kT - \tau))\, d\tau \tag{2.6-26}$$

Even though the functions $h(-(t_0 + kT - \tau))$, $k = 0, \pm1, \pm2, \ldots$, are reversed and time-shifted by t_0, they still form an orthonormal set. Theorem 2.5-1 tells us that the inner products (2.6-26) of the members of any such set with $\eta(t)$ form a set of IID Gaussians with mean 0 and variance $\frac{1}{2}N_0$.

The reverse result holds as well: If the w_k are IID, then there must be some set of T-shifts of the form $h(t_1 + kT + t)$ that are orthogonal. Thus the model in Fig. 2.33 depends precisely on the orthogonality of $h(t)$.

The remainder of this section is devoted to cases of receiver filter mismatch that are important in practice.

Example 2.6-2. Integrate-and-Dump Receiver for a Pulse Train. When the basic transmitter pulse is the simple, square NRZ pulse (see Fig. 2.2), the matched filter response is this square, time reversed and shifted right by $\frac{1}{2}T$; in other words, it is a causal version of the same pulse. The effect of the matched filter is to integrate

[19]The model requires precise knowledge of the symbol interval T and is thus not as useful for receivers that estimate symbol timing (see Chapter 4).

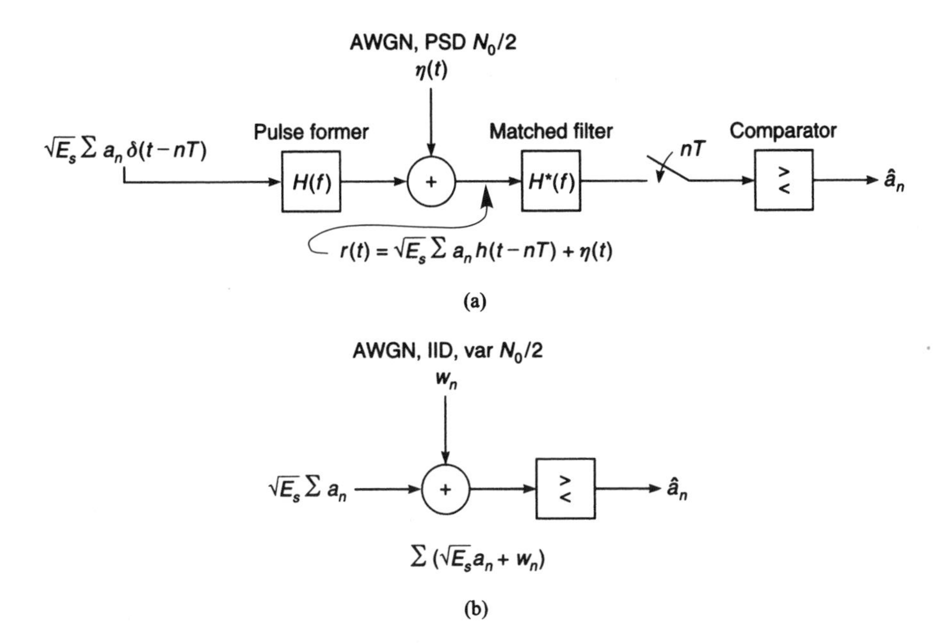

Figure 2.33 Equivalent models for baseband linear modulation over an AWGN channel when $h(t)$ is unit-energy pulse with orthogonal T-shifts: **(a)** continuous time; **(b)** discrete time. For M-ary modulation the threshold comparator has $M-1$ thresholds.

the received signal over the interval $(0, T)$ and pass the result on to the sampler. To prepare for the next pulse, which arrives during $(T, 2T)$, the integrator needs to be emptied out; it then repeats the identical operation as before. Such a receiver is shown for the binary case in Fig. 2.34. A receiver of this type is called an *integrate-and-dump* receiver, or simply an I-and-D. The I-and-D is a maximum likelihood receiver for the NRZ pulse, but above and beyond that, it performs a reasonable and logical operation: It averages the pulse over a symbol interval and compares the result to a threshold. Because of this and because of its simplicity, the I-and-D is often employed when the pulse is not NRZ. Now the receive filter is not a matched filter. When the basic transmitted pulse is confined to one interval, it is not hard to calculate the loss from the mismatch. For example, suppose the arriving transmitted pulses are shifts-by-nT of the drooping pulse

$$v(t) = \begin{cases} Ae^{-t/T}, & 0 < t < T \\ 0, & \text{otherwise} \end{cases} \tag{2.6-27}$$

and the transmission scheme is binary antipodal signaling [i.e., $a_n = \pm 1$ and $s(t) = \Sigma\, a_n \sqrt{E_b}\, v(t-nT)$). To play fair we must normalize $v(t)$ to unit energy by setting $A = \sqrt{2/T(1-e^{-2})}$. The AWGN with density $\frac{1}{2}N_0$ is added to $s(t)$ to produce $r(t) = s(t) + \eta(t)$. The integrator in the I-and-D is matched to the NRZ pulse that would be $1/\sqrt{T}$ over $(0, T)$. Were it to arrive, the sample at the end of an integration would be $a_0\sqrt{E_b}$,

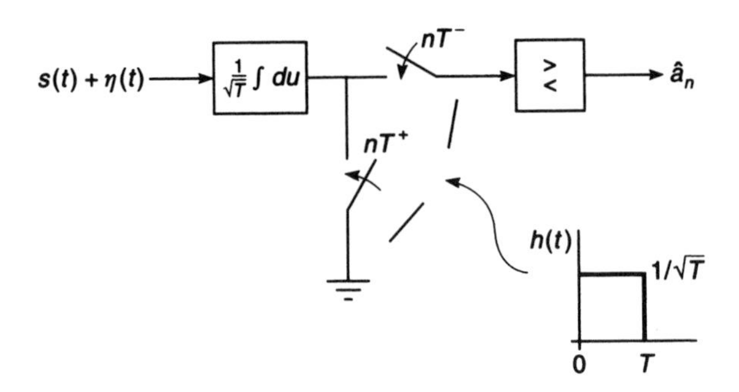

Figure 2.34 Integrate-and-dump receiver, with I-and-D impulse response shown. Switches close just before and after nT.

plus a zero-mean IID Gaussian variate of variance $\frac{1}{2}N_0$, as a consequence of Theorem 2.6-2. With the drooping pulse (2.6-27), the integration in the zeroth interval produces

$$\frac{1}{\sqrt{T}}\int_0^T a_0\sqrt{E_b}v(t)\,dt = \sqrt{\frac{E_b}{T}}\int_0^T a_0Ae^{-t/T}dt$$
$$= \frac{\sqrt{2}(1-e^{-1})}{\sqrt{(1-e^{-2})}}\,a_0\sqrt{E_b} = 0.961a_0\sqrt{E_b}$$

plus the Gaussian variate

$$w_0 = \frac{1}{\sqrt{T}}\int_0^T \eta(t)\,dt$$

By Theorem 2.6-2, w_0 and $w_1, w_2, \ldots$, from all the other intervals are zero-mean IID Gaussians with $\frac{1}{2}N_0$, since the I-and-D circuit impulse response has orthogonal T-shifts and unit energy. In both cases the I-and-D receiver compares the noisy sample to zero. The probability of error is $Q(\sqrt{2E_b/N_0})$ in the NRZ case. In the drooping case, the decision is between $\pm 0.961\sqrt{E_b}$; from Section 2.6.1 this is the probability that w_0 exceeds $D/2$, where $D = 2(.961)\sqrt{E_b} = 1.92\sqrt{E_b}$. Equation (2.6-4) yields $Q(\sqrt{D^2/2N_0}) = Q(\sqrt{1.85E_b/N_0})$, a loss of 0.34 dB in energy efficiency.

Example 2.6-3: No Filter at All—The Sampling Receiver. If the receive filter is set to $H(f) = 1$, what remains is the sampling receiver of Fig. 2.3. If the channel noise is truly white Gaussian, a sample of $s(t) + \eta(t)$ has infinite variance, according to Eq. (2.6-22). Clearly, this is unacceptable and in fact all receivers in AWGN have some sort of filter that at least narrows the noise bandwidth. As an example, we go back to Fig. 2.13c, which shows the eye pattern when a 30% excess-bandwidth binary RC pulse train in AWGN has been prefiltered by a six-pole Butterworth filter. The ratio E_b/N_0 in the channel is 80. The six-pole filter has cutoff frequency $1/T$ hertz, compared to the pulse bandwidth of $0.65/T$ hertz, so that it has almost no effect on the pulses, but from (2.6-22) it reduces the noise bandwidth to $1/T$ and the noise power to about N_0/T. It can be measured off the eye pattern that the standard deviation of a sample (at time 0) is about 12% of the mean value, which is ±1; the actual deviation is about $1/\sqrt{80}$. In terms of the Q function, the error probability (i.e., the probability that the Gaussian variation exceeds the negative of the true sample value) is $Q(\sqrt{80})$. A maximum likelihood receiver would obtain $Q(\sqrt{2E_b/N_0})$, which is $Q(\sqrt{160})$. We see that the sampling receiver with a $1/T$-hertz prefilter suffers a penalty equivalent to a reduction by half in the bit energy E_b. In fact, the cutoff frequency of the true matched filter is about half of $1/T$. Generally, a prefilter whose cutoff is a factor β larger than the matched filter cutoff will lead to a factor β reduction in the energy efficiency of the receiver.

Example 2.6-4: Crude Matched Filtering of an Isolated Pulse. In many receivers, the matched filter is replaced by the crude effect of the bandwidth of the IF amplifier or by a simple filter that follows at baseband. In this example the object is to optimize over the crude filter bandwidth so that the SNR loss at the sampler compared to the case of a true matched filter is as little as possible. The calculation for some simple, practical cases is carried out in Schwartz [17]. The pulse is an isolated NRZ, such as might occur in radar. Then the true matched filter has a sinc(fT) spectrum; the pulse and the spectrum are shown in Fig. 2.2c. First, we look at replacing the filter with an ideal low-pass filter whose bandwidth is B hertz; this also models a sharp-cutoff IF amplifier whose bandwidth is $f_{IF} \pm B$. Figure 2.35 plots the loss in

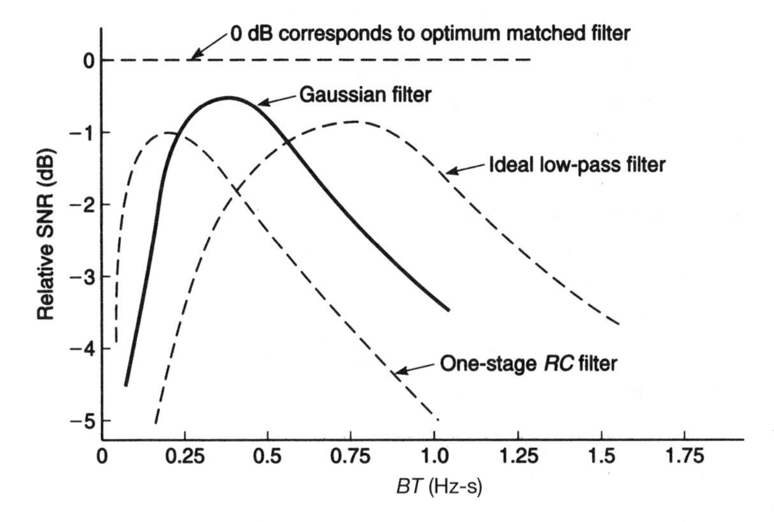

Figure 2.35 Loss in output signal-to-noise ratio (SNR) of mismatched filter compared to perfect matched filter, in decibels versus BT. Here T is symbol time. (From Schwartz [17]. Copyright McGraw-Hill, Inc., 1990; reproduced with permission.)

SNR at the sampler against the bandwidth-symbol time product BT. The best BT is about 0.7, at which a loss of about 0.8 dB occurs. It is of interest that the maximum is very broad. The figure also shows filters with a Gaussian spectrum shape, whose 3-dB bandwidth is BT, and one-pole filters, with transform $H(f) = 1/(1 + j2\pi fRC)$. In the latter, the 3-dB bandwidth in terms of the BT product is T/RC Hz-s. The least loss for these is about 1 dB at $BT = 0.2$ Hz-s. For an extension of this isolated pulse detection to a pulse train, see the Problems.

PROBLEMS

Pulses and Their Properties

2-1. Using Program 2.2-2, or a similar method, create and plot a root RC pulse with excess bandwidth 0.3. Verify that it has unit energy. By numerical integration, show that the pulse is orthogonal to itself at several offsets $T, 2T, \ldots$.

2-2. Using various Programs, create three Nyquist pulses: An RC pulse with excess bandwidth 0.3, another with bandwidth 1.0, and a sinc(·) pulse. Plot the values of the pulses at $nT + 0.1T$, $n = 1, 2, \ldots, 10$. This simulates the effect of a $0.1T$ sampling error. Compare the rate of decay of these values to that theoretically expected [see, e.g., (2.2-6)].

2-3. Without using Theorem 2.2-3, prove that the pulse $\mathrm{sinc}(t/T)$ is orthogonal with respect to shifts of nT. (*Hint:* Execute the calculation in the spectral domain.)

2-4. It is intended that a baseband transmission system use NRZ pulses defined on $[-\frac{1}{2}T, \frac{1}{2}T]$, as shown in Fig. 2.2. Instead, distortions occur in the pulse generation, and the pulse has form $v(t) = \exp(-.5t/T)$ on the same interval. Compute and compare the Fourier transforms $V(f)$ for these two pulses. Carry out the calculation well into the spectral side lobes, and take care that both pulses have the same energy. What asymptotic rates of decay in the side lobes are expected in both cases?

2-5. A root RC pulse with excess bandwidth 0.3 is time truncated to the interval $[-3T, 3T]$. This action creates spectral side lobes that can be unacceptable in practice. Using Fourier theory, estimate the theoretical size and rate of decay of these side lobes. With a program such as 2.2-2, generate and transform such a pulse and verify your estimates.

2-6. The number of derivatives possessed by a pulse is related to the asymptotic rate of decay of the pulse spectrum as $f \to \infty$. Derive this relationship.

2-7. Prove that the convolution of any T-orthogonal pulse with itself creates a pulse that satisfies the Nyquist zero-crossing criterion for spacing nT.

2-8. Devise for your toolbox a program that forms the eye pattern for a signal $y(t)$. MATLAB is recommended. The inputs should be a vector `y` of signal values and `n`, the number of values per symbol interval. Produce plots like those in Figs. 2.12

and 2.13. (To do this easily, obtain `y` from Program `pulsetr` and add Gaussian noise with `rand` to produce a picture like the one shown in Fig. 2.12c or 2.13c.)

2-9. Suppose that a binary pulse-amplitude-modulated (PAM) signal with 30% RC pulses is filtered by a Butterworth filter with cutoff frequency $0.5/T$ Hz. This is the case pictured in Fig. 2.18b. Study the effect of filter order by creating eye patterns for two, four, and eight pole filters. The figure gives the 6-pole case. (*Recommendation:* Use packages similar to those in MATLAB to create the Butterworth filter.)

2-10. An engineer who intends to design a binary PAM system with orthogonal pulses uses RC pulses instead of root RC pulses by mistake. Assume that the pulses have 30% excess bandwidth and that the receive filter is matched to the RC pulse. Plot the impulse response of the whole system, and construct an eye pattern taken at the matched filter output.

Gaussian Noise

2-11. Independent AWGN noise variates η_n with variance $N_0/2$ are added to signal variates x_n which are themselves independent Gaussians with variance σ. The two sequences are independent of each other and all variates have zero mean. The variate $r_n = \eta_n + x_n$ is received.

(a) What is the mean and variance of r_n?

(b) A useful estimate of x_n is the conditional mean $\mathcal{E}[X_n|r_n]$. Find the probability density function of the variable X_n conditioned on r_n.

2-12. Gaussian noise of bandwidth 1 MHz is observed to have mean zero and rms value 100 mV.

(a) If the bandwidth is further restricted to 10 kHz with an ideal low-pass filter, what will the mean and rms values be?

(b) Find the probability that the absolute value of an observation will lie above 20 mV.

(c) Give the normalized autocorrelation function $R(\tau)$ between two observations τ apart in time.

2-13. Wideband Gaussian noise with density $N_0/2$ is filtered by the simple one-pole RC filter whose impulse response is $h(t) = e^{-t/RC}$, $\tau \geq 0$, and 0 otherwise.

(a) Find the mean and variance of a single observation.

(b) Give the autocorrelation $R(\tau)$ between two observations.

Signal Space

2-14. Demonstrate Eq. (2.5-21), that is, show that a translation by z_o minimizes the energy of a signal set, where z_o is the centroid of the signal set.

2-15. Suppose three message signals are sent in one signal space dimension. In vector form they are $s_1 = -a\sqrt{E}$, $s_2 = 0$, and $s_3 = +a\sqrt{E}$. A real value equal to one of these plus AWGN is received.

(a) Give the decision regions for the ML receiver.

(b) Now let the a priori probabilities of the signals be $P[\mathbf{s}_1] = P[\mathbf{s}_2] = \frac{1}{4}$; $P[\mathbf{s}_3] = \frac{1}{2}$. Design the MAP receiver when $E/N_0 = 1$.

(c) Repeat (b) when $E/N_0 = 10$.

2-16. Figure 2.25 shows the decision regions for the ML receiver for a certain signal set $\{\mathbf{s}_1, \mathbf{s}_2, \mathbf{s}_3\}$. Take the radius in the figure to be $\sqrt{E}$ and the actual location of the points to be $\mathbf{s}_1 = \sqrt{E} \angle 67.5°$, $\mathbf{s}_2 = \sqrt{E} \angle 22.5°$, $\mathbf{s}_3 = \sqrt{E} \angle -80°$. Now suppose $P[\mathbf{s}_1] = P[\mathbf{s}_2] = \frac{1}{4}$; $P[\mathbf{s}_3] = \frac{1}{2}$. Draw the decision regions for the MAP receiver when $E/N_0 = 2$.

2-17. Consider the three signals $s_1(t) = 1$ on $[0, 1]$, $s_2(t) = 1$ on $[0, 2]$, and $s_3(t) = 1$ on $[0, 3]$; otherwise, all three are 0. By means of the Gram–Schmidt procedure, derive an orthonormal basis in signal space for these three signals and express the signals as vectors on this basis.

2-18. The standard signal space representation of QPSK is given in Fig. 2.29a. The ML decision regions $\mathcal{S}_1, \ldots, \mathcal{S}_4$ there are bounded by the coordinate system axes. Take the signal energy to be $2E$ as it is in Fig. 2.26c, and assume AWGN.

(a) Let $E/N_0 = 1$, and let $P[\mathbf{s}_2] = P[\mathbf{s}_3] = P[\mathbf{s}_4] = 0.2$ and $P[\mathbf{s}_1] = 0.4$. What are the decision regions for MAP reception?

(b) Repeat with $E/N_0 = 10$.

(c) Repeat with $E/N_0 = 1$ and $P[\mathbf{s}_2] = P[\mathbf{s}_3] = P[\mathbf{s}_4] = 0.01$, $P[\mathbf{s}_1] = 0.97$.

2-19. Consider the standard antipodal binary detection problem, with the vector form signals $\mathbf{s} = \pm a$. The received signal is $r = s + \eta$, where η is the *Laplacian* random variable with PDF

$$p(\eta) = \tfrac{1}{2}\, e^{-|\eta|}, \qquad \text{all } \eta$$

Adapt the argument in Sections 2.5.1–2.5.3 and derive the ML receiver.

2-20. An octal signaling system uses the eight-dimensional orthogonal constellation (Fig. 2.26b shows the two-dimensional case).

(a) Find the normalized square minimum distance of this scheme.

(b) What gain or loss in bit-normalized energy and bandwidth does the scheme have compared to the binary scheme in Fig. 2.26b?

(c) Give an error estimate of the form $AQ(\sqrt{d^2 E_b/N_0})$; that is, give d^2 and A.

2-21. Consider the signal constellation composed of eight points spaced evenly around a circle centered at the origin. (This will be called the 8PSK constellation in Chapter 3; Fig. 3.10 gives a picture.)

(a) Compute its normalized square minimum distance.

(b) Repeat for 16PSK, which is 16 points around the circle.

2-22. Prove the exact formula (2.6-13) for the error probability of the four-point square signal constellation. (*Hint:* Break the constellation into two two-point antipodal signal sets.)

2-23. **(a)** Find the normalized square minimum distance of a constellation consisting of three points arranged in an equilateral triangle. Such an arrangement is called a simplex.

(b) Repeat for four points equilaterally spaced in three dimensions; this is a pyramid with four faces, each of which is an equilateral triangle.

Receivers

2-24. In Problem 2.4, a distorted pulse with an exponential shape replaced an NRZ baseband transmission pulse. The matched filter receiver for the NRZ pulse is the integrate-and-dump receiver. Suppose the I-and-D is used with the distorted pulse. Estimate the loss in effective E_b/N_0 that occurs because of this. Be sure that your distorted pulse has the same energy as the intended NRZ pulse.

2-25. Consider standard baseband binary linear modulation with NRZ pulses of width T. The linear receiver (Fig. 2.7) with its matched filter is used for detection.

(a) Consider a single transmitted pulse $+\sqrt{E_b/T}\,\text{rect}(t/T - \frac{1}{2})$ on the interval $[0, T]$. Sketch the waveform at the output of the matched filter (assume a unity gain filter).

(b) Gaussian noise with PSD $N_0/2$ is present at the filter input. Give the probability density function of the filter output sample at time T.

(c) Now suppose $+\sqrt{E_b/T}\,\text{rect}(t/T - \frac{1}{2}) - \sqrt{E_b/T}\,\text{rect}(t/T - \frac{3}{2})$ is transmitted, in the absence of noise. What will the filter output be? In particular, what is the output at T and $2T$?

2-26. *(Extension of Example 2.6-3)* Consider again the binary NRZ pulse detection in Problem 2–25. The matched filter there is replaced with a simple RC lowpass filter whose impulse response is $h(t) = \sqrt{1/T}\,e^{-t/RC}$, $t \geq 0$, and zero otherwise. This filter is "dumped" at the end of each NRZ pulse interval. A sampler and comparator follows the filter as usual.

(a) The single pulse in Problem 2-25(a) is sent. What is the output of the RC filter?

(b) Where is the best sampling time in $[0, T]$?

(c) Use Eq. (2.6-22) to find the variance in the Gaussian noise component of the sample.

(d) In terms of the Q function, what is the probability of detection error in terms of RC, E_b, and N_0? What dB loss is this compared to optimal matched filter detection?

REFERENCES[20]

1. *J. D. Gibson, *Principles of Analog and Digital Communications,* 2nd ed., Macmillan, New York, 1993.

[20]References marked with an asterisk are recommended as supplementary reading.

2. Members of technical staff, *Engineering and Operations in the Bell System,* Bell Telephone Labs., Murray Hill, NJ, 1977.
3. H. Nyquist, Certain factors affecting telegraph speed, *Bell System Tech. J.,* April 1924, pp. 324–346.
4. H. Nyquist, Certain topics on telegraph transmission theory, *Trans. AIEE,* vol. 47, April 1928, pp. 617–644.
5. R. A. Gibby and J. W. Smith, Some extensions of Nyquist's telegraph transmission theory, *Bell System Tech. J.,* vol. 44, September 1965, pp. 1487–1510.
6. S. Chennakeshu and G. J. Saulnier, Differential detection of $\pi/4$-shifted-DQPSK for digital cellular radio, *IEEE Trans. Vehic. Tech.,* vol. 42, February 1993, pp. 46–57.
7. S. Chennakeshu et al., A trellis-coded modulation scheme for land-mobile radio, in *Coded Modulation and Bandwidth-Efficient Transmission* (Proceedings of the 5th Tirrenia Workshop on Digital Communications, September 1991), E. Biglieri and M. Luise, eds., Elsevier, Amsterdam, 1993, pp. 329–340. See also refs. therein.
8. J. B. Anderson, T. Aulin, and C.-E. Sundberg, *Digital Phase Modulation,* Plenum, New York, 1986, Chapter 4.
9. *S. S. Haykin, *Digital Communications,* John Wiley & Sons, New York, 1988.
10. V. A. Kotelnikov, The theory of optimum noise immunity, PhD Thesis, Molotov Energy Institute, Moscow, January 1947; available under the same name from Dover, New York, 1968 (R. A. Silverman, translator).
11. *J. M. Wozencraft and I. M. Jacobs, *Principles of Communication Engineering,* John Wiley & Sons, New York, 1965.
12. H. L. van Trees, *Detection, Estimation, and Modulation Theory,* Part I, John Wiley & Sons, New York, 1968.
13. *H. Stark, F. B. Tuteur, and J. B. Anderson, *Modern Electrical Communications,* 2nd ed., Prentice-Hall, Englewood Cliffs, NJ, 1988.
14. *C. E. Shannon, Communication in the presence of noise, *Proceedings IRE,* vol. 37, pp. 10–21, 1949; reprinted in *Claude Elwood Shannon: Collected Papers,* N. J. A. Sloane and A. D. Wyner, eds., IEEE Press, New York, 1993.
15. W. B. Davenport, Jr. and W. L. Root, *Random Signals and Noise,* McGraw-Hill, New York, 1958.
16. G. Turin, An introduction to matched filters, Special Matched Filter Issue, *IRE Trans. Information Theory,* vol. IT-6, June 1960, pp. 311–329.
17. *M. Schwartz, *Information Transmission, Modulation, and Noise,* 4th ed., McGraw-Hill, New York, 1990.

CHAPTER 3

CARRIER TRANSMISSION

3.1 MODULATION WITH A CARRIER

In the introduction to Chapter 2, we defined a carrier modulation system as one that translates a baseband-modulated signal to a new frequency band. The baseband signals in Chapter 2 occupy a bandwidth that extends from DC up to about the rate of arrival of the data symbols. Is this signal practical for transmission? A 2400-symbol/s pulse train in a telephone line modem is an audio-bandwidth signal and is ideal for transmission over a telephone line. But sending the same signal by radio would require an antenna hundreds of kilometers long. The signal needs to be translated to a higher frequency. Which frequency depends on many factors: Government regulations intervene; some frequencies are cheaper to use than others; some are only line-of-sight; some work best under water or in deep space. Beyond all this, the realities of electronic design imply that faster pulse trains need to be translated to higher carrier frequencies before they are processed and amplified.

For all these reasons, then, we come to the subject of carrier modulation. A form for a general *bandpass signal* is

$$s(t) = A(t) \cos[\omega_0 t + \psi(t)] \tag{3.1-1}$$

in which an amplitude signal $A(t) \geq 0$ and a phase signal $\psi(t)$ are both low-pass. The radian frequency ω_0, what we will soon call the carrier frequency, sets the center of the bandpass spectrum, whose width is about twice the bandwidth of $A(t)$ and $\psi(t)$. It takes, in general, two low-pass signals to describe a bandpass signal. In the case of a digitally modulated bandpass signal, these two are often used; they form the *amplitude and phase* description of the signal.

An equivalent description is another form,

$$I'(t) \cos \omega_0 t - Q'(t) \sin \omega_0 t \tag{3.1-2}$$

Digital Transmission Engineering, Second Edition. By John B. Anderson

With basic trigonometric identities, one can show that these two are related to $A(t)$ and $\psi(t)$ by

$$I'(t) = A(t)\cos\psi(t), \qquad Q'(t) = A(t)\sin\psi(t) \tag{3.1-3}$$

When a bandpass signal is a digital modulation, it is convenient to refine this description a little further and write (3.1-2) as

$$s(t) = \sqrt{\frac{2E_s}{T}}[I(t)\cos\omega_0 t - Q(t)\sin\omega_0 t] \tag{3.1-4}$$

In this version, E_s will be the average energy devoted to a transmission symbol if

$$\mathcal{E}\left\{\frac{1}{T}\int_T [I^2(t) + Q^2(t)]\,dt\right\} = 1 \tag{3.1-5}$$

in which the integral is over a complete symbol interval. Equation (3.1-5) serves to normalize the signals $I(t)$ and $Q(t)$. Here, I and Q are called the *in-phase* and *quadrature* baseband signals, respectively, and the form (3.1-4) is called the in-phase and quadrature, or simply the baseband, representation of $s(t)$.

We can summarize the conversion between the two carrier modulation forms as follows. For

$$A(t)\cos[\omega_0 t + \psi(t)] = \sqrt{\frac{2E_s}{T}}[I(t)\cos\omega_0 t - Q(t)\sin\omega_0 t]$$

we have

$$I(t) = \frac{A(t)\cos\psi(t)}{\sqrt{2E_s/T}}, \qquad Q(t) = \frac{A(t)\sin\psi(t)}{\sqrt{2E_s/T}} \tag{3.1-6}$$

and[1]

$$\begin{aligned} \psi(t) &= \arctan\left[\frac{Q(t)}{I(t)}\right] \\ A(t) &= \sqrt{\frac{2E_s}{T}}\sqrt{I^2(t) + Q^2(t)} \geq 0 \end{aligned} \tag{3.1-7}$$

The quantity $\sqrt{I^2(t) + Q^2(t)}$ is called the *envelope* of the modulation. It is a normalized measure of the amplitude of the modulated carrier.

Equation (3.1-4) is the standard definition of a carrier digital modulation, and ω_0 is its carrier frequency. Often, the baseband signals $I(t)$ and $Q(t)$ are independent

[1]Here and throughout, arctan is a function of *two* arguments; that is, $\arctan(y/x) = 45°$ when $y = x = 1$, but $-135°$ when $y = x = -1$. Compare the `atan2` function in MATLAB.

data-bearing pulse trains exactly like those in Chapter 2. Modulations of this kind include the phase-shift keying (PSK) and QAM classes that are the subjects of Sections 3.2–3.5. Some modulations are not based on independent pulse trains; an example is the frequency-shift keying (FSK) class in Section 3.6. It is more convenient to describe this class in terms of the amplitude and phase form (3.1-1), although the baseband form (3.1-4) still holds.

A modulation composed of two independent pulse trains in the style of (3.1-4) is called a *quadrature modulation*, which means, etymologically, that it is composed of two signals at right angles. The natural basis in signal space, as we shall see, is the basic pulse $v(t)$ times cos $\omega_0 t$ and times sin $\omega_0 t$, which makes two dimensions for each symbol interval. This compares with one basis signal, $v(t)$, per interval in Chapter 2. Nonquadrature modulations need a more complicated signal space.

A critical element in demodulating a carrier modulation signal is finding synchronization with the carrier cos $\omega_0 t$. This can be thought of as establishing the zero reference for the phase $\psi(t)$ in the amplitude and phase form (3.1-1), or setting out the cos $\omega_0 t$ and sin $\omega_0 t$ signals in a quadrature modulation. In effect, synchronization sets up the signal space basis at the receiver. Since synchronization is a complex subject in itself, we discuss it separately in Chapter 4, and we assume in this chapter that a perfect replica of the carrier signal is available.

The plan of this chapter is first to study PSK modulations in Sections 3.2–3.4 and then to extend them to QAM in 3.5. These systems are the heart of practical transmission today. Then we turn to FSK systems in Sections 3.6 and 3.7. Distortions in transmission are the subject of Section 3.8. Analog implementation is discussed in each section, but the special subject of discrete-time implementation appears in Appendix 3C.

3.2 PHASE-SHIFT KEYING

In phase-shift keying (PSK), the transmission symbols modulate the phase of a carrier. The use of the word *keying* is a holdover from telegraphy, where the instrument of digital modulation was a telegraph key. If there are M different symbols, the carrier may simply hold one of the M phases for each symbol interval, but more likely a shaped pulse of the carrier waveform is transmitted in one of the M phase offsets. If $M = 2$, the modulation is binary PSK, or BPSK; if $M = 4$, it is quaternary PSK, or QPSK. The most common type of PSK is QPSK with shaped carrier pulses. PSK modulations with other alphabets are sometimes used; these are denoted by 8PSK, 16PSK, and so on, with the digit representing the number of phases.

Binary and quaternary PSK are quadrature modulations and can be viewed as linear modulations; that is, their transmitted signal can be viewed as a superposition of independent pulses. In this section, we will focus on these two modulations and will explore their spectra, error probabilities, and other properties. All follow directly from the methods already developed in Chapter 2. Section 3.4 studies several generalizations of PSK, including the extension to more phases. Strictly speaking, PSK with $M > 4$ is neither quadrature nor linear.

3.2.1 Binary PSK

The simplest form of phase-shift keying is BPSK. It consists simply of carrier bursts in one of two opposite phases. With NRZ modulation pulses, the BPSK signal in a symbol interval is just

$$\begin{aligned} &\text{For data } +1: \quad s_1(t) = \sqrt{\frac{2E_s}{T}} \cos \omega_0 t \\ &\text{For data } -1: \quad s_2(t) = \sqrt{\frac{2E_s}{T}} \cos(\omega_0 t + \pi) \end{aligned} \tag{3.2-1}$$

in amplitude-and-phase form and

$$\begin{aligned} &\text{For data } +1: \quad s_1(t) = +\sqrt{\frac{2E_s}{T}} \cos \omega_0 t \\ &\text{For data } -1: \quad s_2(t) = -\sqrt{\frac{2E_s}{T}} \cos \omega_0 t \end{aligned} \tag{3.2-2}$$

in baseband form. In this second form, the in-phase signal $I(t) = \pm 1$ and the quadrature signal $Q(t)$ is always zero. E_s is the energy devoted to one of the ± 1 symbols. Figure 3.1 shows the NRZ-modulated sinusoid that results from the data sequence $+1, -1, +1, +1$.

The NRZ pulse BPSK is also called square-pulse or "pure" BPSK. To construct a signal space basis for it during interval $[(n - \frac{1}{2})T, (n + \frac{1}{2})T]$, we can use the step-by-step Gram–Schmidt procedure in Section 2.5. The procedure gives, through Eq. (2.5-11) the first basis function

$$\phi_1(t) = \sqrt{\frac{2}{T}} \cos \omega_0 t, \qquad \left(n - \frac{1}{2}\right)T < t \leq \left(n + \frac{1}{2}\right)T \tag{3.2-3}$$

and then comes to a halt, because there are no other signals in the interval except for $-\phi_1(t)$. This tells us that the signal space for the modulation spans one dimension per symbol interval. The signals (3.2-1) or (3.2-2) are clearly $\pm\sqrt{E_s}\phi_1(t)$, which means that the vector representing $s_1(t)$ is the scalar $+\sqrt{E_s}$ and the one representing $s_2(t)$ is $-\sqrt{E_s}$. Mathematically, this vector component is the inner product $\int s_i(t) \phi_1(t)\, dt$ in Eqs. (2.5-5)–(2.5-6) in Chapter 2. The signal constellation for this BPSK is Fig. 2.26a; BPSK is thus antipodal signaling.

All antipodal signaling schemes have the error probability (2.6-5), and so we have proved that NRZ pulse BPSK has error probability

$$P_e = Q\left(\sqrt{\frac{2E_b}{N_0}}\right) \qquad \text{(BPSK error probability)} \tag{3.2-4}$$

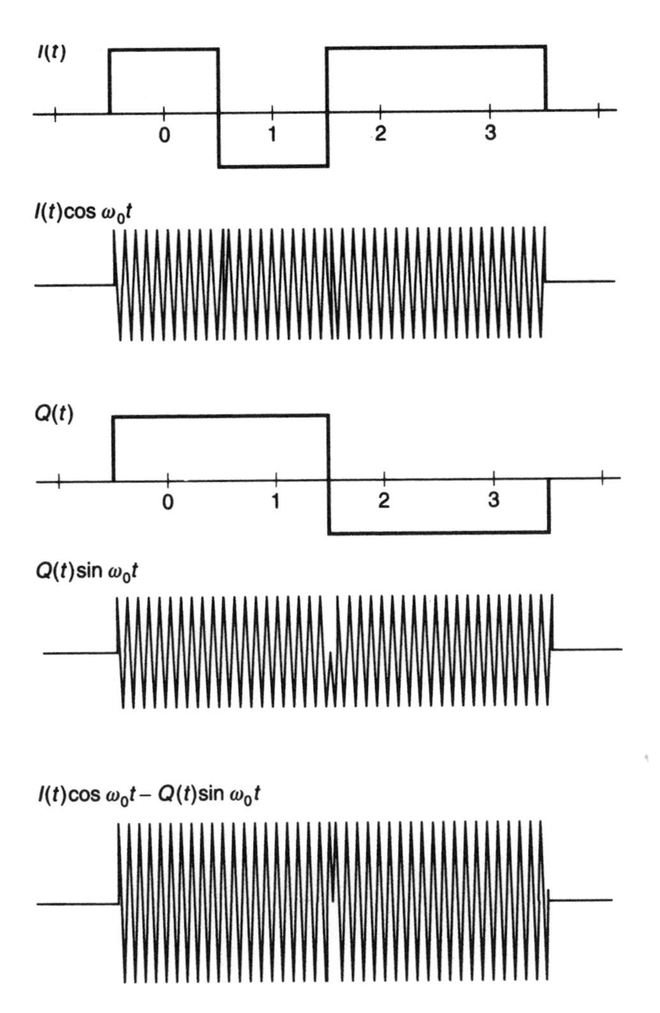

Figure 3.1 Binary and quatenary PSK created from four NRZ pulses. The in-phase data are [+1, –1, +1, +1] and the quadrature data are [+1, +1, –1, –1]. In-phase and quadrature signals combine to form QPSK at the bottom. Top waveform alone if BPSK. Time axis is in symbols.

Note that E_b, the energy per data bit, is also E_s. A plot of (3.2-4) has been given in Fig. 2.28.

Binary PSK with NRZ pulses is easy to understand, but it is seldom practical because the square-pulse modulation has very wide bandwidth. One way to narrow the bandwidth is to use the smooth orthogonal pulses of Section 2.2.3. A set of these pulses $\{v(t - nT)\}$, $n = 0, \pm 1, \pm 2, \ldots$, which is based on $v(t)$, has the property that the inner product of any two pulses is zero (Definition 2.2-2). Suppose we form the pulse train

$$s(t) = \sum_n a_n v(t - nT) \sqrt{2E_s} \cos \omega_0 t \qquad (3.2\text{-}5)$$

This is essentially a summation of weighted time shifts of the basic pulse $v(t) \cos \omega_0 t$.[2] The term $v(t)$, which could be, for instance, $(1/\sqrt{T})\text{sinc}(t/T)$ or the root RC pulse of (2.2-11), shapes the burst of $\cos \omega_0 t$. The new pulse is orthogonal under T-shifts just as surely as $v(t)$ is:

$$\begin{aligned}
&\int_{-\infty}^{\infty} v(t) \cos \omega_0 t \, v(t - nT) \cos \omega_0 t \, dt \\
&\quad = \int \left[\frac{1}{2} + \frac{1}{2} \cos 2\omega_0 t \right] v(t)v(t - nT) \, dt \\
&\quad = \frac{1}{2}(0) + \frac{1}{2} \int v(t)v(t - nT) \cos 2\omega_0 t \, dt \\
&\quad \to 0 \quad \text{as } \omega_0 \to \infty, \qquad n \neq 0
\end{aligned} \qquad (3.2\text{-}6)$$

The first term here is zero because of the orthogonality of $v(t)$; the second is zero simply because $\int f(t) \cos 2\omega_0 t \, dt$ tends to zero as ω_0 grows, for any well-behaved $f(t)$.[3]

Equation (3.2-5) is the baseband form for pulse-shaped BPSK; $Q(t)$ is again zero and the in-phase signal is

$$I(t) = \sum_n a_n \sqrt{T} \, v(t - nT) \qquad (3.2\text{-}7)$$

The normalizing condition (3.1-5) is satisfied if $a_n = \pm 1$ and $v(t)$ is orthonormal. As a candidate for a basis function during the nth interval, we can try the function $\sqrt{2}v(t - nT) \cos \omega_0 t$, which is the basis function (3.2-3) with v instead of a square pulse. The inner product of this function with the entire transmission (3.2-5) is

$$\begin{aligned}
&\int_{-\infty}^{\infty} \sqrt{2}v(t - nT) \cos \omega_0 t \sum_n a_m v(t - mT) \sqrt{2E_s} \cos \omega_0 t \, dt \\
&\quad = 2\sqrt{E_s} \int \left[\frac{1}{2} + \frac{1}{2} \cos 2\omega_0 t \right] v(t - nT) \sum_m a_m v(t - mT) \, dt \\
&\quad = \sqrt{E_s} \int v(t - nT) \sum_m a_m v(t - mT) \, dt \\
&\quad + \sqrt{E_s} \int \cos 2\omega_0 t \, v(t - nT) \sum_m a_m v(t - mT) \, dt
\end{aligned} \qquad (3.2\text{-}8)$$

The first term here is simply $a_n\sqrt{E_s}$, by virtue of Eq. (2.2-8), which expresses the orthogonality of v at baseband. The second term tends to 0 as $\omega_0 \to \infty$, as the second term did in (3.2-6). Thus, our candidate basis function can exactly express the entire contri-

[2] A true time shift would be $v(t - nT) \cos \omega_0(t - nT)$; Eq. (3.2-5) is transmitted in real equipment.

[3] A finite ω_0 causes a small deviation from zero that manifests itself in practical equipment as well.

bution to the transmission from the nth interval, and it is orthogonal to contributions from other intervals. In addition, it has unit energy if v does. Therefore, it is an ideal basis function, and it is clear that there is only one such function needed per interval. The outcome $a_n\sqrt{E_s}$ gives the possible vector components along this axis: $\pm\sqrt{E_s}$.

The foregoing argument shows that orthogonal-pulse BPSK has the same antipodal constellation that NRZ pulse BPSK has and, consequently, the same error expression (3.2-4). From the signal space point of view, they are the same modulation.

An important practical example of an orthogonal smoothed pulse is the root RC pulse in Example 2.2-5. A modulation by +1, –1, +1, +1 with a 30% excess bandwidth pulse is shown for comparison at the top of Fig. 3.2. This picture is Eq. (3.2-5) with $a_n = \pm 1$, $E_s = 1$, and $v(t)$ from (2.2-11). Note how the RC signal rings on in time, compared to the equivalent NRZ modulation.

As far as receivers are concerned, orthogonal-pulse BPSK behaves as does any other orthogonal-pulse modulation: The simple linear receiver in Fig. 2.7 or Fig. 2.33 is a maximum-likelihood detector in Gaussian noise. This time, the matched filter $H^*(f)$ in the receiver is matched to a shaped burst of a sinusoid. Actually, PSK linear receivers are usually realized in an alternate way. Rather than being match filtered at carrier frequencies, the PSK signal is converted to its baseband signals, and the filtering is done against them. This standard receiver is called the *quadrature receiver,* and we will take it up in Section 3.3.

3.2.2 BPSK Spectra

Because BPSK is a linear modulation, its spectrum follows directly from the linear spectrum theorem, Theorem 2.3-1; that is, its spectrum is a scaled version of the spectrum of its underlying pulse. From Section 2.3, the theorem has two requirements. First, the transmission $s(t)$ must be a weighted sum of pulses that all have the same magnitude spectrum. This is clear from (3.2-5), which is a summation of time shifts; the magnitude spectrum is that of the pulse $v(t)\cos\omega_0 t$. Second, the data symbols a_n must be IID with mean zero, something that we assume for all modulations. We can think of binary PSK as converting a baseband spectrum based on $v(t)$ to a bandpass spectrum based on $v(t)\cos\omega_0 t$.

The precise steps in the application of Theorem 2.3-1 are as follows. Let the BPSK signal be $\Sigma a_n v(t-nT)\sqrt{2E_s}\cos\omega_0 t$, in which the unit-energy pulse $v(t-nT)$ appears every T seconds. Let the prototype pulse in the theorem be $w_1(t) = \sqrt{2}v(t)\cos\omega_0 t$, which is again a unit-energy pulse, with transform

$$W_1(f) = \frac{1}{\sqrt{2}}\,[V(f-f_0) + V(f+f_0)], \qquad \omega_0 = 2\pi f_0$$

Take the IID symbols in the theorem to be $\pm\sqrt{E_s}$; their expected square value is E_s. There are N pulses. Equation (2.3-3) times $1/NT$ becomes

$$\frac{E_s}{T}\left[\frac{1}{2}|V(f-f_0)|^2 + \frac{1}{2}|V(f+f_0)|^2\right] \tag{3.2-9}$$

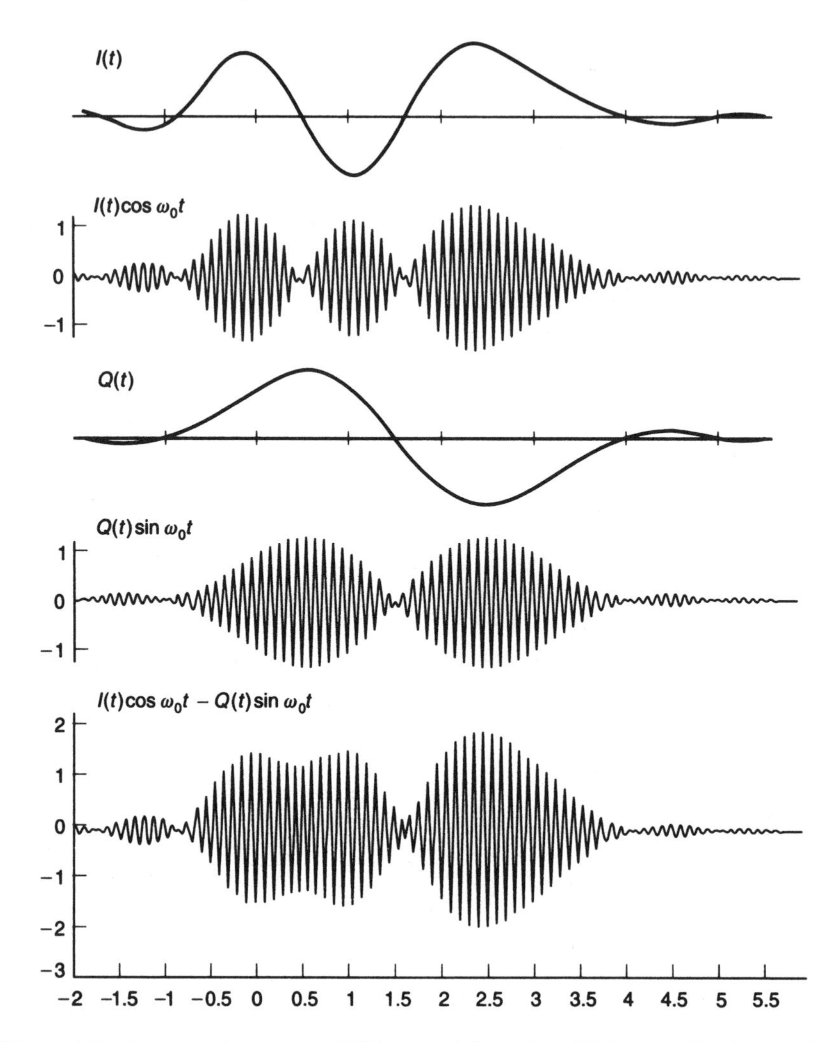

Figure 3.2 Binary and quatenary PSK created from four 30% root RC pulses, with the same in-phase and quadrature data as in Fig. 3.1. Top waveform alone is BPSK: signals combine to form QPSK at bottom. Note that signals extend over considerably more than four intervals; some parts outside [–2, 6] are truncated for clarity. Time axis is in symbols.

which is the general PSD expression for BPSK. It is two replicas of the baseband pulse PSD, one centered at f_0 hertz and the other at $-f_0$.

Example 3.2-1: Sinc Pulse BPSK. The unit-energy sinc pulse with symbol interval T is

$$v(t) = \frac{1}{\sqrt{T}}\ \mathrm{sinc}\,\frac{t}{T}$$

Its transform is $\sqrt{T}\text{rect}(t/T)$, or

$$V(f) = \begin{cases} \sqrt{T}, & -\dfrac{1}{2T} < f < \dfrac{1}{2T} \\ 0, & \text{otherwise} \end{cases}$$

Both $v(t)$ and $V(f)$ are sketched in Fig. 2.2. The baseband PAM scheme based on $v(t)$ and $a_n = \pm 1$ has signal

$$s(t) = \sum_n a_n \sqrt{\frac{E_s}{T}} \operatorname{sinc}\left(\frac{t - nT}{T}\right)$$

and PSD equal to $NE_s|V(f)|^2$, if there are N pulses. The BPSK carrier scheme based on $v(t)$ has signal

$$s(t) = \sum_n a_n \sqrt{\frac{2E_s}{T}} \operatorname{sinc}\left(\frac{t - nT}{T}\right) \cos \omega_0 t$$

The PSD is calculated from (3.2-9) and is sketched in Fig. 3.3. The factor NE_s is not included in the sketch. Both the baseband and the carrier schemes will have error probability $Q(\sqrt{2E_b/N_0})$ with ML detection.

Any orthogonal pulse could be substituted for $(1/\sqrt{T})\operatorname{sinc}(t/T)$ in the example. For example, the root RC pulse in Example 2.2-5 and Eq. (2.2-11) leads to a slightly wider PSD, which Fig. 3.3 also shows for the 30% case. The top RF signal in Fig. 3.2 is a piece of a signal with this spectrum. Another example is the NRZ pulse case in Fig. 3.4. The decibel scale on the vertical axis helps to show the important details. Because of the relationship in (3.2-9), the NRZ plot in Fig. 2.9 is the same as the right half of the right replica in Fig. 3.4; the remaining parts of the PSD are translations and reflections of 2.9.[4]

Carrier PSDs are more often plotted in decibels versus normalized frequency above the carrier, $(f - f_0)T$, so that they take on the appearance of the baseband plot. A similar convention is applied to power-out-of-band plots. However, it must be remembered that a carrier modulation has an upper and lower sideband, so that a carrier signal power bandwidth is twice that of the baseband signal. To help distinguish the two cases, the double-sideband carrier bandwidth is called the *RF bandwidth*. For example, the 99% RF bandwidth of NRZ pulse BPSK is $20.4/T$, compared to $10.2/T$ for a sequence of baseband NRZ pulses.

3.2.3 Quaternary PSK

Quaternary PSK is the most common form of phase-shift keying. With the basics we have just developed for binary PSK, QPSK is a straightforward generalization.

[4]The distortion in the symmetry of Fig. 3.4 is due to the finite carrier frequency, which is $10/T$, a typical value in practice.

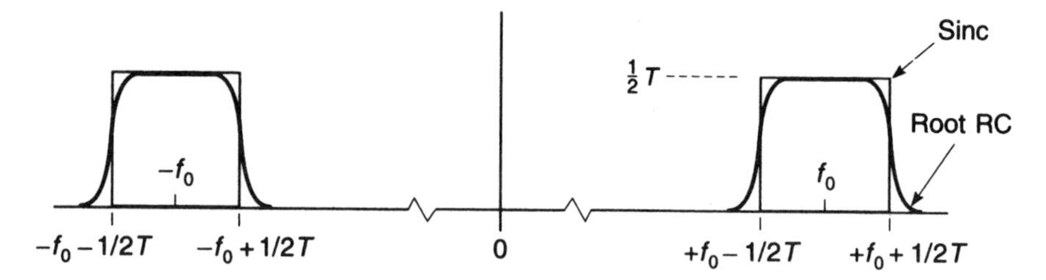

Figure 3.3 Idealized power spectral densities for BPSK for sinc and root RC pulses; T = symbol time.

In Section 3.4, we will in turn generalize QPSK to produce quadrature amplitude modulation. Thus QPSK is at the heart of much of modulation theory.

QPSK can usefully be thought of as two-dimensional BPSK. Whereas the BPSK in (3.2-2) or (3.2-5) has an $I(t)$ baseband signal but no $Q(t)$, QPSK has both, as two independent signals. The general baseband form for QPSK for a unit-energy pulse v is

$$s(t) = \sum_n a_n^I v(t - nT) \sqrt{2E_s} \cos \omega_0 t - \sum_n a_n^Q v(t - nT)\sqrt{2E_s} \sin \omega_0 t \quad (3.2\text{-}10)$$

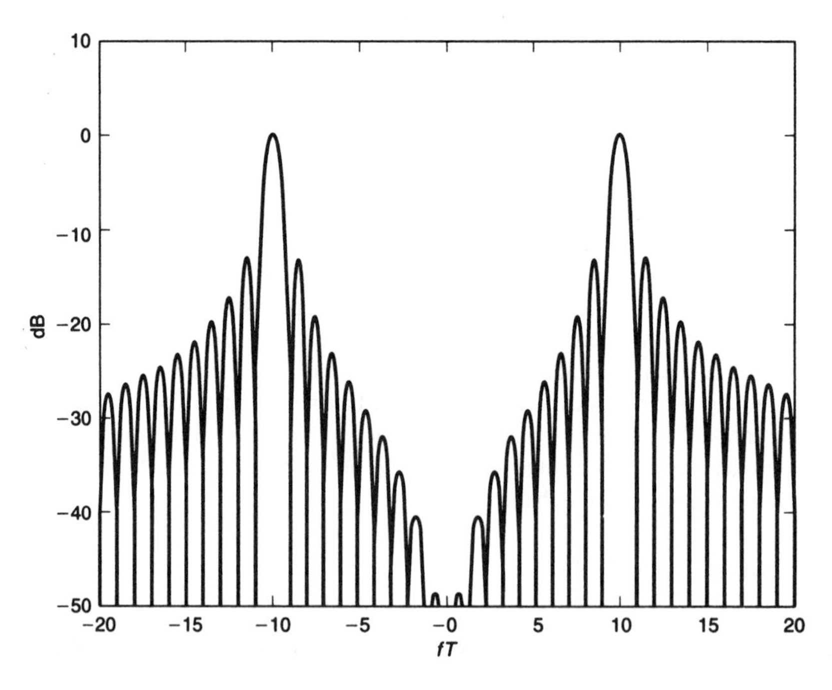

Figure 3.4 Actual power spectral density of NRZ pulse BPSK and QPSK versus units of $1/T$ for carrier frequency $10/T$. Nulls appear at $f = f_0 + n/T$, n an integer. For QPSK, $T_b = \frac{1}{2}T$.

in which $a_n^I = \pm 1/\sqrt{2}$ and $a_n^Q = \pm 1/\sqrt{2}$ are independent sequences of transmission symbols and E_s, as usual, is the average energy per symbol interval. The expression for BPSK in (3.2-5) consists of the first term alone (with a ± 1 symbol alphabet to preserve the energy normalization). The second term is the same form except for the sin $\omega_0 t$. These two terms are, in fact, orthogonal to each other because the integral of one against the other is zero as $\omega_0 \to \infty$. In effect, (3.2-10) is two independent BPSKs, with a 90° phase shift between them.

We can write this signal in the standard quadrature modulation form (3.1-4) as follows:

$$s(t) = \sqrt{\frac{2E_s}{T}}\,[I(t)\cos\omega_0 t - Q(t)\sin\omega_0 t]$$

where

$$I(t) = \sum_n a_n^I \sqrt{T}\,v(t-nT), \qquad Q(t) = \sum_n a_n^Q \sqrt{T}\,v(t-nT) \tag{3.2-11}$$

Here $I(t)$ and $Q(t)$ satisfy the energy normalization condition (3.1-5) if v is unit energy and the symbols are $\pm 1/\sqrt{2}$. One set of these symbols, a_n^I, is carried in the in-phase "channel," while the other set, a_n^Q, is carried in the quadrature channel.

What has just transpired is the standard description of QPSK, but it is not really clear how the modulation lives up to its idea, fourfold phase shifting. To see this, consider just the pulse $v(t - nT)$ from the nth interval. By the conversion formula (3.1-7), the amplitude and phase form of (3.2-11) with a single I and Q pulse is

$$\psi(t) = \arctan\frac{Q(t)}{I(t)} = \arctan\left[\frac{a_n^Q \operatorname{sgn}[v(t)]}{a_n^I \operatorname{sgn}[v(t)]}\right]$$

$$A(t) = \sqrt{\frac{2E_s}{T}}\sqrt{\frac{1}{2}Tv(t-nT)^2 + \frac{1}{2}Tv(t-nT)^2} = \sqrt{2E_s}\,|v(t-nT)|$$

with sgn(x) the sign of x. Depending on the two transmission symbols, the phase here is one of the values 45°, 135°, 225°, 315°. We can write the amplitude-and-phase form as

$$\sqrt{2E_s}\,v(t-nT)\cos(\omega_0 t + \tfrac{1}{4}\pi + \tfrac{1}{2}i\pi), \qquad i = 0, 1, 2, 3 \tag{3.2-12}$$

which is four phase shifts of the basic sinusoidal pulse.

When v is the NRZ pulse, $v(t - nT) = 1/\sqrt{T}$ in the nth interval, Eq. (3.2-11) becomes the baseband form

$$s(t) = \sqrt{\frac{2E_s}{T}}\,[a_n^I \cos\omega_0 t + a_n^Q \sin\omega_0 t] \tag{3.2-13}$$

in the nth interval and the amplitude and phase form

$$s(t) = \sqrt{\frac{2E_s}{T}} \cos(\omega_0 t + \tfrac{1}{4}\pi + \tfrac{1}{2}i\pi) \tag{3.2-14}$$

Table 3.1 summarizes the relationships among the data bits, phases, and transmission symbols in both NRZ and general QPSK. In the NRZ case, the signals $I(t)$ and $Q(t)$ simply hold the transmission symbol values a^I and a^Q throughout an interval, while in the general case, these scale a pulse. It can be seen that the binary data symbols map to the signs of the transmission symbols, with binary 0 mapping to "+" and binary 1 mapping to "–". Furthermore, the first binary maps to the I-channel sign while the second one maps to the Q-channel sign: The I-channel thus "carries" the first data bit and the Q-channel the second. These conventions are arbitrary, since we could, in principle, give the binaries any two names and specify any mapping to the four QPSK signals. The conventions chosen here make QPSK easier to understand in the long run; in particular, they underscore the idea that QPSK is two parallel BPSKs.[5]

We can find a signal space basis and constellation for QPSK in the nth interval by the following sequence of steps; v is any unit-energy T-orthogonal pulse.

1. By taking the inner product of $v(t - nT) \cos \omega_0 t$ and the $s(t)$ in the general form (3.2-11), discover that

$$(s(t), v(t - nT) \cos \omega_0 t) = a_n^I \sqrt{\tfrac{1}{2}E_s}$$

   The calculation follows the model of (3.2-8), with some additional $\sin \omega_0 t \cos \omega_0 t$ terms that tend to zero with ω_0. Consequently, $v(t - nT) \cos \omega_0 t$ is a good basis function for the nth interval.

2. By taking the inner product of $-v(t - nT) \sin \omega_0 t$ and $s(t)$, discover that

$$(s(t), -v(t - nT) \sin \omega_0 t) = a_n^Q \sqrt{\tfrac{1}{2}E_s}$$

   The calculation is again similar to (3.2-8). Find that $v(t - nT) \cos \omega_0 t$ and $v(t - nT) \sin \omega_0 t$ are orthogonal as $\omega_0 \to \infty$. Thus $-v(t - nT) \sin \omega_0 t$ is a good second basis function for the nth interval.

3. Since the two functions found so far can express any form that $s(t)$ takes in the nth interval, conclude that no more basis functions need to be found. Normalize the functions, and conclude that the orthonormal basis is

$$\{\phi_1(t), \phi_2(t)\} = \{\sqrt{2}v(t - nT) \cos \omega_0 t, -\sqrt{2}v(t - nT) \sin \omega_0 t\} \tag{3.2-15}$$

   for the nth interval.

[5]The mapping of binary 0 to "+" is confusing to some. A major reason for the map is that 0 and + act as the identity elements in the binary field and the system of sign changes, respectively.

TABLE 3.1 Equivalences of Angles, Shifts, Data Bits, and Transmission symbols for QPSK

Phase Shift, i	Angle (deg)	As Binaries	a_n^I	a_n^Q
0	45	00	$+1/\sqrt{2}$	$+1/\sqrt{2}$
1	135	10	$-1/\sqrt{2}$	$+1/\sqrt{2}$
2	225	11	$-1/\sqrt{2}$	$-1/\sqrt{2}$
3	315	01	$+1/\sqrt{2}$	$-1/\sqrt{2}$

Note: $I(t) = a_n^I$ and $Q(t) = a_n^Q$ for NRZ pulses.

4. By finding the inner products $(s(t), \phi_1(t))$ and $(s(t), \phi_2(t))$, conclude that the first and second signal space vector components for $s(t)$ are (as $\omega_0 \to \infty$)

$$\mathbf{s} = (a_n^I \sqrt{E_s}, a_n^Q \sqrt{E_s}) = (\pm\sqrt{\tfrac{1}{2}E_s}, \pm\sqrt{\tfrac{1}{2}E_s}) \tag{3.2-16}$$

By summing the squares of these components, find that the energy of all four possible signals is E_s; hence the average energy is also E_s.

The constellation that represents (3.2-16) is shown in Fig. 3.5; this figure is the square constellation of Fig. 2.26c, which is, in turn, a two-dimensional antipodal constellation. We could conclude the last from the fact that QPSK is two independent BPSKs. Figure 3.5 plots the constellation points in terms of the average energy per data bit, which is given by $E_b = \frac{1}{2}E_s$, since the signal carries two data bits in each signaling interval.

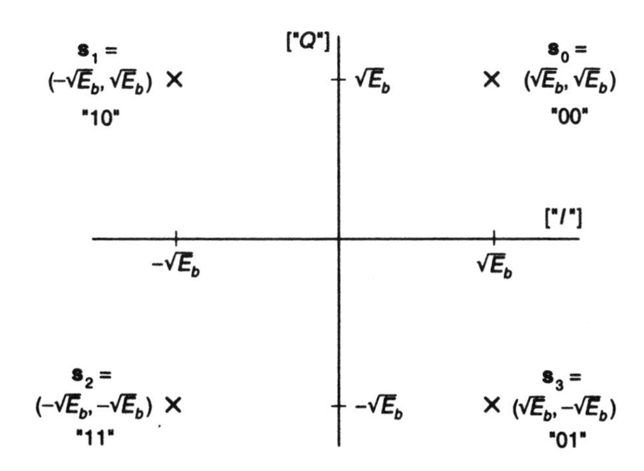

Figure 3.5 Constellation of points in signal space for QPSK showing vector components and corresponding binary data symbols. Axes represent the size of first and second vector components.

From signal space theory, we can directly deduce the error probability of QPSK, just from the fact that it has a square constellation. The calculation was done at length in Section 2.6, with the result that in terms of E_b, the probability is

$$p_e \approx 2Q\left(\sqrt{\frac{2E_b}{N_0}}\right) \tag{3.2-17}$$

or precisely (2.6-13), which is

$$p_e = 2Q\left(\sqrt{\frac{2E_b}{N_0}}\right) - Q^2\left(\sqrt{\frac{2E_b}{N_0}}\right)$$

A plot of (2.6-13) is given later in Fig. 3.17.

The power spectral density of QPSK again follows directly from the spectrum theorem, Theorem 2.3-1, and is, in fact, identical to the PSD of BPSK, Eq. (3.2-9). This time, some of the pulses are time shifts and some are 90° phase shifts. Either way, all the pulses have the same magnitude spectrum, and the steps leading to the PSD (3.2-9) follow through the same, except that there are now $2N$ pulses with energy $\frac{1}{2}E_s$ rather than N pulses with energy E_s.

Even though QPSK has the same spectrum shape as BPSK, it carries twice as many data bits per interval as BPSK. Measured in hertz per data bit carried, its bandwidth is half as great. *Quaternary PSK has twice the bandwidth efficiency of BPSK.* Since its error probability is about the same, QPSK is therefore preferred to BPSK in most applications.

Example 3.2-2: Sinc Pulse QPSK. We can produce sinc-pulse QPSK by extending Example 3.2-1. The unit-energy sinc pulse is still $v(t) = (1/\sqrt{T})\text{sinc}(t/T)$. The QPSK scheme based on this is

$$s(t) = \sqrt{\frac{2E_s}{T}}\,[I(t)\cos\omega_0 t - Q(t)\sin\omega_0 t]$$

with

$$I(t) = \sum_n a_n^I \,\text{sinc}\left(\frac{t-nT}{T}\right), \qquad Q(t) = \sum_n a_n^Q \,\text{sinc}\left(\frac{t-nT}{T}\right)$$

when written in the standard quadrature form (3.2-11). Recall that the transmission symbols here are $\pm 1/\sqrt{2}$. The signal constellation in each interval is Fig. 3.5 and the error probability is Eq. (2.6-13). The power spectrum is Fig. 3.3, as it was in Example 3.2-1.

An example of a 30% root RC QPSK waveform is shown in Fig. 3.2. This waveform consumes 650 kHz of RF bandwidth to carry 1 Mb/s data. Sinc pulse QPSK consumes 500 kHz; 30% root RC BPSK consumes 1.3 MHz.

3.2.4 Envelopes and *I/Q* Plots

An important element in the design of RF circuitry is how much the power of the RF sinusoid is expected to vary. Efficient RF amplifiers have difficulty following RF amplitude changes and the resulting amplitude distortion confuses the detection of the signal. Earlier we defined the quantity $\sqrt{I^2(t)+Q^2(t)}$ as the envelope of a modulated sinusoid. The envelope is effectively the local amplitude of the RF signal.

For a BPSK signal, the envelope boils down to the absolute value of the pulse train $\Sigma a_n v(t-nT)$ that modulates $\cos \omega_0 t$. For QPSK and for other modulations with both $I(t)$ and $Q(t)$, the relation to the modulating data is more complex. A useful way to show the relation is with the *I/Q plot* of Fig. 3.6. This is a plot of $Q(t)$ versus $I(t)$ as t evolves. A modulation with a constant envelope traces out a perfect circle, but most pulse modulations wander through a complex path, as in the figure. The radial square distance of an *I/Q* point, which is the square value of the envelope, is the local power of the signal at that time. Transitions through the center of the plot are the hardest for a power amplifier to follow because the short-term signal power passes close to zero. With NRZ pulse QPSK, the signal spends the entire signal interval at one of four spots, and the *I/Q* plot becomes, basically, the constellation plot in Fig. 3.5.

Figure 3.7 plots the distribution, or histogram, of the envelope in Fig. 3.6 over time. Plots such as Fig. 3.7 help in the design of transmitter amplifiers because they show what fraction of time the signal exceeds certain limits on power.

3.3 QUADRATURE TRANSMITTER AND RECEIVER

Phase-shift keying gives the occasion to introduce two basic circuits in digital communication, the quadrature transmitter and receiver. We shall discuss them here in terms of QPSK, but they find application in all of carrier modulation.

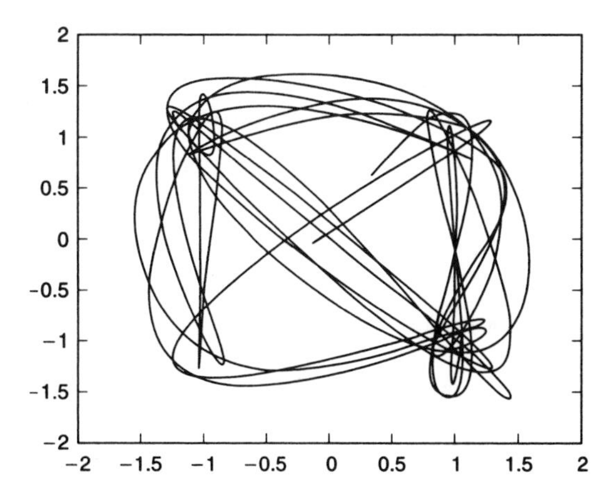

Figure 3.6 An *I/Q* plot for 30% root RC pulse QPSK; plot traces 40 random bits in *I* and in *Q*.

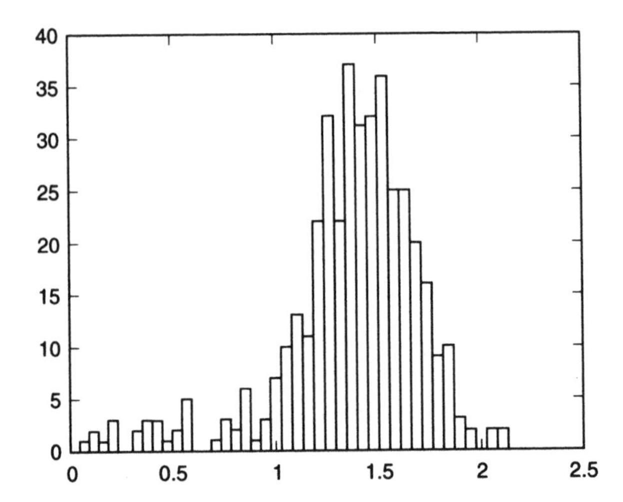

Figure 3.7 Histogram of envelope samples for *I/Q* plot in Fig. 3.6; 10 samples per interval, 400 total.

Quadrature Transmitter Circuit. An implementation of the basic QPSK equation [Eq. (3.2-11)] is shown in Fig. 3.8. This circuit is called the *quadrature transmitter*. It consists of multiplications of $I(t)$ and $Q(t)$ by $\cos \omega_0 t$ and $\sin \omega_0 t$. The circuit carries out the standard conversion in (3.1-7) from the baseband to the amplitude and phase form of a carrier modulation. It may thus be used to generate any carrier signal.

At high frequencies, the multipliers in the quadrature transmitter form a small piece of hardware called a double-balanced mixer. At low frequencies the circuit is usually realized as software. In either case, some earlier processing creates $I(t)$ and $Q(t)$, from which may come any desired passband signal.

Quadrature Receiver. The standard ML receiver for QPSK is the *quadrature receiver* shown in Fig. 3.9. The receiver is the carrier version of the linear receiver

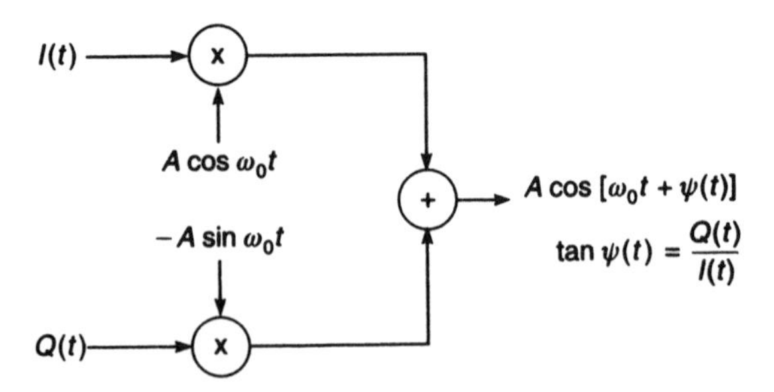

Figure 3.8 Quadrature transmitter circuit. In standard QPSK, take A as $\sqrt{2E_s/T}$.

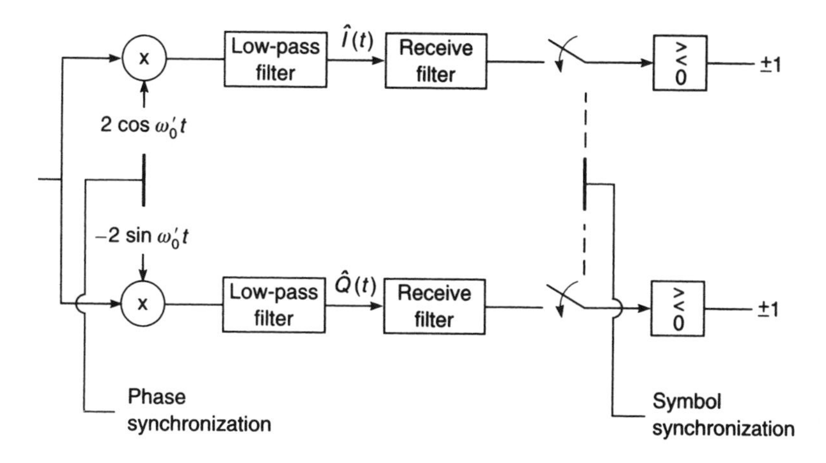

Figure 3.9 Quadrature receiver, applied to QPSK.

of Fig. 2.7. Like that receiver, it is ML when the transmitted pulse train is made from orthogonal pulses. Also like that receiver, it finds wide application with other pulses and signals.

The receiver in Fig. 3.9 begins with a multiplication of the noisy input signal

$$r(t) = \sqrt{\frac{2E_s}{T}}[I(t)\cos\omega_0 t - Q(t)\sin\omega_0 t] + \eta(t) \tag{3.3-1}$$

by $\sin\omega_0' t$ and $\cos\omega_0' t$. These sinusoids are supplied by a carrier synchronizer circuit like those in Chapter 4, and for now we will assume that $\omega_0 = \omega_0'$. The modulated signal in (3.3-1) is a standard quadrature signal of the form (3.1-4), and if the signal is QPSK, $I(t)$ and $Q(t)$ are the independent pulse trains in (3.2-11). The $\eta(t)$ is additive noise, and in this chapter we will assume that it is white Gaussian noise with spectral density $\frac{1}{2}N_0$. The hardware that performs the multiplication is another double-balanced mixer. After the mixer comes a pair of low-pass filters. Taking the top rail of the receiver, we can see that the filter input is

$$2\sqrt{\frac{2E_s}{T}}[I^2(t)\cos^2\omega_0 t - Q(t)\sin\omega_0 t\cos\omega_0 t] + 2\eta(t)\cos\omega_0 t$$

$$= \sqrt{\frac{2E_s}{T}}[I(t)(1+\cos 2\omega_0 t) - Q(t)\sin 2\omega_0 t] + 2\eta(t)\cos\omega_0 t$$

Similar trigonometry gives the lower filter input. After the two low pass filters what remains is

$$\hat{I}(t) \triangleq \sqrt{\frac{2E_s}{T}}\, I(t) + \eta_I(t), \qquad \hat{Q}(t) \triangleq \sqrt{\frac{2E_s}{T}}\, Q(t) + \eta_Q(t)$$

It can be seen that $\hat{I}(t)$ and $\hat{Q}(t)$ are estimates of $I(t)$ and $Q(t)$ consisting of a scaled $I(t)$ and $Q(t)$ plus noise waveforms $\eta_I(t)$ and $\eta_Q(t)$. These are given by $2\eta(t) \cos \omega_0 t$ and $-2\eta(t) \sin \omega_0 t$, respectively. It can be shown that they are Gaussian noises with density N_0, band-limited to the filter bandwidth.[6]

With $\hat{I}(t)$ and $\hat{Q}(t)$ extracted, the receiver becomes identical to the baseband Gaussian noise receiver in Section 2.6. What follows next in Fig. 3.9 can be a bank of matched filters, an integrate-and-dump, or some other processor. In particular, if the transmitter pulse trains are made up of orthogonal pulses, inserting the linear receiver components in Fig. 2.33 provides an ML detector for the pulses, and the AWGN channel model in the figure is valid as well. Figure 3.9 is a complete linear receiver if the Receive Filter block is replaced by the filter $H^*(f)$ matched to the pulse-shaping filter $H(f)$. Most QPSK detectors are of the quadrature receiver type and have the structure of Fig. 3.9, although the filter may not be closely matched to the transmission pulse. Note that the receiver needs both carrier phase and symbol timing signals; obtaining these is the subject of Chapter 4.

We have already proven in Section 3.2 that the error probability of orthogonal-pulse QPSK is $Q(\sqrt{2E_b/N_0})$ by applying a signal space analysis to the RF signals. As an alternate derivation, we can directly analyze the baseband signals $\hat{I}$ and $\hat{Q}$ in Fig. 3.9 as follows. In the top rail, $\hat{I}(t)$ is a binary PAM pulse train plus additive Gaussian noise of spectral density N_0. The actual energy per binary symbol in the pulse train is

$$\int_{\mathcal{T}} \mathcal{E}\left[\frac{2E_s}{T} I^2(t)\right] dt = \frac{2E_s}{T} \int \mathcal{E}[I^2(t)]\, dt$$

where the expectation is over all the signals that may appear in the symbol interval $\mathcal{T}$. By the standard normalizing condition (3.1-5),

$$\frac{1}{T}\int \mathcal{E}[I^2(t)]dt = \frac{1}{2}$$

if half the original RF signal energy lies in each of the I and Q signals. The energy per bit in $\hat{I}(t) - \eta_I(t)$ is thus E_s, the symbol energy factor from the original signal. Recall that we characterize the SNR in a channel, whether passband or baseband, as the actual energy per data bit divided by twice the noise power spectral density. In the top rail this ratio is then $E_s/2N_0$. As in the two-signal derivation in Section 2.6, this SNR with antipodal, orthogonal binary pulses and an ML detector leads to the error probability $Q(\sqrt{2(E_s/2N_0)})$. In the original signal E_s was $2E_b$, where E_b is the data bit energy in the original signal. The error probability in the top rail may thus be written $Q(\sqrt{2E_b/N_0})$, which is the BPSK error result (3.2-4).

If the entire quaternary symbol is to be correct, then both the I and Q decisions need to be. They are independent because η_I and η_Q are independent, and the error probability becomes the expression (2.6-13). Although Chapter 2 assumes truly

[6]This is shown in Appendix B; η_I and η_Q are Gaussian and uncorrelated as $\omega_0 \to \infty$ and are hence independent.

white noise, we can assume that the low-pass filter (LPF) block in Fig. 3.9 and any receiver bandpass filters that come before the figure have bandwidth larger than the signal and therefore do not damage it. Only the noise *density* matters at the Receive Filter block.

The quadrature receiver will come up repeatedly in the rest of the chapter because it is a basic detection circuit. In part, this is because so many modulations simply send their data in independent $I(t)$ and $Q(t)$ signals. But on a more fundamental level, any passband signal may be broken down into equivalent I and Q signals, and the quadrature receiver is a system for estimating these. The signals may be complicated and dependent on each other; in this case, some further processing will be required to find the data in them. The next sections illustrate these ideas.

3.4 GENERALIZATIONS OF PHASE-SHIFT KEYING

In this section we will extend the idea of PSK in a number of directions. Most of these extensions aim at making the signal easier to synchronize or easier to handle electronically. First, we extend QPSK to a higher number of phases.

3.4.1 General PSK

In the general case, a PSK signal takes on one of M phases; in amplitude and phase form, MPSK looks like[7]

$$\sum_n \sqrt{2E_s}\, v(t-nT)\cos\left(\omega_0 t + \frac{2i_n\pi}{M}\right), \qquad i_n = 0, \ldots, M-1 \qquad (3.4\text{-}1)$$

in which v is unit energy and E_s is the energy in each M-ary symbol interval. The baseband form is again (3.2-11), but now the two transmission values a_n^I and a_n^Q there are *dependent*. For example, when $i_n = 0$ in (3.4-1), which means 0° phase, $a_n^I = 1$ and $a_n^Q = 0$; when $i_n = 1$ and the phase is $2\pi/M$ rads, $a_n^I = \cos(2\pi/M)$ and $a_n^Q = \sin(2\pi/M)$. All eight values of the symbol pair for the case of 8PSK are plotted in Fig. 3.10a. Plotted in the plane this way, they form a set of M points around a circle.

M-ary PSK is a linear modulation in the sense that if we let the elements in a standard pulse train like (2.2-1) be complex numbers, then (3.4-1) may be rewritten as the real part of

$$\sum_n \sqrt{2E_s}\, v(t-nT)e^{j2i_n\pi/M}\, e^{j\omega_0 t} \qquad (3.4\text{-}2)$$

which is evidently a pulse train. Here is an example of a notational convenience that appears often in communication theory, writing carrier signals as complex numbers. In complex form, the standard I and Q form (3.2-11) is

[7]The $\frac{1}{4}\pi$ offset in (3.2.12), which allowed QPSK to be viewed directly as two independent BPSKs, is dropped here, because independence is no longer at issue.

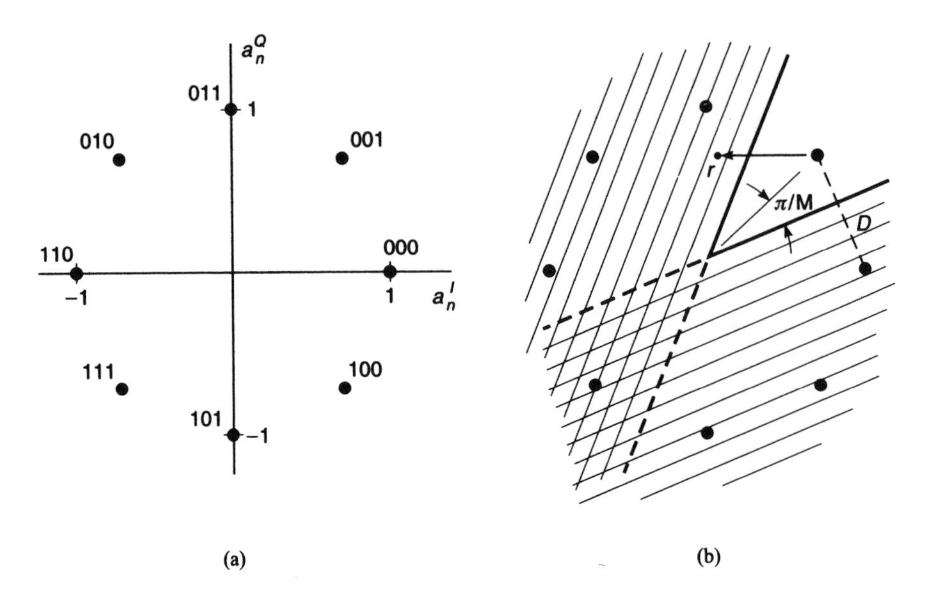

Figure 3.10 **(a)** Transmission symbols for 8PSK; labels show corresponding data bits. **(b)** Calculation of symbol error probability for MPSK for 8PSK case. Minimum distance is D; circle radius is $\sqrt{E_s}$.

$$\mathrm{Re}\left\{\sqrt{\frac{2E_s}{T}}[I(t) + jQ(t)]e^{j\omega_0 t}\right\} \tag{3.4-3}$$

It needs to be kept in mind that there is nothing "unreal" about a carrier signal; the complex notation is only a method of keeping the books. Here, it provides a way of seeing that M-ary PSK might be viewed as a linear pulse train. Whether the insight is of any use depends on whether MPSK *acts* like a pulse train, which, as we will see, it does.

The usual Gram–Schmidt process derives the signal space basis for M-ary PSK in the nth interval. Taking $\sqrt{2}v(t - nT)\cos\omega_0 t$ as the initial basis function, one finds that the only other basis function for the interval is $-\sqrt{2}v(t - nT)\sin\omega_0 t$; in other words, the basis is (3.2-15), the same as the one for QPSK. The signal vector components in the nth interval are still

$$s = (a_n^I\sqrt{E_s},\ a_n^Q\sqrt{E_s})$$

as in (3.2-16), but the transmission symbol values now lie around a unit circle as in Fig. 3.10. What this shows is that MPSK is a quadrature modulation, since it is based on the $\sin\omega_0 t/\cos\omega_0 t$ basis functions, and that the $\sin\omega_0 t$ and $\cos\omega_0 t$ terms in the baseband form are affected independently by white Gaussian noise. Yet the terms themselves no longer track independently of each other.

The power spectral density of M-ary PSK follows from the linear spectrum theorem, Theorem 2.3-1, and is again the spectrum of $v(t)$. The reason is that all the elements in the real part of the sum (3.4-2) have the same amplitude spectrum and the weighting coefficients average out to zero and are independent from interval to interval. Because of the outcome of the spectral calculation and the Gram–Schmidt process, we are justified in thinking of MPSK as a linear modulation.

The symbol error probability of M-ary PSK in white Gaussian noise follows from the signal space theory in Sections 2.5 and 2.6. We first sketch the precise calculation for an ML receiver and then give a simple approximate formula based on the minimum-distance concept. The calculation is sketched in Fig. 3.10b which illustrates the example of 8PSK. For a given transmitted MPSK point, the receiver decision region is a wedge-shaped piece of the plane, whose angular size is $2\pi/M$ radians. A signal vector $\boldsymbol{r}$ received in the remaining region of the plane causes an M-ary symbol error whose probability is the integral of the two-dimensional Gaussian density of the noise $\boldsymbol{\eta}$ over this region. The region is shaded in Fig. 3.10b and the arrow is an $\boldsymbol{\eta}$. From the symmetry of the constellation and of the noise, the error probability must be the same for all M transmitted points. Some further thought will reveal that only the *angle* of the received $\boldsymbol{r}$ matters. A precise error calculation based on this fact is given in Proakis [1], Section 5-2.

A surprisingly accurate estimate to the error probability comes from the minimum-distance estimate (2.6-17). As illustrated in Fig. 3.10b, the signal space minimum distance between two adjacent points is

$$D = 2\sqrt{E_s}\,\sin\left(\frac{\pi}{M}\right)$$

From (2.6-16) and the fact that $E_s = E_b \log_2 M$, the normalized minimum distance is

$$d_{\min} = \frac{D}{\sqrt{2E_b}} = \sin\left(\frac{\pi}{M}\right)\sqrt{2\log_2 M} \tag{3.4-4}$$

By substituting this into (2.6-17), we get

$$p_e \leq 2Q\left[\sin\left(\frac{\pi}{M}\right)\sqrt{2\left(\frac{E_b}{N_0}\right)\log_2 M}\right] \tag{3.4-5}$$

where the factor $2K/M$ is 2. An alternate way to derive (3.4-5) is to realize that one of the multiples of Q in (3.4-5) is exactly the integral of the horizontally shaded area in Fig. 3.10b, while the other is the vertically shaded area; the inequality appears because a small area in the third quadrant is counted twice.

By comparing (3.4-5) to (2.6-5), we see that the energy loss of MPSK compared to antipodal signaling is a factor $(\log_2 M)\sin^2(\pi/M)$. Table 3.2 summarizes these theoretical losses for PSKs up to $M = 16$ and gives the actual energy requirements to achieve a symbol error rate of 10^{-5}; the latter all lie about 0.3 dB above the distance-based estimates. It is clear that the energy requirement of MPSK grows rapid-

TABLE 3.2 Comparison of Bandwidth and Energy Requirements of M-ary PSK and Differential PSK

M	Relative Bandwidth	Asymptotic Loss (dB)	Actual E_b/N_0 Needed for $P_e = 10^{-5}$ (dB)
2 (BPSK)	B	0	9.6
2 (DPSK)	B	0	10.3
4(QPSK)	$\frac{1}{2}B$	0	9.9
4 (4DPSK)	$\frac{1}{2}B$	2.3	12.1
8	$\frac{1}{3}B$	3.6	13.6
16	$\frac{1}{4}B$	8.2	18.1

B is a conventional bandwidth such as 99% bandwidth.

ly with M, but since an M-ary symbol carries $\log_2 M$ bits, the signal bandwidth needed per data bit drops too, as $1/\log_2 M$. This saving is a valuable advantage, but it comes with an energy cost, and what we see here is a trade-off between bandwidth and energy.

In large measure, this trade-off is a fact of life in communication theory, but we can still ask whether the MPSK trade-off is the most efficient one. As Chapter 3 continues, we will find schemes with a more efficient trade-off; for this reason, PSKs with more than eight phases are seldom encountered in practice. The symbol error rate of MPSK is compared to a number of competitors later in Fig. 3.17.

Gray Coding. The previous discussion has focused on symbol error rate, but what happens to the individual data bits that may be carried in the M-ary symbol is important as well. A sensible mapping of the data bits to the symbol would be one that minimizes the number of bits in error when the most common demodulator errors occur. These are the ones that decide in favor of an adjacent decision region. Our version of QPSK in Fig. 3.5 carries out such a mapping; all adjacent decision errors lead to a single error in the bit pairs, and none lead to a double error. A mapping that minimizes adjacent decision bit errors is called a *Gray coding* of the data bits. A Gray coding for octal PSK is somewhat trickier to design, and one example is shown in Fig. 3.10b.

3.4.2 Offset QPSK

Offset QPSK is another scheme that exploits the independence of the in-phase and quadrature signals in a PSK modulation. It is simply QPSK, except that one of the baseband signals is delayed, or offset, by half a symbol interval. The technique is illustrated for NRZ pulse trains in Fig. 3.11, in which the quadrature signal $Q(t)$ is delayed by $\frac{1}{2}T$; the in-phase and quadrature data symbols are the same ones used in Fig. 3.1, $\{1, -1, 1, 1\}$ and $\{1, 1, -1, -1\}$.

The offset idea is motivated by the fact that 180° phase transitions never occur in an offset scheme. Rather, twice as many transitions occur that are limited to 0 and ±90°. With smoothed-pulse QPSK or with QPSK that has been channel filtered, it is the 180° transitions that lead to electronic circuit difficulties. Figure 3.12 compares

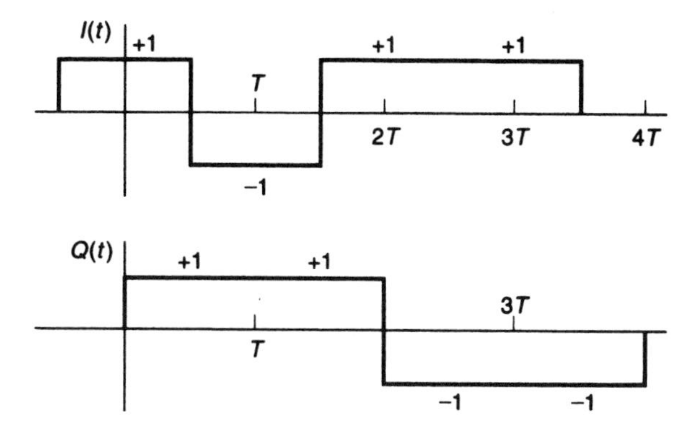

Figure 3.11 Illustration of offset QPSK for data symbols in Fig. 3.1; $Q(t)$ is delayed $\frac{1}{2}T$, compared to Fig. 3.1.

the 30% root RC QPSK signal in Fig. 3.2 to an offset QPSK signal that carries the same data. It can be seen that the offset signal has less envelope variation, especially near the 180° phase reversal at $1.5T$, but also, for instance, during the ringing before and after the main signal burst. An I/Q plot will show the same reduction. Figure 3.13, based on transmission of 10 I and Q symbols, compares QPSK and offset

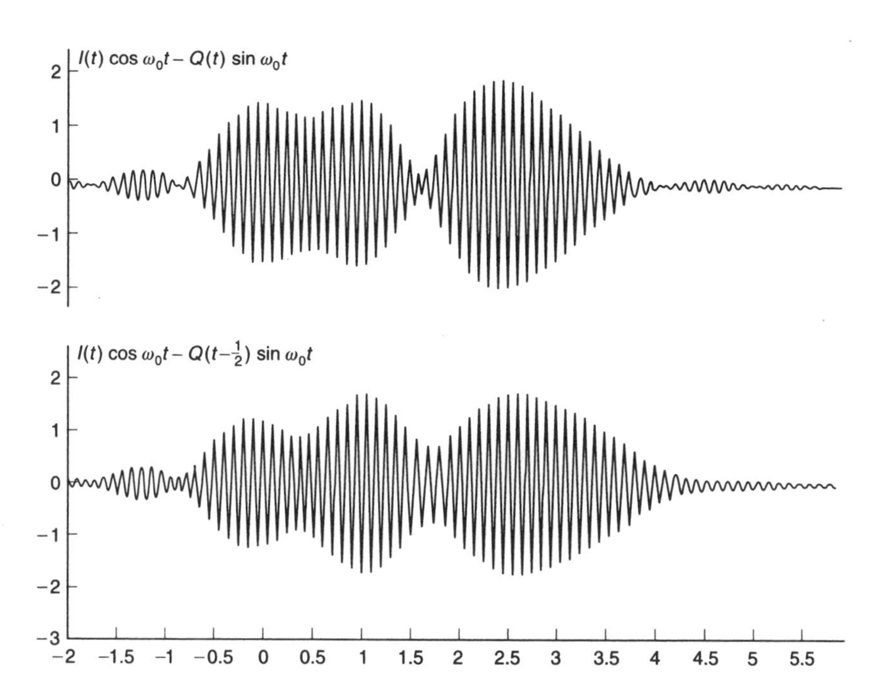

Figure 3.12 Comparison of QPSK **(top)** and offset QPSK **(bottom)** with same data and 30% root RC pulses. Data as in Figs. 3.1, 3.2, and 3.11.

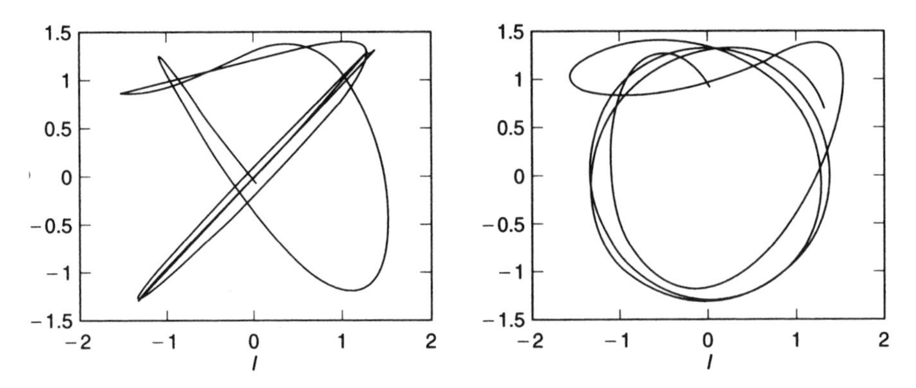

Figure 3.13 The I/Q plots for QPSK **(left)** and offset QPSK **(right)** for I data (+ – + – + – – + + –) and Q data (+– + – + + + + – +). There are five 180° phase changes in the left plot.

QPSK; the offset plot tends to trace a rough circle, while ordinary QPSK repeatedly crosses through the origin.

Offset QPSK has the same error probability in white Gaussian noise as QPSK, as a consequence of the I and Q channel independence, because the error probability is still $Q(\sqrt{2E_b/N_0})$ in each channel. Because the two pulse trains still satisfy the requirements of the linear spectrum theorem (Theorem 2.3-1) in the same way, the power spectrum of offset QPSK is also identical to that of QPSK. Offset QPSK and QPSK illustrate schemes that look the same from a theoretical point of view, but behave quite differently in practice.

3.4.3 Differential PSK

There are actually two concepts under the heading of differential phase-shift keying and they are easily confused with each other: In differential *encoding*, the data are mapped to the standard PSK signals in a different way; in differential *detection*, a new kind of receiver detects the data, whatever the encoder mapping was.

A differentially encoded modulator is one where a data symbol dat_k is the *difference* between two transmitted symbols. If we let these encoded symbols be enc_k and enc_{k-1} and take all as binaries, then

$$\text{enc}_k = \text{dat}_k \oplus \text{enc}_{k-1} \tag{3.4-6a}$$

or, equivalently,

$$\text{dat}_k = \text{enc}_k \oplus \text{enc}_{k-1} \tag{3.4-6b}$$

where $\oplus$ means modulo 2 addition. Table 3.3 illustrates binary data sent by a differential binary modulator. The first row is a binary data stream and the second is the differential encoding of the data. A 1 in the top stream means that the symbol below differs from the one before it in row two. The encoded stream must start with a known initial symbol enc_0 in order to start the scheme. In the table, the first data symbol is 0, and $\text{enc}_0 = 0$, so enc_1 must be the same, namely 0. The symbol dat_2 is 1, and so enc_2 must switch to 1.

The second row in Table 3.3 is converted in the third row to the transmission values $a_0, a_1, \ldots$, and these are applied to the pulse train formula (3.2-5) to create a BPSK in the usual way. The fourth row is the set of phases ψ_k in the amplitude and phase description of the BPSK. A formula for this is $\psi_k = \psi_{k-1} + \frac{\pi}{2}[1 - (a_k a_{k-1})]$.

The differential encoding idea is easily extended to QPSK: The four data bit pairs 00, 01, 11, 10 might, for example, map to 0°, 90°, 180°, 270° *plus* the previous transmitted phase (which is one of 45°, 135°, 225°, 315°), rather than to the values in Table 3.1. Differential encoding fits in naturally with modulations based on frequency shifts, as we will see in Section 3.6, and mitigates difficulties that come up in Chapter 4 with carrier phase synchronization.

A differential demodulator is quite a different matter. This is a receiver structure, rather than a method of encoding or decoding symbols. A differential demodulator works by comparing the present received signal to a previous part of the signal in such a way that it extracts the difference between them. A straightforward NRZ-pulse application of the principle is shown in Fig. 3.14. A noisy phase-modulated signal enters at the left and is multiplied by a one-symbol-delayed version of itself. After filtering by an LPF, what remains, by simple trigonometry, is the cosine of the phase difference between symbols, $\cos(\psi_k - \psi_{k-1})$, plus a zero-mean noise term, $\eta'(t)$. The parts within the dashed box are the standard filter/sampler/comparator components from Chapter 2.

The possible signal phases and the resulting low-pass filter outputs are listed in Table 3.4. It can be seen that the filter puts out +1 when the last two phases are the same and −1 otherwise; that is, with the usual correspondence $-1 \leftrightarrow$ data 1 and $+1 \leftrightarrow$ data 0, the filter output is the modulo-2 sum of the last two symbols.

Differential PSK (DPSK) is the name given to binary PSK with *both* this differential detection and the differential encoding. The latter produces enc_k according to Eq. (3.4-6a). The detection produces $\text{enc}_k \oplus \text{enc}_{k-1}$, which, by Eq. (3.4-6b), is directly dat_k without further computation. This can be demonstrated by applying the second form to row two of Table 3.3, to produce row one. The scheme has the advantage that a detection error will not automatically propagate, as it does in many

TABLE 3.3 Example of Differential Encoding of Binary PSK with Initial Phase of 0 rad

Data bits, dat_k		0	1	1	1	1	0	1	0	1	....
Bits after differential encoding, enc_k	0	0	1	0	1	0	0	1	1	0	....
Transmission symbol a_n, carried by $I(t)$		+1	−1	+1	−1	+1	+1	−1	−1	+1	....
Carrier phase (rad), ψ_k		0	π	0	π	0	0	π	π	0	....

Note: First entry in row 2 is the initial symbol enc.

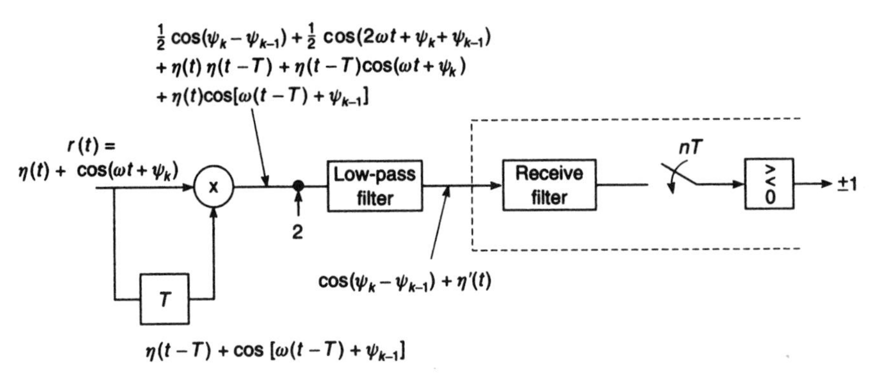

Figure 3.14 Differential PSK receiver structure, showing signals during the kth symbol interval.

differential schemes. Furthermore, it can be shown that the probability of an initial error after a period of correct detection is

$$p_e = \tfrac{1}{2} e^{-E_b/N_0} \tag{3.4-7}$$

The proof is lengthy because the noise $\eta'(t)$ is not simply white additive noise. It may be found in Proakis [1] or Simon et al. [2]. By applying the Q-function estimate (2.6-3) for the BPSK error probability (3.2-4), we find that it too tends in logarithm to the same form (3.4-7). That is, BPSK, QPSK, and DPSK all have about the same error probability as a function of E_b/N_0 as E_b/N_0 grows. With such a good error probability and its freedom from a separate phase synchronizer, DPSK has been an attractive modulation when its wide bandwidth is acceptable.

The idea of differential detection may be extended to four-phase and higher PSK transmission, although 4DPSK with a receiver of the form in Fig. 3.14 requires asymptotically 2.3 dB more E_b/N_0 for the same error probability compared to QPSK. The actual E_b/N_0 requirements of DPSK and 4DPSK at $p_e = 10^{-5}$ appear in Table 3.2. The error probability may be improved by special receivers that observe more than two symbols, and these are described in Ref. 2.

Differential detection is easily extended to orthogonal shaped pulses by matching the receive filter in Fig. 3.14 to the pulse in the usual way. With square pulses the filter and sampler form an integrate-and-dump circuit.

TABLE 3.4 Operation of DPSK in Fig. 3.14

Last Two Data Bits	Sent Carrier Phase ψ_k (rad) If $\psi_{k-1} = 0$	Sent Carrier Phase ψ_k (rad) If $\psi_{k-1} = \pi$	Advance in Transmitted Phase $\psi_k - \psi_{k-1}$ (rad)	LPF Output at Receiver
Same	0	π	0	+1
Different	π	0	π	−1

3.5 QUADRATURE AMPLITUDE MODULATION

We have seen that PSK transmissions can be represented as constellations of points in the I/Q plane. For the simpler PSKs, the points are scaled versions of the transmission symbols a^I and a^Q, which are in turn ± 1 data values. Quadrature amplitude modulation (QAM) is the extension of this idea of directly mapping data to I/Q points. Now a^I and a^Q will carry multilevel rather than binary data. We will describe some of these schemes and calculate their spectra and error rates.

Actually the term QAM has come to mean several things in communications engineering, depending on the context. As a generic term, it refers to transmission via mapping data to points in an independent I/Q plane—that is, to one that is independent from the planes for other symbol intervals. The constellations in Fig. 3.15 illustrate this more general idea. In other contexts, QAM refers to a specific kind of constellation, namely, a *rectangular* one formed by sending independent M-ary PAM waveforms in I and Q. This must necessarily create a rectangular array like the ones in Fig. 3.16. QPSK is rectangular QAM with binary ($M = 2$) pulse trains in each dimension. Figure 3.16b shows the most common rectangular scheme, 16QAM, which has quaternary 4PAM transmission in I and Q.

The standard form of rectangular QAM with M-ary pulse trains is (3.2-11) with M-ary a^I and a^Q; that is,

$$s(t) = \sqrt{\frac{2E_s}{T}}[I(t) \cos \omega_0 t - Q(t) \sin \omega_0 t]$$

$$I(t) = C_o \sum_n a_n^I \sqrt{T} v(t - nT) \tag{3.5-1}$$

$$Q(t) = C_o \sum_n a_n^Q \sqrt{T} v(t - nT)$$

where

$$a_n^I, a_n^Q \in \{\pm(M-1), \pm(M-3), \ldots, \pm 1\}$$

Here $v(t)$ is the usual orthonormal pulse and the transmission symbols are taken as odd integers. The latter is a common convention, which we adopt, but it does mean that a constant C_0 needs to be introduced in order to make $I(t)$ and $Q(t)$ satisfy the unit-energy normalization (3.1-5). This done, E_s becomes the QAM symbol energy, the average energy in one of the M^2 constellation points.

The value $M = 4$ creates 16QAM, a scheme in which a^I and a^Q takes values in the set $\{-3, -1, 1, 3\}$ and $C_0 = 1/\sqrt{10}$ (see Lemma 3.5-3 to follow). The actual transmission symbols, meaning the values that scale the pulses $\sqrt{T}v(t - nT)$, are $\{-3/\sqrt{10}, -1/\sqrt{10}, 1/\sqrt{10}, 3/\sqrt{10}\}$. For example, the constellation point $(a^I, a^Q) = (-1, 3)$ in Fig. 3.16b maps to the data bits 1001; the corresponding baseband pulses for the signal interval are

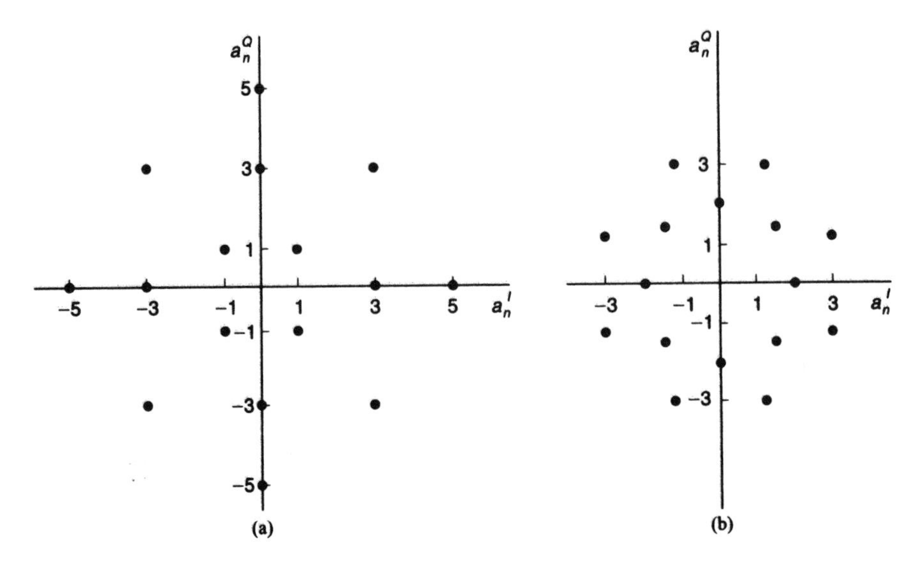

Figure 3.15 Two nonrectangular QAM constellations: **(a)** V.29 modem standard; **(b)** double-circle constellation (outer radius 3.16; inner radius 2).

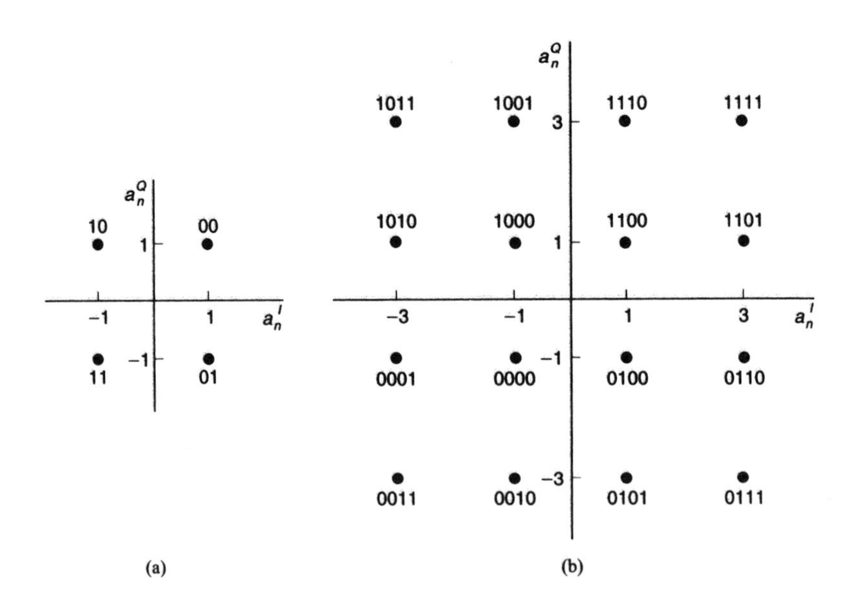

Figure 3.16 Two rectangular QAM constellations: **(a)** 4QAM; **(b)** 16QAM, the V.32 ALT modem standard. Values of a_n^I and a_n^Q are not normalized but are set to simple integers instead.

$$I: \quad \left(\frac{-1}{\sqrt{10}}\right)\sqrt{T}v(t-nT)$$

$$Q: \quad \left(\frac{3}{\sqrt{10}}\right)\sqrt{T}v(t-nT)$$

and the RF pulse is

$$\sqrt{2E_s}\left[\frac{-1}{\sqrt{10}}\, v(t-nT)\cos\omega_0 t - \frac{3}{\sqrt{10}}\, v(t-nT)\sin\omega_0 t\right]$$

This pulse adds to those from other intervals to form the transmitted signal.

In Fig. 3.5 (QPSK) and Fig. 3.10 (8PSK), the constellation points are also the transmission symbol values, but it is more common in QAM to use the unnormalized integers.

Figure 3.15a is a 16-point constellation that is not rectangular. The points take positions on the usual integers, and the average square value of the points is 13.5, which means that C_0 is $1/\sqrt{13.5}$ if E_s is to be the energy in a 16-ary symbol interval. The constellation here is that of the CCITT V.29 modem standard. We will find that it has worse error performance in AWGN than does Fig. 3.16b, 16QAM, but it is easier to synchronize, and if the channel suddenly worsens, it can be modified more easily into a simpler constellation with more noise resistance.

3.5.1 QAM Spectra and Error Performance

Because QAM signals are a linear sum of independent pulses with the same magnitude spectrum, the QAM power spectral density is the same as that of its pulse, from the linear spectrum theorem (Theorem 2.3-1). The argument follows the same pattern that it did in Section 3.2 for BPSK and QPSK. All the time shifts of the two basic pulses, $v(t)\cos\omega_0 t$ and $v(t)\sin\omega_0 t$, have the same magnitude spectrum. The pulse amplitudes a_n^I and a_n^Q are zero mean and independent from time to time and of each other, since QAM constellations ordinarily are symmetric and the points are selected in an IID fashion.

Although QAM is like BPSK and QPSK in that it carries the spectrum of its pulse, its bandwidth *per data bit* can be much less, since there are many more points in the constellation. If B is the bandwidth of BPSK according to some measure and there are M^2 points in the QAM, the bandwidth per data bit of the QAM is $B/\log_2 M^2$. This savings in bandwidth is the chief reason for using QAM.

The error performance of a QAM depends primarily on the minimum distance among points in its constellation. Before looking at the distance question, it will be helpful to collect together some distance properties of quadrature signals. These greatly simplify QAM calculations, and the proofs demonstrate some manipulations that are common in I/Q signal analysis.

The first states that in calculating distances between two carrier signals, we need only consider their baseband signals.

Lemma 3.5-1. Let $s_1(t)$ and $s_2(t)$ be carrier signals of the form

$$s_k(t) = \sqrt{\frac{2E_s}{T}}[I_k(t) \cos \omega_0 t - Q_k(t) \sin \omega_0 t], \qquad k = 1, 2$$

where I and Q are well-behaved signals.[8] Define the differences between the in-phase and quadrature signals as

$$\Delta I(t) \triangleq I_1(t) - I_2(t), \qquad \Delta Q(t) \triangleq Q_1(t) - Q_2(t) \tag{3.5-2}$$

Then in the limit $\omega_0 \to \infty$ the energy in the difference between $s_1(t)$ and $s_2(t)$ satisfies

$$\int [s_1(t) - s_2(t)]^2 \, dt = \frac{E_s}{T} \int [\Delta I(t)^2 + \Delta Q(t)^2] \, dt \tag{3.5-3}$$

Proof. Subtracting one carrier signal from the other, we get the difference energy

$$\begin{aligned}\int [s_1(t) - s_2(t)]^2 \, dt &= \frac{2E_s}{T} \int [\Delta I(t) \cos \omega_0 t - \Delta Q(t) \sin \omega_0 t]^2 \, dt \\ &= \frac{2E_s}{T} \int [\Delta I(t)^2 \cos^2 \omega_0 t - 2\Delta I(t)\Delta Q(t) \cos \omega_0 t \sin \omega_0 t \\ &\quad + \Delta Q(t)^2 \sin^2 \omega_0 t] \, dt\end{aligned}$$

Applying standard trigonometric identities to this gives

$$\frac{2E_s}{T} \int \left[\frac{1}{2} \Delta I(t)^2 + \frac{1}{2} \Delta Q(t)^2 + W(t) \right] dt \tag{3.5-4}$$

Here $W(t)$ consists of a number of trigonometric terms of the form $f(t) \sin 2\omega_0 t$ or $f(t) \cos 2\omega_0 t$; a basic theorem of calculus states that the integral of such a term tends to zero as $\omega_0 \to \infty$ for any well-behaved $f(t)$. Thus (3.5-4) tends to

$$\frac{E_s}{T} \int [\Delta I(t)^2 + \Delta Q(t)^2] \, dt$$

as $\omega_0 \to \infty$, which gives (3.5-3). This proves the lemma.

An important point to notice is that I and Q affect the square distance between two signals *independently*. That is, we can make two carrier signals distinguishable by piling up distance in either the in-phase or the quadrature dimensions or in both.

[8] I and Q need only be square-integrable bandpass functions.

Our second lemma concentrates on QAM signals and shows that for these we need consider only the transmission symbols a^I and a^Q in finding the distance. Here a^I and a^Q may take any real values, not just the standard integer values.

Lemma 3.5-2. Now let $I_1(t), I_2(t), Q_1(t), Q_2(t)$ be made up of weighted orthonormal pulses according to

$$I_k(t) = C_0 \sum_n a_n^I(k) \sqrt{T} v(t - nT), \qquad k = 1, 2$$
$$Q_k(t) = C_0 \sum_n a_n^Q(k) \sqrt{T} v(t - nT), \qquad k = 1, 2 \tag{3.5-5}$$

Define the symbol differences at symbol time n to be

$$\Delta a_n^I \triangleq a_n^I(1) - a_n^I(2), \qquad \Delta a_n^Q \triangleq a_n^Q(1) - a_n^Q(2) \tag{3.5-6}$$

so that the baseband signal differences are

$$\Delta I(t) = C_0 \sum_n \Delta a_n^I \sqrt{T} v(t - nT)$$
$$\Delta Q(t) = C_0 \sum_n \Delta a_n^Q \sqrt{T} v(t - nT) \tag{3.5-7}$$

Then the baseband energy difference in (3.5-3) is

$$\frac{E_s}{T} \int [\Delta I(t)^2 + \Delta Q(t)^2] \, dt = E_s C_0^2 \sum_n [(\Delta a_n^I)^2 + (\Delta a_n^Q)^2] \tag{3.5-8}$$

Proof. The proof is a straightforward exercise with orthonormal functions. By substituting (3.5-7), we get the following for the left-hand side of (3.5-8):

$$\frac{C_0^2 E_s}{T} \int \Bigg[\left(\sum_n \sqrt{T} \, \Delta a_n^I v(t - nT) \right) \left(\sum_{n'} \sqrt{T} \Delta a_{n'}^I \, v(t - n'T) \right)$$
$$+ \left(\sum_n \sqrt{T} \, \Delta a_n^Q v(t - nT) \right) \left(\sum_{n'} \sqrt{T} \, \Delta a_{n'}^Q v(t - n'T) \right) \Bigg] dt$$
$$= C_0^2 E_s \int \Bigg[\sum_n \sum_{n'} \Delta a_n^I \Delta a_{n'}^I \, v(t - nT) v(t - n'T)$$
$$+ \sum_n \sum_{n'} \Delta a_n^Q \Delta a_{n'}^Q \, v(t - nT) v(t - n'T) \Bigg] dt$$
$$= C_0^2 E_s \sum_n \sum_{n'} [\Delta a_n^I \Delta a_{n'}^I + \Delta a_n^Q \Delta a_{n'}^Q] \int v(t - nT) v(t - n'T) \, dt$$

Because the time shifts of $v(t)$ are orthonormal, the integral here is unity if $n = n'$ and zero otherwise. Thus, only the "diagonal" terms survive, and the result is the right-hand side of (3.5-8). This proves the lemma.

The next lemma calculates the average symbol energy represented by a QAM constellation, which we will need in order to find the error performance. The energy so found is the energy in *one* use of the constellation; that is, in one joint in-phase and quadrature pulse. If there are N pulses and all are orthogonal, the total average energy is N times this number. Once again, a^I and a^Q take any real values, and all constellation points are equally likely.

Lemma 3.5-3. Consider a general QAM orthogonal-pulse signal of the form (3.5-1) that consists of just the nth pulse. The average energy of this signal is

$$E_s C_0^2 \sum_{a^I, a^Q} P[a^I, a^Q][(a_n^I)^2 + (a_n^Q)^2] \tag{3.5-9}$$

where the sum is over all the joint values that a^I and a^Q take. In particular, the average energy of the QAM pulse is E_s if C_0 satisfies

$$C_0 = \left(\frac{1}{\sum_{a^I, a^Q} P[a^I, a^Q][(a_n^I)^2 + (a_n^Q)^2]} \right)^{1/2} \tag{3.5-10}$$

Proof. The proof follows easily from Lemmas 3.5-1 and 3.5-2 by considering a particular pair of signals in (3.5-3). Take the first signal to be the QAM signal with just one pulse and transmission symbols (a^I, a^Q); the second signal is simply zero. Then ΔI and ΔQ in (3.5-3) are I and Q for the first signal, and similarly for Δa^I and Δa^Q in (3.5-8). Consequently, the energy of the signal with (a^I, a^Q) is $E_s C_0^2[(a_n^I)^2 + (a_n^Q)^2]$, and its expected value is (3.5-9).

Now we are ready to calculate some example error probabilities. Even though the signals are carrier rather than baseband, the method is still the signal space analysis of Section 2.6. The basis functions for the nth QAM interval are again (3.2-15). Asymptotic estimate (2.6-17), which is typically a tight upperbound, gives

$$p_e \leq \frac{2K}{\mu} Q\left(\sqrt{\frac{d_{\min}^2 E_b}{N_0}}\right) \tag{3.5-11}$$

Here $d_{\min}$ is the nearest-neighbor distance in the constellation, K is the number of distinct pairs at this distance, and μ is the number of points in the constellation. As always, the normalized $d_{\min}$ is $D_{\min}/\sqrt{2E_b}$, in which $D_{\min}$ and E_b are as calculated from the constellation at hand.

Example 3.5-1: The V.32 ALT[9] Modem Constellation. The modulation in this telephone line standard is rectangular 16QAM. The carrier frequency is 1800 Hz and the symbol rate is 2400 Hz; consequently, the scheme carries 9600 b/s. We will ignore many finer details of V.32 ALT and treat it simply as rectangular 16QAM. To estimate the error rate, we need to calculate $D_{\min}$ and E_b off the constellation, which is Fig. 3.16b. From these two comes the minimum distance. The actual values that E_s and C_0 take in the QAM signal will not matter. Each point in Fig. 3.16b occurs with probability $\frac{1}{16}$; there are four points with energy 2, eight with energy 10, and four with energy 18. From Lemma 3.5-3, then, the average energy of the nth symbol is

$$E_s C_0^2 \tfrac{1}{16}[4(2) + 8(10) + 4(18)] = E_s C_0^2(10)$$

The average energy per data bit is

$$E_b = \frac{E_s}{\log_2 16} = E_s C_0^2(10/4)$$

The closest that any pair of points in Fig. 3.16b lies is 2; from Lemmas 3.5-1 and 3.5-2, the square distance between the signals that correspond to such points is

$$D_{\min}^2 = E_s C_0^2[(\Delta a_n^I)^2 + (\Delta a_n^Q)^2] = E_s C_0^2(2^2)$$

Consequently, the normalized minimum distance is

$$d_{\min}^2 = \frac{D_{\min}^2}{2E_b} = \frac{E_s C_0^2 4}{2E_s C_0^2(10/4)} = 0.8$$

The error probability estimate (3.5-11) for this QAM is then

$$3Q\left(\sqrt{\frac{0.8\, E_b}{N_0}}\right) \tag{3.5-12}$$

since the number K of distinct point pairs at the minimum distance is 24 and there are $\mu = 16$ points in the constellation. This estimate averages over all the constellation points. As was discussed in Section 2.6, a more refined estimate is possible if it is known that a particular point $\mathbf{s}_i$ was transmitted. This estimate is $N_i Q(\sqrt{0.8E_b/N_0})$, in which N_i is the number of nearest neighbors to $\mathbf{s}_i$. In Fig. 3.16b, N_i is 2, 3, or 4, depending on the point. To conclude the example, we calculate the bandwidth of this 16QAM. If B is the bandwidth of the carrier pulse, the bandwidth consumption of the scheme is $B/\log_2 16 = \frac{1}{4}B$ hertz per data bit. For a practical modem, we can take a 30% excess bandwidth root RC pulse (see Example 2.2-5), which has a baseband band-

[9]These designations refer to ISO (International Standards Organizations) protocols for modems.

width of $1.3/2T$ Hz, with T the symbol time. After carrier modulation, the RF bandwidth will be twice this, centered on the carrier location 1800 Hz. The RF bandwidth is thus

$$\frac{2(1.3)}{2(1/2400)} = 3120 \text{ Hz}$$

running from 240 Hz up to 3360 Hz. This is intended to occupy the passband of a voice telephone channel.

The error probability in (3.5-12) is about 4.0 dB worse in a good channel than the precise expression for QPSK in (2.6-13), meaning that 16QAM needs about 4.0 dB higher E_b for the same error probability. The minimum distances in the two expressions differ by 4.0 dB (2 versus 0.8) and we can ignore the leading factors (2 and 3) when E_b/N_0 is high. However, 16QAM consumes half the bandwidth per data bit that QPSK does. Here is the basic trade-off presented by QAM. For the most part, trade-offs of bandwidth for energy like this are a fact of nature and cannot be avoided.

The precise symbol error probability of rectangular 16QAM is plotted in Fig. 3.17, along with the probabilities for a number of other schemes. Except for BPSK, QPSK, and DPSK, numerical integration is required for all these probabilities; the method for QAM is summarized in Ziemer and Peterson [4]. An accurate plot of the data bit error rate is even harder, because the data bit mapping needs to be accounted for (a precise rendering appears in Ref. 2). An estimate such as (3.5-12) is much easier to generate; it errs, typically, by less than 0.3 dB.

It is possible to improve the QAM minimum distance a little by rearranging the points into a ball rather than a square. This *shaping gain,* as it is called, occurs because signal space points arranged uniformly in a sphere make the most efficient use of energy. Some examples of optimized 16-ary constellations appear in Gibson [3]. It is also possible to fare a lot worse than rectangular 16QAM through a foolish arrangement of points: As Fig. 3.17 shows, the 16PSK constellation is about 4 dB worse than rectangular 16QAM. The 16PSK would be considered only if a constant envelope were needed.

It is not difficult to write down an asymptotic law for the energy efficiency of rectangular QAM. Suppose that the I and Q channels each carry the usual M-ary symbol alphabet $\{\pm1, \pm3, \ldots, \pm(M-1)\}$. The average symbol energy $E^{(M)}$ with an M-ary I and Q alphabet turns out to be

$$E^{(M)} = \tfrac{2}{3}(M^2 - 1) \qquad (3.5\text{-}13)$$

The calculation method consists of summing the squares of constellation points, just as in Example 3.5-1, and applying some standard series formulas.[10] The square

[10]A faster but more subtle calculation is as follows: Apply Lemma 3.5-3 to the M-ary I channel acting alone; by symmetry, the Q-channel has the identical energy. The two energies add.

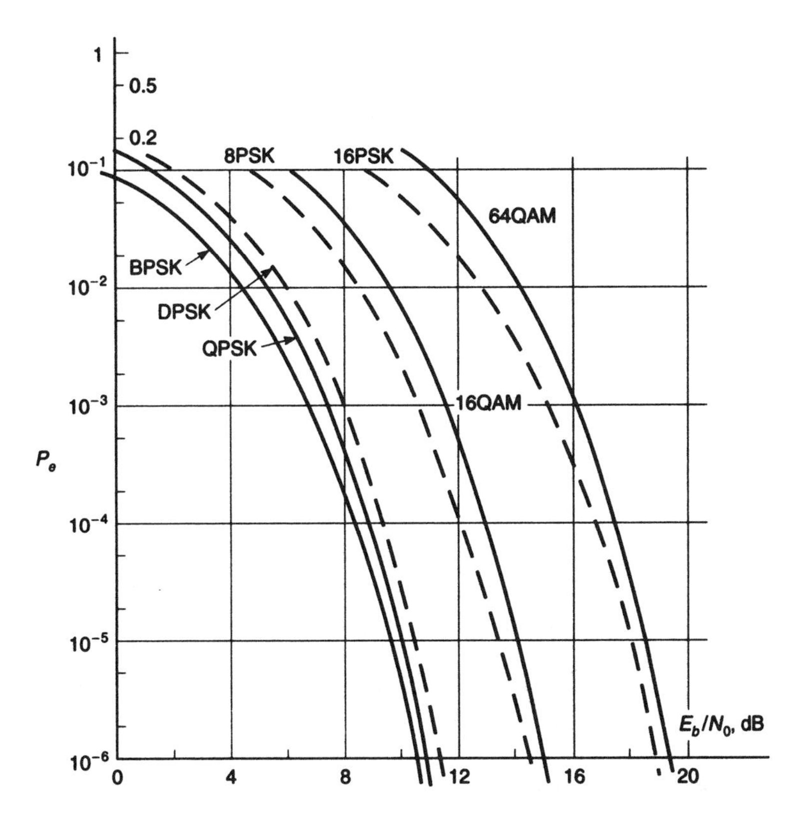

Figure 3.17 Actual symbol error probabilities of selected modulations: BPSK, QPSK, DPSK, 8PSK, 16PSK, and rectangular 16QAM and 64QAM. Data bit error rates differ somewhat from these curves.

distance between nearest neighbors is always 4, as it is in Fig. 3.16b, no matter what M is. So, the normalized square distance of M^2QAM is

$$d^2_{\min} = \frac{4}{2E_b} = \frac{4 \log_2 M^2}{2(2/3)(M^2 - 1)} = \frac{6 \log_2 M}{M^2 - 1} \tag{3.5-14}$$

We can focus on the practical case where $M = 2^m$; that is, m bits are carried in each of I and Q. Now the square distance becomes

$$\frac{6m}{4^m - 1} \tag{3.5-15}$$

The expression tends to zero with m as $(\frac{1}{4})^m$. This means that the minimum distance of rectangular QAM drops asymptotically 6 dB per bit carried in I or Q. The actual figures for $m = 1, 2, 3, 4$ (which give QPSK, 16QAM, 64QAM, and 256QAM, respectively) are losses of 0, 4, 8.4, and 13.3 dB, compared to the QPSK distance.

These losses occur because the average energy of QAM rapidly grows for a fixed distance between points; consequently, the normalized minimum distance in (3.5-11) rapidly drops. In a telephone line with 256QAM, for example, an E_b/N_0 of roughly 9.6 + 13.3, about 23 dB, would be required for a symbol error rate near 10^{-5} in this 19.2 kb/s modem. Can it work well? The needed value of E_s/N_0 is 9 dB higher, a total of 32 dB, since $E_s = 8E_b$. Good-quality telephone lines have 35–40 dB, but it is clear that if the SNR momentarily drops, a 256QAM modem will begin to make errors.

Some further theoretical analysis can be done to extend (3.5-15) to general QAM constellations. The result is that, asymptotically, the normalized minimum distance of QAM drops about 3 dB for each doubling of the points in the constellation, so that about 3 dB more energy is required for the same error probability. In fact, one can show that the capacity of the underlying Gaussian channel model [Eq. (6.1-14)] also grows at the rate of one bit per doubling of the energy.

We conclude with an example of a nonrectangular QAM constellation. As with Example 3.5-1, this one is part of an international modem standard. Modem standards contain a great many details with which we will not concern ourselves at this moment. For example, the binary data may be randomized and differentially encoded across successive constellations; the mapping to a constellation point is further complicated by Gray coding; the constellation itself may change under certain conditions. All of these have some effect on the error rate. A readable introduction to these details appears in Webb and Hanso [5].

Example 3.5-2: The V.29 Modem Constellation. As in Example 3.5-1, the V.29 telephone line standard specifies a 2400-Hz symbol rate and a 16-point constellation, so that 9600 b/s are carried. However, the carrier frequency is 1700 Hz this time, and the constellation is the nonrectangular one in Fig. 3.15a. There are four points each with energy 2, 9, 18, and 25, and each point has probability $\frac{1}{16}$, so that (3.5-9) yields

$$E_s C_0^2 \tfrac{1}{16}[4(2) + 4(9) + 4(18) + 4(25)] = E_s C_0^2(13.5)$$

for the average symbol energy. The average bit energy is then $13.5/\log_2 16 = 3.375$, times $E_s C_0^2$. Just as in Example 3.5-1, the least-square distance between constellation points is $D_{\min}^2 = 4E_s C_0^2$, although this time there are only eight distinct point pairs that have this distance. The normalized square distance is therefore

$$d_{\min}^2 = \frac{D_{\min}^2}{2E_b} = 0.59$$

and the error probability estimate (3.5-11) is

$$\frac{16}{16} Q\left(\sqrt{\frac{0.59\, E_b}{N_0}}\right) \qquad (3.5\text{-}16)$$

Compared to rectangular 16QAM, (3.5-16) implies an energy loss of 10 $\log_{10}(0.8/0.59) = 1.3$ dB at high E_b/N_0, although the small leading factor makes up somewhat for this in a poor channel. The 1.3-dB loss stems from the lack of compactness in the constellation. The arrangement for V.29 was chosen because the modem can revert easily to an eight- or four-point subconstellation in case the channel worsens. With 30% root RC pulses, the V.29 RF bandwidth is 3120 Hz, the same as Example 3.5-1, but with the 1700-Hz carrier location, the occupancy now runs from 140 Hz to 3260 Hz.

3.5.2 Partial-Response Signaling

A linear modulation produces a new constellation for each symbol interval, and until now we have treated these constellations as independent. Carrier pulses can interfere with each other, just as the baseband pulses did in Section 2.4. Sometimes the intersymbol interference is undesired—for example, a channel filter—but other times ISI may be introduced intentionally into a QAM scheme. *Partial-response signaling* (PRS) is a name given to this technique. The usual motivation for PRS is modification or reduction of the signal bandwidth.

The introduction of ISI to form a partial-response signal may be performed in practice at baseband or at carrier frequencies, and the two techniques are equivalent, but the processing is normally at baseband, so we will work there. To begin, imagine that the data symbol sequence $\{a_n\}$ is applied to the discrete-time filter $H(z)$ before it reaches the pulse shaper $V(f)$, as shown in Fig. 3.18. This is in contrast to the usual QAM signal generation, in which $\{a_n\}$ is applied directly to $V(f)$ to generate an orthogonal pulse train. The filter $H(z)$ is order-m and finite-response with z-transform $h_0 + h_1 z^{-1} + \cdots + h_m z^{-m}$. We will focus on the generation of the in-phase signal $I(t)$ in the standard QAM form (3.5-1), but we could as well be generating the standard PAM baseband signal (2.2-1). The response of the top system in Fig. 3.18 is then

$$I(t) = \sqrt{T} C_0 \sum_n c_n v(t - nT) \qquad (3.5\text{-}17)$$

$\{a_n\}$ → $H(z)$ → $C_0\sqrt{T}V(f)$ → $I(t)$

(a)

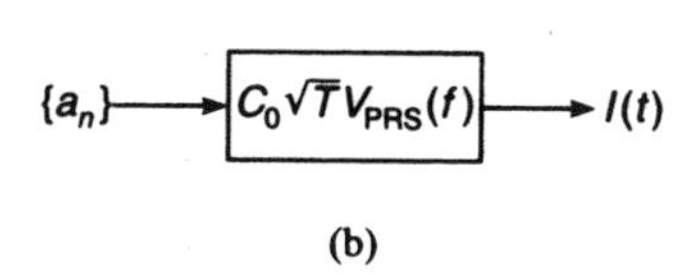

(b)

Figure 3.18 Two methods to generate baseband partial-response signals: **(a)** discrete-time prefilter; **(b)** via modified pulse.

in which

$$c_n = \sum_{i=0}^{m} h_i a_{n-i}$$

is an element in the output sequence with transform $A(z)H(z)$. By rearranging the sums in (3.5-17), we get

$$\begin{aligned} I(t) &= \sqrt{T}C_0 \sum_n \sum_{i=0}^{m} h_i a_{n-i} v(t-nT) \\ &= \sqrt{T}C_0 \sum_n \sum_{i=0}^{m} a_n h_i v(t-(n+i)T) \\ &= \sqrt{T}C_0 \sum_n a_n \, v_{\text{prs}}(t-nT) \end{aligned} \tag{3.5-18}$$

in which

$$v_{\text{prs}}(t) = \sum_{i=0}^{m} h_i v(t-iT)$$

These manipulations show that a partial-response signal may be generated in two equivalent ways, by a discrete-time filtering of the data by $H(z)$ or by standard QAM with the new pulse $v_{\text{prs}}(t)$. Both are shown in Fig. 3.18.

Partial-response signaling began in the early 1960s with the introduction by Lender of a system called *duobinary* signaling (see, e.g., Ref. 6). The technique was extended by Kretzmer (see Ref. 7, among others). Partial-response signaling is sometimes called correlative encoding. A readable summary of traditional PRS was written by Pasupathy [8] in 1977. In the 1980s and 1990s, PRS was greatly generalized, however, to the point where it is now considered a form of coded modulation. It is known, for example, how to optimize $H(z)$ to obtain the best combination of signal bandwidth and signal energy [9], and $H(z)$ may have poles rather than just zeros. Codes for magnetic recording can be viewed as an extension of the PRS idea to nonlinear $H(z)$.

For an example of PRS, we will take Lender's original technique. In the duobinary scheme, $H(z) = 1 + z^{-1}$. That is, c_n in (3.5-17) is the sum of the present and the previous data symbol. As originally envisioned, the scheme encodes binary ± 1 data and the orthogonal pulse $v(t)$ was the standard unit-energy sinc pulse in Example 2.2-2. The pulse $v_{\text{prs}}(t)$ in the equivalent form (3.5-18) is thus

$$v_{\text{prs}}(t) = \frac{1}{\sqrt{T}} \left[\text{sinc}\left(\frac{t}{T}\right) + \text{sinc}\left(\frac{t-T}{T}\right) \right] \tag{3.5-19}$$

which has Fourier transform

$$V_{\text{prs}}(f) = \begin{cases} 2\sqrt{2}\, e^{-j\pi f T} \cos(\pi f T), & 0 \le |f| < \dfrac{1}{2T} \\ 0, & \text{otherwise} \end{cases}$$

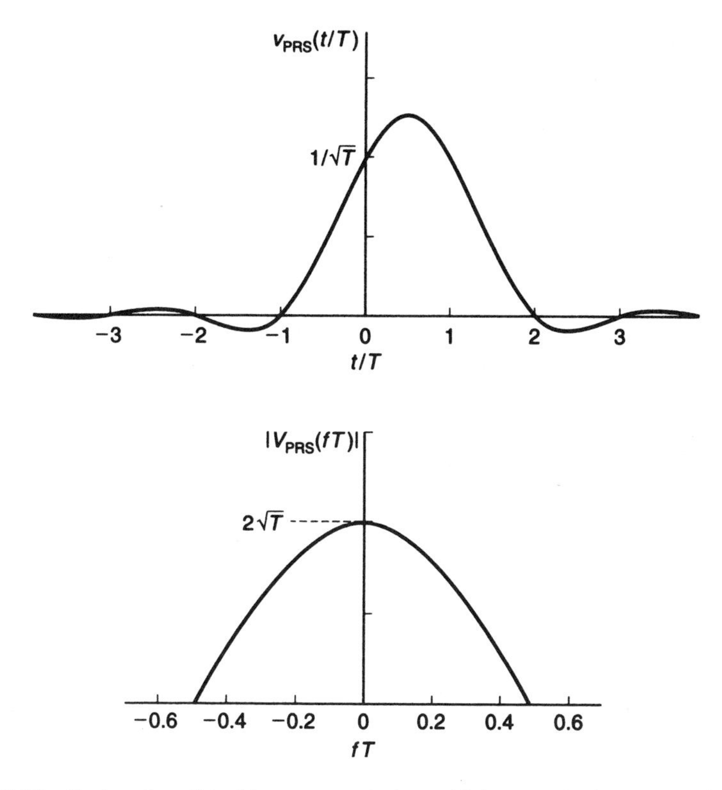

Figure 3.19 Basic pulse of duobinary transmission, with its magnitude spectrum $|V_{PRS}(f)|$. Pulse has energy 2.

Figure 3.19 shows $v_{prs}(t)$ and its transform. If we leave out the normalizing constant C_0, the complete baseband signal will be

$$I(t) = \sum_n a_n \frac{1}{\sqrt{T}} \left[\operatorname{sinc}\left(\frac{t - nT}{T} \right) + \operatorname{sinc}\left(\frac{t - T - nT}{T} \right) \right] \qquad (3.5\text{-}20)$$

Since the data $\{a_n\}$ are ± 1, the value of $I(t)$ at the symbol times nT will be one of the values -2, 0, $+2$; this is easiest to see from the form (3.5-17). Some calculation shows[11] that the average energy in (3.5-20) is $2N$ if there are N symbols. This contrasts with ordinary sinc pulse modulation, which would have energy N, and the reason tracks back to the energy of $v_{prs}(t)$, which is 2. In other words, the duobinary techniques doubles the average signal energy.

The technique also doubles the minimum square distance of the signaling scheme. The demonstration of this follows the signal space methods that we have

[11]Using the method in Theorem 2.3-1, one gets that $I(t)$ with N data symbols has energy $2N$. The hypotheses of the theorem are satisfied with v_{prs} because the PRS scheme is still linear modulation; however, v_{prs} has energy 2.

developed, but the dependence between symbol intervals makes it an exercise in coded modulation theory, which is beyond the scope of this chapter (see Section 7.1). For now we can draw the conclusion that both the square distance and the energy have doubled, so that the normalized distance remains unchanged. Consequently, the asymptotic error probability of duobinary PRS is the antipodal signaling expression $Q(\sqrt{2E_b/N_0})$.

Since the duobinary signal is formed by linear modulation, the linear spectrum theorem (Theorem 2.3-1) applies, and the power spectrum of the signal is the square of the transform shown in Fig. 3.19. That is, a baseband duobinary scheme has PSD $|V_{\text{prs}}(f)|^2$, scaled by the signal energy, and a carrier duobinary scheme has PSD $\frac{1}{2}|V_{\text{prs}}(f+f_o)|^2 + \frac{1}{2}|V_{\text{prs}}(f-f_o)|^2$. These have the same bandwidth as the original modulation with just the sinc $v(t)$, and the normalized distance is the same. So what is the advantage of duobinary signaling? To see the answer, we need to look more carefully at the pulse $v_{\text{prs}}(t)$. By studying Example 2.2-5, it can be concluded that the transform $V_{\text{prs}}(f)$ is the transform of a double-width root RC pulse with 100% excess bandwidth; consequently, the duobinary pulse must be (2.2-11) with $2T$ and $\alpha = 1$. This reduces to

$$v_{\text{prs}}(t) = \frac{4}{\sqrt{T}\pi} \frac{\cos(\pi t/T)}{1-(2t/T)^2} \tag{3.2-21}$$

The slow time roll-off at the rate T/t that the original $v(t)$ pulse had has been replaced here by a much faster $(T/t)^2$ roll-off. This was Lender's original intention: to obtain a friendlier pulse roll-off, while keeping the same bandwidth and energy efficiency as sinc pulse transmission.

At the receiver, a matched filter followed by a sampler/comparator (as in Fig. 2.7 or 3.9) will produce, in general, an ML estimate of the sequence $\{c_n\}$. This is obvious from realization (3.5-17), so long as $v(t)$ is orthogonal. In theory, one can obtain the data $\{a_n\}$ from the $\{c_n\}$ by the formula $a_n = c_n - a_{n-1}$. A problem with this, however, is error propagation: Once an error occurs in an a_n, later a_n will be wrong with high probability. Lender eliminated this propagation with a precoding method. An intermediate sequence $\{b_n\}$ is formed from the data symbols according to the rule $b_n = a_n b_{n-1}$; the duobinary sequence $c_n = b_n + b_{n-1}$ is formed in the usual way. The symbols here are all ± 1 with ordinary arithmetic.[12] At the receiver, the following decoding rule is applied to the sampler/comparator outputs:

$$a_n = \begin{cases} +1, & \text{if } c_n = \pm 2 \\ -1, & \text{if } c_n = 0 \end{cases} \tag{3.5-22}$$

By exhausting all the cases, it can be shown that this decoding is not affected by earlier errors in decoding the data sequence.

An alternate decoder that eliminates error propagation is the Viterbi algorithm in Chapter 6.

[12]Alternately, all symbols may be taken in $\{0, 1\}$ and the operations recast in an equivalent modulo-2 form.

TABLE 3.5 Traditional Partial-Response Schemes

Class	$H(z)$	Comments
I	$1 + z^{-1}$	Duobinary
II	$1 + 2z^{-1} + z^{-2}$	
III	$2 + z^{-1} - z^{-2}$	
IV	$1 - z^{-2}$	Zero at DC
V	$-1 + 2z^{-2} - z^{-4}$	Zero at DC

Several other types of PRS are now in use, and these are listed in Table 3.5. A "class" naming system has grown up for these; duobinary, for example, is called class I partial response signaling. Some of these schemes have special properties. It can be shown that the class IV and V schemes have *spectral nulls* at DC [8]. As baseband signaling, these schemes are useful in an undersea cable, in which power for the repeaters is carried at DC and the data-bearing signal must be kept away from nearby frequencies. An application to carrier signaling occurs in "transparent tone in band" systems, in which a small pilot tone is transmitted at the carrier frequency location; the tone is used to drive the carrier synchronization circuit and must be free of interference from the data-bearing signal.

Some of the schemes in the table reduce the transmission bandwidth. Indeed, this is an object in present research on more sophisticated PRS schemes. An interesting theoretical point is that no PRS scheme can have higher normalized minimum distance with a given data symbol set than does the system with $H(z) = 1$.[13] The square distance with binary ± 1 data is limited, for example, to 2. A worthwhile object, then, is to find schemes at a given narrow bandwidth with distance as close to this limit as possible.

3.6 FREQUENCY-SHIFT KEYING

All of the foregoing schemes can be thought of as linear modulations composed of pulses that are independent from interval to interval. Frequency-shift keying (FSK) modulations work by shifting the frequency of a carrier in response to the data. The FSK signals, like all carrier signals, can be broken down into I and Q components, but the signals do not consist of independent pulses, nor are I and Q in general independent. Therefore, FSK is neither linear nor quadrature. We need different, more sophisticated methods to analyze FSK. Detection is also more complex.

The FSK schemes are important because they are constant-envelope and at the same time have bandwidth and error probability on a par with pulse-shaped QPSK. We will begin with the pure form, where the data simply switch the frequency, and then turn to some extensions. Section 3.7 continues with some detector designs.

[13]The distance here is called the the *antipodal signal distance*. See further discussion in Section 7.1.

3.6.1 Basic FSK

A general form for all FSKs is

$$s(t) = \sqrt{\frac{2E_s}{T}} \cos\left(\omega_0 t + \frac{a_n h\pi (t - nT)}{T} + \psi_n\right), \qquad nT \le t \le (n+1)T \quad (3.6\text{-}1)$$

As before, T is the time devoted to one transmission symbol,[14] E_s is the average symbol energy, and a_n is the nth transmission symbol. In a binary FSK, a_n takes values in $\{-1, +1\}$, in a quaternary FSK, $a_n \in \{-3, -1, +1, +3\}$, and so on. The second term in the cosine argument is unique to FSK: It represents a linear shift of phase, totaling $a_n h\pi$ radians, across the nth symbol interval. The constant h is the *modulation index,* the constant of proportionality between the symbols and the total phase shift. For example, with binary FSK and $h = \frac{1}{2}$, the total shift will be $\pm\frac{1}{2}\pi$ radians. Here h plays a role similar to the modulation index β in analog FM; in what amounts to digital FM, h/T is the frequency shift for a unit symbol. The third factor in (3.6-1), ψ_n, is used to ensure phase continuity between intervals and may or may not be present.

The transmitted frequencies in pure-form FSK thus take the values $\{\omega_0 + a_n h\pi/T\}$ in radians per second, with $a_n \in \{\pm 1, \pm 3, \ldots\}$. The FSK signals are categorized by how widely these frequencies vary:

1. In *wideband* FSK, $h \gg 1$ and $\psi_n = 0$. Because the frequencies are far apart, they hardly affect one another, and the FSK is modeled by the orthogonal signaling constellation of Fig. 2.26b, with one dimension in signal space for each frequency. The error probability is $Q(\sqrt{E_b/N_0})$ from (2.6-6), and the bandwidth is about twice the largest frequency deviation.
2. In *discontinuous-phase narrowband* FSK, $0 < h < 1$ and again $\psi_n = 0$. The phase offset begins from zero at the beginning of each interval, and there is generally a discontinuity at the beginning of the next interval. See Fig. 3.20. In fact, the intervals of this FSK are independent, and it is in some ways a linear modulation. Although the bandwidth is narrower than wideband FSK, this FSK is seldom used because its bandwidth can be narrowed further by removing the phase discontinuities.
3. In *continuous-phase narrowband* FSK, $0 < h < 1$ and ψ_n is chosen so that the carrier phase is continuous at every interval boundary. The continuity is guaranteed by setting

$$\psi_n = \pi h \sum_{i<n} a_i \qquad (3.6\text{-}2)$$

A common name for these signals is continuous-phase FSK, abbreviated CPFSK. The signals have excellent error and bandwidth performance, but both of these depend on h in a complicated way.

[14]Note that in FSK analysis the symbol intervals are not centered at nT.

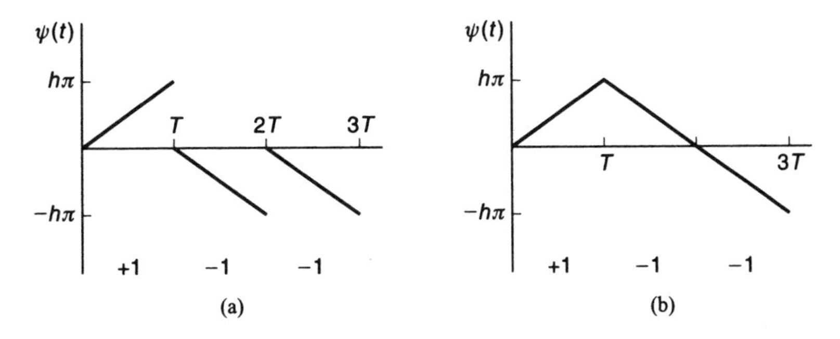

Figure 3.20 Excess phase $\psi(t)$ for **(a)** discontinuous-phase narrowband FSK and **(b)** continuous-phase FSK with the same data.

Since FSK is not a linear modulation, it is not very useful to describe it with pulses. A better way is to work with the excess phase relative to the carrier, the last two terms in (3.6-1), which we will denote $\psi(t)$. A plot of this is shown in Fig. 3.20a for a category 2 signal and in 3.20b for category 3, in both cases for the symbols +1, −1, −1. From these pictures it is clear that the type 2 signals have discontinuities and the type 3 signals do not and that each symbol causes a linear excursion of $h\pi$ radians.

In addition to plotting $\psi(t)$, we can make a plot of *all* possible phase trajectories and get a plot like Fig. 3.21. A plot like this is called a *phase tree,* and this one is for the CPFSK signal of Fig. 3.20b. Phase trees are a very useful analysis tool.

It is straightforward to obtain signal distances from $\psi(t)$. The hard way to do this is to construct an orthogonal basis for the signal space with the Gram–Schmidt procedure: In each interval the signals take on M frequencies and a variety of shifts by

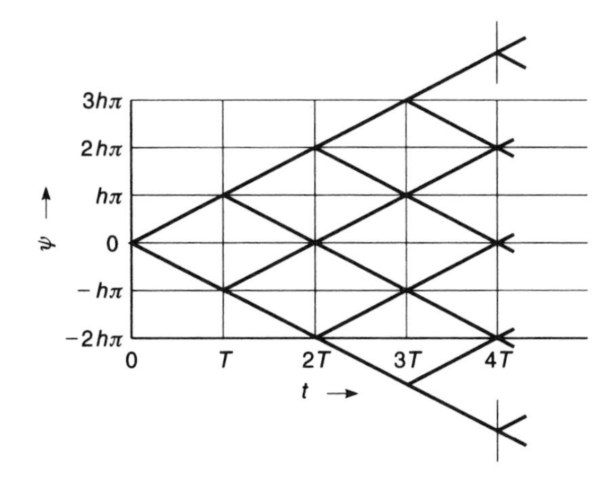

Figure 3.21 Phase trajectories of all CPFSK signals, starting from zero excess phase at time zero.

multiples of $h\pi$; a sine and cosine for each frequency will suffice, four dimensions altogether for binary FSK. But the basis functions are not simple and the signal constellation is not illuminating. A simpler way is to use the Parseval identity, Eq. (2.6-18), and skip the signal space creation entirely. We can derive a simple distance formula that works for any pair of FSK signals. Let the signals be $s_1(t) = \sqrt{2E_s/T} \cos[\omega_0 t + \psi^{(1)}(t)]$ and $s_2(t) = \sqrt{2E_s/T} \cos[\omega_0 t + \psi^{(2)}(t)]$, where $\psi^{(1)}(t)$ and $\psi^{(2)}(t)$ are the first and second excess phases. From Parseval, the square distance between them is

$$\|\mathbf{s}_1 - \mathbf{s}_2\|^2 = \int [s_1(t) - s_2(t)]^2 \, dt$$

which, in the manner of the proof of Lemma 3.5-1, reduces eventually to

$$\frac{2E_s}{T} \int [1 - \cos[\psi^{(1)}(t) - \psi^{(2)}(t)]] \, dt \tag{3.6-3}$$

as ω_0 becomes large. Thus the square distance depends only on the *phase difference*. This useful formula works for any constant-envelope signals, even for PSK. For example, it gives a square distance of $(2E_s/T)[T - 2T \cos(\frac{1}{2}\pi)] = 2E_s$ for two QPSK signals $\frac{1}{2}\pi$ apart for one interval; when normalized, this leads to $d^2_{\min} = 2$.

To find the minimum distance for CPFSK [category 3] with (3.6-3), we need to sort through the different signal pairs to find the least-distant ones. One such pair is shown in Fig. 3.22; the two signal phases trace a length $-$ T diamond shape, created by a $\psi^{(1)}(t)$ that is driven by symbols $+1, -1$ and a $\psi^{(2)}(t)$ driven by $-1, +1$. Succeeding transmission symbols can take any values, but they will not help distinguish the signals in the interval $[0, 2T]$. Many similar diamonds can be seen in Fig. 3.21. Each corresponds to two signals separated by this same distance. It can be shown (see Ref. 10) that the diamond figure is the one leading to the minimum distance for M-ary CPFSK with index $h \leq \frac{1}{2}$; for binary CPFSK, the diamond holds for $h \leq 1$. To compute the distance of the diamond, we apply (3.6-3) to the interval $[0, T]$ and double the results; this gives

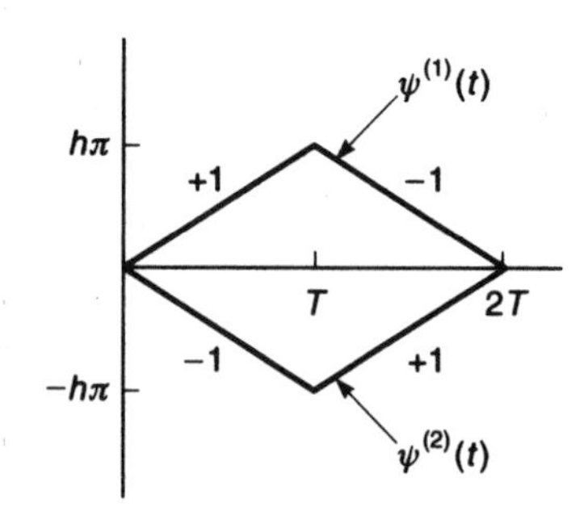

Figure 3.22 Excess phase for two CPFSK signals lying at minimum distance for most kinds of CPFSK.

$$
\begin{aligned}
D_{\min}^2 &= 2\frac{2E_s}{T}\int_0^T\left[1-\cos\left(\frac{2\pi ht}{T}\right)\right]dt \\
&= 4E_s\left[1-\frac{\sin 2\pi h}{2\pi h}\right]
\end{aligned}
\tag{3.6-4}
$$

Since $E_s = E_b \log_2 M$, the normalized square distance $d_{\min}^2 = D_{\min}^2/2E_b$ is

$$
d_{\min}^2 = 2\log_2 M\left[1-\frac{\sin 2\pi h}{2\pi h}\right]
\tag{3.6-5}
$$

Figure 3.23, adapted from Ref. 10, plots the minimum square distance $d_{\min}^2$ for CPFSK and $0 < h < 1$. Much of the plot is (3.6-5). It can be seen that much higher distances are available than with PSK or QAM. M-ary CPFSK with $M = 4$ or more offers a doubling or tripling of the basic QPSK distance of 2. Even binary CPFSK reaches a peak of 2.43, as shown in Example 3.6-1. These distances are to be placed in the usual approximate error formula, $p_e \approx Q(\sqrt{d_{\min}^2 E_b/N_0})$.[15]

The minimum distance just calculated introduces a new analytical difficulty, that the distance depends on more than one transmission symbol. The diamond shape extended over two symbols. There exist CPFSKs whose minimum distance stems from triple-symbol differences, and certain bandwidth-efficient coded modulations have distances that depend on very long symbol sequences. The calculation of these distances is a technical subject we will delay until Section 7.1. In Section 3.7, we will come to a related subject, building detectors for CPFSK, and here, too, the receiver must observe and compute over several intervals.

3.6.2 FSK Spectra

The spectrum of CPFSK is not easy to compute, because the modulation is not linear. Fast, general numerical methods for nonlinear modulation appear in Refs. 1 and 10, and writing a program for it is a profitable and not too strenuous exercise. Closed-form formulas exist for some nonlinear modulations, but these are often of legendary size. A formula for the power spectral density of M-ary CPFSK is small enough to be repeated here. First, define the quantities

$$
c = \frac{(1/M)\sin M\pi h}{\sin \pi h}
$$

$$
g_i = [fT - (2i - M - 1)h/2]\pi, \qquad i = 1, \ldots, M
$$

$$
a_{ij} = \frac{\cos(g_i + g_j) - c\cos(g_i + g_j - 2\pi fT)}{1 - 2c\cos(2\pi fT) + c^2}, \qquad i, j = 1, \ldots, M
$$

Then the PSD is

$$
\mathcal{E}[|S(f)|^2] = G(f - f_0) + G(f + f_0)
$$

[15]Computation of the coefficient in front of Q is an advanced topic. It is generally 2 or less.

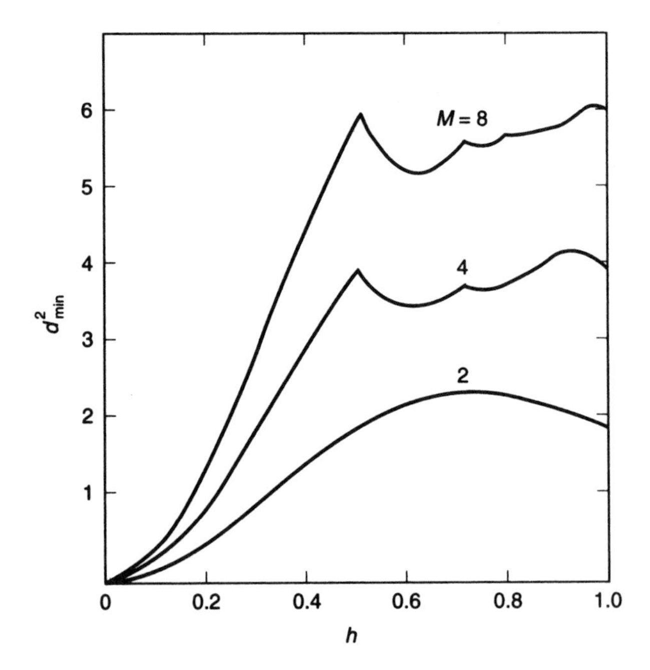

Figure 3.23 Square minimum distance versus modulation index h for binary, quaternary, and octal CPFSK.

where

$$G(f) = \frac{2}{M} \sum_{i=1}^{M} \left[\frac{(1/2)\sin^2 g_i}{g_i^2} + \frac{1}{M} \sum_{j=1}^{M} \frac{a_{ij} \sin g_i \sin g_j}{g_i\, g_j} \right] \tag{3.6-6}$$

Here, f is in hertz, but note that in every instance it appears as fT, a symbol-normalized quantity with dimension hertz-second. A spectrum like $G(f)$, which is the same form as $G(f-f_0)$ but centered at 0 instead of f_0, is called a *baseband* spectrum.

Figure 3.24 shows the spectral densities of several binary and quaternary CPFSKs. Both horizontal axes are scaled to fT_b in hertz-seconds per bit, so that the bandwidth consumption per actual data bit can be easily compared ($fT_b = \frac{1}{2}fT$ in the quaternary case). At the same time that the PSDs are varying in this complex pattern, the CPFSK distances are gyrating as in Fig. 3.23.

Judging both energy and bandwidth efficiency at once can be challenging, and it is often necessary, since both are critical in the choice of a modulation. An energy–bandwidth plot such as Fig. 3.25 is one way to carry out the comparison. In the figure, bandwidth per data bit is the vertical axis and normalized distance in decibels relative to QPSK is the horizontal one; the horizontal axis in effect measures energy efficiency per bit, since a gain in distance is directly a saving in E_b through the approximate error expression $Q(\sqrt{d^2_{\min} E_b/N_0})$. Good schemes in Fig. 3.25 are lower and to the left. It is relatively easy to see what a good h and M might be in CPFSK, for

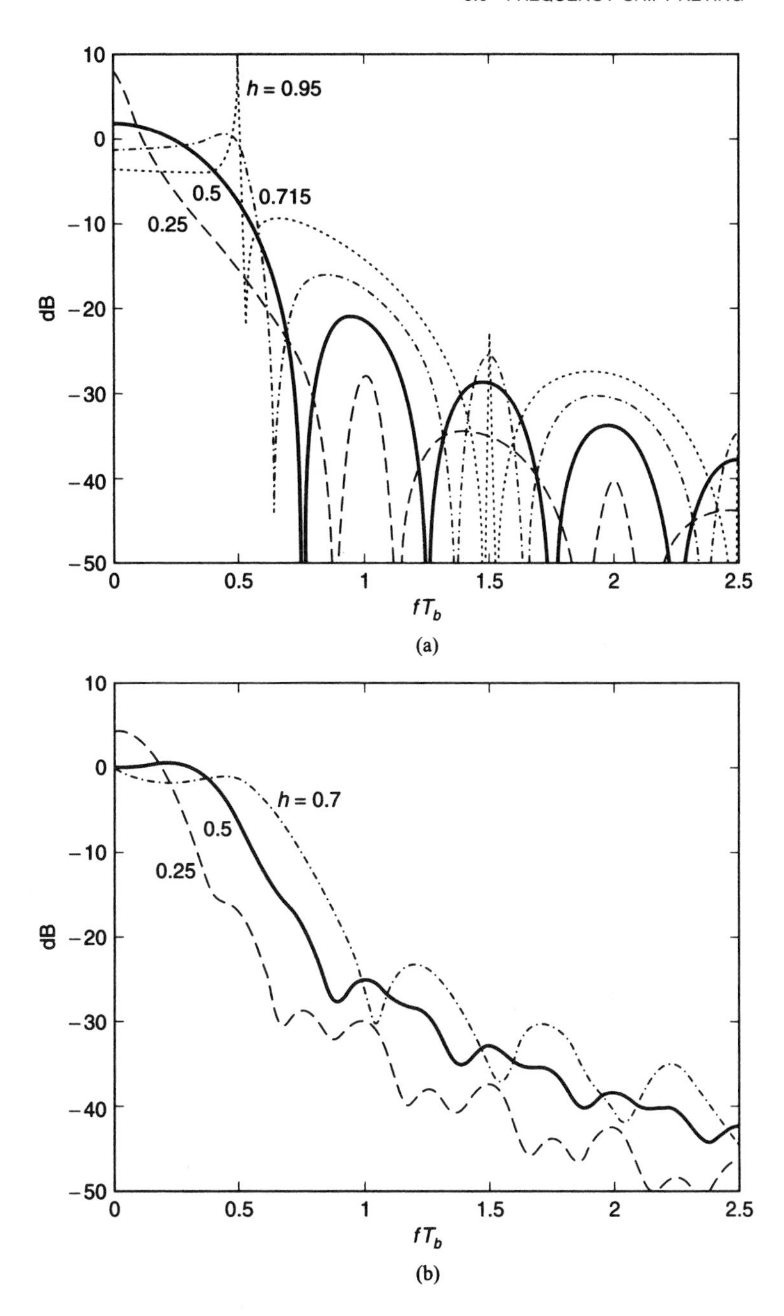

Figure 3.24 The PSDs of CPFSK for several modulation indices of practical interest: **(a)** binary CPFSK; **(b)** quaternary CPFSK. Both plots are normalized to unit data bit power; frequency axes are normalized to data bit time T_b. Note that RF bandwidth is twice the values shown here.

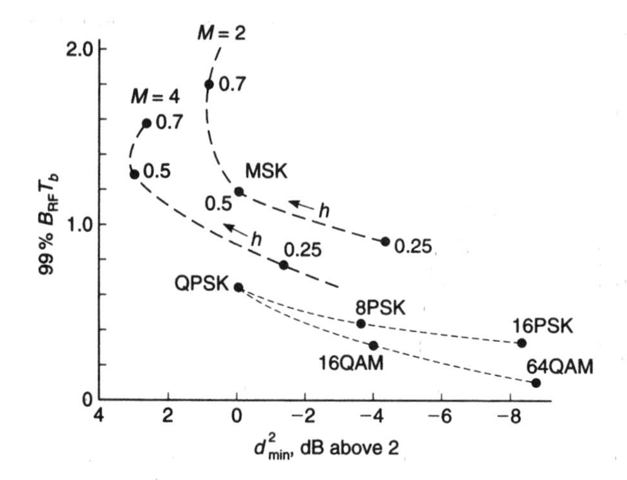

Figure 3.25 Energy–bandwidth efficiency of CPFSK and other selected schemes plotted as 99% RF power bandwidth and minimum distance above 2 in decibels. Both are normalized to data bit rate. For CPFSK, performance runs along trajectories shown as h varies.

a desired energy or bandwidth. Quaternary CPFSK is always better than binary, but plots for $M > 4$ show little further improvement (see Ref. 10, Chapter 5).

3.6.3 FSK Examples

Next we turn to some concrete examples of FSK modulation.

Example 3.6-1: Binary and Quaternary CPFSK with Small Index. By reducing the modulation index h, a CPFSK with reduced bandwidth may be obtained. From Fig. 3.24a the 99% RF power bandwidth of binary CPFSK with index $h = 0.25$ is about 0.9 Hz-s/bit (twice the baseband 0.45), and from Fig. 3.24b a quaternary CPFSK with the same h has 99% bandwidth about the same.[16] Their energy efficiency, however, is quite different. From Eq. (3.6-5), the 2CPFSK has $d^2_{\min} = 0.73$, while the 4CPSFK has $d^2_{\min} = 1.45$, a 3-dB improvement. By comparison, QPSK has minimum distance 2 and, with 50% root RC pulses, about the same bandwidth per data bit. Thus, QPSK has somewhat better energy efficiency, but 4CPFSK has a constant envelope. The 2CPFSK is constant envelope but has low energy efficiency. Thus quaternary CPFSK is important in narrowband applications that need a constant envelope.

Example 3.6-2: Energy-Efficient Binary CPFSK. By optimizing (3.6-5) over h, we can find the binary CPFSK with highest minimum distance. The optimum oc-

[16]A good approximation to the 99% power bandwidth is the largest frequency at which the PSD is 20 dB below its peak.

curs at $h = 0.715$ with distance 2.43, an improvement of 0.85 dB over the benchmark QPSK distance 2. This result, discovered in the early 1970s, shows that even binary CPFSK can have higher energy efficiency than any linear modulation. However, (3.6-6) shows that the 99% bandwidth of this CPFSK is about 1.8 Hz-s/bit, three times that of a good QPSK.

Example 3.6-3: Minimum-Shift Keying. This scheme, abbreviated MSK, is the name given to binary CPFSK with index $\frac{1}{2}$. The name stems from the fact that MSK is the smallest index ("minimum shift") CPFSK with the full benchmark square distance 2. An alternate name for the scheme is fast FSK, which alludes to the same fact. The power spectrum of MSK appears in Fig. 3.24a as a solid line. The general baseband spectrum formula (3.6-6) reduces, after some struggle, to the simple form

$$G(f) = \frac{16}{\pi^2}\left(\frac{\cos 2\pi fT}{1 - 16f^2T^2}\right)^2 \tag{3.6-7}$$

Minimum-shift keying has a first spectral null at $fT_b = 0.75$ away from the carrier and at every 0.5 thereafter. While MSK's main lobe is wider than NRZ pulse QPSK's (compare Fig. 2.9 or Fig. 3.4 to Fig. 3.24a), its side lobes diminish much faster than QPSK's; the 99% RF power bandwidth of MSK is about 1.20, while QPSK exceeds 10. These facts have made MSK the preferred alternative to QPSK when the signals must have constant envelope. An interesting fact about MSK is that it actually *is* a form of shaped QPSK when viewed the right way. Specifically, MSK is the offset QPSK of Section 3.4, with a special pulse shape and twice the symbol interval. We can give the outline of a derivation as follows. An offset QPSK signal has the form $I(t)\cos\omega_0 t - Q(t - \frac{1}{2}T_q)\sin\omega_0 t$, in which T_q is the QPSK symbol interval. Now let $T_q = 2T$ and define the special pulse shape

$$v(t) = \begin{cases} \dfrac{1}{\sqrt{T}}\cos\left(\dfrac{\pi t}{2T}\right), & -T \le t \le T \\ 0, & \text{otherwise} \end{cases} \tag{3.6-8}$$

This is the positive half of a sinusoid, centered at zero. We can remove the $T = \frac{1}{2}T_q$ shift from the quadrature signal by noting that

$$u(t) \triangleq v(t - T) = \begin{cases} \dfrac{1}{\sqrt{T}}\sin\left(\dfrac{\pi t}{2T}\right), & 0 \le t \le 2T \\ 0, & \text{otherwise} \end{cases}$$

Denote the offset QPSK transmission symbols as b_n^I and b_n^Q; then the QPSK baseband signals are

$$\begin{aligned} I(t) &= \sum b_n^I v(t - nT_q) \\ Q(t - T) &= \sum b_n^Q u(t - nT_q) \end{aligned} \tag{3.6-9}$$

Figure 3.26 shows an example of these signals. We show that (3.6-9) is an MSK signal by converting to amplitude and phase form. First, perform the conversion over the time interval $[(n-\frac{1}{2})T_q, nT_q]$ in the figure. From (3.1-7), the signal excess phase is

$$\psi(t) = \arctan\left(\frac{Q(t)}{I(t)}\right) = \arctan\left(\frac{b^Q_{n-1}\sin[[t-(n-1)T_q]\pi/2T]}{b^I_n\cos[(t-nT_q)\pi/2T]}\right)$$

$$= \arctan\left[\frac{b^Q_{n-1}}{b^I_n}\tan\left(\frac{(t-nT_q)\pi}{2T}\right)\right]$$

$$= 2\pi + \frac{[(-1)/(+1)](t-nT_q)\pi}{2T}$$

This is a linear increase of phase from $\frac{3}{2}\pi$ to 2π radians at a slope corresponding to an index of $\frac{1}{2}$. (It is easiest to see the quadrants and the start and stop phases by noting the signs and values of I and Q in Fig. 3.26.) Second, we can perform the conversion over the other half of the T_q interval, namely, $[nT_q, (n+\frac{1}{2})T_q]$. Now

$$\psi(t) = \arctan\frac{b^Q_n\sin[(t-nT_q)\pi/2T]}{b^I_n\cos[(t-nT_q)\,\pi/2T]}$$

$$= \frac{[(-1)/(+1)](t-nT_q)\pi}{2T}$$

This is a linear decrease of phase from 2π down to $\frac{3}{2}\pi$, at the index $\frac{1}{2}$ slope. Some contemplation of Fig. 3.26 will show that slopes are always $\pm\frac{1}{2}$ and that there can be

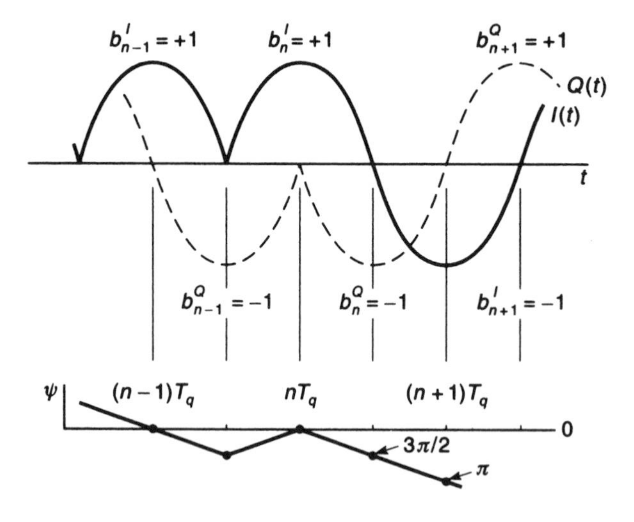

Figure 3.26 Example of baseband signal in offset QPSK construction of MSK. Here b^I_n and b^Q_n denote offset QPSK transmission symbols. The MSK phase is shown at bottom.

no phase jumps because there are no jumps in I or Q. The excess phase is thus an MSK signal. The symbols b_n are not, however, the symbols a_n, in definition Eq. (3.6-1). The mapping between a's and b's is rather complicated. In brief, the a's need to be differentially encoded to produce the ±1 stream $v_n = a_n/a_{n-1}$; even v_n are mapped to $b^I_{n/2}$ and odd v_n are mapped to $b^Q_{(n+1)/2}$; after offset QPSK detection, the a's are recovered via the differential decoding $a_n = v_n/v_{n-1}$. The MSK/Offset QPSK equivalence is admittedly not easy to demonstrate, but it is of some practical importance because the offset circuits are an easy way to implement the modulation. For a fuller explanation of the equivalence, see Ref. 2 or 13.

Example 3.6-4: Radioteletype Wideband FSK. This venerable service still finds shipboard and radio amateur use. RTTY, as it is called, is asynchronous data transmission via binary wideband FSK [our category 1] at a speed usable by mechanical teletypewriters. Dozens of formats exist, but we focus on a common older one called 60 Speed TTY that is based on a symbol interval $T = 0.022s$ and on transmission of one of two RF frequencies, 170 Hz apart. This ±85-Hz frequency shift works out to a modulation index of $h = (85)(0.022) = 1.87$. A common detector circuit is composed of filters that resonate at the two frequencies. These work as crude matched filters for the orthogonal signal constellation in Fig. 2.26b; the frequencies here are not totally orthogonal, and they get less so in other RTTYs that have a faster symbol rate, but the detector still works reasonably well. An even older standard had a ±425-Hz deviation; h for this one is 9.35. An excellent reference for older FSK schemes is the *ARRL Handbook* [14].

3.6.4 CPM Coded Modulation

Continuous-phase modulation (CPM) codes were the first widely studied class of coded modulations. These codes are an extension of, and were originally inspired by, CPFSK. As a general concept, a code word is a sequence of symbols with a pattern. A receiver looks for one of the patterns known to be in the set of code words. If it cannot find one, it knows there is a transmission error, and it seeks a legal pattern close to the received sequence. A coded modulation is a set of code words where the "symbols" are pieces of a modulated signal. In CPFSK, the pieces received must form one of the patterns in Fig. 3.21. The critical aspect of that picture, if the index and the starting point are known, is that all the paths through it are continuous: The CPFSK signal piece during an interval must start where the previous one left off and end where the next one takes off. This property of memory, or patterning, in the signal leads directly to the narrow bandwidth and high distance of CPFSK.

CPM codes extend this idea in two related ways: by adding a higher order of memory and by smoothing the phase transitions. The first primarily improves the minimum distance, while the second narrows the bandwidth. Most CPM codes have signals that are defined by an extension to (3.6-1) that looks like

$$s(t) = \sqrt{\frac{2E_s}{T}} \cos\left[\omega_0 t + 2\pi h \sum_n a_n q(t - nT)\right] \tag{3.6-10}$$

The excess phase term in (3.6-10) amounts to a convolution of the transmission symbols . . . , a_{n-1}, a_n with the function $q(t)$, so that CPM codes are a kind of convolutional code in phase. The *phase response* function $q(t)$ has a form like the examples shown in Fig. 3.27: $q(t) = 0$, $t < 0$, and $q(t) = \frac{1}{2}$, $t > LT$. The requirements ensure that the signal phase is continuous if $q(t)$ continuous. The $q(t)$ in Fig. 3.27b gives directly CPFSK and Eq. (3.6-1); the next $q(t)$, in 3.27c, widens the memory to $2T$ and rounds the corners in 3.27b. The $q(t)$ in 3.27d extends these ideas much further and belongs to a modulation called Gaussian MSK (GMSK); it will be easier to explain GMSK after some material in the next section. Figure 3.27a generates PSK or, more precisely, differential PSK.

The smoothness of the phase response $q(t)$ is crucial because it sets the rate of decay of the spectral side lobes, which in turn governs how much a modulation will

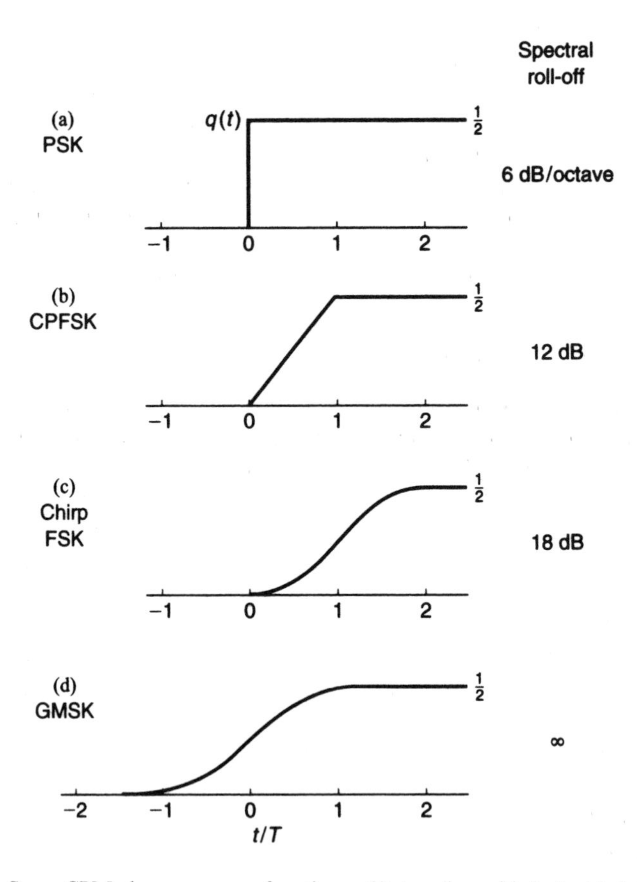

Figure 3.27 Some CPM phase response functions $q(t)$, together with their side-lobe roll-off rates. Responses begin at $t = 0$ except for GMSK, which is centered at zero.

interfere with transmissions nearby in the spectrum. Suppose we measure the smoothness of $q(t)$ by the number c of continuous derivatives that it has. Then Baker [15] has shown formally that as f tends away from the carrier frequency, the PSD obeys asymptotically the relation

$$\mathcal{E}[|S(f)|^2] \approx |(f-f_0)T|^{-(2c+4)} \tag{3.6-11}$$

The corresponding roll-off rates in decibels per octave change in $f-f_0$ are listed in Fig. 3.27. With PSK, or with any other phase-discontinuous modulation, $q(t)$ itself is discontinuous and only its first integral has continuity; this sets $c = -1$ and implies that the asymptotic roll-off of any discontinuous-phase modulation is $|(f - f_0)T|^{-2}$, or 6 dB/octave. For CPFSK, $c = 0$, which translates to 12 dB/octave. These rates are clear in Figs. 2.9, 3.4, and 3.24. Baker's formula and phase continuity are not at all theoretical curiosities. In digital mobile radio, the edge of a neighboring channel can start two octaves above the main lobe of the previous channel, and the formula correctly predicts that CPFSK will cause $6 \times 2 = 12$ dB less interference at the next band edge than NRZ-QPSK. The lower CPM modulations in Fig. 3.27 will be even more effective.

References 1–4 introduce CPM, and Refs. 10–12 present a wealth of detail about individual codes. Each widening of the active part of $q(t)$ improves the minimum distance and narrows the spectral main lobe; each order of smoothing reduces the side lobes in accordance with (3.6-11). Another kind of CPM codes, called *multi-h* codes, work by advancing h cyclically through a set of different indices. The effect is to introduce another kind of memory into the signals. CPM signals of whatever kind always have a constant envelope.

3.7 FSK DETECTORS

A receiver for FSK is more complex than one for PSK or QAM, because it needs to observe more than one signal interval if it is to be maximum likelihood. Still, simple receivers are available, when the lower performance can be tolerated. Simplicity can be achieved by observing a single interval, or by working without an exact phase reference, or by detecting frequency instead of phase changes.

3.7.1 FSK Maximum-Likelihood Detection

No matter how complicated the FSK signals, it is always an option, at least in theory, to set up the bank-of-matched-filters receiver in Fig. 2.32, or its equivalent, the bank of correlators in Fig. 2.30. If the signal is observed for L intervals from a known starting point, there will be M^L filters in the bank. These are sketched out in phase and time in Fig. 3.21 for the case of CPFSK. In this pure matched filter receiver, each filter is matched to a different phase-modulated RF signal.

Precisely how many filters are truly needed is a complex issue. Eventually, a symbol stops affecting the detection of later symbols, and L, the observation width, need not be longer than this width of symbols. For example, with most CPFSKs, including MSK, L can be 2 if the detection starts from a known point in the signal. If the start is unknown, ML detection is still possible, but L must be 3 or more. If the observation is just one symbol, as it is in PSK and QAM, performance is much worse: The detector sees the nearest-neighbor pair in Fig. 3.22 only over $[0, T]$, and from Eq. (3.6-3) the square distance is only half the true minimum distance. The single-interval detector thus suffers a loss of 3 dB in the energy needed to produce a given error probability.

In addition to the length issue, it is often possible to reduce the matched filtering to a few components that repeat often in the set of signals. For example, if h is a simple fraction, parts of Fig. 3.21 will soon be the same as other parts modulo 2π. But this is an advanced topic.

Although we have begun the discussion with detection at RF, FSK receivers usually work at baseband with the signals $I(t)$ and $Q(t)$. We have already seen that any bandpass signal may be expressed in these terms. With FSK, $I(t)$ and $Q(t)$ are certainly not linear pulse trains, but using the conversion formula (3.1-6) we can still set them down as

$$I(t) = \cos \psi(t), \qquad Q(t) = \sin \psi(t) \tag{3.7-1}$$

where $\psi(t)$ is the FSK excess phase.

With I and Q from (3.7-1), an FSK signal may be generated by the standard quadrature transmitter of Fig. 3.8, with $A = \sqrt{2E_s/T}$. The standard quadrature receiver in Fig. 3.9 recreates $\hat{I}(t)$ and $\hat{Q}(t)$. To complete a baseband ML receiver, a bank of matched filters or other ML detector replaces the right half of Fig. 3.9, since the receive filter and sampler there are no longer adequate. Distances can be computed directly from I and Q differences by using Lemma 3.5-1.

3.7.2 Coherency in FSK Detection

Until now, we have assumed that a detector knows the starting point of the signals that it observes. Since FSK and CPM signals are constant-envelope, this means that the detector has an accurate phase reference, which is to say that it knows the zero point of the excess phase in definition (3.6-1). It also knows the symbol interval boundaries. Such a receiver is called a *coherent detector*.

The opposite extreme is a *noncoherent* detector, a receiver that makes no use of a phase reference at all.[17] One such receiver is the *discriminator* detector. A discriminator is a circuit that observes the instantaneous frequency of a signal. This frequency is the time derivative of the total phase in (3.6-1) or (3.6-10), which for FSK is

$$\omega_0 + \frac{a_n h \pi}{T} \quad \text{(rad/s)}, \qquad nT \le t \le (n+1)T \tag{3.7-2}$$

[17]The term *incoherent* is also used.

and more generally for CPM is

$$\omega_0 + 2\pi h \sum_n a_n g(t - nT) \qquad \text{(rad/s)} \tag{3.7-3}$$

in which $g(t) \triangleq dq(t)/dt$ is defined to be the *frequency pulse* of the CPM signal. Since the phase response $q(t)$ starts and ends with constant sections, $g(t)$ is a true pulse and (3.7-3) is a pulse train modulated by the data $\{a_n\}$ and offset by the constant ω_0. A phase offset in the signal acts as a constant during the differentiation and is ignored. Discriminators are easy to build in hardware or software and can easily be designed to ignore the DC offset in (3.7-3) and any other offsets that vary more slowly than the modulation.

The notion of a varying offset brings up a subtlety in noncoherent detection, whether or not the unknown phase reference also varies. If it does not vary, or changes very slowly, then receivers can be built [2, 12] that essentially duplicate ML detection; that is, they act as if the phase reference were known. Partly, this stems from the fact that FSK and CPM schemes carry the data in the change of phase rather than in the absolute phase; partly, the receiver identifies the offset along with the data as it works.

If the reference varies too much, a discriminator detector will need to be employed, but now the error probability will be considerably worse than the ML probability. An analysis of the new error is technically daunting even in the Gaussian case, because the noise is no longer Gaussian at the discriminator circuit output. Added to this, it is not easy to derive the optimal receiver filter that comes after the discriminator. A good rule of thumb is that discriminators lead to 3-dB energy degradation compared to ML detection.

A third kind of coherency occurs in *partially coherent* detection. Now the phase reference is partly known. For example, its probability distribution may be known but not the exact value of the reference at a given time. A particular case is the estimation of a phase reference from a signal with additive Gaussian noise: In many estimation circuits, the AWGN disturbance converts to a noise in the reference that has a Tikhonov distribution. It is possible to derive good receivers based on this distribution [2].

3.7.3 Digital FM View

A strong motivation to use discriminator detection is the fact that ordinary analog FM receivers already have one built in. It is only necessary to add a pulse detection circuit in order to detect FSK or CPM. This has been a strong driver on the development of digital mobile radio, which had to evolve from an analog FM system already in place. A way to look at this kind of FSK system is to think of it as *digital FM*. The system begins with a pulse train like (3.7-3), made up of smoothed pulses, which is basically the linear modulation pulse train of Sections 3.2–3.5. This drives an FM modulator. A frequency modulation detector recreates the pulse train, from which the data are detected, again in the manner of a linear modulation. A workable system is thus created out of spare parts.

A digital FM system can transmit any CPM signal, and some of these signals have very attractive spectrum and energy properties. Perhaps the best known one, aside from CPFSK, is Gaussian MSK, or GMSK, first proposed by Murota and Hirade [16]. The frequency pulse of GMSK has the Gaussian-integral shape

$$g(t) = \frac{1}{T}\left[Q\left(\frac{t + T/2}{\sigma T}\right) - Q\left(\frac{t - T/2}{\sigma T}\right)\right] \tag{3.7-4}$$

in which

$$\sigma = \frac{\sqrt{\ln 2}}{2\pi B_b}, \qquad 0 < B_b < \infty$$

and B_b is a parameter that sets the signal bandwidth. The modulation index h in (3.7-3) is $\frac{1}{2}$. The name Gaussian MSK comes from the fact that GMSK results when the ordinary MSK frequency pulse train, which is an NRZ train, passes through a low-pass filter whose impulse response has the Gaussian shape $\exp(-t^2/2\sigma^2)$. It can be shown that the 3-dB-down bandwidth of this filter is B_b hertz. This is, of course, not the final bandwidth of the modulation itself, but $2B_b/T$ hertz is a reasonable estimate of the GMSK 99% RF power bandwidth when $B_b < 1$.

Thus GMSK is filtered square-pulse digital FM. Figure 3.28 shows the frequency pulses at several values of B_b. When B_b is large, $g(t)$ is almost the square pulse of MSK; when B_b is small, the pulse spreads out over 3–5 symbol intervals, and there will be considerable intersymbol interference among the pulses. Upon integrating $g(t)$, we get the phase response pulse $q(t)$ for use in (3.6-10), and this is the CPM expression of GMSK. The phase pulse for $B_b = 0.3$ appears in Fig. 3.27d. The CPM

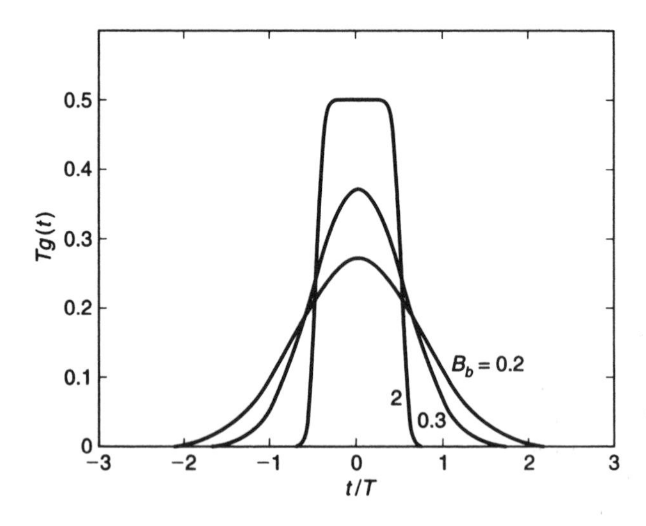

Figure 3.28 Some GMSK frequency pulses $g(t)$ for bandwidth parameter $B_b = 0.2, 0.3, 2$.

expression for GMSK reminds us that a full maximum-likelihood CPM receiver does exist and is a higher-cost higher-performance alternative to the discriminator.[18] The CPM analysis also leads to an interesting conclusion about the GMSK spectrum: $q(t)$ has infinitely many continuous derivatives, and from the Baker formula (3.6-11) the GMSK side lobes must therefore fall off extremely rapidly.

Example 3.7-1: GMSK Pulse for GSM Digital Cellular Telephones. The European GSM[19] digital cellular standard, announced in 1986, replaced earlier analog FM standards. GSM is a time-division multiple-access system that transmits bursts of 271-kb/s GMSK over several carrier ranges—for example, 890–960 MHz. The GMSK has bandwidth factor $B_b = 0.3$; its frequency pulse appears in Fig. 3.28 and its phase pulse appears in Fig. 3.27d. Figure 3.29 shows an approximate baseband power spectrum; there are virtually no side lobes, and the 99% RF bandwidth is about 250 kHz. The GSM system divides its spectrum allocation into 200-kHz-wide RF subchannels with a GMSK transmission in each one. In rough numbers, the GMSK spectrum is down 18 dB at the next subchannel band edge and 50 dB at the edge after that. In reality, the limit to side-lobe reduction is set not by the modulation design, but by the discrete-time signal processing that generates the pulses (see Appendix 3C). Figure 3.30 shows an eye diagram for the signal at the output of a discriminator. The eye is closed by about half compared to ordinary MSK. Data symbols are cheaply detected by taking the sign of the signal at times nT. A good reference for GSM and GMSK is Steele [17].

3.8 TRANSMISSION IMPAIRMENTS

All the foregoing chapter has assumed that only white Gaussian noise corrupts a transmitted signal, which otherwise arrives in pristine condition. Sadly, this is seldom the case, but we can temper this sorrow with the knowledge that communication engineers derive a handsome income from fixing these damaged signals.

We can summarize the many impairments to signals as follows. Some of these are better explained in other chapters, and the rest will form this section.

1. *Noise.* This has been discussed heretofore. We usually design as if it were white Gaussian. If a filter affects the noise within the signal bandwidth before the signal plus noise reach the detector, then the detector design must change if it is to remain ML (discussed in Section 3.8.1).
2. *Channel Filters.* Filters usually occur in or before a physical channel, before the point where the noise appears. The filter may model time dispersion in a

[18]A good CPM receiver is not too expensive, but it is challenging to design well; one alternative is to approximate the infinite-width GMSK pulse with a width-4 pulse. Gains of 5 dB or more over the discriminator have been reported.

[19]*Groupe Speciale Mobile*, the European mobile radio working group that devised the standard; GSM cellular telephone has been the standard after the mid-1990s in Europe and many other regions.

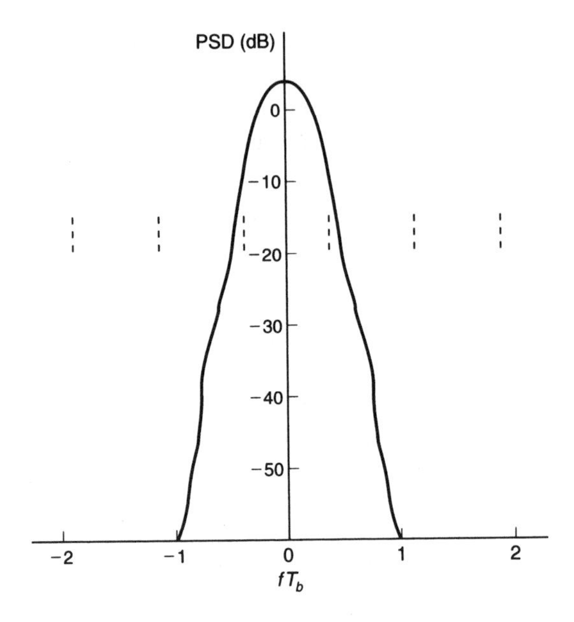

Figure 3.29 Power spectral density of GMSK with $B_b = 0.3$. The 200-kHz channel spacing of GSM cellular telephone is shown.

physical channel. They thus damage the signal but do not modify the noise (Section 3.8-1 and Chapter 5).

3. *Cochannel Interference (CCI).* This occurs when part of the in-phase signal appears in the detector where the quadrature signal should be and vice versa. The CCI can be caused by an asymmetrical channel filter (discussed in Section 3.8-1) or by inaccurate phase and symbol synchronization (Chapter 4 and Section 2.4).

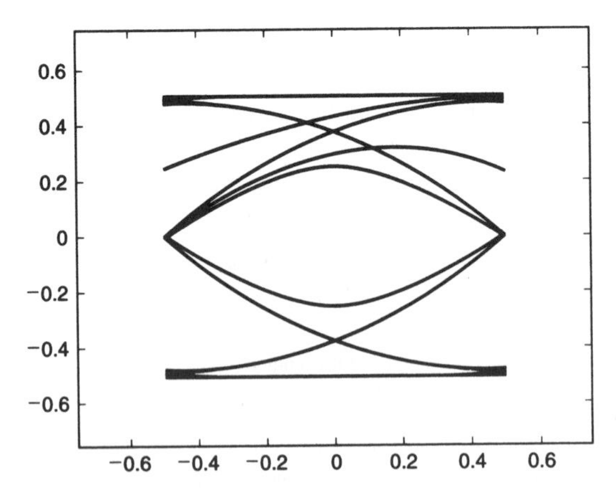

Figure 3.30 Eye diagram at discriminator output for GMSK with $B_b = 0.3$.

4. *Adjacent Channel Interface (ACI).* This is sidebands from another signal in a nearby channel.
5. *Intersymbol Interference (ISI).* This is contribution from symbols before or after the present one in a pulse train. It is usually caused by channel dispersion or filtering, but may stem from symbol timing error (Sections 3.8-1 and 2.4; Chapter 5).
6. *Phase Error.* Here the carrier signal fed to the quadrature receiver has a phase error, which leads to CCI (Chapter 4).
7. *Symbol Timing Error.* The timing that drives the receiver filter samplers is in error, which leads to ISI (Chapter 4 and Section 2.4).
8. *Fading.* There is a short but massive drop in signal power, accompanied often by a loss of phase and symbol synchronization and a change in channel dispersion (Section 5.4).
9. *Nonlinearity.* Usually, this occurs in the transmitter power amplifier. Nonlinearity changes the signal spectrum and creates intermodulation distortion, in addition to distorting amplitudes (Section 3.8-2).
10. *AM–PM Conversion.* Many power devices convert amplitude shifts into phase shifts (Section 3.8-2).
11. *Implementation Error.* In discrete-time processors, insufficient sampling rate and arithmetic precision can raise the detector error rate (Appendix 3C).

This is a daunting list. But perhaps the greatest design challenges arise from malevolent *interactions* among effects when more than one is present. In transmission impairments, the whole is truly more than the sum of the parts.

3.8.1 Filtering

A filter can affect a carrier-modulated signal at different points in a communication link. Figure 3.31 summarizes the possibilities that exist in the great majority of systems. The most common place for a filter is inside the transmitter; the filter may carry out the baseband pulse shaping or it may restrict the transmission outside a certain band in accordance with government regulation. Sometimes the restriction occurs out in the physical channel, intentionally or otherwise. It commonly happens that time dispersion occurs in the channel, a smearing of the channel impulse re-

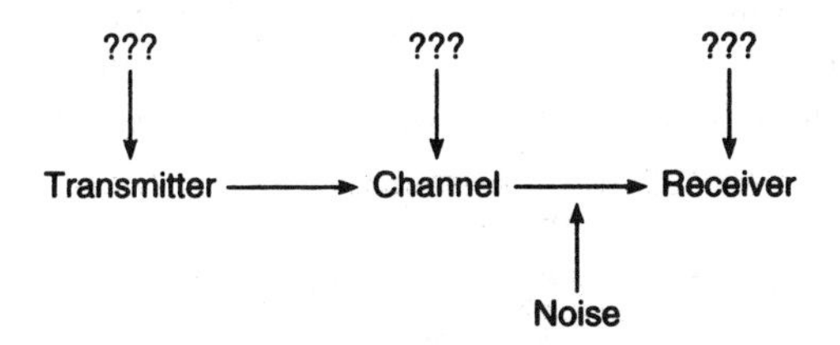

Figure 3.31 Different locations for filter impairment in transmission system.

sponse by reflections, refraction, and other physical phenomena. Over short time spans, at least, dispersion can be modeled by a filter with the same impulse response. Next in the figure comes the noise. Despite the fact that it is often called "channel" noise, noise in a link usually traces to the front end of the receiver, which lies after the physical channel, and, normally, after any filter impairment. While all receivers have front-end and IF filters, these normally serve to keep unwanted signals out and do not invade the band of the desired signal. Occasionally, a filter inside the receiver affects both the signal and the noise, but this is abnormal. All these filters should not be confused with matched filters or with the "Receive Filter" in a quadrature receiver circuit, both of which are part of the detection.

In filter impairment problems, the analysis is usually carried out with baseband equivalent filters and signals. We have already seen in Section 3.1 that any bandpass signal may be written in the form

$$s(t) = I(t) \cos \omega_0 t - Q(t) \sin \omega_0 t \tag{3.8-1}$$

in which $I(t)$ and $Q(t)$ are the low-pass in-phase and quadrature baseband components of the signal. In the same way, the impulse response of a bandpass filter may be written as

$$h(t) = h^I(t) \cos \omega_0 t - h^Q(t) \sin \omega_0 t \tag{3.8-2}$$

Appendix A shows a number of results about baseband equivalent circuits, the chief one being that the baseband components of the filter output $y(t)$ are related to those of the input by

$$\begin{aligned} y^I(t) &= \tfrac{1}{2}[I(t) * h^I(t) - Q(t) * h^Q(t)] \\ y^Q(t) &= \tfrac{1}{2}[I(t) * h^Q(t) + Q(t) * h^I(t)] \end{aligned} \tag{3.8-3}$$

This equivalence is shown graphically in Fig. 3.32. Appendix A also explains the conversion back and forth between h, h^I, and h^Q.

Noise in a bandpass system may also be represented by baseband components. The bandpass noise waveform is

$$\eta(t) = \eta^I(t) \cos \omega_0 t - \eta^Q(t) \sin \omega_0 t \tag{3.8-4}$$

in which $\eta^I(t)$ and $\eta^Q(t)$ are baseband component noise waveforms. The facts about these are summarized in Appendix B. Fortunately, when the noise background is AWGN and the filtering inside the receiver does not invade the signal passband, the facts are very simple: *Background AWGN with density* $\tfrac{1}{2}N_0$ *may be expressed as two AWGN baseband components with density* N_0.

Symmetrical Bandpass Filters. Most bandpass filters are symmetrical about the carrier frequency, that is, their response Fourier transform is

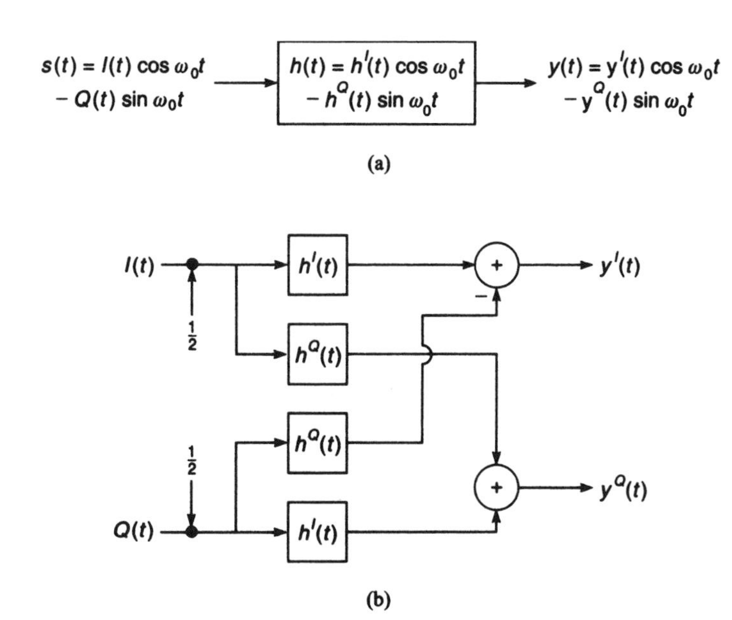

Figure 3.32 **(a)** Bandpass and **(b)** baseband equivalent filter representations.

$$H(f) = \tfrac{1}{2}[H_B(f - f_0) + H_B(f + f_0)] \tag{3.8-5}$$

in which $H_B(f)$ is a legal low-pass filter with bandwidth $[-\frac{1}{2}B, \frac{1}{2}B]$. In consequence, the magnitude of $H(f)$ is symmetrical about f_0 and $-f_0$ with RF bandwidth B, and the phase is antisymmetrical about the same. An illustration is shown in Fig. 3.33. For example, Butterworth and elliptical bandpass filters are low-pass filters frequency-shifted according to (3.8-5).[20] With symmetric filtering, the baseband equivalent of the bandpass filter has its identical shape but at baseband, and it affects the signals $I(t)$ and $Q(t)$ independently. This is shown in Appendix A. Here lies the motivation for the baseband analysis method: In most problems the baseband filters look like the bandpass ones and the noise is still AWGN. Most of what we need to know has already been covered in the baseband filter discussion in Section 2.4

Section 2.4 shows how typical baseband filters affect pulse trains like the $I(t)$ and $Q(t)$ that occur in PSK and QAM carrier modulation. As Figs. 2.18 and 2.19 show, low-pass filtering at baseband closes the eye pattern and creates intersymbol interference in the samples after the receive filter/sampler. A symmetric bandpass filter acting on a carrier signal creates the identical effect on $I(t)$ and $Q(t)$.

Filters also cause delay, which may need compensation itself and will lead to

[20]RF circuit designers make use of several such lowpass-to-bandpass conversions and call the results for all of them by the low-pass name. We will assume that the filter is realized by (3.8-5).

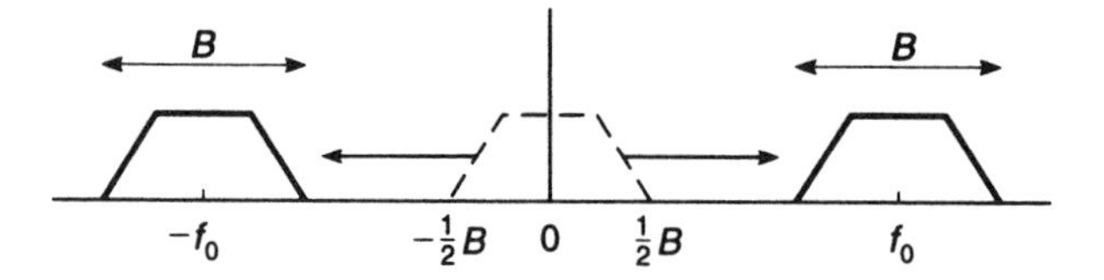

Figure 3.33 Typical symmmetric bandpass filter spectrum.

particular difficulty if it does not affect all signal frequencies equally. A way to measure this is to compute the *group delay* of the filter, defined as

$$\tau_g = -\frac{d\theta(\omega)}{d\omega} \tag{3.8-6}$$

where $\theta(\omega)$ is the phase of $H(f)$ at $\omega = 2\pi f$. A group delay plot is shown in Fig. 3.34 for the Butterworth filter with cutoff $0.5/T$ hertz in Example 2.4-1. The delay is roughly flat and about $1.4T$ out to $0.4/T$ hertz; this delay is reflected in Fig. 2.17 and is subtracted in the corresponding eye diagrams. A signal with spectral components outside $0.4/T$ will suffer not only delay but distortion as well. It is important in signal filter design to make the group delay as flat as possible. Since a flat delay is a linear phase, an equivalent filter requirement is linear phase.

When constant-envelope carrier signals are bandpass-filtered, the result is envelope variation, even though none existed before. This can be demonstrated by con-

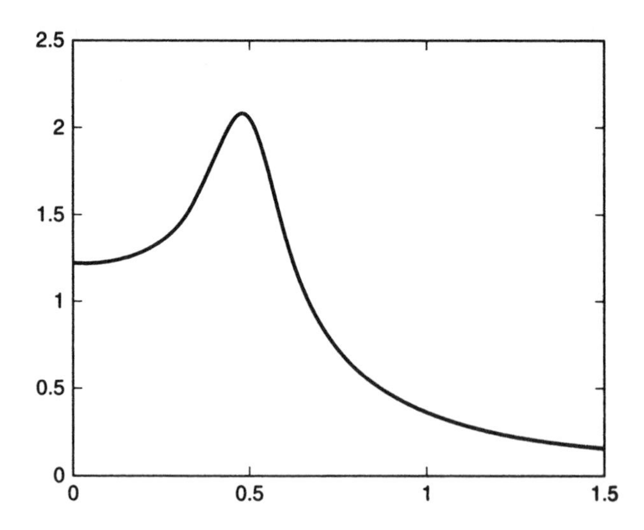

Figure 3.34 Group delay versus frequency normalized to symbol time for six-pole Butterworth low-pass filter with 3-dB cutoff frequency $0.5/T$. See Example 2.4-1 for filter description.

verting carrier signal and filter to baseband, filtering, and converting back, as in Fig. 3.32, but the outcome is rather difficult to quantify. Figure 3.35 shows oscilloscope photos of NRZ-BPSK and MSK waveforms that were subjected to increasingly narrow filtering. As the filter narrows, the envelope varies more, and the effect is more severe with BPSK. Filter-induced envelope variation destroys the very reason for using constant-envelope signals and so needs to be avoided with FSK, CPM, and like signals.

Asymmetric Bandpass Filters. An important fact shown in Appendix A is that filters that are not symmetric about f_0 and $-f_0$ lead to *crosstalk* between the I and Q components of the data-bearing signal. This means that some of $I(t)$ in the signal (3.8-1) appears in the filter output (3.8-3) Q-component and some of $Q(t)$ appears in the output's I-component. The effect is also called co-channel interference (CCI). In a quadrature modulation that carries two pulse trains, the trains are no longer independent. A condition equivalent to asymmetry, proven in Appendix A, is that the filter quadrature component $h^Q(t)$ does not equal zero. This is also clear from Fig. 3.32.

Asymmetric filters can arise in several ways. An improperly tuned receiver will place a carrier signal off-center in the bandpass, and if the filter is not wideband enough, the signal will see asymmetrical filtering. Receiver IF filters with imperfect

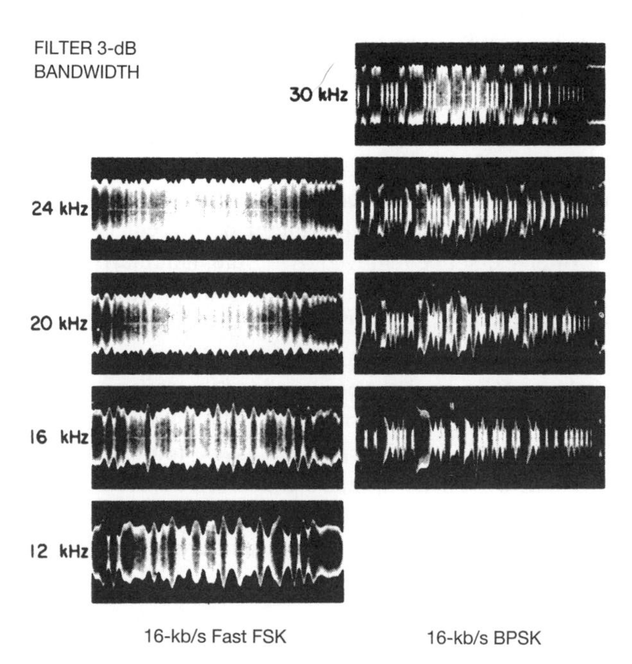

Figure 3.35 Variations in envelope caused by filtering MSK **(left)** and BPSK **(right)**. Symbol rate is 16 kHz; filter is a Chebychev with 3-dB RF bandwidth as shown. (Courtesy of J. L. Pearce, Communications Research Centre, Government of Canada, Ottawa.)

components, and particularly ceramic crystal filters, can lack symmetry and lead to crosstalk.

A particularly nasty effect can occur when an asymmetric filter comes after the additive noise and the filter asymmetry invades the signal band. As shown in Appendix B, the baseband equivalent noise components are neither white nor uncorrelated. The design of an ML detector for this situation is possible but considerably more complicated. An account is found in the classic text by Van Trees [18].

Deriving the components $h^I(t)$ and $h^Q(t)$ for a measured asymmetric filter response $h(t)$ is not difficult and is described in Appendix A.

3.8.2 Other Signal Impairments

Amplifier Nonlinearity. One of the inescapable trade-offs in RF power amplifier design is that the more efficient an amplifier is, the more it responds nonlinearly to differing amplitudes in the sinusoidal input. A class A amplifier is linear but has a 20–30% efficiency; a single power device in class B is 40–60% efficient. Class C amplifiers can be more than 90% efficient but are strongly nonlinear, and they are really useful at this efficiency only with constant-envelope modulations. These facts pose an interesting design dilemma in a power-supply-limited situation, such as a satellite downlink or a hand-held telephone. On the one hand, constant-envelope transmission can be employed with a class C amplifier that converts nearly all the available DC power into RF power. On the other hand, a pulsed linear modulation can be used and will be more bandwidth-efficient, or simpler, or (with QPSK) more energy efficient. But the necessary linear power amplifier will convert much less DC power to RF. The net result, typically, is a loss of 2–4 dB in the received E_b/N_0.

Consequently, many links contain amplifier nonlinearity. The nonlinearity creates signal harmonics, which can generally be filtered out, and intermodulation products between different carriers that might be present in the same amplifier. These products often cannot be removed. As a rule, nonlinearity also leads to *spectral spreading,* a widening in the bandwidth of the amplified signal. One classic example occurs in digital telephony, and it shows how spreading can occur without an explicit amplifier nonlinearity. Digital telephones contain a compander before the analog-to-digital conversion, whose purpose is to improve intelligibility by raising weak speech sounds and limiting strong ones. Although the analog-to-digital conversion rate is typically 8 kHz, the speech must be limited considerably below 8 kHz/2 = 4 kHz because the band-limited signal will be spread out again by the nonlinearity.

Figure 3.36 shows what happens when a unit-energy BPSK signal with 30% root-RC pulses passes through the nonlinearity

$$g(x) = [1 - e^{-|x|}]\,\mathrm{sgn}(x) \tag{3.8-7}$$

The bandwidth of the signal is $0.65/T$ hertz on either side of the carrier going in (solid line) but develops considerable side lobes coming out (dashed line) and of

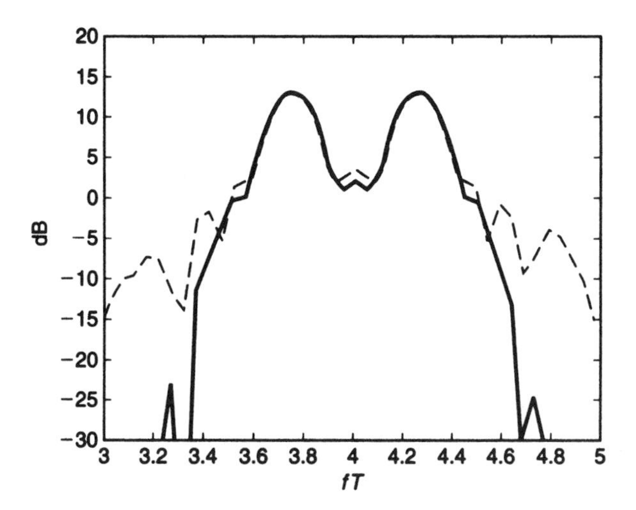

Figure 3.36 Spectral spreading of 30% root RC BPSK transmission of symbols + − − + +, caused by nonlinearity of Eq. (3.8-7). Carrier locations is at $fT = 4$ Hz-s.

course can no longer be thought of as made of clean root RC pulses. An exponential rule like (3.8-7) applies to many power amplifiers. An important amplifier for satellite use is the traveling-wave tube (TWT), and the actual relation between input and output power for one of these is shown in Fig. 3.37.

A particularly difficult impairment occurs when filtering creates or worsens envelope variations, which are then distorted by a nonlinearity.

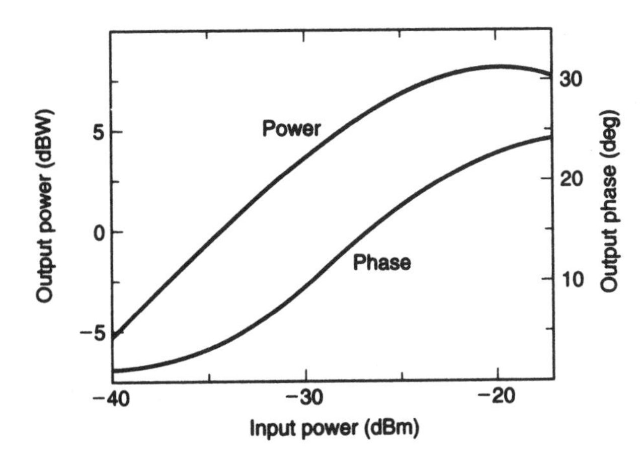

Figure 3.37 Plot of output versus input power and output phase versus input power for typical TWT amplifier. (Reproduced from Ref. 20 by permission. Copyright 1981 by John Wiley and Sons.)

AM-to-PM Conversion. Another eccentricity of power amplifiers is a tendency to convert changes in the amplitude of a sinusoid into changes in its phase, a phenomenon called AM-to-PM conversion. In effect, stronger sinusoids move through the device faster and acquire what appears to be a phase advance. This plays havoc with phase modulations. TWTs have large AM-to-PM conversion, and the effect for a standard tube is shown in Fig. 3.37. A pulse-shaped QPSK that undergoes 4:1 amplitude shifts could sustain a 20°–25° gratuitous phase shift with this amplifier. To reduce nonlinearity and AM–PM, TWTs often run in the lower part of their input/output characteristic in *back-off* mode—that is, at reduced power. Further information about TWTs in satellites may be found in Ref. 20.

A link that has all three interacting impairments—filters, nonlinearity, and phase error—is a heavy challenge.

PROBLEMS

Phase-Shift Keying

3-1. (a) Show that $\sqrt{2/T}\,\mathrm{sinc}(t/T)\cos\omega t$ has unit energy in the limit of $\omega \to \infty$.

(b) Show that the BPSK basis function $\sqrt{2}v(t)\cos\omega t$ has unit energy in the same limit whenever $v(t)$ does.

(c) Show that the function $\sqrt{2}v(t)\cos\omega t$ is orthogonal to T-shifts whenever $v(t)$ is, again in the same limit.

3-2. Verify that the energy of the signal in (3.1-4) is E_s whenever (3.1-5) holds.

3-3. Repeat Problem 2–5 for an RF root RC pulse; that is, let the pulse $\sqrt{2}v(t)\cos\omega t$ be time truncated to the interval $[-3T, 3T]$, where $v(t)$ has unit energy and is orthogonal to T-shifts. Find the pulse power spectrum and estimate the rate of decay of the spectral side lobes.

3-4. Create an I/Q plot similar to Fig. 3.6, but let the pulse be100% root RC instead of 30%. Comment on the differences between the two plots.

3-5. Design a standard quadrature receiver for 8PSK; that is, create a design like that in Fig. 3.9, but replace the threshold blocks with suitable comparisons and logic, so that the receiver decision is ML for the 8PSK constellation in Fig. 3.10.

3-6. Consider the NRZ-pulse BPSK in (3.2-2). The values of T and ω_0 and the interval for t are chosen so that the '+' cosine burst consists of exactly five cycles, namely the signal $\sqrt{2E_s/T}\sin\omega_0 t$, $0 \le t \le 10\pi/\omega_0$.

(a) Give the bandpass (not baseband) matched filter that is matched to the signal for data +1.

(b) For an isolated five-cycle burst input, what is the output of this matched filter?

(c) What is the best time for the sampler?

3-7. Prove that the offset QPSK realization of MSK has constant envelope (see Fig. 3.26).

3-8. Prove that carrier on–off keying modulation has symbol error probability (2.6-6); that is, $p_e = Q(\sqrt{E_b/N_0})$. (*Hint:* It is sufficient to work with the simple binary signal set $\{0, \sqrt{E_b} \cos \omega_0 t\}$ in each symbol interval.)

3-9. Write a program that computes the true symbol error probability of 8PSK by numerical integration. Compare the results at several p_e to the bound (3.4-5).

QAM and Its Distance

3-10. Design a 16QAM system that carriers 48 kbit/s with a 200-kHz carrier. Root RC pulses are used with 30% excess bandwidth.

(a) List the 16 actual signals in the first signal interval.

(b) Give the PSD of the modulation.

(c) Give the most accurate symbol error probability estimate of the form $AQ(\sqrt{d^2 E_b/N_0})$; that is, give A and d^2.

(d) Give a data bit assignment to the 16 constellation points which will minimize the BER in a good channel.

3-11. Data bits are to be transmitted two at a time with a signal set of size four in AWGN. The five signal sets listed below are available. Compare them in terms of bandwidth and approximate error probability. In every case, the coefficient b is set so that the average energy per bit is the same and T is the symbol interval.

(i) $\pm b\sqrt{2/T} \cos \omega_0 t \pm b\sqrt{2/T} \sin \omega_0 t$

(ii) $\pm b\sqrt{2}\sqrt{2/T} \cos \omega_0 t$ and $b\sqrt{2/T} \cos \omega_0 t \pm b\sqrt{2/T} \sin \omega_0 t$

(iii) $b\sqrt{2}\sqrt{2/T} \cos(\omega_0 t + \theta)$, with $\theta = \pm\pi/4, \pm 3\pi/4$

(iv) $\pm b\sqrt{2/T} \cos \omega_0 t \pm b\sqrt{2/T} \cos(\omega_0 t + 2\pi t/T)$

(v) $\pm b\sqrt{2/T} \cos(\omega_0 t + t\pi/T) \pm b\sqrt{2/T} \cos(\omega_0 t + 2\pi t/T)$

3-12. Find the Euclidean square minimum distance of 64QAM by repeating the steps in Example 3.5-1 for this case. Check your result against Eq. (3.5-14).

3-13. (a) Give the bandwidth reduction and energy growth of 64QAM, compared to 16QAM. (Use (3.5-14) to help with the energy estimation).

(b) Give an error estimate of the form (3.5-11) for 64QAM.

(c) A telephone line modem with this 64QAM is to be designed. Suppose the standard telephone channel has $E_s/N_0 = 30$ dB, the modem carrier frequency is 1700 Hz, and the QAM pulse is 30% root RC. Estimate the probability of symbol error for the modem and how many bits per second it can carry.

3-14. The V.29 modem standard includes a provision whereby if the modem finds itself with a bad channel E_s/N_0, it can reduce the 16-point constellation in Example 3.5-2 down to the innermost eight points or to the innermost four.

(a) Calculate the normalized Euclidean square minimum distances in these two cases.

(b) If $E_s/N_0 = 13$ dB, estimate the symbol error probability by a Q-function argument for all three operation modes.

Frequency-Shift Modulations

3-15. Prove that the Euclidean square distance between two constant-envelope sinusoids with phases $\psi^{(1)}(t)$ and $\psi^{(2)}(t)$ satisfies Eq. (3.6-3).

3-16. **(a)** By means of Eq. (3.6-5), find an expression for the limiting form of the square distance of 2CPFSK as its index $h \to 0$. It is sufficient to assume that the minimum distance stems from the diamond-shaped event in Fig. 3.22.

(b) Now let h grow. At what h does another event replace the diamond one as the minimum-distance causing event?

3-17. In one interval, standard binary CPFSK with index $h = \frac{1}{4}$ can exhibit eight signals with upward tending phase and eight with downward phase (see Fig. 3.21).

(a) Create a signal space basis for these signals.

(b) Give the vectors that represent the 16 signals. (*Note*: Fourteen are phase rotations of the other two!)

(c) Verify (b) by using it to compute the distance of the minimum-distance achieving diamond event in Fig. 3.22.

3-18. Suppose a four-signal FSK system with symbol time T uses the signals $\pm\cos \omega_0 t$ and $\pm\cos(\omega_0 t + \pi h t/T)$.

(a) By the Gram–Schmidt process, find an orthonormal set of signal space basis functions for these four signals. If needed, simplify your calculations by assuming that $\omega_0 \to \infty$.

(b) Sketch the signal constellation for the first symbol interval if $h = \frac{1}{4}$.

3-19. **(a)** For the four-signal constellation in Problem 3–18, design a matched filter receiver with filters that are based on the signals themselves, not the basis signals.

(b) Design a second ML receiver based instead on the basis functions found by the problem.

3-20. Write a program for your toolbox which evaluates the CPFSK power spectrum (3.6-6). Check it against Fig. 3.24 and compute the 2CPFSK spectra for index $h = 0.1$ and 0.01.

3-21. Consider wideband binary FSK with symbol time T. The two signals are cosines at frequencies $\omega_0 \pm \pi h/T$. The receiver is two filters, matched to these signals.

(a) What is the correlation between the signals when $h = 3$? When $h = 3.1$? Normalize the correlation by $2E_b$ in both cases.

(b) In the $h = 3.1$ case, what appears at the two matched-filter sampler outputs when a signal is sent? Estimate how much this affects the noise immunity of the receiver.

3-22. MSK is continuous-phase binary FSK with index $h = \frac{1}{2}$. Its minimum distance is $d^2_{\min} = 2$. Let the channel have AWGN.

(a) Suppose the signal is known up to time nT. Show that two symbols—that is, the signal up to $(n + 2)T$—are needed if MSK is to achieve the asymptotic error probability $Q(\sqrt{d^2_{\min} E_b/N_0})$.

(b) Now suppose nothing is known before nT except the phase reference. Show that the signal must be observed for three symbols, up to $(n + 3)T$, to obtain the same error probability.

Impairments and Noise

3-23. Analyze the effects of band limitation on the RF envelope in the following way. Consider NRZ pulse QPSK modulation, which has constant envelope and whose baseband power spectrum sinc(fT) is the one in Fig. 2.9. Limit this spectrum. The new modulation will be linear, but will have a new, possibly longer pulse, and its envelope $\sqrt{I^2(t) + Q^2(t)}$ will no longer be constant. Generate 40 symbol intervals of this signal from random data and make a histogram of envelope samples similar to Fig. 3.7. Compare the effect of two bandwidth limitations:

(a) Truncate the NRZ spectrum to its first null ($|fT| = 1$), inverse-transform to find the pulse, and plot the envelope histogram.

(b) Repeat, truncating to the third null ($|fT| = 3$).

3-24. Repeat Problem 3–23 for the constant-envelope modulation scheme MSK, described in Example 3.6-3. [*Hint:* Like NRZ-pulse QPSK, MSK can be thought of as linear modulation with pulse (3.6-8); truncate the spectrum (3.6-7) and proceed as before to find envelope histograms.]

3-25. Analyze the effect of amplitude limitation on the spectrum in the following way. Consider the baseband 30% root RC pulse signal in Example 2.2-5 and Fig. 2.8. With no amplitude limitation, its power spectrum is the 30% RC one in Fig. 2.2. Introduce limitation directly on the signal produced by data +1, –1, +1, +1, –1, –1, compute the new signal's spectrum, and compare to the RC spectrum. Consider two amplitude limiters:

(a) The hard limiter defined by

$$v_{\text{out}} = \begin{cases} v_{\text{in}}, & |v_{\text{in}}| < 1.1 \\ 1.1, & \text{otherwise} \end{cases}$$

(b) The soft limiter defined by (sgn(x) is the sign of x)

$$v_{\text{out}} = \begin{cases} v_{\text{in}} & |v_{\text{in}}| < 1 \\ \text{sgn}(v_{\text{in}})[|v_{\text{in}}| + 0.15\,[1 - e^{-(|v_{\text{in}}|-1)}]], & \text{otherwise} \end{cases}$$

3-26. A receiver has a 30% root RC matched filter implemented as a bandpass filter centered at the carrier f_0. AWGN passes through this, so that it has a raised-cosine PSD with the shape shown in Fig. 3.3 and passband density $\frac{1}{2}N_0$ W/Hz. Let the result be $\eta(t)$.

(a) Find the variance of a sample of $\eta(t)$.

(b) Find and sketch the autocorrelation $R_{\eta\eta}(\tau)$ for η.

(c) According to Eq. (3B-1), η may be expressed in the form

$$\eta(t) = \eta^I(t) \cos \omega_0 t - \eta^Q(t) \sin \omega_0 t, \qquad \omega_0 = 2\pi f_0$$

where η^I and η^Q are low pass processes with parameters given in Appendix B. Give the PSDs of η^I and η^Q. What are the variances of samples of these processes?

3-27. Write a program for your toolbox that generates samples of low pass band-limited Gaussian noise. Let the inputs be the passband PSD $N_0/2$, the bandwidth B, and the sample frequency f_s. Demonstrate your program for the parameters in Fig. 3.41c. (An extension of this problem is 5-25 in Chapter 5.) (*Hint:* MATLAB is recommended, with Butterworth filtering of order 6 or so; start with the routine `[b,a] = butter(6,2*ctf/fs)`, in which `6` is the filter order, B is taken as `ctf`, the cutoff frequency, and `fs` is the sample frequency.)

3-28. An element in some receivers is a squaring circuit. From an input $r(t)$, it produces $r^2(t)$.

(a) Suppose the input is the standard I/Q-form signal $s(t)$ in (3.1-4) plus the flat-PSD bandpass AWGN in Fig. 3.40, with double-sideband bandwidth B and density $\frac{1}{2}N_0$. If the noise is represented as $\eta(t) = \eta^I \cos \omega_0 t - \eta^Q \sin \omega_0 t$, give an expression for the output in terms of η^I, η^Q, and the baseband signal components $I(t)$ and $Q(t)$.

(b) Set the signal $s(t)$ to zero. Find the PSD of the squared noise $\eta^2(t)$ that remains.

(c) Replace the signal so that $r(t) = s(t) + \eta(t)$. Find the mean value of $r^2(t)$.

3-29. A discriminator detector works by taking the derivative of its input signal. AWGN that is present will be differentiated as well. Suppose the AWGN spectrum is the flat-PSD bandpass one in Fig. 3.40. Give the PSD of the differentiated noise. How does the total noise power depend on B? [*Hint:* A certain filter $H(f)$ performs differentiation; use this $H(f)$.]

REFERENCES[21]

1. *J. G. Proakis, *Digital Communications,* 3rd ed., McGraw-Hill, New York, 1995.
2. M. K. Simon, S. M. Hinedi, and W. C. Lindsey, *Digital Communication Techniques,* Prentice-Hall, Englewood Cliffs, NJ, 1995.
3. *J. D. Gibson, *Principles of Digital and Analog Communications,* 2nd ed., Macmillan, New York, 1993.
4. *R. E. Ziemer and R. L. Peterson, *Digital Communications and Spread Spectrum Systems,* Macmillan, New York, 1985.

[21]References marked with an asterisk are recommended as supplementary reading.

5. *W. Webb and L. Hanzo, *Quadrature Amplitude Modulation,* Pentech/IEEE Press, Piscataway, NJ, 1994.
6. A. Lender, The duobinary technique for high speed data transmission, *IEEE Trans. Commun. Technol.,* vol. 82, 1963, pp. 214–218.
7. E. R. Kretzmer, Generalization of a technique for binary data communication, *IEEE Trans. Commun. Technol.,* February 1966, pp. 67–68,
8. S. Pasupathy, Correlative coding: A bandwidth-efficient signaling scheme, *IEEE Commun. Mag.,* vol. 15, July 1977, pp. 4–11.
9. A. Said and J. B. Anderson, Bandwidth-efficient coded modulation with optimized linear partial response signals, *IEEE Trans. Information Theory,* vol. IT-44, March 1998, pp. 701–713.
10. *J. B. Anderson and A. Svensson, *Coded Modulation Systems,* Kluwer-Plenum, New York, 2003.
11. F. Xiong, *Digital Modulation Techniques,* Artech House, Boston, 2000.
12. J. B. Anderson, T. Aulin, and C.-E. Sundberg, *Digital Phase Modulation,* Plenum, New York, 1986.
13. S. Pasupathy, Minimum shift keying: A spectrally efficient modulation, *IEEE Commun. Mag.,* vol. 17, July 1979, pp. 14–22.
14. M. J. Wilson, ed., *ARRL Handbook,* American Radio Relay League, Newington, CT, 1988 and other years.
15. T. J. Baker, Asymptotic behavior of digital FM spectra, *IEEE Trans. Commun.,* vol. COM-22, 1974, pp. 1585–1594.
16. K. Murota and K. Hirade, GMSK modulation for digital mobile radio telephony, *IEEE Trans. Commun.,* vol. COM-29, 1981, pp. 1044–1050.
17. R. Steele, ed., *Mobile Radio Communications,* Pentech/IEEE Press, Piscataway, NJ, 1992.
18. H. L. Van Trees, *Detection, Estimation, and Modulation Theory,* Vol. I, John Wiley & Sons, New York, 1968.
19. A. Papoulis, *Probability, Random Variables, and Stochastic Processes,* 2nd ed., McGraw-Hill, New York, 1984.
20. V. K. Bhargava, D. Haccoun, R. Matyas, and P. Nuspl, *Digital Communications by Satellite,* John Wiley & Sons, New York, 1981.

APPENDIX 3A: BASEBAND REPRESENTATION OF BANDPASS FILTERS

Just as bandpass signals can be expressed by baseband I and Q components, so also can bandpass filters. To do so is a conceptual convenience because it reduces signaling and filtering to the simple low-pass case of Chapter 2. Signal processors are often much easier to construct at baseband as well. This appendix shows how to make the conversions back and forth between the baseband and bandpass implementations.

First, we represent a bandpass filter response. We have seen in Section 3.1 that any bandpass signal may be written as

$$x(t) = x^I(t)\cos\omega_o t - x^Q(t)\sin\omega_o t \tag{3A-1}$$

in which $x^I(t)$ and $x^Q(t)$ denote the in-phase and quadrature baseband signals. In the same way, we may represent the impulse response of a bandpass filter as

$$h(t) = h^I(t)\cos\omega_o t - h^Q(t)\sin\omega_o t \tag{3A-2}$$

since it, too, is a bandpass signal. By bandpass signal is meant a signal whose spectrum is nonzero in an RF bandwidth B hertz that is centered at a location f_0 hertz, where $\omega_0 = 2\pi f_0$ and $f_0 > \frac{1}{2}B$; this is portrayed in Fig. 3.38. Time-limited, physical signals cannot satisfy this condition exactly, but it is a virtually perfect approximation of reality.

As with any filtering process, the filter output is $y = h * x$, the convolution of $h(t)$ and $x(t)$. We can state our main result as follows.

Theorem 3A-1. For bandpass x and h, the convolution y may be written as

$$y(t) = \tfrac{1}{2}(x^I * h^I - x^Q * h^Q)\cos\omega_o t - \tfrac{1}{2}(x^I * h^Q + x^Q * h^I)\sin\omega_o t \tag{3A-3}$$

where x^I, x^Q, h^I, and h^Q are defined in (3A-1) and (3A-2).

Proof. The proof needs three Fourier transform pairs, which we state first:

$$g(t)\cos 2\pi f_0 \leftrightarrow \tfrac{1}{2}[G(f-f_0) + G(f+f_0)] \tag{3A-4a}$$

$$g(t)\sin 2\pi f_0 \leftrightarrow \frac{1}{2j}[G(f-f_0) - G(f+f_0)] \tag{3A-4b}$$

$$g(t)\sin 2\pi f_0 \cos 2\pi f_0 \leftrightarrow \frac{1}{4j}[G(f-2f_0) - G(f+2f_0)] \tag{3A-4c}$$

To begin, take the formal convolution of x and h, to get

$$\begin{aligned} y = x * h = {} & x^I\cos\omega_o t * h^I\cos\omega_o t + x^Q\sin\omega_o t * h^Q\sin\omega_o t \\ & - x^Q\sin\omega_o t * h^I\cos\omega_o t - x^I\cos\omega_o t * h^Q\sin\omega_o t \end{aligned} \tag{3A-5}$$

The rest of the proof works by examining each term in turn in (3A-5). Taking the second term as an example, find its Fourier transform to be

$$\frac{1}{2j}[X^Q(f-f_0) - X^Q(f+f_0)]\frac{1}{2j}[H^Q(f-f_0) - H^Q(f+f_0)] \tag{3A-6}$$

Here, the transform pair (3A-4b) is used. Because of the bandpass condition on x and h, (3A-6) reduces to

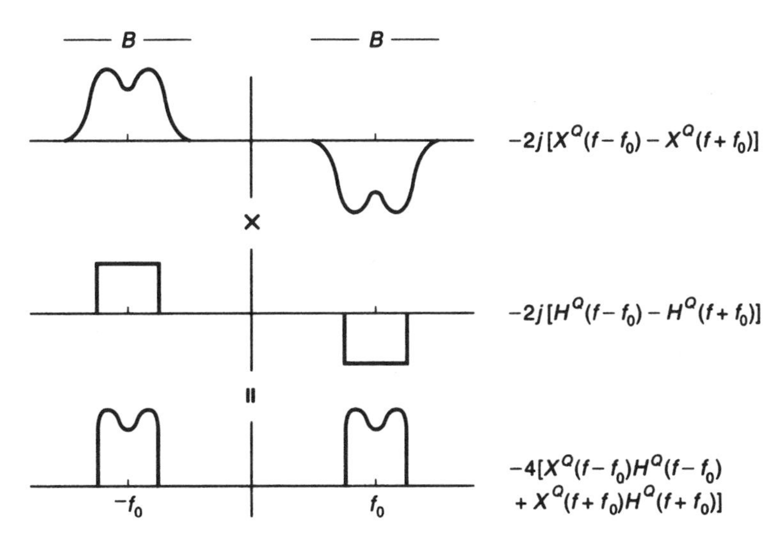

Figure 3.38 Illustration of critical step in Theorem 3A-1. Here H is an ideal bandpass filter.

$$\left(\frac{1}{2j}\right)^2 [X^Q(f-f_0)H^Q(f-f_0) + X^Q(f+f_0)H^Q(f+f_0)] \tag{3A-7}$$

Figure 3.38 illustrates how this works. From transform pair (3A-4a) with $g = x^Q * h^Q$, get that (3A-7) has inverse transform

$$-\tfrac{1}{2}x^Q * h^Q \cos \omega_0 t$$

By the same sort of argument, the first term in (3A-5) becomes $\frac{1}{2}x^I * h^I \cos \omega_0 t$ through two uses of transform pair (3A-4a). By using pair (3A-4c) and then (3A-4b), one converts the third and fourth terms in (3A-5) to $-\frac{1}{2}x^Q * h^I \sin \omega_0 t$ and $-\frac{1}{2}x^I * h^Q \sin \omega_0 t$. The total of all four terms in (3A-3), which proves the theorem.

The equivalence embodied in (3A-3) is diagrammed in Fig. 3.32. It says that I and Q components of the bandpass output $y(t)$ are

$$y^I = \tfrac{1}{2}(x^I * h^I - x^Q * h^Q), \qquad y^Q = \tfrac{1}{2}(x^I * h^Q + x^Q * h^I) \tag{3A-8}$$

with

$$y(t) = y^I(t) \cos \omega_0 t - y^Q(t) \sin \omega_0 t$$

An immediate consequence of (3A-8) is that if $h^Q(t)$ is not zero, there is *crosstalk* between x^I and x^Q; that is, some of x^I appears in y^Q and some of x^Q appears in y^I. When $h^Q(t)$ is zero, the filter response has no quadrature component. Transforming the response (3A-2) then gives just

$$H(f) = \tfrac{1}{2}[H^I(f - f_0) + H^I(f + f_0)] \tag{3A-9}$$

Since $h^I(t)$ is a real, low-pass signal, its transform $H^I(f)$ has symmetric absolute value about $f = 0$ and antisymmetric phase, and consequently the same symmetry exists about f_0 and $-f_0$ in $H(f)$. The low-pass $H^I(f)$ is simply moved out to $\pm f_0$.

It remains to show how to obtain the components $h^I(t)$ and $h^Q(t)$ from a known $h(t)$. This is important because one may want to implement a desired $h(t)$ at baseband rather than bandpass, or one may need to quantify signal crosstalk via (3A-8). A signal processor that directly produces h^I and h^Q from h is the standard quadrature demodulator circuit in Fig. 3.9. After the double-balanced mixer appear the two signals

$$\begin{aligned} 2h(t) \cos \omega_0 t &= h^I(t) + h^I(t) \cos 2\omega_0 t - h^Q(t) \sin 2\omega_0 t \\ -2h(t) \sin \omega_0 t &= h^Q(t) - h^Q(t) \cos 2\omega_0 t + h^I(t) \sin 2\omega_0 t \end{aligned} \tag{3A-10}$$

if the input to the mixer is simply (3A-2). After the low-pass filters, it is clear that h^I and h^Q remain on the top and bottom rails. To calculate these without reference to the demodulator, define $H'(f)$ to be the right, positive frequency half of the spectrum $H(f)$. Then

$$\begin{aligned} H^I(f) &= H'(f + f_0) + H'(-(f - f_0)) \\ H^Q(f) &= j[H'(f + f_0) - H'(-(f - f_0))] \end{aligned} \tag{3A-11}$$

The calculation is illustrated in Fig. 3.39.

Equation (3A-11) shows incidentally that an $H(f)$ symmetrical about f_0 and $-f_0$ implies no crosstalk, since $H^Q(f)$ is then zero. This demonstrates that crosstalk occurs *if and only if* the bandpass filter spectrum is asymmetrical.

An alternate way to characterize bandpass filter responses is by complex notation. In this method, x, h, and y are expressed by the complex numbers

$$\tilde{x}(t) = x^I(t) + jx^Q(t), \qquad \tilde{h}(t) = h^I(t) + jh^Q(t), \qquad \tilde{y}(t) = h^I(t) + jy^Q(t)$$

in which the real parts for each are the in-phase components and the imaginary parts are the quadrature components. The actual bandpass signals are given by

$$x(t) = \mathrm{Re}\{\tilde{x}(t) \exp(j2\pi f_0 t)\}$$

and similarly for $h(t)$ and $y(t)$. In this notation, filtering is expressed by

$$\tilde{y}(t) = \tfrac{1}{2}\tilde{h}(t) * \tilde{x}(t)$$

that is, the filter output is directly the convolution of $\tilde{h}$ and $\tilde{x}$. It needs to be stressed, as it was in Section 3.4, that the complex notation here is only an alternate method of bookkeeping. All the results in this appendix follow through in either system. Many texts, including Proakis [1], carry out the details.

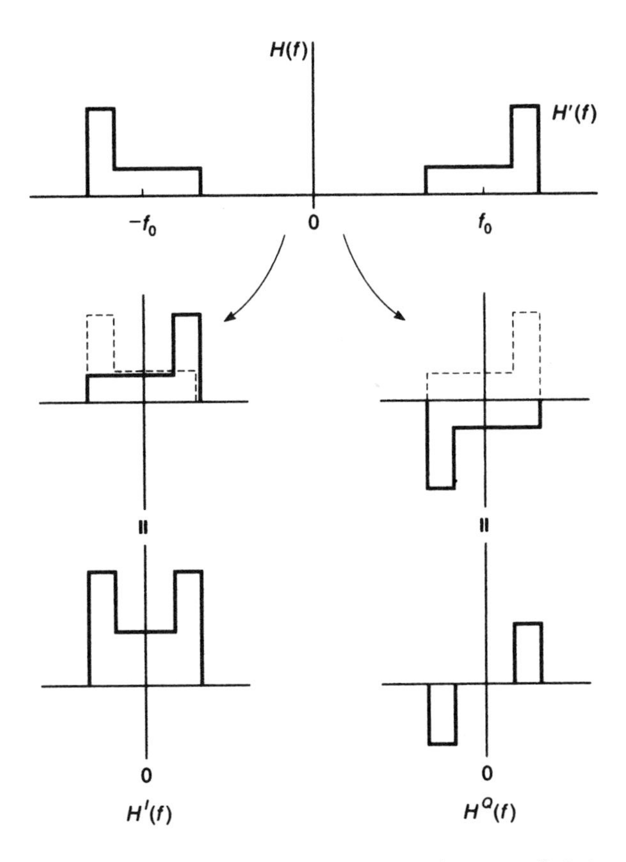

Figure 3.39 Calculation of baseband filter components $H^I(f)$ and $H^Q(f)$ from given bandpass filter $H(f)$ for asymmetrical case.

APPENDIX 3B: BANDPASS NOISE

Appendix 3A extended the notion of baseband I and Q components to bandpass filtering, and we now extend the notion to noise in bandpass systems. Ordinarily, white Gaussian noise added to a bandpass signal translates to white Gaussian noise components that add independently to the signal's baseband components. But in some circumstances, the noise components end up being not white or not independent.

Just as with any other bandpass signal, an outcome $\eta(t)$ of a bandpass noise process may be expressed as

$$\eta(t) = \eta^I(t) \cos \omega_0 t - \eta^Q(t) \sin \omega_0 t, \qquad \omega_0 = 2\pi f_0 \tag{3B-1}$$

in which $\eta^I(t)$ is the in-phase baseband component and $\eta^Q(t)$ is the quadrature component and both of these are *low-pass*. We collect results about the probability distribution of these components in the following theorem.

Theorem 3B-1. Let $\eta(t)$ be a wide-sense stationary (WSS) process with zero mean, whose power spectral density $S_{\eta\eta}(f)$ is zero outside an RF bandwidth B hertz, where $f_0 > \frac{1}{2}B$. Let the notation $R_{xx}(\tau)$ denote the autocorrelation of x and $R_{xy}(\tau)$ denote the cross-correlation of the processes x and y. Then

1. $\eta(t)$ and $\eta^Q(t)$ are WSS with zero mean
2. $$R_{IQ}(\tau) = -R_{QI}(\tau) \tag{3B-2}$$
3. $$R_{II}(\tau) = R_{QQ}(\tau) \tag{3B-3}$$
4. $$R_{\eta\eta}(\tau) = R_{II}(\tau) \cos \omega_0\tau + R_{QI}(\tau) \sin \omega_0\tau \tag{3B-4}$$
5. $$S_{\eta\eta}(f) = \frac{1}{2}[S_{II}(f-f_0) + S_{II}(f+f_0)] + \frac{1}{2j}[S_{QI}(f-f_0) - S_{QI}(f+f_0)] \tag{3B-5}$$

Proof. We will summarize the steps in proving these results and leave the details to a good stochastic processes text such as Papoulis [19]. Result (1) follows from the fact that $\eta(t)$ is a WSS process and $\eta^I(t)$ and $\eta^Q(t)$ can be obtained from it via the demodulation process at the end of Appendix 3A. Results 2–4 all follow from the stationarity of $\eta(t)$. Specifically, we can write $R_{\eta\eta}(\tau)$, which we know to be a function of τ only, as

$$\begin{aligned}\mathcal{E}[\eta(t)\, \eta(t+\tau)] = \mathcal{E}\{&[\eta^I(t) \cos \omega_0 t - \eta^Q(t) \sin \omega_0 t] \\ &\times [\eta^I(t+\tau) \cos \omega_0(t+\tau) - \eta^Q(t+\tau) \sin \omega_0(t+\tau)]\}\end{aligned}$$

From the stationarity of $\eta^I(t)$ and $\eta^Q(t)$, the right-hand side reduces to

$$\begin{aligned}&R_{II}(\tau) \cos \omega_0 t \cos \omega_0(t+\tau) + R_{QQ}(\tau) \sin \omega_0 t \sin \omega_0(t+\tau) \\ &- R_{QI}(\tau) \sin \omega_0 t \cos \omega_0(t+\tau) - R_{IQ}(\tau) \cos \omega_0 t \sin \omega_0(t+\tau)\end{aligned}$$

By means of trigonometric identities, this reduces to

$$\begin{aligned}&\tfrac{1}{2}[R_{II}(\tau) + R_{QQ}(\tau)] \cos \omega_0\tau - \tfrac{1}{2}[R_{IQ}(\tau) - R_{QI}(\tau)] \sin \omega_0\tau \\ &+ \tfrac{1}{2}[R_{II}(\tau) - R_{QQ}(\tau)] \cos \omega_0(2t+\tau) - \tfrac{1}{2}[R_{QI}(\tau) + R_{IQ}(\tau)] \sin \omega_0(2t+\tau)\end{aligned} \tag{3B-6}$$

Since (3B-6) can be a function only of τ, the last two lines must not depend on t, and consequently (3B-2) and (3B-3) must hold. This leaves only the first line, which, together with (3B-2) and (3B-3), demonstrates (3B-4). The power spectral density of $\eta(t)$ is the Fourier transform of (3B-4), and this is simply (3B-5). This completes the proof.

When $R_{IQ}(\tau) = 0$ for all τ, the baseband components of the bandpass noise are uncorrelated. The next theorem shows that this happens precisely when the bandpass power spectral density is symmetrical about f_0 and $-f_0$. (This symmetry notion is explained in Appendix 3A.) Recall that power spectral densities are always real and nonnegative.

Theorem 3B-2. The cross correlation $R_{QI}(\tau) = 0$, all τ, if and only if $S_{\eta\eta}(f) > 0$ is symmetrical about f_0 and $-f_0$.

Proof. If $R_{QI}(\tau) = 0$, then the cross-spectral density $S_{QI}(f) = 0$ for all f, since these are transform pairs. From (3B-5), it follows that

$$S_{\eta\eta}(f) = \tfrac{1}{2}[S_{II}(f-f_0) + S_{II}(f+f_0)] \tag{3B-7}$$

Here, $S_{II}(f)$ is a power spectral density and is hence symmetrical about zero and non-negative; the expression in (3B-7) is therefore symmetric about f_0 and $-f_0$ and non-negative.

To prove the reverse implication, assume that $R_{QI}(\tau) \neq 0$. Because $R_{QI}(\tau)$ is real, $S^*_{QI}(f) = S_{QI}(-f)$. If $S_{QI}(f)$ has a real part, it becomes imaginary when its frequency shifts $S_{QI}(f-f_0)$ and $S_{QI}(f+f_0)$ are placed in the formula (3B-5). Then $S_{\eta\eta}(f)$ must acquire an imaginary part if S_{QI} satisfies the bandpass condition $f_0 > B$. But $S_{\eta\eta}(f)$ is then not a power spectral density, and consequently $S_{QI}(f)$ must be pure imaginary. Its frequency shifts $S_{QI}(f-f_0)$ and $S_{QI}(f+f_0)$ are therefore antisymmetric about f_0 and $-f_0$. When inserted in (3B-5), their contribution becomes real and they destroy the symmetry already present in $S_{II}(f-f_0)$ and $S_{II}(f+f_0)$. Thus $S_{\eta\eta}(f)$ is not symmetrical about f_0 and $-f_0$. This completes the proof.

If a WSS process $y(t)$ is linear filtered by $H(f)$ in order to produce $\eta(t)$, the power spectral densities of $y(t)$ and $\eta(t)$ always satisfy

$$S_{\eta\eta}(f) = |H(f)|^2\, S_{yy}(f) \tag{3B-8}$$

An important case occurs when $y(t)$ is white, so that $S_{\eta\eta}(f)$ becomes the power spectrum of $H(f)$. In particular, a bandpass H symmetrical about f_0 and $-f_0$ produces noise with the same PSD symmetry whose I and Q components by Theorem 3B-2 are therefore uncorrelated. Conversely, asymmetrical filters produce correlated components. Just as asymmetrical filters introduce crosstalk into I and Q signals, they introduce correlation into noise.

Just as with other bandpass signals, the analysis of bandpass noise may be carried out with complex numbers with the same outcomes; see Refs. 1 and 19.

When bandpass noise is Gaussian, then so are the baseband components and linear filter outputs. Uncorrelated Gaussians are independent. In the common practical case of white Gaussian background noise, therefore, symmetrical bandpass filters produce independent baseband noise components.

White Gaussian noise is a mathematical abstraction that closely models the effect of noise in many practical receivers. Truly white noise cannot be expressed in baseband components, but this noise cannot really exist in any case. An accurate and effective model is to imagine that AWGN with spectral density $\frac{1}{2}N_0$ has been prefiltered by a brick wall bandpass filter whose RF bandwidth is B hertz centered at f_0. Here, B is wide enough so that the desired signal can pass through without significant distortion. We can then express both the signal and the noise in baseband components. The signal components are as in (3A-1). The noise components are distributed as specified in Theorem 3B-1: $\eta^I(t)$ and $\eta^Q(t)$ are zero-mean WSS processes and from (3B-5), $S_{II}(f)$ and $S_{QQ}(f)$ must satisfy

$$S_{II}(f) = S_{QQ}(f) = \begin{cases} N_0, & -\frac{1}{2}B \leq f \leq \frac{1}{2}B \\ 0, & \text{otherwise} \end{cases} \tag{3B-9}$$

Because $\eta^I(t)$ and $\eta^Q(t)$ are filtered Gaussians, they themselves must be Gaussian. In summary, a baseband processor at bandwidth $[-\frac{1}{2}B, \frac{1}{2}B]$ sees AWGN with density N_0, twice that of a bandpass processor. This bandpass/baseband noise equivalence is expressed in Fig. 3.40.

Figure 3.41 shows examples of Gaussian bandpass noise. Only the ratio B/f_0 matters in these pictures. In the top example $B/f_0 = 1$, while in the second it is 0.25, which leads to a strongly sinusoidal waveform. It can be shown that the amplitude of these sinusoids at a predetermined time t is Rayleigh distributed and their phase is uniformly distributed [19]. The actual baseband components of the narrowband waveform appear at the bottom of the figure; these change very slowly.

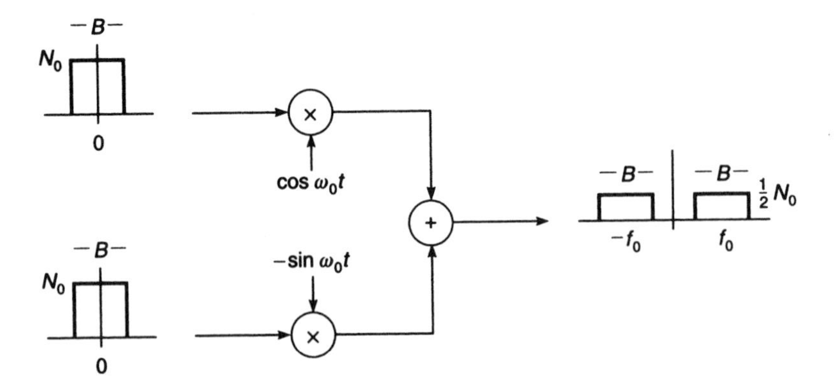

Figure 3.40 Conversion between Gaussian baseband noise components with power spectral density N_0 and bandpass Gaussian noise with PSD $\frac{1}{2}N_0$.

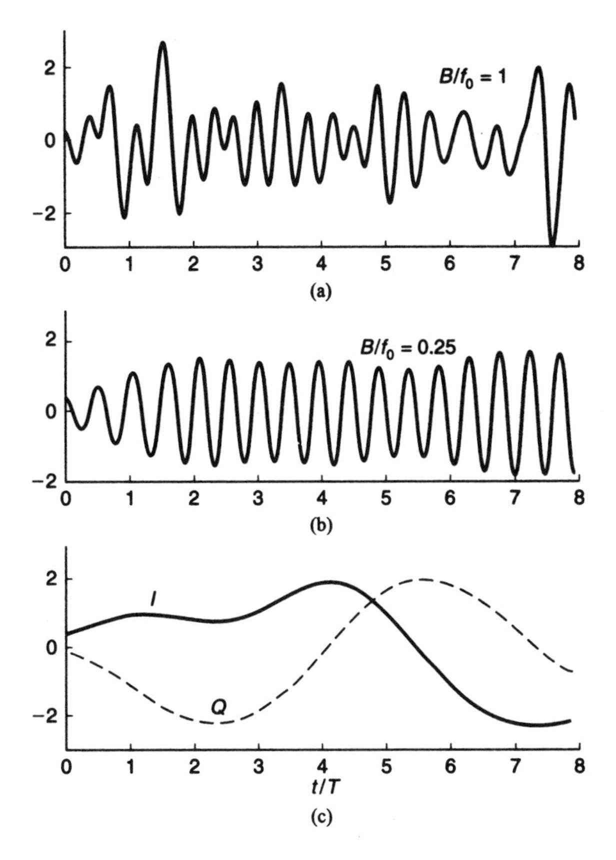

Figure 3.41 Examples of bandpass noise: **(a)** RF bandwidth 2 Hz centered on $f_0 = 2$ Hz; **(b)** RF bandwidth 0.5 Hz centered on 2 Hz; **(c)** baseband compontents that produced middle waveform. Butterworth six-pole filter. Passband sample variances normalized to 1.

The autocorrelation function of the Gaussian process in (3B-9) is

$$R_{II}(\tau) = R_{QQ}(\tau) = \frac{N_0 \sin \pi B \tau}{\pi \tau} \tag{3B-10}$$

Observe that a sequence of noise samples taken at the rate twice $\frac{1}{2}B$ hertz are uncorrelated, since $R(k/B) = 0$ for all integer k. The samples are thus independent as well.

APPENDIX 3C: DISCRETE-TIME IMPLEMENTATION AND SIMULATION

In the present day, all but high-speed modems are implemented with programmable processors. It often happens that we wish to test a communication link by computer simulation. In either case, we need a discrete-time implementation that mimics the continuous-time processing in Chapter 3, and we present three alternatives in this appendix. Before starting, it is useful to distinguish the ideas of simulation and implementation. When a system is simulated, some large part of it mimics functions but does not actually carry them out and consequently does not test for imperfections in those parts. The first alternative below illustrates this well, because it contains no actual pulses, no channel, and no matched filters. The last alternative simply constructs a modem and simulates nothing, except possibly the channel noise. The second alternative lies somewhere in between.

(i) The Symbol AWGN Model. This method replaces pulse train generation, the channel, and the matched filter/sampler with a memoryless AWGN channel model invoked once per symbol. The method is the same with bandpass or low-pass signals and has been given already in Fig. 2.33. It reduces the system to transmission symbols that scale a perfect baseband orthogonal pulse sequence, which passes through an AWGN channel with noise density $\frac{1}{2}N_0$ to a perfectly synchronized matched filter. The overall effect, as shown in Section 2.6, is that an independent zero-mean Gaussian variate with variance $\frac{1}{2}N_0$ is added to each transmission symbol value. Later processing may work on the noisy symbol sequence—as, for example, in PRS detection.

Method (i) cannot simulate any but an AWGN channel, nor can it test any of the impairments in Section 3.8, other than the noise. It cannot easily test nonorthogonal pulses or modulation schemes that are not linear. Its main use is to simulate modems that are part of a larger system, one that might be a network or have error-correcting coding or a particular data source such as speech.

(ii) The Signal Space Method. In this approach, the actual transmitted waveforms, be they pulses or nonlinear modulation signals, baseband or bandpass, are realized as signal space vectors $\boldsymbol{s}_i$, in principle exactly as in Sections 2.5–2.6. An AWGN channel is again assumed, and its effect is simulated by adding a zero-mean variate with variance $\frac{1}{2}N_0$ to each component in the signal vector. The ML receiver can be implemented in one of two ways. Remaining in the vector space, the receiver can compute distances from signal and signal-plus-noise vector components, as in Eq. (2.5-19), and find the closest signals. Or the receiver can re-create the physical received signal from its noisy vector components and apply that to a matched filter implementation. Many other non-ML receivers are imaginable.

Method (ii) works with any signal set and can test synchronization and many kinds of waveform impairments. It demands a Gaussian channel, but in principle the noise need not be white. The advantage of the signal space method is the simple and direct way that it mimics the effect of the Gaussian channel. The disadvantage is that a Gram–Schmidt process is required to find the basis signals, and inner products must be performed to find the signal vector components. The basis signals can be very complicated.

(iii) Full Discrete-Time Implementation. We will assume in what follows that a complete discrete-time transmitter–receiver set is desired. Method (iii) is not a simulation at all but is a direct implementation of all desired parts of a modem. Signals are discrete-time samples of the real analog signals, and filters are realized in discrete time. Potentially any signal set can be used and any impairment tested.

A major decision in a discrete-time modem is the sampling rate. Since a carrier signal would have to be sampled at twice the carrier frequency, and preferably higher, bandpass processing is relatively seldom implemented in discrete time. An exception is telephone line modems, whose carrier frequency is nearly the same as their baseband bandwidth. More commonly, bandpass and baseband signals are converted back and forth by some means, and the discrete-time implementation deals with baseband components.

A rule of thumb for sampling rate is that it should be high in the transmitter and low in the receiver. The transmitter must create pure, clean signals, and this requires a high sampling rate and high precision. Another rule of thumb states that the spectral side lobes of a signal generated by digital processing cannot lie lower than 6 dB down for each bit of precision. In order for signals not to interfere with neighboring channels, considerable precision may be required. (An alternative is to append a sharp analog filter after the transmitter.) In the receiver, precision is much less important because distance measurement can be rough. A sampling rate of $10/T$ is commonly used, where T is the symbol rate, but rates as low as $2/T$–$4/T$ have been used without affecting the detector error rate.

A second challenge in setting up the implementation is modeling the white Gaussian channel. Several derivations exist, but the one illustrated in Fig. 3.42 is an easy one to understand. Since the sampling rate is some f_s, we imagine that AWGN of density $\frac{1}{2}N_0$ has first been applied to an ideal low-pass filter whose bandwidth is $[-\frac{1}{2}f_s, \frac{1}{2}f_s]$. The noise waveform is sampled at f_s hertz. Because of the filter, there is

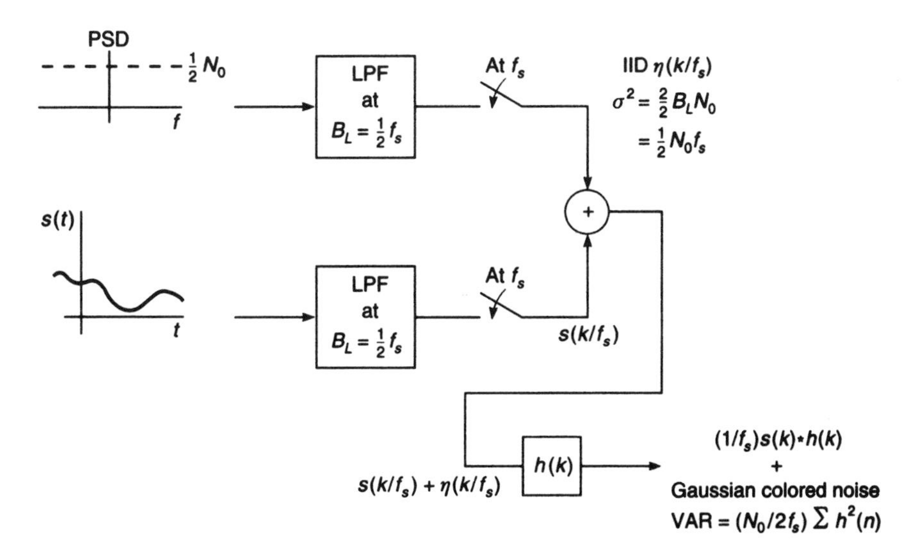

Figure 3.42 Modeling of Gaussian noise in discrete-time modem implementation.

no aliasing, and the filtered noise samples must be Gaussian and zero mean and have variance $\sigma^2 = (\frac{1}{2}N_0)(\frac{1}{2}f_s)2 = \frac{1}{2}N_0 f_s$; from Eq. (3B-10), these variates are also independent. The signal $s(t)$ is also sampled at f_s, where now the signal bandwidth must be less than $\frac{1}{2}f_s$. To summarize, an analog AWGN channel with $\frac{1}{2}N_0$ may be replaced by signal samples, to which are added independent Gaussians with variance $\sigma^2 = \frac{1}{2}N_0 f_s$.

Consider a long sample sequence $s(1), \ldots, s(N)$ with sample variance var(s). The average energy E_s in $s(t)$ is

$$E_s \approx \frac{T\sum_{k=1}^{N} s^2(k)(1/f_s)}{N/f_s} = \frac{T}{N}\sum_{k=1}^{N} s^2(k) \tag{3C-1}$$

where T is the symbol time. This is T var(s), if we assume the samples $s(k)$ have zero mean. Thus, for M-ary signaling and noise samples with σ

$$\frac{E_b}{N_0} = \frac{T\,\mathrm{var}(s)}{(2\sigma^2/f_s)\log_2 M} \tag{3C-2}$$

Observe that a reduction of T by, for example, 10 requires an increase in $s(t)$ by $\sqrt{10}$ if E_b/N_0 is to retain the same value.

If the next element is a filter $h(t)$, let its impulse response sequence, sampled at f_s, be $h(0), h(1), \ldots$. The analog convolution $s(t) * h(t)$ is approximated at $t = n/f_s$ by

$$s * h \text{ at } t = \int s(t-u)h(u)\,du \approx y(n) \triangleq \frac{1}{f_s}\sum_k s(n-k)h(k) \tag{3C-3}$$

Filtering the noise sequence produces colored variates with variance

$$\frac{N_0 f_s}{2}\left[\sum h^2(k)\right]\frac{1}{f_s^2} = \frac{N_0}{2f_s}\sum h^2(k) \tag{3C-4}$$

For example, $h(k)$ could be a matched filter, and then the actual SNR at the output at time n/f_s would be $y^2(n)$ divided by (3C-4).

CHAPTER 4

SYNCHRONIZATION

It is all too easy to forget that digital communication cannot work without synchronization. In reality a digital system depends on an entire chain of synchronizers. In earlier chapters we have seen that all digital transmission, whether baseband or carrier, hangs on a framework of symbol times T, $2T$, $3T$, . . . , which must be set up. This kind of synchronization is called *symbol timing*. With carrier modulation, an even more basic synchronization is needed, this time to the sinusoidal carrier itself. Without this *carrier* synchronization, the I and Q signal orthogonal basis cannot be set up, ML detection is impossible, and the next step in the chain, the symbol timing, usually cannot be performed. Beyond symbol timing, the synchronization chain extends forward in a series of steps we can lump together into *frame,* or *word* timing. In PCM, for example, data bits are formed into PCM words, words into frames, and frames may be multiplexed into superframes. All of this structure requires demarcation.

Synchronization is not a new subject. Nineteenth-century AC power generators were synchronized with the equivalent of the first-order loop we will discuss. Synchronization is also not a simple subject, and it requires some different ways of thinking, but it is based on distinct principles. We will begin with these and then turn directly to practical circuits for carrier synchronization in Section 4.2. All of these depend on the same basic block, the phase-lock loop, the analysis of which is the subject of Sections 4.3 and 4.4. With that mastered, we extend the carrier synchronization ideas to symbol timing and frame timing in the final sections.

4.1 SOME SYNCHRONIZER PRINCIPLES

Before getting into the details of synchronizer circuits, we need to think about what they need to do and what the challenges they face are. For now we will concentrate on carrier phase synchronization. Here are some basic principles of synchronization.

The Product of a Synchronizer Is a Spectral Line. A phase synchronizer produces a pure sinusoid (the carrier or a relative of it), and symbol and frame synchronizers produce the equivalent of a ticking clock; ideally, the clock is without artifact and distortion. It is clear from our study of carrier modulation spectra that often they contain no spectral lines or if they do, other signal components are present as well. A phase synchronizer can be thought of as creating a spectral line, or its time equivalent, a pure sinusoid, where none existed before or as filtering other parts of a signal away, so that only a clean spectral line remains. Symbol and frame synchronizers have an equivalent task.

Disturbances in the Clock or Carrier Phase. There are many of these, and we can break them down into those that must be tracked, because the phase and timing of the transmission move with them, and those that must be ignored, because they are unrelated. It is useful to break these changes down into step changes and linear, or "ramp," changes in the phase. A mobile radio transmission to a moving vehicle provides a good example of steps and ramps. As the vehicle moves out from behind a building, it encounters a new dominant path to the transmitter, and the signal phase changes nearly stepwise to the phase of the new path. When the vehicle changes speed, a phase ramp occurs because the Doppler effect shifts the apparent transmitter frequency. Some concrete mobile radio examples appear in Section 4.5.

It is a mistake to attribute all carrier phase shifts to motion and to poor circuit design. All oscillators produce at least a little phase noise—that is, a little random jiggling in their generated phase. When the noise is in the transmitter, it needs to be tracked; when it is in the receiver, it needs to be ignored. But even when all the equipment is perfect and unmoving, phase shift can still occur. In a classic calculation it can be shown that the sun rising on a buried coaxial cable between New York and Chicago increases the phase of a 300-Mb/s data transmission enough to produce 200 extra bits at the receiver, even with perfect oscillators at both ends. The culprits here are the sun and the speed of propagation in a warming dielectric. The only option is to adjust to the phase change.

Disturbances that are unrelated to the signal phase must be ignored by the synchronizer. The most important of these is the noise that accompanies the signal. Another kind of noise, called data noise, traces back to the signal itself and is a byproduct of the signal processing done in the synchronizer. We will take it up in Section 4.4. A third kind of disturbance is momentary loss of signal, as occurs in a fading channel.

Synchronization Takes Signal Energy. Another way of stating this principle is that if synchronization were somehow provided for free, the apparent E_b/N_0 in the transmission would seem to be higher. In many communication links a separate pilot tone or clock provides the phase reference. Everyday examples are telephone channel banks and FM stereo broadcasting. Here the energy consumed is obvious:

It is the energy of the pilot. If the pilot were not needed, the energy could be applied to the data symbols instead. A less obvious example is a link that is *self-synchronizing*. Here there is no pilot and the phase timing references are derived from the data signal itself; as a consequence, they are not perfect. The result is data errors in larger numbers than would be expected from the E_b/N_0 figure alone and a poorer bit error rate curve. To compensate, one must augment E_b/N_0, and here lies the lost signal energy.

The Quantity of Energy Grows with Signal Instability. Unstable links—those with rapid variation in their phase reference—need more signal energy in order to stay synchronized. The heart of a synchronizer is a phase-lock loop structure that tries to track the phase reference while ignoring the unrelated disturbances such as noise. As we will study in Section 4.4, a fundamental trade-off exists in this tracking: The faster the structure responds to changes in the reference, the more it responds to noise. A fast-responding loop will track a step change in the signal phase with good accuracy, but it also responds to the channel noise; if the signal E_b/N_0 drops, as it might during a signal fade, more of what the loop sees is noise and the noise begins to dominate the loop response. A designer has only two basic remedies now: Slow the loop's response or increase the signal E_b/N_0. Only the second alternative maintains a fast-tracking loop.

As a signal's phase becomes more unstable, the dilemma here grows, because faster and faster loop response is needed. It is clear that eventually the signal energy E_b must be augmented. In fact, some combinations of signal energy and instability are simply unusable, because the receiver cannot be synchronized. A mobile cellular radio channel working at a given symbol rate will waver with increasing violence as the vehicle speed grows, until the synchronization fails. The remedies are to slow the data rate, so that the apparent speed of the synchronizer grows, or to improve the transmitter power, so that the noise will not bother a faster-responding loop.

Analog and Digital Loops. Every synchronizer loop design has two basic implementations: a continuous-time one and a discrete-time or "digital" one. The synchronizer is really a control system, either analog or discrete time, that seeks to minimize the error between a local, controlled oscillator and an outside reference oscillator. Most often, the choice between an analog or digital implementation depends on the data rate. With slow data—for example, a cellular telephone or a computer modem—digital implementation based on software or perhaps on a special-purpose chip is generally best. At high speeds, for example, with a satellite modem or a high-capacity optical fiber, only an analog implementation is possible.

The analyses of analog and discrete-time loops differ in several ways. We will begin with the analog one and then extend it in Sections 4.6–4.7 to an analysis for the discrete case.

Acquisition Versus Tracking. There are two modes to synchronizer operation. At first, the synchronizer has only a rough idea of the target's frequency, be it a carrier frequency, a symbol rate, or a frame rate. It needs to sweep its local oscillator over a wide range, a mode of operation called the *acquisition* mode. Once the local oscillator is close to the right frequency, the synchronizer switches to the *tracking* mode and, hopefully, locks to the exact reference phase. If the synchronizer loses track, it returns to the acquisition mode, an act called a resynch. Once back in the tracking mode, it is said to be "synchronized," but despite the static sound of this word, the synchronizer is constantly adapting to a shifting phase reference.

The acquisition mode may employ different circuitry, more suitable to its function, but often it uses the tracking-mode loop, reparameterized so that it is wider ranging. In some systems, the initial phase or timing is provided by the system, so that the acquisition mode consists of identifying this information.

4.2 PHASE SYNCHRONIZER CIRCUITS

The heart of most synchronizers is the phase-lock loop (PLL). The parts of the PLL may be scattered around the synchronizer and may appear as different circuit elements, but they are nonetheless almost always present. We will study three basic carrier phase synchronizers: the raised-power loop, the remodulation loop, and a more complex synchronizer called the Costas loop. Symbol time and frame synchronizers are treated in Section 4.7–4.8.

Two generic versions of the PLL are shown in Figs. 4.1a and 4.1b. The signals themselves appear in Fig. 4.1a, while the signal phases appear in Fig. 4.1b. The PLL consists of a summing junction, called the phase detector, a loop filter that sets the PLL's dynamics, and a local oscillator. By convention, this oscillator is called a voltage-controlled oscillator (VCO), a reference to a particular hardware implemen-

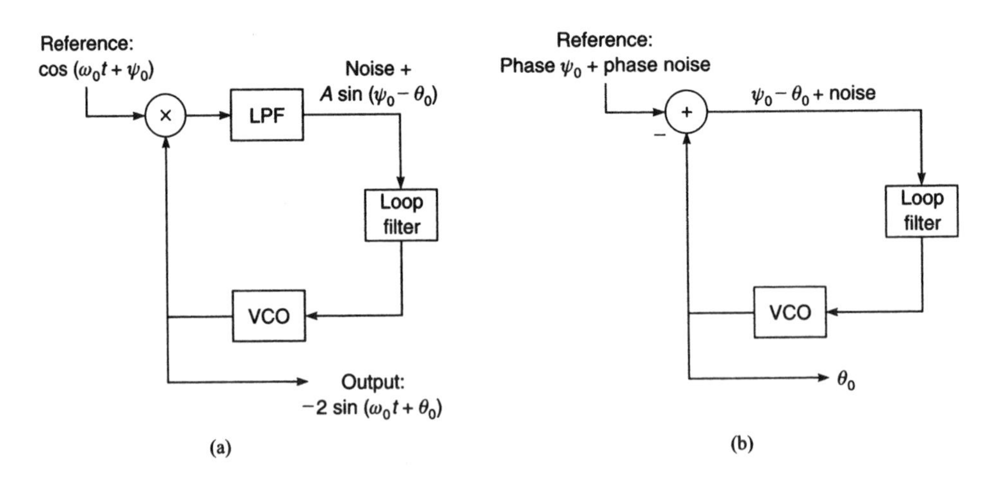

Figure 4.1 Two generic PLLs: **(a)** version with sinusoids and multiplier/LPF phase detector; **(b)** view showing phase offsets only, with a linear phase detector.

tation; the VCO produces an oscillation whose frequency is proportional to the loop filter output. The phase detector output, called variously the phase difference signal or the error signal, is a signal proportional to the difference between the phase at the input to the PLL and the VCO phase. As in Fig. 4.1a, it is not always linearly related to the difference.

The PLL has its origins in control theory, where the input to the PLL is called the *reference* signal. We will borrow this and other control engineering terminology in what follows. The purpose of the PLL is to force the error signal to zero and lock the local phase in the VCO to the external reference phase. The challenge here is that the reference constantly shifts and may be obscured by noise. The output of the PLL is the VCO oscillation.

When the external phase is obscured by noise, the PLL is also attempting an approximate solution to a classic problem of estimation theory: What is the ML estimate of the phase? Rigorous solutions for this problem are given in Ziemer and Peterson [4] and van Trees [8]. Simple PLL circuits similar to Fig. 4.1 generally have performance close to the ML circuit.

A real PLL can put out various oscillating signals, not just sinusoids. We will show them as sinusoidal in what follows. In a low-speed application, however, the oscillations will be digital-electronic square waves, the VCO will be a square-wave generator (most likely, a multivibrator), and the phase detector will count elapsed time between transitions in two square waves. But the dynamics of the PLL are similar.

4.2.1 Raised-Power Loops

The first principle of synchronization is obtain a clean spectral line from a signal that does not contain one. In signal processing, a classic means to this end is to raise the signal to a power, and this is the most common design method for synchronizers. Consider, for example, an NRZ pulse BPSK signal:

$$\begin{aligned} s^2(t) &= \frac{2E_s}{T}\cos^2[\omega_o t + \psi(t)] \\ &= \frac{E_s}{T} + \frac{E_s}{T}\cos[2\omega_o t + 2\psi(t)], \qquad \psi(t) = 0, \pi \qquad (4.2\text{-}1) \\ &= \frac{E_s}{T} + \frac{E_s}{T}\cos(2\omega_o t) \end{aligned}$$

The original signal $s(t)$ is phase-modulated by the random data signal $\psi(t)$ and consequently contains no lines in its spectrum, but it is clear that the squared signal in (4.2-1) contains a DC component and a pure sinusoid at frequency $2\omega_0$. Its modulation has been removed. A frequency division by 2 on the sinusoid will provide the carrier.

The case of pulse-shaped BPSK is somewhat harder. If we represent the modulating pulse train as in Chapter 3 by $I(t)$, the squared signal becomes

$$
\begin{aligned}
s^2(t) &= \frac{2E_s}{T}[I(t)\cos\omega_0 t]^2 \\
&= \frac{E_s}{T}I^2(t) + \frac{E_s}{T}I^2(t)\cos(2\omega_0 t)
\end{aligned}
\tag{4.2-2}
$$

Now the second term is a pure sinusoid at $2\omega_0$ multiplying a nonnegative slowly varying signal. What is the spectrum of this term? On an intuitive level, the term has to be an AM-modulated signal—that is, a spectral impulse plus some sort of bandpass sidebands. Thinking more analytically, we can express a long-duration $I^2(t)$ in the form

$$
I^2(t) = \kappa_{\text{AV}} + \kappa(t) \tag{4.2-3}
$$

in which $\kappa_{\text{AV}} > 0$ is the average value of the nonnegative $I^2(t)$ and $\kappa(t)$ is a signal with average value zero. The transform of (4.2-3) is $2\pi\kappa_{\text{AV}}\delta(\omega) + K(\omega)$. Thus the second term in (4.2-2) contains a spectral line of weight $(E_s/T)\pi\kappa_{\text{AV}}$ at radian frequency $2\omega_0$.

The transform of (4.2-2) contains one other term in the neighborhood of frequency $\pm 2\omega_0$, namely, $\frac{1}{2}(E_s/T)K(\omega \pm 2\omega_0)$, which acts to disturb a circuit that tries to lock to the spectral line. The signal $\kappa(t)$ is called *data noise,* because it is a disturbance and it stems from the data signal itself. We will return to the subject of data noise in Section 4.4.

Figure 4.2 combines Fig. 4.1a and the idea of squaring into a single synchronizer called a *squaring loop*. The original PLL appears in the dashed box. The input to the synchronizer is the modulated signal $A(t)\cos(\omega_0 t + \psi_0 + \psi(t))$, which has a con-

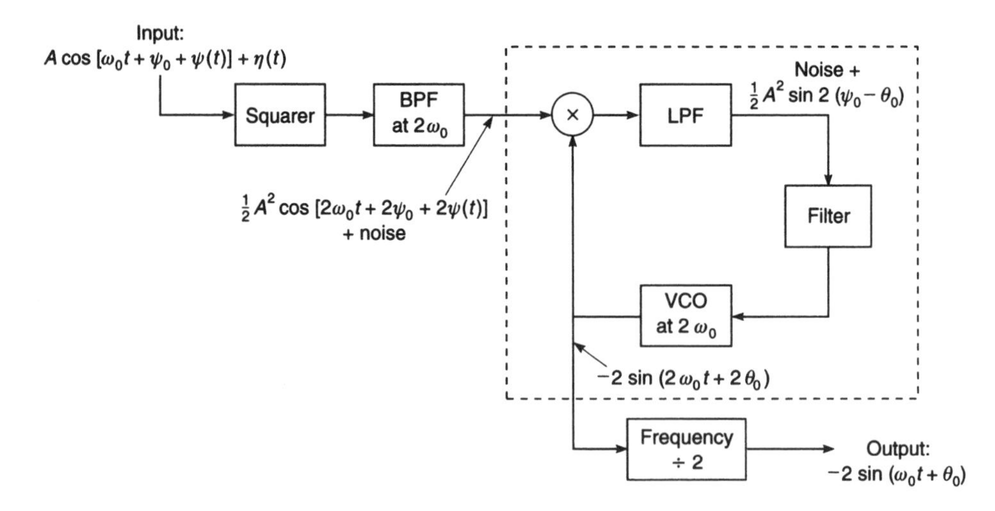

Figure 4.2 Squaring loop synchronizer, based on generic PLL of Fig. 4.1a. Data phase $\psi(t)$ assumed to be π or 0.

stant phase offset ψ_0, and the output is the VCO output after a frequency is divided by 2. Observe that the PLL itself runs at twice the carrier frequency. For now, let us assume that the data-carrying phase is an NRZ pulse taking values 0 or π. Then the squaring loop works as follows. The combination of the squarer and the bandpass filter (BPF) centered at $2\omega_0$ doubles the data-bearing signal phase. Since $2\psi(t)$ is one of 0 or 2π, the result is $A^2 \cos(2\omega_0 t + 2\psi_0)$, in which ψ_0 is the slowly varying offset. Noise survives the squaring process and enters the PLL. In the PLL, a VCO oscillates close to frequency $2\omega_0$ and produces the output $-2 \sin(2\omega_0 t + 2\theta_0)$, which has its own slowly varying phase offset $2\theta_0$. The multiplication of this with the PLL input produces the phase difference signal $\frac{1}{2}A^2 \sin(2\psi_0 - 2\theta_0)$ plus some higher-frequency terms, which are removed by a low-pass filter. [A little trigonometry demonstrates this easily; the $-2 \sin(\cdot)$ form of the VCO output is solely to facilitate the trigonometry.]

The purpose of the synchronizer is to force the VCO offset θ_0 to the signal offset ψ_0, so that the VCO and the signal carrier have identical phases. The loop filter regulates the dynamics of the process and attempts to average out and ignore any noise. A phase difference signal that averages out positive will instruct the VCO to increase phase, which tends to cancel the positive difference. A negative difference has the opposite effect.

The most common of the raised-power synchronizers is the fourth-power loop or "frequency quadrupling" loop, an example of which is shown in Fig. 4.3. The input signal is raised to its fourth power and bandpass-filtered around $4\omega_0$ so that only a sinusoid at four times the phase remains; a PLL running at $4\omega_0$ locks to this phase, and the synchronizer output is the VCO output after frequency is divided by 4. Other than these factors of 4, the circuit is the same as the squaring loop in Fig. 4.2.

Variants of the fourth-power loop are effective with many modulations. With NRZ pulse QPSK, the operation is particularly easy to trace. Since the data-bearing

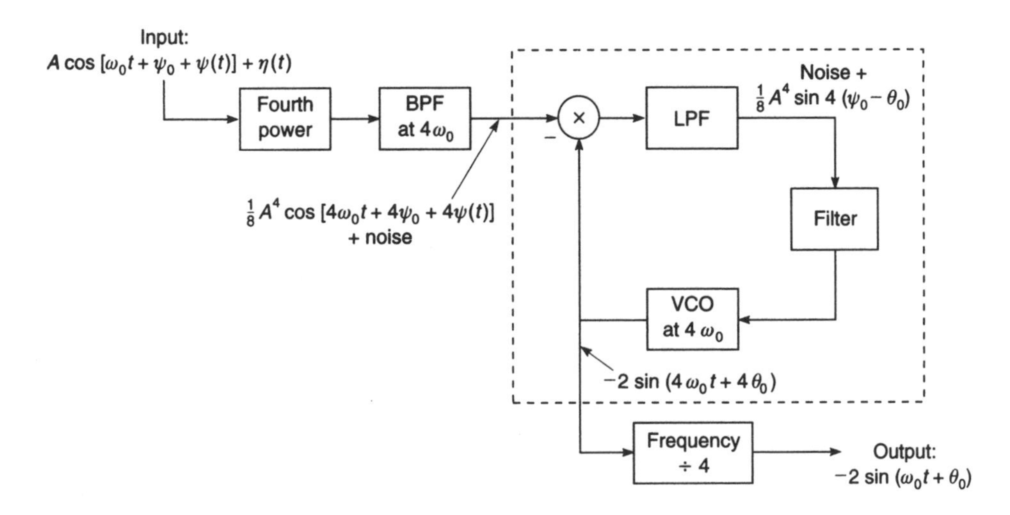

Figure 4.3 Fourth-power synchronizer, again based on Fig. 4.1a.

phase is $\psi(t) = \frac{1}{2}n\pi + \frac{1}{4}\pi$ at all times, n an integer, the data phase is cleaned away and the reference signal for the PLL is $-\frac{1}{8}A^4 \cos(4\omega_0 t + 4\psi_0)$. After the PLL's phase detector, the phase difference signal is $\frac{1}{8}A^4 \sin(4\psi_0 - 4\theta_0)$. The PLL tries to force this signal to zero, thereby attempting to match its phase offset $4\theta_0$ to the $4\psi_0$ offset of the reference signal at its input.

When the modulating pulse is not an NRZ or the modulation is, for instance, 16QAM instead of QPSK, the PLL reference signal is not a steady $\cos(4\omega_0 t + 4\psi_0)$, and the phase difference signal has other components besides $\sin(4\psi_0 - 4\theta_0)$. Nonetheless, the synchronizer still works passably well.

4.2.2 Remodulation and Costas Loops

Although the raised-power strategy will very often produce a spectral line for the PLL, the data noise components become more and more severe as the modulation departs from simple NRZ pulse PSK. With some kinds of signaling (e.g., CPM signals), raising to simple powers may not produce a line at all. In these cases, the remodulation loop may be effective. This loop is also called a decision feedback loop.

Figure 4.4 shows this synchronizer. There is no raising to a power. Instead, the receiver does its best to detect the data symbols, and these are modulated anew with the VCO output; the result is combined with the delayed original signal in the multiplier/low-pass filter to produce a phase error signal, which in turn controls the VCO. Note how elements of the receiver and even a "transmitter" are mixed with PLL components inside the loop. The pure-form PLL is gone, although the struc-

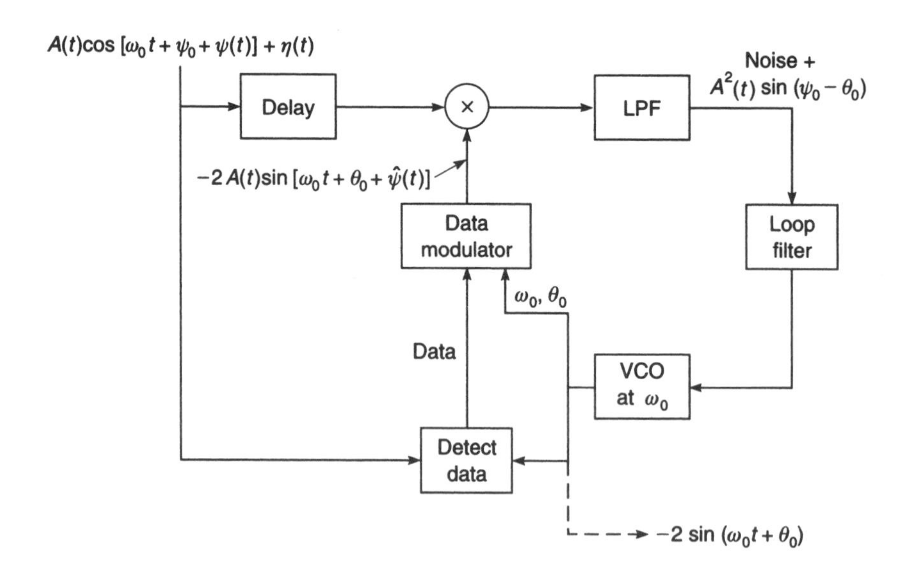

Figure 4.4 Remodulation loop. Data detection now resides inside loop. Signal amplitude has been written as $A(t)$ to emphasize that pulse shaping may vary A.

ture as a whole continues to work as a PLL. The structure's output is $-2 \sin(\omega_0 t + \theta_0)$, as usual, although this signal is probably not needed outside the loop.

While beating a reconstructed signal against a received signal might seem an obvious strategy, the remodulation loop suffers from an important shortcoming: The delay of the receiver lies inside its loop. Delay is the mortal enemy of any control loop. The detector must create at least a one-symbol delay, even if the transmission is simple PSK. By comparison, the raised-power loop instantly compares its VCO signal to the receiver input and adapts, subject only to the designed-in dynamics of the loop. A fast-acting loop can adjust to a phase disturbance in less than a symbol interval, thereby avoiding a detection error in that interval.

A more serious case is when the detector includes a further delay, such as a Viterbi decoder. Studies [1] have shown that the overall receiver generally works better if early, unreliable data are given to the remodulation loop, rather than delayed, reliable data.

The remodulation loop introduces the concept of simultaneous estimation of phase and data in the same circuit. A related idea underlies the Costas loop, first proposed in 1956 [2]. A loop of the Costas type instantaneously feeds back the baseband signals $I(t)$ and $Q(t)$ to a multiplier or similar simple circuit, which produces an error signal that drives the VCO. There is no detection delay in the loop.

Figure 4.5 illustrates the Costas idea with a variant that synchronizes shaped-pulse QPSK and that, outside the loop, estimates the data as well.[1] It is easier to follow the loop operation at first by assuming NRZ pulses. It can be seen that when the loop is reasonably near lock, the signal on the top rail after the LPF is close to the data symbol value a_n^I and the signal on the bottom rail is close to a_n^Q. The limiter box enforces this by ignoring small perturbations in the signal, which could be the opposite rail symbol (if the loop is not locked), or the shaping of the pulse, or simply channel noise. The cleaned-up ± 1 levels a_n^I and a_n^Q feed a network that produces from the received baseband signals a phase difference signal $2 \sin(\psi_0 - \theta_0)$. This signal, working with the loop filter and VCO, effectively mimics the PLL of Fig. 4.1a.

4.2-3 *S*-Curves and Lock Points

The purpose of all these loops is to lock an internal oscillator to the phase of an external oscillator. What forces this lock is the phase difference signal, which acts as the control signal in the loop. A plot that shows how this control signal depends on the actual phase difference is called an *S-curve*. Several *S*-curves are shown in Fig. 4.6a–c. The first one, Fig. 4.6a, plots the effect of the multiplier/low-pass filter phase detector in Figs. 4.2–4.5, for which the phase difference signal output will be $\sin(\Delta\phi)$ when the input signal phases differ by $\Delta\phi$. The *S*-curves take their name from the *S*-on-its-side shape of this classic case. The dot at the origin of the *S* denotes a *stable lock point*: A positive phase difference generates a positive control signal to the VCO, which increases its frequency and closes the phase difference toward the lock point; similarly, a negative phase difference reduces the VCO fre-

[1]The estimate is not maximum likelihood; however, the circuit can be modified to produce one.

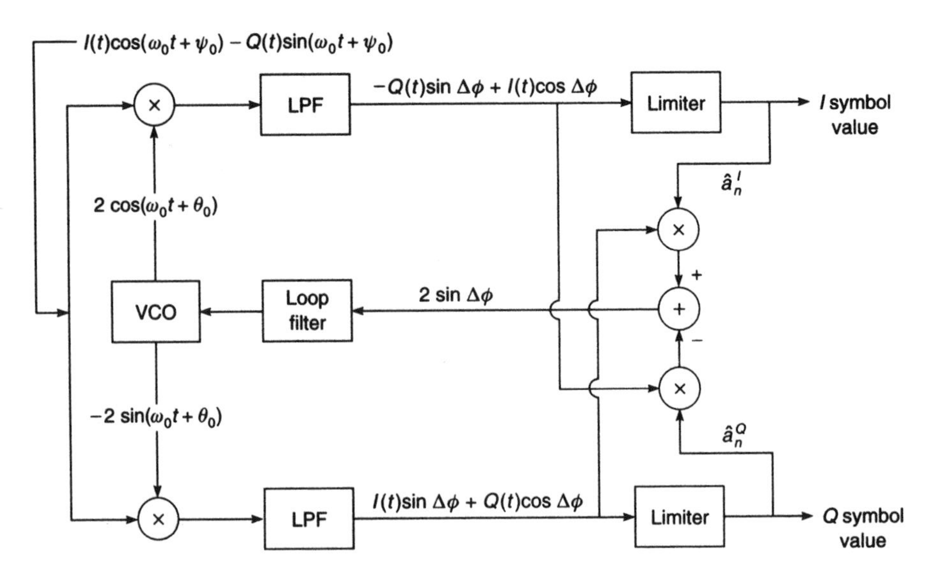

Figure 4.5 A QPSK Costas loop. Both data and phase are obtained, $I(t)$ and $Q(t)$ are shaped-pulse trains, and limiter extracts sign of its input. $\Delta\phi = \psi_0 - \theta_0$.

quency and drives the control signal upward along the curve to the origin. The effect is shown by arrows along the S-curve.

The square blocks in Fig. 4.6a locate *unstable* lock points along the S-curve. Here the behavior is reversed from that just described, and the loop tends to move away from the lock point. However, the control signal is zero precisely at the point, so that the loop freezes, and nothing happens, at least initially. An analogy to this situation is a pencil balanced on its point. Eventually, a random disturbance nudges

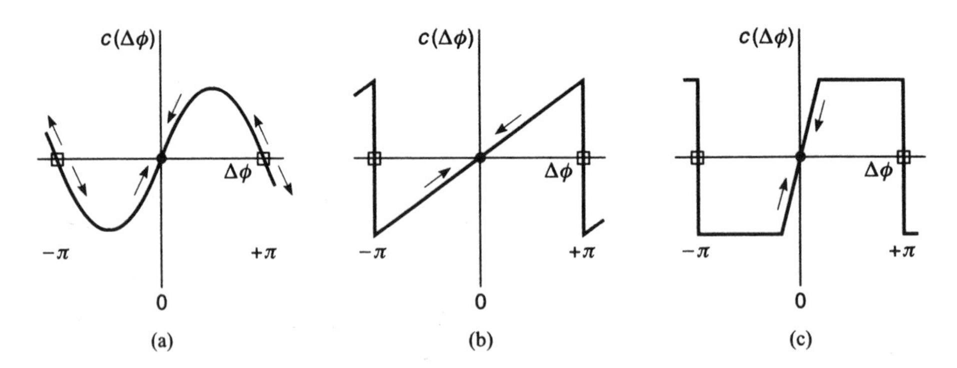

Figure 4.6 Examples of S-curves showing loop control signal $c(\Delta\phi)$ versus $\Delta\phi$, reference phase minus VCO phase: **(a)** multiplier/LPF S-curve; **(b)** linear S-curve; **(c)** fast-lock S-curve.

the loop away from the unstable lock, and it progresses along the curve to a stable point, although the process may take some time.

Figures 4.6b and 4.6c show other types of *S*-curves. Figure 4.6b represents a linear phase detector, that is, a phase difference signal that is linearly proportional to the actual phase difference. This is the situation in the PLL model in Fig. 4.1b. The linear *S*-curve is the natural one for a discrete-time or a software-implemented loop, in which the phase detector is likely to simply count time between zero crossings of the VCO and the input oscillations. It is worth noting that at any reasonable phase error, the sinusoidal *S*-curve (Fig. 4.6a) is essentially a linear *S*-curve; its chief difference is that the restorative force it applies does not keep growing with large phase errors. Figure 4.6c is a variant of the linear *S*-curve with a faster lock at small phase errors.

We need to distinguish between the *S*-curve of a simple PLL, which we have just been discussing, and the *S*-curve of a synchronizer. The fourth-power loop of Fig. 4.3 contains a PLL as a subsystem, but the PLL runs at four times the carrier frequency. In terms of the phase at the synchronizer input and output, the PLL phase difference is $\Delta\phi = 4\psi_0 - 4\theta_0$ and the control signal is $\sin(4\psi_0 - 4\theta_0)$, neglecting a constant multiplier. Replotting Fig. 4.6a against the phase difference $\psi_0 - \theta_0$ gives Fig. 4.7. There are now stable lock points at $-\frac{1}{2}\pi$, $\frac{1}{2}\pi$, and π, as well as at 0. In other words, the synchronizer as a whole can lock equally well to the carrier phase plus any multiple of $\frac{1}{2}\pi$ radians. This is the *phase ambiguity* problem.

Some study of the Costas loop will show that it suffers from the same ambiguity, even though its VCO runs at the carrier frequency. Phase ambiguity afflicts most synchronizer circuits. The usual cures for it are (*i*) to transmit special symbols at the beginning of a data frame that inform the receiver to which $\frac{1}{2}\pi$ multiple the synchronizer has locked, or (ii) to differentially encode the data so that it is contained in the phase difference from symbol to symbol, rather than in the absolute phase.

4.2.4 Effects of Error in the Phase Reference

Without an accurate carrier phase reference, a receiver cannot establish an accurate signal space coordinate system for detection. For quadrature demodulation, with *I*

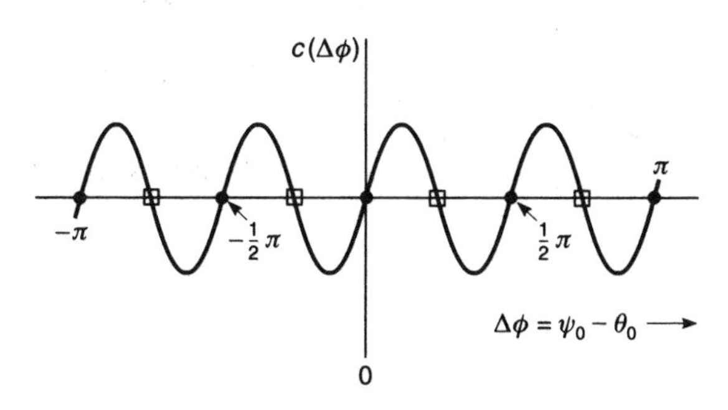

Figure 4.7 An *S*-curve for a fourth-power synchronizer: VCO control signal versus synchronizer phase error.

and Q components, Fig. 4.8 shows what goes wrong when a static phase error $\Delta\phi = \psi_0 - \theta_0$ occurs in QPSK detection. The receiver attempts to detect with the rotated coordinate system indicated by the dashed lines; these lines become the mistaken ML detection regions. With no phase error, the shortest noise vector that causes detection error has length $\sqrt{E_b}$, shown as d on the figure, but with a phase error $\Delta\phi$ the geometry in the figure shows that noises as short as

$$|A| = d(1 - \tan\Delta\phi)\cos\Delta\phi \tag{4.2-4}$$

may lead to error.

A precise error calculation runs as follows. For correct detection under Gaussian noise, the added noise must lead to a point inside the dashed receiver decision region. The projections of the noise on any two orthogonal directions are independent, and if we take the directions A and B shown on the figure, we obtain

$$\begin{aligned} P[\text{QPSK error}] &= 1 - P[\text{correct detection}] \\ &= 1 - \left[1 - Q\left(|A|\sqrt{\frac{2}{N_0}}\right)\right]\left[1 - Q\left(|B|\sqrt{\frac{2}{N_0}}\right)\right] \\ &\approx Q\left(|A|\sqrt{\frac{2}{N_0}}\right) + Q\left(|B|\sqrt{\frac{2}{N_0}}\right) \\ &= Q\left((\cos\Delta\phi - \sin\Delta\phi)\sqrt{\frac{2E_b}{N_0}}\right) \\ &\quad + Q\left((\cos\Delta\phi + \sin\Delta\phi)\sqrt{\frac{2E_b}{N_0}}\right) \end{aligned} \tag{4.2-5}$$

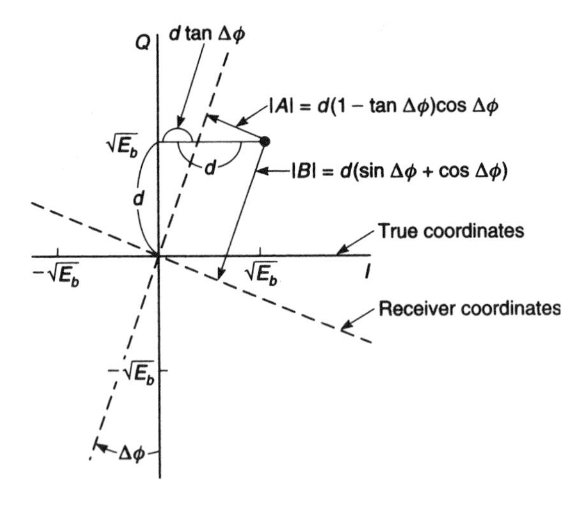

Figure 4.8 Effect on QPSK of coordinate system rotated by $\Delta\phi$. Minimum distance of QPSK is $2d = 2\sqrt{E_b}$. Upper right point is transmitted.

This compares with $2Q(\sqrt{2E_b/N_0})$ for QPSK with a perfect phase reference. A similar calculation can be made for any QAM modulation.

A simpler calculation, left to the reader, shows that the probability of error of BPSK with a carrier reference error of $\Delta\phi$ is

$$P_e = Q\left(\cos \Delta\phi \sqrt{\frac{2E_b}{N_0}}\right) \tag{4.2-6}$$

At large E_b/N_0, (4.2-5) and (4.2-6) imply that the loss in minimum distance in decibel terms is

$$\begin{aligned} &\text{QPSK:} \quad 20 \log_{10} (\cos \Delta\phi - \sin \Delta\phi) \quad \text{(dB)} \\ &\text{BPSK:} \quad 20 \log_{10} \cos \Delta\phi \quad \text{(dB)} \end{aligned} \tag{4.2-7}$$

For example, a carrier reference error of 0.1 rad (5.7°) leads to a loss of 1.0 dB for QPSK, but only 0.04 dB for BPSK. A careful QPSK calculation at $P_e = 10^{-5}$ with (4.2-5) yields a loss of 0.7 dB. An increase in E_b by these amounts will be necessary at high E_b/N_0 to make up for the synchronization error.

Equation (4.2-7) generally overestimates the loss for QPSK. Some actual measured losses, taken from Ziemer and Ryan [3], appear in Fig. 4.9. It is clear that

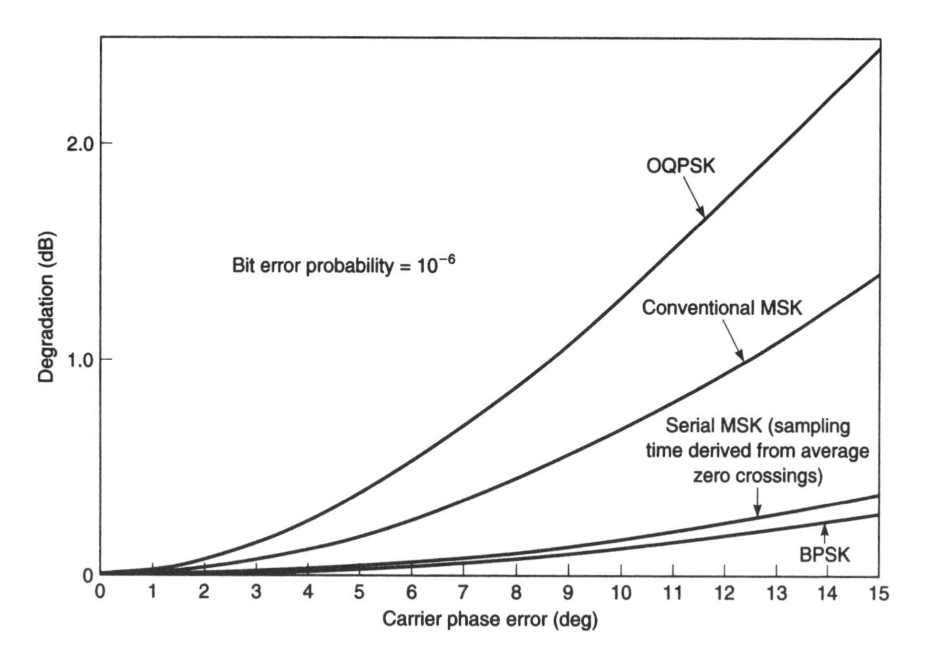

Figure 4.9 Demodulator degradation in terms of extra E_b/N_0 required versus synchronizer phase error at BER = 10^{-6} for several types of modulation. (From Ref. 3. Copyright IEEE. Used by permission.)

QPSK is more sensitive to synchronization error than competing simple modulations.

All of the foregoing has assumed a static phase error during detection. When the disturbance to the synchronizer is channel noise or data noise, $\Delta\phi$ is actually a random variable. One can still get good estimates for the damage to P_e by substituting the standard deviation of $\Delta\phi$ for $\Delta\phi$ in (4.2-4)–(4.2-7).

4.3 THE ANALOG PHASE-LOCK LOOP

We have seen that synchronizers explicitly contain a PLL or at least act like they do. A framework for the analysis of this important circuit is set up in this section. The plan is to write the dynamical equations for the PLL and then derive the responses of the PLL to the common phase transients and, in Section 4.4, to noise. With these we can design an acceptable synchronizer.

4.3.1 PLL as a Control System

Although we use the PLL in a synchronization system, it is really a control system: a loop that controls a VCO phase so that it tracks an external signal phase. A control engineering text such as Ref. 9 is a useful aid in understanding what follows. Figure 4.10 expresses the PLL of Fig. 4.1b in terms of a unity-feedback control system. All the variables shown are Laplace transforms of phase signals. The input to the system is $R(s)$, the *reference phase*. In the fourth-power loop of Section 4.2, for example, the reference phase as a time-domain signal was the slowly varying offset $4\psi_0(t)$. The term $C(s)$ represents the *controlled variable*; in a synchronizer this is the VCO phase, and in the fourth-power loop it is $4\theta_0(t)$. The *error,* or actuating signal, $E(s)$ is the difference between the reference and the controlled signals. Finally, $G(s) = C(s)/E(s)$ is the open-loop transfer function for the system; $G(s)$ includes the loop filter, the workings of the VCO, and other circuits we might add inside the PLL.

An important function for us is the closed-loop transfer function $H(s)$, the ratio $C(s)/R(s)$ between the controlled phase variable and the reference phase variable.

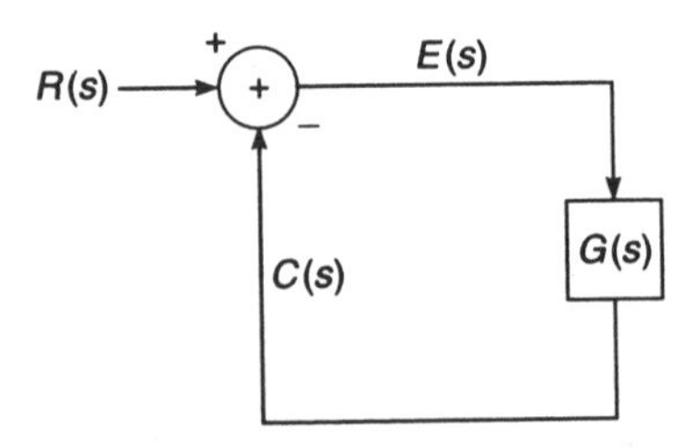

Figure 4.10 Linearized PLL as unity-feedback control system. All variables are transforms of slowly varying phase signals in Fig. 4.1b.

The VCO phase, which is fed to the rest of the receiver, thus has transform $C(s) = H(s)R(s)$. Some manipulation shows that

$$\frac{C(s)}{R(s)} = \frac{G(s)}{1 + G(s)} \triangleq H(s) \qquad \text{(closed-loop transfer function)} \qquad (4.3\text{-}1)$$

This follows from the fact that $[R(s) - C(s)]\, G(s) = C(s)$. Another important function is the *error transfer function* $1 - H(s)$, which is the ratio $E(s)/R(s)$ between the phase error and the phase reference:

$$\frac{E(s)}{R(s)} = \frac{1}{1 + G(s)} = 1 - H(s) \qquad \text{(error transfer function)} \qquad (4.3\text{-}2)$$

Since the converged loop has zero error, it must be that $H(0) = 1$.

The loop in Fig. 4.10 is called a *linearized loop,* because it assumes that the PLL phase detector is the linear S-curve in Fig. 4.6b. If the PLL really contains a multiplier/LPF phase detector, then the assumption amounts to making the approximation $\sin(\phi_0(t) - \theta_0(t)) \approx \phi_0(t) - \theta_0(t)$, which is accurate at any reasonable phase error.[2] If the PLL really contains a linear detector, there is still an assumption, this time that the phase error lies in $(-\pi, \pi)$.

In control engineering, the convention is to express the open-loop transfer function in the pole-zero form

$$G(s) = \frac{K(1 + T_1 s)(1 + T_2 s) \cdots}{s^m (1 + T_a s)(1 + T_b s) \cdots} \qquad (4.3\text{-}3)$$

Here the factor K is called the DC gain of the loop, or simply the *loop gain.* Each factor in the numerator and denominator corresponds to a zero and pole in the s-plane, respectively. The integer m plus the number of factors in the denominator is the *order* of the control system. The value of m itself denotes the *type* of the system, type 0, type 1, and so on. We will be interested in types 1 and 2.

For Type 1. A constant phase error $E(s)$ leads to a constant rate of change in the controlled variable, the VCO phase $C(s)$; for a step change in the reference phase $R(s)$, the eventual, steady-state error $E(s)$ is zero.

For Type 2. A constant phase error leads to a constant acceleration in the controlled variable; for a change in the slope of the reference phase, the steady-state error $E(s)$ is zero.

In other words, a type 1 PLL can adjust with zero eventual error to a step phase change, but it cannot "catch" a ramp phase change. On the other hand, a type 2 PLL

[2] A Fokker–Planck analysis has been devised that makes the sin(·) approximation unnecessary, but it has been applied only to loops with simple dynamics. See Ref. 4, p. 268.

can adjust to a phase ramp with zero error. The proof of these statements is a consequence of the final value theorem of Laplace transforms, which states that

$$\lim_{t\to\infty} q(t) = \lim_{s\to 0} sQ(s) \tag{4.3-4}$$

where $q(t)$ and $Q(s)$ are a transform pair. For example, consider the error response of a type 2 system to a ramp change in the reference signal. The reference transform is therefore $R(s) = r_0/s^2$, in which r_0 is the slope of the ramp in radians per second; that is, r_0 is the radian frequency offset of the reference. From (4.3-2) and (4.3-3), $sE(s)$ tends to

$$\lim_{s\to 0} \frac{sR(s)}{1+G(s)} = \lim_{s\to 0} \frac{sr_0/s^2}{1+K/s^2} = 0$$

as $s \to 0$. So, the steady-state phase error, $e_{ss}(t)$, is zero. We can expect that a type 2 PLL will eventually catch up with a Doppler or other frequency shift in a carrier phase.

PLL Open-Loop Transfer Function. The open-loop transfer function $G(s)$ of a PLL [Eq. (4.3-3)] consists of the loop filter $F(s)$, the VCO, and possibly other circuit elements, all in cascade. We will write all these as unity-gain circuit elements, combine their actual gains into the single DC gain factor K in (4.3-3), and count K as part of the filter $F(s)$. The gain K is a critical factor in PLL design, and it may in fact be necessary to insert a gain block in the loop in order to provide it. The VCO, as a circuit block, is a device that puts out $-2\sin(\omega_0 t + \theta_0)$, the signal shown in Fig. 4.1a, in which the excess phase θ_0 is a slowly varying signal that satisfies

$$\frac{d\theta_0}{dt} = K_{\text{vco}} e_{\text{vco}}(t) \tag{4.3-5}$$

That is, the VCO deviation in frequency from ω_0 is proportional to its input e_{vco}. It is not necessary that the VCO oscillation be sinusoidal; the important point is that its excess phase satisfies (4.3-5).[3] The phase θ_0 here is the controlled variable C in Fig. 4.10, and if we take the transform of (4.3-5), the result is

$$sC(s) = K_{\text{vco}} E_{\text{vco}}(s) \tag{4.3-6}$$

With (4.3-6) in hand, we can write the open-loop transfer function $G(s)$. The signal controlling the VCO is the error signal $e(t)$ filtered by $F(s)$. Thus $E_{\text{vco}}(s) = KF(s)E(s)$. Combining this with (4.3-6), we get

$$G(s) = \frac{C(s)}{E(s)} = \frac{KF(s)}{s} \tag{4.3-7}$$

Here the VCO gain has been absorbed into the factor K.

[3]As an electronic device, a VCO has a center, or "rest," frequency, and a DC bias may be required to hold the VCO at ω_0; if so, e_{vco} should be thought of as a deviation from that bias.

Although a great many PLL transfer functions are conceivable, we will concentrate on four common ones. The parameters and transfer functions for these are summarized in Table 4.1.

1. *The First-Order Loop.* The loop filter is a gain block. Here, $G(s) = K/s$, so that the closed-loop response has a single pole.
2. *The Two-Pole Loop.* The loop filter has a single pole, so that the closed-loop response is second order but lacks an s-plane zero.
3. *The Double-Integral Loop.* The loop filter has a pole at 0 and a single zero. This is the only type 2 second-order loop; it tracks a frequency shift with zero steady-state error.
4. *The General Second-Order Loop.* The filter has one pole and one zero.

4.3.2 Phase Transient Responses

In what follows, we will analyze the responses of loop kinds 1–3 to phase transients.

The First-Order Loop. When the loop filter $F(s)$ is simply a constant, $G(s) = K/s$; the closed-loop transfer function is

$$H(s) = \frac{K}{s + K} \tag{4.3-8}$$

and the error transfer function is

$$\frac{E(s)}{R(s)} = 1 - H(s) = \frac{s}{s + K} \tag{4.3-9}$$

Referring to Eq. (4.3-3), we see that the loop is type 1. Consequently, it responds to a step change in the reference phase with steady-state error zero. In fact, we can easily solve for the unit step response:

$$\mathcal{L}^{-1}\left[\frac{H(s)}{s}\right] = u(t)[1 - e^{-Kt}] \tag{4.3-10}$$

Here $u(t)$ denotes the unit step function. The error response $u(t)$ is

$$e(t) = \mathcal{L}^{-1}\left[\frac{1 - H(s)}{s}\right] = \mathcal{L}^{-1}\left[\frac{1}{s + K}\right] = u(t)e^{-Kt} \tag{4.3-11}$$

The final value theorem gives that the steady-state error with a unit ramp input is $e_{ss} = 1/K$, which shows that the error due to a frequency shift cannot be reduced independently of the gain, an unpleasant conclusion. This is a motivation to investigate higher-order loops.

TABLE 4.1 Transfer Functions and Loop Parameters for PLLs of First and Second Order

Loop Type	Loop Filter $F(s)$	Open-Loop Transfer Function $G(s)$	Define ω_n, ζ as Functions of K, T_1, T_2, T_a		Closed-Loop Transfer Function $H(s)$	Error Transfer Function $1-H(s)$	Noise Equivalent Bandwidth B_N (Hz)
			ω_n (rad/s)	ζ			
First order	1	K/s	K	—	$\frac{K}{s+K}$	$\frac{s}{s+K}$	$\frac{1}{4}K$
Two-pole	$\frac{1}{1+T_a s}$	$\frac{K}{s(1+T_a s)}$	$\sqrt{\frac{K}{T_a}}$	$\frac{1}{2\sqrt{KT_a}}$	$\frac{\omega_n^2}{D(s)}$	$\frac{s^2+2\zeta\omega_n s}{D(s)}$	$\frac{K}{4}$ or $\frac{\omega_n}{8\zeta}(1)$
Double integral	$\frac{(1+T_1 s)}{s}$	$\frac{K(1+T_1 s)}{s^2}$	$\sqrt{K}$	$\frac{1}{2}T_1\sqrt{K}$	$\frac{2\omega_n s+\omega_n^2}{D(s)}$	$\frac{s^2}{D(s)}$	$\frac{\omega_n}{8\zeta}+\frac{\omega_n\zeta}{2}(1)$
General second order	$\frac{(1+T_1 s)}{(1+T_a s)}$	$\frac{K(1+T_1 s)}{s(1+T_a s)}$	$\sqrt{\frac{K}{T_a}}$	$\frac{1}{2}\sqrt{\frac{K}{T_a}}\left(T_1+\frac{1}{K}\right)$	$\frac{(2\zeta\omega_n-\omega_n^2/K)s+\omega_n^2}{D(s)}$	$\frac{s^2+(\omega_n{}^2/K)s}{D(s)}$	$\frac{\omega_n}{8\zeta}+\frac{\omega_n}{8K^2\zeta}(2\zeta K-\omega_n)^2$

Note: K is the loop gain parameter; T_1 and T_a are the pole-zero form open-loop parameters; ω_n (the natural frequency) and ζ (the damping factor) are parameters for the closed-loop transfer function in polynomial form; $D(s) = s^2 + \zeta\omega_n s + \omega_n^2$. (1). Note ω_n here is in radians per second even though B_N is in hertz.

Two-Pole Loop. The remaining PLLs that we will discuss are all second order, and the traditional parameters to describe any such system are ω_n, the natural frequency, and ζ, the damping coefficient. The denominator of the second-order closed-loop transfer is $D(s) = s^2 + 2\zeta\omega_n s + \omega_n^2$, which implies that two poles are located at

$$s = -\zeta\omega_n \pm \omega_n\sqrt{\zeta^2 - 1} \tag{4.3-12}$$

When $\zeta < 1$, the system is said to be underdamped; it has complex poles and its response will be oscillatory. When $\zeta > 1$, the system is overdamped, has two real poles, and does not oscillate. Figure 4.11 shows the step response of a PLL whose $H(s)$ consists of just these two poles and no zero, along with the step response of a first-order loop with the same open-loop gain K. It is clear that the underdamped cases adapt faster to the step, but they overshoot the step by a considerable margin when the damping is low. In many control applications, $\zeta = 1/\sqrt{2} = 0.707$ is taken as a good compromise between overshoot and slow adaptation. For reference, we give the step response of the two-pole PLL:

$$\mathcal{L}^{-1}\left[\frac{H(s)}{s}\right] = 1 - \frac{e^{-\zeta\omega_n t}}{\sqrt{1-\zeta^2}} \sin(\omega_n\sqrt{1-\zeta^2}\, t + \cos^{-1}\zeta) \tag{4.3-13}$$

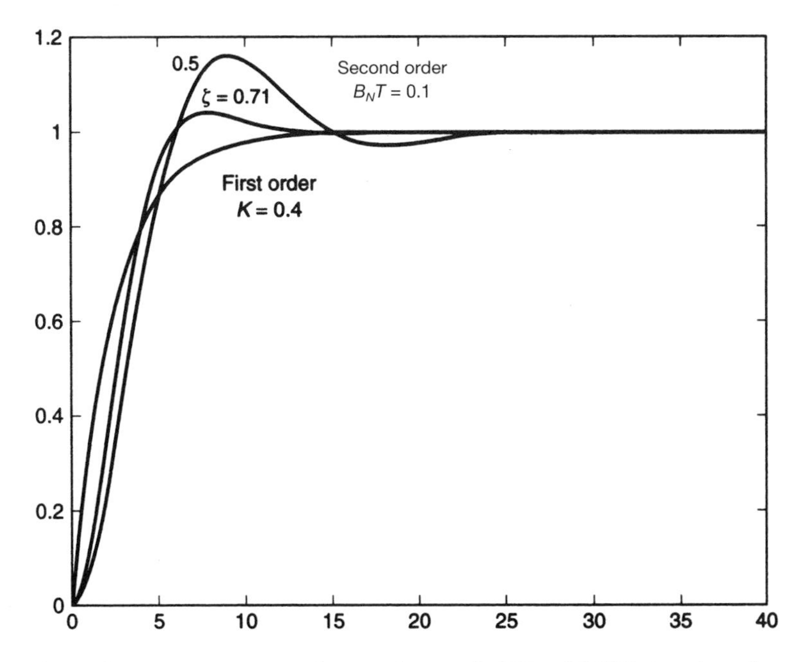

Figure 4.11 Some step responses of one- and two-pole PLLs. All PLLs here have the same gain parameter K and the same noise equivalent bandwidth.

The error response to a step is the negative of the second term. It can be shown that the peak overshoot in the unit step response is

$$1 + \exp\left(\frac{-\zeta\pi}{\sqrt{1-\zeta^2}}\right) \tag{4.3-14}$$

This is 1.04 at $\zeta = 0.707$.

Although ω_n and ζ are the traditional parameters in second-order system analysis, they are not the most convenient ones for PLL design. We will soon see that the bandwidth of the closed-loop response is what sets most of the PLL's response. Specifically, the parameter that we will need is the *noise-equivalent bandwidth*

$$B_N \triangleq \frac{1}{|H(0)|^2} \int_0^\infty |H(j2\pi f)|^2 \, df \tag{4.3-15}$$

Figure 4.12 illustrates the calculation. The idea is simple: B_N is the single-sided bandwidth in hertz of an ideal low-pass filter with the same DC gain that passes the same energy as $H(\cdot)$. The noise response of a PLL depends on this single parameter B_N.

Returning to the two-pole PLL, we find that the one-pole loop filter $K/(1 + T_a s)$ leads to the formulas in Table 4.1; particularly, definition (4.3-15) applied to the closed-loop transfer function $\omega_n^2/D(s)$ leads to the value $B_N = \omega_n/8\zeta$. The details of these calculations are left to the reader. However it is defined, the two-pole PLL requires two parameters. These can be the pole position T_a and the loop gain K or they can be ω_n and ζ. Instead, we will work with B_N and ζ, where B_N characterizes the PLL's noise response and ζ is the sole remaining design parameter.

Figures 4.13a and 4.13b show the step error responses of the two-pole loop for $B_N = 0.2/T$ and $0.05/T$, respectively. The free parameter is the damping coefficient ζ. In the sequel we will work with $B_N = 1/T$ as well, but the response of the two-pole PLL in this case is essentially instant. In both plots, it is clear that ζ in the range 0.707–1 are the best damping coefficients. Their error response is essentially finished at $0.5/B_N T$ symbols.

Before continuing, we give a routine for computing the important responses of the two-pole loop.

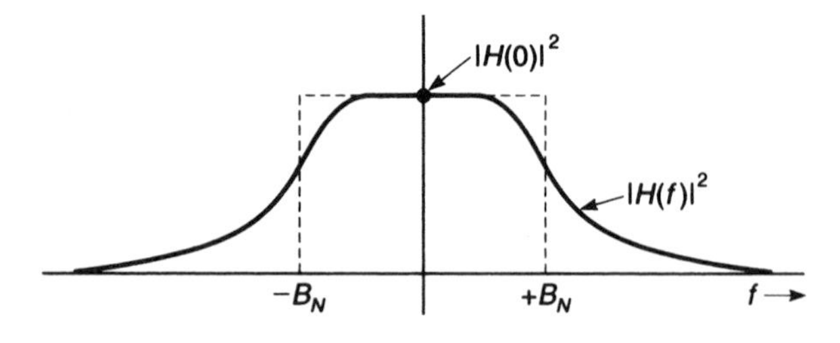

Figure 4.12 Concept of noise-equivalent bandwidth. Areas under two functions are the same.

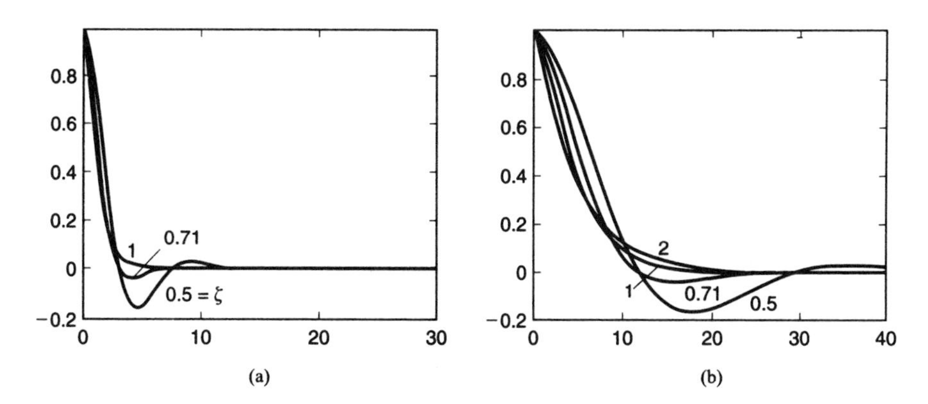

Figure 4.13 Step error responses of two-pole PLL with different damping factors: **(a)** bandwidth 0.2 Hz-s/symbol; **(b)** bandwidth 0.05 Hz-s/symbol. Horizontal axis in length-T symbol intervals.

Program 4.3-1: Error Responses of the Two-Pole PLL. The following MATLAB program computes and plots the two-pole loop error responses to a unit step and unit ramp input. The program is based on the MATLAB function `step(num,den,t)`, which computes the step response at a vector of times `t` for the analog system polynomial with numerator $N(s)$, represented by `num`, and denominator $D(s)$, represented by `den`. The system polynomial in question is the error transfer function

$$1 - H(s) = \frac{s^2 + 2\zeta\omega_n s}{s^2 + 2\zeta\omega_n s + \omega_n^2}$$

in Table 4.1. The program accepts as input the noise equivalent bandwidth `bn`, the damping coefficient `zeta`, and the times `t`. From these it finds `wn`, the natural frequency, and then `num` and `den`. The function `step(num,den,t)` directly gives the step error response `estep`. The ramp error response `eramp` is the step response with numerator $N(s)/s$ instead of $N(s)$:

```
%         Set up numerator and denominator from bn and zeta
wn = 8*zeta*bn;              %Find natural frequency by Table 4.1
num = [1 2*wn*zeta 0];
den = [1 2*wn*zeta wn^2];
%
%         Find step error response
[estep,x,t] = step(num,den,t);
plot(t,estep)
%         Find ramp error response
nrp = [0 1 2*wn*zeta];    %Set up ramp numerator
[eramp,x,t] = step(nrp,den,t);
plot(t,eramp)
```

As an example, set `bn = .2` and `zeta = .5`. Then

```
wn = .8
num = [1 .8 0]
den = [1 .8 .64]
```

and the curve in Fig. 4.13a results. For the unit ramp error response, `nrp` becomes `[0 1 .8]`, and the curve in Fig. 4.14 is the outcome.

Before proceeding further, a word about the scaling of this figure and those that follow. Figure 4.13a and 4.13b are time responses of a PLL, but the x axis is in symbol times. We introduce the idea of symbol time by replacing t with t/T on the x axis, so that time is measured in symbols. We can go one step further and replace ω_n with $\omega_n T$ and B_N by $B_N T$. Observe, for example, how Eq. (4.3-13) remains unchanged if ω_n is replaced by $\omega_n T$ and t by t/T. A single generic response plot will be the result for all T when the $B_N T$ product is fixed and the time axis is scaled to t/T. As an example, two receivers, one with $T = 1$ s/symbol and $B_N = 0.5$ Hz and the other with a 1-MHz symbol rate and $B_N = 500$ kHz, both have the same normalized bandwidth, namely, $B_N T = 0.5$ Hz-s/symbol, and the same PLL response measured in symbol times.

For both bandwidths in the figure, it is clear that significant VCO phase error occurs when a large reference phase step appears. About $0.4/B_N T$ symbol times are needed before the error drops to 20% of the step size. This means that the PLL bandwidth is not really large enough if steps of, say, 45° to 90° are expected, although it may be enough for smaller steps. Probably, the object of the design is a target error probability in the demodulator; then the tolerable PLL bandwidth will depend on the size of the phase steps, how often they appear, and how accurate a VCO phase the particular modulation scheme needs. We will take up some practical design examples in Section 4.5.

When a unit phase ramp appears at the input of the two-pole PLL, the VCO response is similar to Fig. 4.14, which shows the case of $B_N T = 0.2$. The best damping

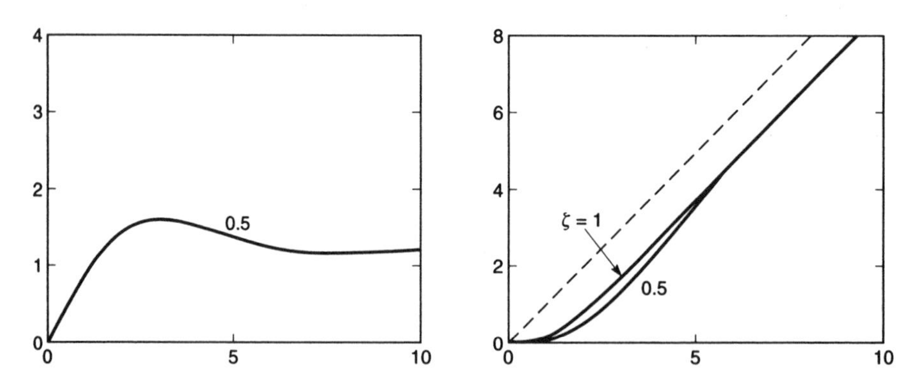

Figure 4.14 Ramp response **(right)** and ramp error response **(left)** for two-pole PLL versus symbol intervals: $B_N T = 0.2$ Hz-s/symbol.

here is again in the range 0.707–1, but an obvious fact about the figure is the appearance of a steady-state phase error. The responses at other bandwidths are similar in shape, except that the steady-state error grows worse as the bandwidth falls. We can easily derive this from the final value theorem. From (4.3-4) and the proper closed-loop entry in Table 4.1, we find for a unit-per – T ramp excitation $1/s^2T$ that

$$\begin{aligned} e_{ss} &= \lim_{s \to 0} \frac{sE(s)}{s^2 T} = \frac{2\zeta}{\omega_n T} \\ &= \frac{1}{4\, B_N T} \end{aligned} \qquad (4.3\text{-}16)$$

This predicts, accurately, that $e_{ss} = 1.25$ in Fig. 4.14. That is, a frequency shift of 1 rad/symbol in the PLL input leads to a 1.25-rad (72°) steady-state VCO error. The error may be reduced only by widening the PLL bandwidth, but this will be unacceptable if the phase reference signal is too noisy.

Double-Integral Loop. In order for a system to have zero steady-state response to a ramp, it must be a type 2 system. There is only one type 2 second-order loop, the double integral (DI) loop. Its transfer functions and parameters are shown in Table 4.1. In terms of the generic PLL in Fig. 4.1b, the DI loop is defined by the loop filter $F(s) = K(1 + T_1 s)/s$, but it is probably easier to think of it in terms of Fig. 4.15, which consists of a filter $F'(s) = K(1 + T_1 s)$ and an *integrator* followed by the VCO. In physical terms, this PLL works by driving the VCO with the integral of the smoothed error signal. A steady-state error signal leads to a linearly growing VCO input, which creates a parabolic VCO phase, which can catch a ramp change in the reference.

It is easy to show that there are no other type 2 loops of second order. We will proceed by allowing the loop to be third order and seeing what is required to bring it down to second. Restricting the general open-loop expression (4.3-3) to type 2 and third order and then substituting it into (4.3-1), we get

$$H(s) = \frac{K(1 + T_1 s)(1 + T_2 s)}{s^2(1 + T_a s) + K(1 + T_1 s)(1 + T_2 s)}$$

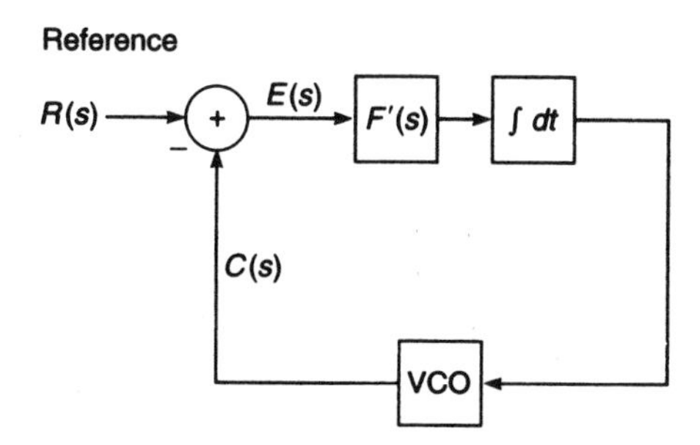

Figure 4.15 Block diagram of double-integral PLL. To be second order, the loop requires that $F'(s) = K(1 + T_1 s)$.

For a second-order loop, T_a must be zero, which leaves

$$H(s) = \frac{K(1 + T_1 s)(1 + T_2 s)}{s^2 + K(1 + T_1 s)(1 + T_2 s)}$$

In a PLL, we do not want feed-through from the input to the output, which means that T_2 here must be zero. What is left is

$$G(s) = \frac{K(1 + T_1 s)}{s^2}, \qquad H(s) = \frac{K(1 + T_1 s)}{s^2 + K(1 + T_1 s)} \tag{4.3-17}$$

The two parameters K and T_1 can be exchanged for ω_n and ζ as shown in Table 4.1, and the expressions for $H(s)$ and $E(s)$ there are the result.

Figures 4.16a–c show the step error response of the DI loop for our three standard noise bandwidths, $B_N T = 1, 0.2, 0.05$. The response at the wide bandwidth rapidly converges to zero and is completely satisfactory. At bandwidth 0.2, a higher damping in the range 1–2 seems to be preferred, but otherwise Figs. 4.13a and

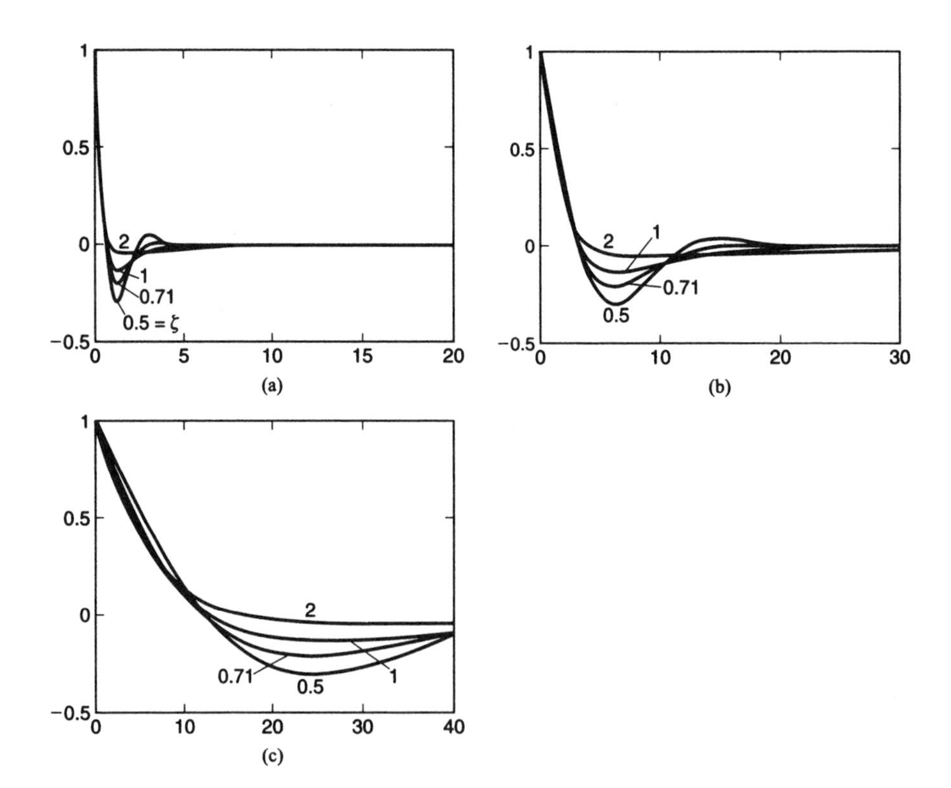

Figure 4.16 Step error responses for double-integral PLLs versus symbol intervals for several damping factors: **(a)** bandwidth $B_N T = 1$; **(b)** $B_N T = 0.2$; **(c)** $B_N T = 0.05$.

4.16b seem to match up well, which shows that the switch to a type 2 PLL has not cost anything in step response. At the narrow bandwidth 0.05, a similar conclusion holds, but both the two-pole and the DI loops respond so slowly now that their step response is probably unsatisfactory unless the steps are small.

Figures 4.17a–c show the DI error response to a unit ramp at the same bandwidths. Now the best damping lies in the range 0.707–1, which conflicts somewhat with the best damping for a step excitation. The wideband ($B_NT = 1$) PLL tracks the ramp well, showing significant phase error during just one symbol interval, even for a major frequency shift. The narrower PLLs fare poorly with the ramp; significant phase error appears over about $2/B_NT$ symbols, and the peak error from the unit ramp grows as $\approx 1/4B_NT$. This is no better than the error (4.3-16) for the two-pole loop, but it is temporary at least and not steady state. If the PLL must have a narrow bandwidth, the allowed frequency shifts will have to be small.

These responses were computed with the following program.

Program 4.3-2: Error Responses of the Double-Integral PLL. This MATLAB program computes and plots the double-integral PLL error responses to the step and

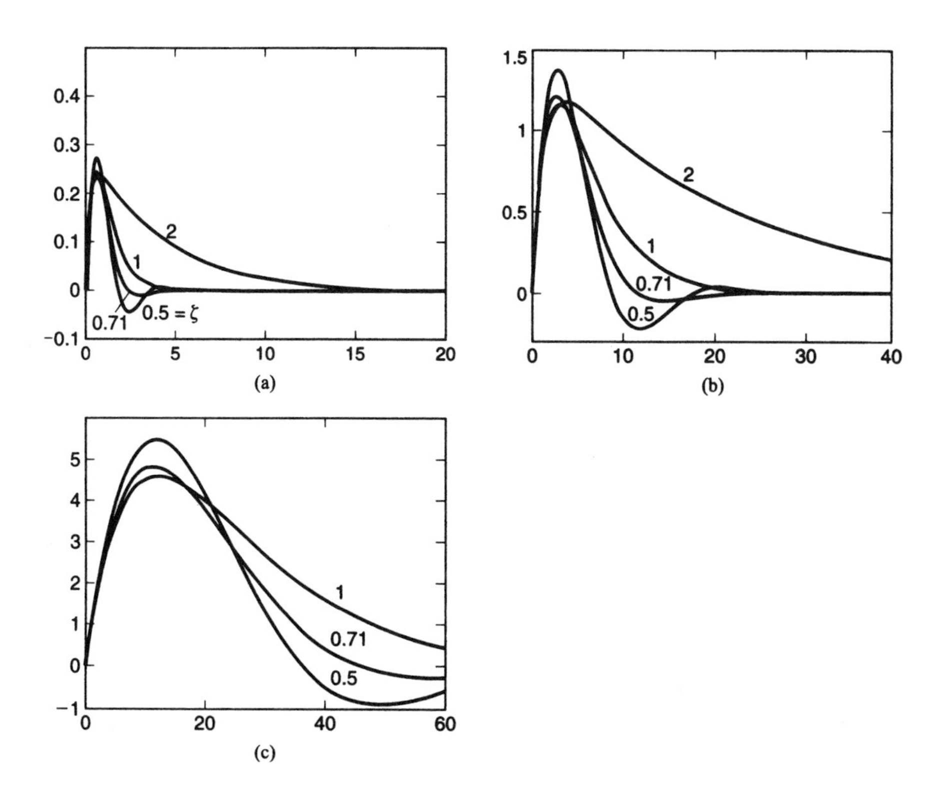

Figure 4.17 Ramp error responses for double-integral PLLs versus symbol intervals for several damping factors: **(a)** bandwidth $B_NT = 1$; **(b)** $B_NT = 0.2$; **(c)** $B_NT = 0.05$.

unit ramp input. The program is similar to Program 4.3-1 and takes the same input, the noise equivalent bandwidth `bn` and damping coefficient `zeta`. The outputs are `estep` and `eramp`. This time the error transfer function polynomial is, from Table 4.1,

$$\frac{s^2}{s^2 + 2\zeta\omega_n s + \omega_n^2}$$

The denominator lines are the same as in Program 4.3-1.

```
%         Set up numerator and denominator from bn and zeta
wn = bn/(.5*zeta+1/(8*zeta)); %Natural frequency by Table 4.1
num = [1 0 0];
den = [1 2*wn*zeta wn^2];
%         Find step error response
[estep,x,t] = step(num,den,t);
plot(t,estep)
%         Find ramp error response
nrp = [0 1 0];                          %Ramp numerator
[eramp,x,t] = step(nrp,den,t);
plot(t,eramp)
```

Setting `bn = .2` and `zeta = .5`, we get

```
wn = .4
num = [1 0 0]
den = [1 .4 .16]
```

and the curve in Fig. 4.16b. For the unit ramp error response, `nrp` becomes `[0 1 0]`, and the curve in Fig. 4.17b is the outcome.

4.3.3 Responses During Acquisition

As mentioned in Section 4.1, the PLL has only a rough idea of the reference signal frequency when it is in the acquisition mode. The PLL will often not behave as a linear system in this mode, and so a detailed study of acquisition uses special, sometimes complex methods. We will summarize some high points here and take a simplified approach. Given an initial frequency offset between the VCO and the reference, the bandwidth and loop gain of a PLL determine whether it will reach lock and how long the process will take.

A PLL can deviate from the tracking mode in several ways, and we begin by defining these. The PLL may go through quite a nonlinear behavior on its way to lock. There may be cycle slips—that is, instants when the phase error moves beyond the range $(-\pi, \pi)$; the PLL now does not converge in a single direction toward lock, and instead it sometimes slips by 2π radians. In the worst case, the multiplication of the VCO and reference sinusoids leads to a beat frequency error signal with an average DC component; this hopefully drives the VCO over time in the right direction. The maximum range of reference frequency over which eventual lock up[4] can occur is called the *capture range*. Analyses of capture range exist, but it is often easier simply to measure the range by testing the PLL.

An easier quantity to predict is the *hold-in* (or *lock-in*) range, which can be defined as the widest reference frequency range over which the PLL stays locked without cycle slips, given that it was in lock to start. The hold-in range only estimates the capture range, but it is often a good, easily found estimate. For all type 1 loops (including, e.g., first-order and two-pole), the hold-in range directly stems from the PLL loop gain K by the following argument. With a linear phase detector[5] the PLL phase error must lie in the range $-\pi < e(t) < \pi$; otherwise, a cycle slip occurs. If the locked-to reference signal suddenly moves to frequency $f_A + f_{\text{vco}}$ hertz, we can model the PLL system input as a phase ramp with slope $2\pi f_A$ radians per second, whose transform is $2\pi f_A/s^2$. From the error response formula (4.3-2), the PLL error response transform is

$$E(s) = \frac{2\pi f_A/s^2}{1 + G(s)}$$

where $G(s)$ is the open-loop general form (4.3-3) with $m = 1$. From the final value theorem (4.3-4) and Table 4.1, the steady-state phase error is

$$e_{\text{ss}} = \lim_{s\to 0} sE(s) = \lim_{s\to 0} \frac{2\pi f_A}{s[1 + G(s)]} = \frac{2\pi f_A}{K} \qquad \text{(rad)} \qquad (4.3\text{-}18)$$

Since $|e_{\text{ss}}| < \pi$, we have the hold-in condition

$$|f_A| < \tfrac{1}{2}K \qquad \text{(Hz)} \qquad (4.3\text{-}19)$$

which is a function solely of the loop gain.

For first-order and two-pole loops, the loop bandwidth B_N is $\frac{1}{4}K$ hertz (see Table 4.1), so that (4.3-19) becomes the very useful relationship

$$|f_A| < 2\,B_N \qquad \text{(Hz)} \qquad (4.3\text{-}20)$$

Consider, for example, phase synchronization by a quadrupler for 1-Mb/s QPSK. A normalized bandwidth $B_N T = 0.05$ translates to a hold-in range of

$$2B_N = \frac{2(0.05)}{T} = 2(0.05)500\ \text{kHz} = 50\ \text{kHz}$$

[4]For analysis, we assume the reference phase is a fixed target. In fact, it slowly varies in applications, and so it is more correct to say that the loop moves toward a tracking mode rather than a lock-up.
[5]With a multiplier/LPF phase detector, replace π in what follows by 1.

above and below the VCO center frequency. Since the VCO runs at 4 times carrier frequency, this is a ±12.5-kHz variation in the carrier. A temperature-compensated crystal oscillator can achieve this tolerance at 1 GHz, but not much higher.

The double-integral PLL has zero steady-state phase error with a ramp, but (4.3-16) shows that its maximum error on the way to zero error is about $1/4B_NT$ radians for a unit-per-T ramp. For this to be less than π with a $|f_A|$-Hz reference error, relation (4.3-20) again must hold. Thus all the PLLs discussed in this chapter have a hold-in range equal to plus or minus *twice their noise bandwidth*.

Very often in loop design we need a small bandwidth, which is in conflict with the need for a wide hold-in range. A special loop filter of form $F(s) = (1 + as)/(1 + bs)$, called a lead-lag filter, leads to a PLL whose hold-in range is not fixed by B_N. See further discussion in Refs. 10, 11, and 18.

A challenging problem in PLL analysis is the pull-in time, the time until the loop is solidly in the tracking mode. Pull-in time is strongly affected by operating conditions such as noise, and it is particularly difficult to estimate when the PLL moves through nonlinear operation. Gardner [10] gives the rule of thumb $3/K$ seconds for this time, which works out to $0.75/B_N$ for the first-order and two-pole loops.

4.4 PLL RESPONSES TO NOISE

Were it not for noise, PLLs would all be wideband and there would be little challenge in their design. With noise we come to the real trade-off and to the possibility that an acceptable trade-off may not exist. The important types of noise are the background noise in the channel, data noise, which is introduced into the PLL by the modulated signal itself, and oscillator phase noise. The PLL should ignore the first two as much as it can. It should track the last one, if it can.

4.4.1 Gaussian Channel Noise

We will assume that the transmitted signal is corrupted by additive white Gaussian noise with two-sided density $\frac{1}{2}N_0$. This is, in any case, the most common practical situation, and a sufficiently accurate analysis of the PLL phase response is not too difficult. The derivation is rather long but the outcome is simple, so we will give it at the outset: The PLL phase error is a nearly Gaussian variable. Its variance obeys the simple relation

$$\sigma_\theta^2 = \frac{B_NT}{E_s/N_0} \tag{4.4-1}$$

where E_s/N_0 is the standard symbol-energy-to-noise ratio, and B_NT is the symbol-normalized PLL bandwidth, in hertz-second/symbol. Equation (4.4-1) simply states that the noise-induced jitter in the VCO phase is the ratio of the normalized bandwidth to the signal SNR.

The analysis that follows applies to a PLL that observes a noisy but unmodulated signal. It thus applies to synchronizers such as the remodulation and Costas loops but not directly to raised-power synchronizers, which pass the signal plus noise first through a nonlinearity. We will add a modification later to account for that. The analysis makes only one approximation: that the VCO phase offset θ_0 is *quasi-static*. Practically, this means that the channel SNR is high and the loop bandwidth is narrow, so that θ_0 changes slowly.[6] With the aid of Fig. 4.18, the analysis goes as follows. Somewhere in the receiver, a bandpass filter, denoted "Front End Filter" in the figure, reduces the AWGN $\eta(t)$ to bandwidth B hertz centered around the carrier frequency. The signal is unaffected by the filter; the noise is reduced to a Gaussian bandpass noise $\eta_{\mathrm{FE}}(t)$, which has power $(2B)(\frac{1}{2}N_0) = BN_0$. This noise can be expressed in the quadrature noise form

$$\eta_{\mathrm{FE}}(t) = \eta_{c,\mathrm{FE}}(t) \cos \omega_0 t - \eta_{s,\mathrm{FE}}(t) \sin \omega_0 t$$

in which $\eta_{c,\mathrm{FE}}$ and $\eta_{s,\mathrm{FE}}$ are independent Gaussian low-pass noises with power spectral density N_0 in the range $(-\frac{1}{2}B, \frac{1}{2}B)$ hertz and zero elsewhere. (For more see Appendix 3B.) After the signal plus $\eta_{\mathrm{FE}}(t)$ passes through the multiplier/LPF, the result is

$$\begin{aligned} &[A \cos(\omega_0 t + \psi_0) + \eta_{c,\mathrm{FE}}(t) \cos(\omega_0 t) - \eta_{s,\mathrm{FE}}(t) \sin(\omega_0 t)][-2 \sin(\omega_0 t + \theta_0)] \\ &\qquad = A \sin(\psi_0 - \theta_0) - \eta_{c,\mathrm{FE}}(t) \sin \theta_0 + \eta_{s,\mathrm{FE}}(t) \cos \theta_0 \qquad (4.4\text{-}2) \\ &\qquad\quad + \text{bandpass terms removed by LPF} \end{aligned}$$

Only simple trigonometric identities are used here. After the LPF, we can think of the total signal as $\sin(\psi_0 - \theta_0) - \eta'(t)$, in which the noise-free phase difference signal $\sin(\psi_0 - \theta_0)$ is corrupted by the noise

$$\eta'(t) = \frac{1}{A} \eta_{c,\mathrm{FE}}(t) \sin \theta_0 - \frac{1}{A} \eta_{s,\mathrm{FE}}(t) \cos\theta_0 \qquad (4.4\text{-}3)$$

Recall that in the noise-free PLL analysis, the factor A from the signal was incorporated into the loop gain K. The effect of continuing to do this now is to bring a factor $1/A$ into the noise (4.4-3).

The noise $\eta'(t)$ is Gaussian, since it is a weighted sum of two white independent Gaussians, it is zero mean, and its power spectral density is consequently $(1/A^2)(\sin^2 \theta_0 + \cos^2 \theta_0)N_0 = N_0/A^2$ in the range $(-\frac{1}{2}B, \frac{1}{2}B)$. In its response to this noise, the multiplier/LPF PLL will behave exactly like the model shown in Fig. 4.19. The noise response alone will be

$$C(s) = H(s)N'(s) \qquad (4.4\text{-}4)$$

[6]Without these assumptions, the VCO phase takes on the Tikhonov distribution (see Ref. 5). The Tikhonov converges to the Gaussian distribution as the SNR grows or B_N narrows.

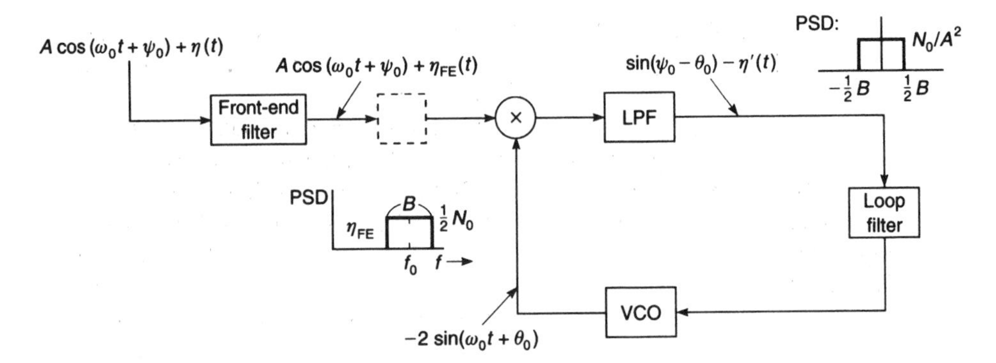

Figure 4.18 Standard multiplier/LPF phase-lock loop of Fig. 4.1a, with noise variables added. Receiver front-end filter reduces AWGN $\eta(t)$ to a Gaussian noise $\eta_{FE}(t)$ of bandwidth B; $\eta'(t)$ is noise excitation to loop. Nonlinearity may be added later at dashed box.

in which $N'(s)$ is the transform of the Gaussian process outcome $\eta'(t)$. From stochastic process theory, $C(s)$ is the transform of a Gaussian process whose PSD is $(N_0/A^2)|H(j2\pi f)|^2$. We take this process as a jitter component $\theta(t)$ in the VCO phase, and a sample has variance

$$\sigma_\theta^2 = \int_{-\infty}^{\infty} \frac{N_0}{A^2} |H(j2\pi f)|^2 \, df$$

From the noise equivalent bandwidth definition (4.3-15), we may write (note $H(0) = 1$)

$$\sigma_\theta^2 = \frac{N_0}{A^2} 2B_N \tag{4.4-5}$$

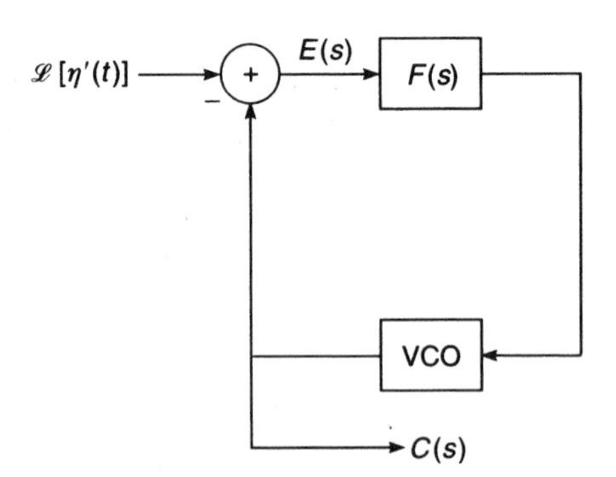

Figure 4.19 Equivalent loop to multiplier/LPF. PLL of Fig. 4.1a, considering noise alone. Shown in Laplace transform domain.

This jitter adds to the basic VCO offset θ_0, which varies much more slowly, according to its own separate dynamic.

The argument leading to (4.4-5) still imagines that the PLL views a simple sinusoid. We need to add in the fact that the sinusoid carries symbols in a framework of symbol times. In the standard modulated signal, the factor A^2 is $2E_s/T$, which sets E_s equal to the signal energy per symbol. Putting this in (4.4-5) yields the relationship (4.4-1).

It is important to realize that the symbol energy and not the data bit energy appears in (4.4-1). That is, synchronization depends on the physical energy of the transmission symbol, not on the energy allotted to a customer data symbol. The distinction can be painful for a transmission with ordinary parity-check coding, as shown in the following example. (Coding is treated in Chapter 6.)

Example 4.4-1: PSK with and without Coding. Consider three transmission systems that all run at the approximate data bit error rate 10^{-5}: QPSK, BPSK, and BPSK with a powerful rate $\frac{1}{2}$ parity-check code that provides a 6-dB coding gain. The channel is characterized by Gaussian noise and phase jumps of significant size. The ratio E_b/N_0 for the QPSK system will need to be about 10 dB in order to guarantee the bit error rate (BER), which means that E_s/N_0 is about 13 dB (since $E_s = 2E_b$). For the BPSK, both E_s/N_0 and E_b/N_0 need to be about 10 dB. For the coded BPSK, $E_s = RE_b = \frac{1}{2}E_b$; furthermore, the 6-dB coding gain means that E_b/N_0 needs to be only $10/10^{6/10} = 2.5$; consequently, $E_s/N_0 = 1.25$. Table 4.2 lists the normalized PLL bandwidths according to (4.4-1) that are needed to assure some values of the rms jitter σ. From our earlier study of transient responses (see Figs. 4.13, 4.14, 4.16, and 4.17), it was clear that we would like to have B_NT in the range 0.5–1, if at all possible, and this does seem reasonable for QPSK and uncoded BPSK. (BPSK can withstand twice the VCO error that QPSK can, so a wider tolerance is acceptable for BPSK.) But the coded BPSK system needs a much slower PLL, and so we can expect trouble with transient responses. The culprit here is that the coded system needs little physical signal energy and the data bit energy that it does need is spread over two physical symbol intervals. These thoughts are continued in Example 4.5-2 of the next section.

One topic remains in extending this straight PLL analysis to an actual synchronizer: the effect of a nonlinearity in the dashed box in Fig. 4.18. With a signal-plus-noise

TABLE 4.2 Normalized PLL Bandwidths Required for Phase Jitter σ

System	B_NT at Selected Phase Jitter Values		
	$\sigma = 0.1$ rad	$\sigma = 0.2$ rad	$\sigma = 0.3$ rad
QPSK	0.2	0.8	1.8
BPSK	0.1	0.4	0.9
BPSK/coding	0.013	0.05	0.11

input, a nonlinearity tends to worsen the SNR. In a raised-power synchronizer with a squarer, a good rule is that the SNR at the squarer output is 3 dB worse than that at the input. By taking the cases of a constant signal $s(t)$ or a pure sinusoid, it is not difficult to see that the output $s^2(t) + 2\eta(t)s(t) + \eta^2(t)$ might obey this. In the fourth-power case, the worsening may be as much as 6 dB. These losses translate through (4.4-1) directly into a worsening of the VCO jitter power by the same ratio. They are referred to generally as *squaring losses*. For a further discussion, see Refs. 4 and 6.

4.4.2 Phase Noise

Oscillators, unfortunately, do not put out a pure tone. The phase of their oscillation always varies randomly, at least to a small degree—a random process called phase noise. This process needs to be distinguished from oscillator *drift,* which is a slow change in the center, or "nominal," frequency of the noise process. In most cases where it is significant, phase noise appears in the transmitter oscillator or in receiver mixer oscillators leading up to the synchronizer circuit; in such cases, it constitutes a bona fide shift in the carrier that must be tracked. The synchronizer bandwidth will need to be wide enough to allow sufficiently agile tracking. Only when phase noise appears in the synchronizer VCO does it need to be ignored, in which case the synchronizer bandwidth might need to be reduced. Occasionally, these needs for large or small bandwidth conflict with the need to ignore other kinds of noise or to follow transients.

The usual design philosophy with phase noise is to design the synchronizer against other types of disturbance, such as channel or data noise, and then specify oscillators whose phase noise is insignificant in comparison. It becomes more difficult to do this as the symbol rate $1/T$ diminishes in proportion to the oscillator frequency f_0. What is important is the ratio of the two, $(1/T):f_0$. For example, telephone line data transmission by standard QAM employs $1/T = 2400$ Hz. The transmission passes through various mixing stages in the local telephone network, and if the highest of these is, say, at 2 MHz, then the ratio $1/Tf_0$ is 2400 Hz/2 MHz = 0.0012. If the same transmission passes through an 800-MHz cellular radio link, the ratio reduces to 3×10^{-6}. Let Δf be the deviation from f_0 due to phase noise. Then the ratio needs to be significantly greater than $|\Delta f|/f_0$. To a first approximation, the bandwidth of phase noise grows in proportion to the oscillator frequency, meaning that $|\Delta f|/f_0$ is constant. Thus, reducing $1/Tf_0$ eventually leads to trouble.

The phase noise properties of a number of oscillators appear in Fig. 4.20, which is taken from Ref. 7. The plot shows the noise power spectral density in decibels versus the deviation from the oscillation center, expressed as $\log_{10}[|\Delta f|/f_0]$. Better oscillators appear to the lower left. The sources of phase noise are complex and not well understood, but in simple terms we can say that there are two. The first produces a flat noise floor at frequencies far from the oscillation center and may be thought of as caused by the random motion of charge carriers. Near the center, the noise density begins to rise, and it eventually assumes an inverse third-power law, $1/|\Delta f|^3$. This second source is called frequency flicker noise. An everyday oscillator in Fig. 4.20, the 100-MHz crystal oscillator (shown dashed), provides a useful ex-

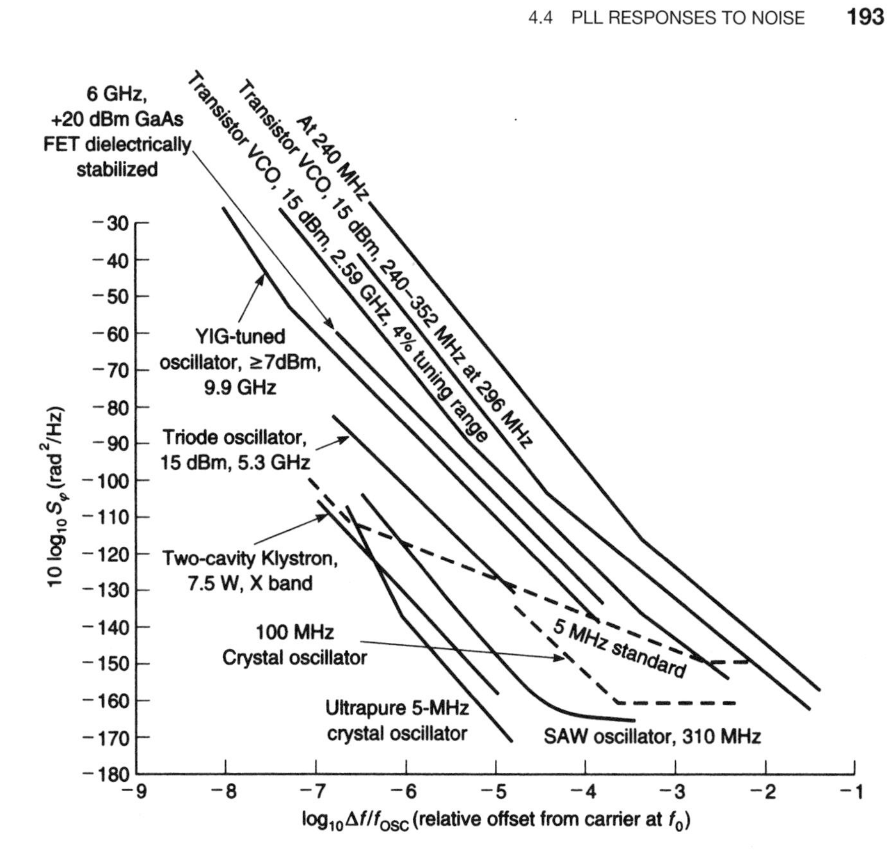

Figure 4.20 Power spectral densities (in dB) of phase noise of selected oscillators. (From Ref. 7. Copyright W. F. Egan, 1981. Used by permission.)

ample. The flat noise floor lies at about –160 dB and ranges inward to within 1 part in 10,000 of f_0; that is, the noise is flat and very weak outside the range 100 MHz ±10 kHz. Closer in to f_0, the noise spectrum rises at the rate of 30 dB/decade.

The simplified mathematical law here is useful, but since the integral near f_0 of a density of the form $1/|f-f_0|^3$ is infinite, the law clearly falls apart if pushed too far. We may content ourselves with the idea that at f/f_0 very close to 1, oscillators should be thought of as stochastic processes with a peaked spectral density. Measurements show that the 100-MHz crystal ceases to act like a pure oscillator for $|\Delta f|/f_0$ less than perhaps 10^{-7} or 10^{-8}, which is 1–10 Hz on either side of 100 MHz. For either of the 2400-Hz symbol rate examples, the central phase noise peak will be present and will create a variation in the carrier phase observed by the synchronizer, but so long as the synchronizer bandwidth exceeds the effective variation bandwidth, the synchronizer will successfully track.[7] An exception would occur if

[7]Actually, the instability in an uncompensated crystal oscillator is larger than $|\Delta f/f_0| = 10^{-6}$ because of long-term drift, but temperature compensation improves this.

the synchronizer bandwidth had to be very narrow because of channel noise (see, e.g., Example 4.5-3), so that it could not track variations whose frequencies lay in the range $|\Delta f|/f_0 < 10^{-8}$; then it would be necessary to widen the loop bandwidth and make a compromise between tracking the carrier phase noise and ignoring the channel noise. Further discussion of such compromises appears in Ref. 4.

4.4.3 Data Noise

In Section 4.2, Eqs. (4.2-1)–(4.2-3), we saw that the squaring of an NRZ pulse BPSK signal produces a $\cos(2\omega_0 t)$ term alone in the region near $2\omega_0$ but that squaring a shaped-pulse BPSK produces, in addition, a disturbance term in that region. A PLL, working as part of a raised-power synchronizer, will be disturbed by the new term. Even a remodulation loop will suffer a similar disturbance when the modulation lacks a constant envelope. This is clear from Fig. 4.4, where the time-varying signal envelope $A(t)$ leads to a time-varying phase error signal $e(t) = A^2(t)\sin(\psi_0 - \theta_0)$, even when the phase difference is constant. Generically, these disturbances are called data noise.

Dealing with data noise is a complex subject, and there are probably several techniques to think about for each modulation that produces noise. We will treat the subject by exploring the two most common examples, pulse-shaped BPSK and QPSK. Some important techniques that arise are the measurement of data noise, the use of limiter circuits, and removing noise with a narrowband PLL. The QPSK example is particularly challenging, and the details of it form Appendix 4A.

We take first the BPSK case. The pulse-train-modulated BPSK signal $I(t)\cos(\omega_0 t + \psi_0)$ is applied to the circuit in Fig. 4.2. Here, $I(t)$ is a superposition of a train of shaped pulses that carry the binary data. After the squarer/band pass filter in the circuit, what remains is the term $\frac{1}{2}I^2(t)\cos(2\omega_0 t + 2\psi_0)$, in which ψ_0 is the slowly varying reference phase offset. As an example for data noise removal, BPSK is very much a special case because we can remove all the data variation from this signal with a limiter. This is a high-gain saturating amplifier that chops off the tops of all the sine wave excursions in the squarer/BPF output, to produce a square wave with the same crossings as $I^2(t)\cos(2\omega_0 t + 2\psi_0)$. If this square wave passes through another bandpass filter centered at $2\omega_0$, the result will be the sinusoid $A\cos(2\omega_0 t + 2\psi_0)$ with the square-wave corners cleaned away. (The factor A is set by the limiter.) The combination of the limiter and the filter is a common circuit element called a *bandpass limiter*. When its output is applied to the PLL in Fig. 4.2, the result is the clean error signal $A\sin 2(\psi_0 - \theta_0)$, which is mostly free of data noise.

Now suppose the modulation is the standard QPSK signal $I(t)\cos(\omega_0 t + \psi_0) - Q(t)\sin(\omega_0 t + \psi_0)$, in which $I(t)$ and $Q(t)$ are independent binary pulse trains and ψ_0 is the usual phase offset. In a typical case, I and Q are made up of raised-cosine (RC) pulses. With simple trigonometric identities, the standard signal may be written as

$$s(t) = A(t)\cos\left[\omega_0 t + \psi_0 + \tan^{-1}\left(\frac{Q(t)}{I(t)}\right)\right] \tag{4.4-6}$$

in which

$$A(t) = \sqrt{I^2(t) + Q^2(t)}$$

After the fourth power/BPF circuit block in Fig. 4.3, only the term

$$\frac{A^4(t)}{8} \cos\left[4\omega_0 t + 4\psi_0 + 4\tan^{-1}\left(\frac{Q(t)}{I(t)}\right)\right] \tag{4.4-7}$$

remains, with the many other fourth-power terms having been blocked by the BPF. At this point, we could apply (4.4-7) to a PLL, or we could pass it through a bandpass limiter that would remove the $\frac{1}{8}A^4(t)$ factor, and then apply the remaining sinusoid to a PLL. It often does not matter much which is done, although as a general rule it is a good idea to remove data-driven envelope variation when possible. We will insert the bandpass limiter and remove the variation. Consequently, the multiplier at the input of the PLL in Fig. 4.3 performs the operation

$$[-2\sin(4\omega_0 t + 4\theta_0)] \cos\left[4\omega_0 t + 4\psi_0 + 4\tan^{-1}\left(\frac{Q(t)}{I(t)}\right)\right]$$

$$= \sin\left[4(\psi_0 - \theta_0) + 4\tan^{-1}\left(\frac{Q(t)}{I(t)}\right)\right] + \text{bandpass terms}$$

The LPF after the multiplier removes the bandpass terms, and with trigonometric identities we can rewrite the remaining term as

$$\sin 4(\psi_0 - \theta_0) \cos\left[4\tan^{-1}\left(\frac{Q(t)}{I(t)}\right)\right] + \cos 4(\psi_0 - \theta_0) \sin\left[4\tan^{-1}\left(\frac{Q(t)}{I(t)}\right)\right] \tag{4.4-8}$$

The first term here appears to be a phase error measure $\sin 4(\psi_0 - \theta_0)$, multiplied by a noisy factor; the last term is a factor $\cos 4(\psi_0 - \theta_0)$ hopefully near 1, multiplied again by a noisy factor. From these we must obtain an indication of phase error. The success of that depends on an analysis of the terms, and that appears in Appendix 4A. For QPSK, the jitter due to data noise satisfies $\sigma_\theta^2 \approx B_N T$. This is a rough estimate, but it is usually close enough to use in a successful design.

4.5 DESIGN EXAMPLES

We turn now to three synchronizer examples that show the roles played by noise and transients in different designs. Most often, one of these dominates the problem and the other may be ignored. Sometimes, as in the second example, several disturbances play competing roles, and then it is necessary to find a design that balances the responses so that the sum of them is minimized. If the result leads to an unacceptable error rate, the remaining alternative is generally to increase the signal

E_b/N_0. Here we see the acting-out of some of the principles discussed at the beginning of the chapter: the trade-off in loop design between noise and transient response, the loop's need for signal energy, and the relation between energy and signal stability.

The focus here is on analog loop design and, except for Example 4.5-3, on carrier phase tracking. The principles, however, apply generally: There is always a ticking clock to be tracked, whose ticks may be missing or obscured by noise, and a physical time needed to track changes within a certain accuracy.

Example 4.5-1: Pulse-Shaped QPSK in a Noise-Free Channel. In this example, QPSK with RC pulse shaping is to be transmitted at the standard "T1" data bit rate, 1.544 Mb/s. The carrier frequency is 1 MHz. The channel is a relatively short coaxial cable; as a consequence, the channel is almost free of noise and distortion. The entire transmission is multiplexed together out of successive bursts of data from different origins, and there may be phase shifts as large as 0.3 rad between adjoining bursts. An everyday example is a high-speed data link between workstations in the same building. We are to design a quadrupling-loop synchronizer.

The hallmarks of this design problem are the lack of channel noise, a set of moderate phase transients, and pulse shaping, from which data noise will arise. A synchronizer need only smooth the data noise and track the 0.3-rad transients such that the error in its output phase lies comfortably inside the range $(-\frac{1}{4}\pi, \frac{1}{4}\pi)$. Theoretically, a QPSK demodulator with good symbol timing will then never make errors. As for the data noise, we will assume that the RC pulse shaping has excess bandwidth factor $\alpha = 0.3$ and use the measurements of the noise obtained in Appendix 4A. The conclusion there is that the signal-to-data-noise ratio after the phase quadrupling is close to unity. Specifically, the quadrupler output is interpreted as a mean value $\mu \sin 4(\psi_0 - \theta_0)$, plus noise of power σ_n^2; the SNR seen by the loop is $(\mu/\sigma_n)^2$. At phase error $\psi_0 - \theta_0 = 0.1$, $(\mu/\sigma_n)^2$ was approximately unity, and we will assume that it remains near there at other phase errors.

To summarize, here are the system specifications.

Carrier: 1 MHz, subject to 0.3-rad transients

$1/T_s = 1.544$ MHz/2 = 0.772 MHz

Modulation: QPSK, RC pulses, $\alpha = 0.3$ (yielding double-sideband RF bandwidth 1.0 MHz, centered at 1 MHz)

Error rate: nominally zero

Synchronization: quadrupler loop (Fig. 4.3)

Carrier phase error $|\psi_0 - \theta_0| < 0.6$ rad at all times

Solution

(i) *Tracking the Transient Response.* Since there is no appreciable frequency transient (phase ramp) to track, we will investigate a two-pole PLL; its step response may be faster than other designs. Since the specification on the output phase (0.6 rad) is larger than the transient expected at the input, any

PLL design will work, so long as its damping is not too small. However, the transient response will add to the data noise response.

(ii) *Data Noise.* The real purpose of the PLL in this synchronizer is to average out the data noise. Our strategy will be to keep the data noise response very small so that the total phase error will never exceed 0.6 rad; consequently, we will attempt to reduce the SNR from near unity at the PLL input to a σ_θ of about 0.06 rad at the synchronizer output. This places the noise response at a small fraction of the transient response. In a quadrupler loop, the output phase error is one-fourth of the VCO phase error, so σ_θ for the VCO can be as large as 0.24. This σ_θ implies a PLL bandwidth of roughly σ_θ^2, or more precisely, from (4A-3),

$$B_N T = \left(\frac{\mu}{\sigma_n^2}\right)\sigma_\theta^2 B_{eff} T \approx (1)(0.24)^2(1.5) \approx 0.086 \text{ Hz-s/symbol} \qquad (4.5\text{-}1)$$

(iii) The step response of such a loop is like Fig. 4.13b, but a little faster. With a damping of $\zeta = 1$, a transient of 0.3 rad before the quadrupler will create a synchronizer output transient of 0.1–0.3 rad for about five symbol intervals. On top of this will lie a small data noise component, and so the design seems satisfactory. From Table 4.1, the PLL, with VCO centered on 4 MHz, is described as follows:

$$\begin{aligned} \omega_n &= 8\zeta B_N = 8(1)(0.086)(0.772 \text{ MHz}) \approx 500000 \text{ rad/s} \\ H(s) &= \frac{(500000)^2}{s^2 + 2(500000)s + (500000)^2} \end{aligned} \qquad (4.5\text{-}2)$$

the gain parameter K and T_a are

$$\begin{aligned} K &= 4B_N = 4(0.086)(0.772 \text{ MHz}) \approx 250{,}000 \\ T_a &= \frac{1}{4K\zeta^2} = \frac{1}{(4)(250000)(1)} = 1 \times 10^{-6} \end{aligned} \qquad (4.5\text{-}3)$$

and the loop filter is $250000/(1 + 10^{-6}s)$. Figure 4.21 is a possible implementation of the loop filter.

Comments. The lack of channel noise is an important factor in this synchronizer design. With noise and phase error at the same time, the QPSK demodulator will have an unacceptable error rate; a wideband loop will be needed to reduce the PLL's transient error enough. But this conflicts with the requirement to smooth the data noise. In addition, the fourth-power box in Fig. 4.3 would create squaring noise. The only alternative would be adding a lot of transmission power, which would reduce the solution to the one here.

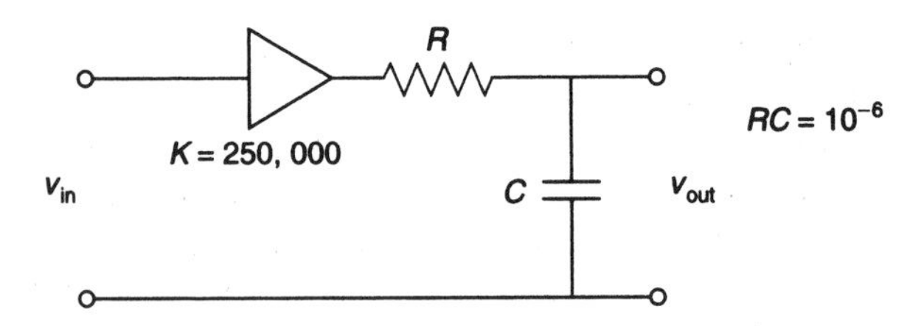

Figure 4.21 Simple circuit implementing loop filter in solution for Example 4.5-1. VCO and phase detector constant assumed to be unity.

Example 4.5-2: A Deep-Space BPSK Link. In this example, the object is to transmit data from a deep-space vehicle such as the recent Voyager probe that visited the outer planets. Because of great distance and weak solar power, the transmission power that reaches Earth is very low. Here E_b/N_0 is kept high enough only by reducing the symbol rate and employing an error-correcting code. On the other hand, there is little competition for bandwidth in outer space, so that NRZ pulse BPSK is a good choice, with its wide bandwidth but easier synchronization. Aside from a high level of Gaussian noise, the only impairment expected in the transmission is a phase ramp caused by changes in vehicle motion and by oscillator temperature shifts.

Here are some initial specifications.

Modulation: NRZ pulse BPSK, $1/T_s = 1200$ symbols/s

Carrier: 4 GHz, subject to frequency transients up to ±200 Hz

Error-correcting code: rate $\frac{1}{4}$, coding gain 6 dB (yielding data rate 300 b/s)

Synchronization: squaring loop (Fig. 4.2)

Bit error rate: 10^{-5}

A sketch of the overall receiver appears in Fig. 4.22.

We will investigate the design of two-pole and double-integral squaring loops for carrier synchronization of the BPSK. A design needs to resolve the conflict between tracking the phase ramp disturbance, which requires a wide bandwidth, and ignoring the channel noise, which requires a narrow bandwidth. What carrier phase

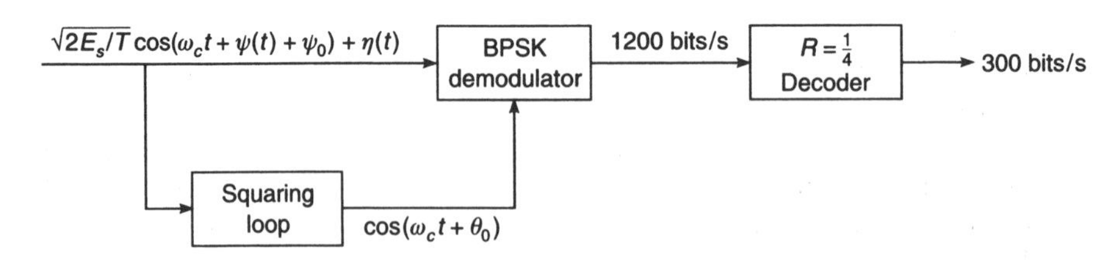

Figure 4.22 Deep-space BPSK receiver block diagram.

error can the demodulator tolerate? With no channel noise, an NRZ pulse binary demodulator can theoretically tolerate an error up to $\pm\frac{1}{2}\pi$ rad. With noise, and a phase error of 0.9 rad, Eq. (4.2-7) predicts that 4 dB more energy is needed to maintain a given error rate. A reasonable starting point might be to specify as follows, at the synchronizer output:

$$\begin{aligned} |\text{error}| \text{ due to ramp} &< 0.6 \text{ rad} \\ \sigma_\theta \text{ due to AWGN jitter} &< 0.3 \text{ rad} \end{aligned} \tag{4.5-4}$$

Then the typical error under a worst-case ramp will be about 0.9 rad. With this view, we will need to augment E_b by 4 dB to make up for synchronizer error. The phase error at the output of a squaring loop is half that of the PLL it contains, so that the specifications on the PLL are 1.2 and 0.6 rad instead of the 0.6 and 0.3 in (4.5-4).

First Solution Attempt

(i) We need to compute E_s, the symbol energy available for synchronization in the underlying BPSK. Were it not for coding, the error rate of 10^{-5} would be attained approximately with $E_s/N_0 = E_b/N_0 = 10$. With a rate $\frac{1}{4}$ code, four BPSK symbols represent one data bit, so that $E_s = \frac{1}{4}E_b$; furthermore, the coding gain reduces E_b by another 6 dB, so that $E_s/N_0 = \frac{1}{4}(10^{-0.6})(10) = 0.628$, or –2.0 dB. Finally, we need to add back the 4 dB that overcomes the effect of synchronizer error, which will give $E_s/N_0 = (.628)(10^{0.4}) = 1.58$.

(ii) With this E_b/N_0, we can solve (4.4-1) for a PLL bandwidth that meets the jitter specification. Before the noise reaches the PLL, it is augmented by 3 dB or so in the squarer (the "squaring loss"), which reduces E_s/N_0 at the PLL input by half. From (4.4-1), we obtain

$$B_N T = \sigma_\theta^2 \left(\frac{E_s}{N_0} \frac{1}{2} \right) = (0.6)^2(1.58)\left(\frac{1}{2} \right) = 0.28 \text{ Hz-s/symbol} \tag{4.5-5}$$

(iii) Next we can solve for a PLL bandwidth that meets the ramp disturbance specification in (4.5-4). The ramp itself has worst-case slope S set by the 200-Hz carrier frequency error, which is

$$S = \frac{2\pi(200 \text{ Hz})}{1200 \text{ symbols/s}} = \pi/3 \text{ rad/symbol}$$

After the squarer, this is doubled to $2\pi/3$. For a two-pole loop to attain a steady-state phase error of 2(0.6) rad, Eq. (4.3-16) tells us that

$$B_N T = \frac{S}{4e_{ss}} = \frac{2\pi/3}{4(1.2)} = 0.44 \text{ Hz-s/symbol} \tag{4.5-6}$$

A DI loop has zero steady-state error, but (see Section 4.3) its transient error will reach the same value, 1.2 rad.

(iv) Now to design the PLL. Unfortunately, the bandwidths (4.5-5) and (4.5-6) are in conflict; that is, the bandwidth needed to reject the noise ($B_N T = 0.28$) is too small to follow the expected phase ramps. One approach is to go ahead and design the PLL at the wider bandwidth. The DI loop seems the most sensible because its ramp error is at least only transient. From Fig. 4.17, a damping near 1 seems reasonable. From Table 4.1, $\zeta = 1$ and $B_n = 0.44/T$ mean that

$$\omega_n = \frac{B_N}{\zeta/2 + 1/8\zeta} = \frac{(0.44)(1200)}{0.625} = 845 \text{ rad/s}$$
$$H(s) = \frac{2(845)s + (845)^2}{s^2 + 2(845)s + (845)^2} \tag{4.5-7}$$

The VCO is centered at 8 GHz.[8] Figure 4.23 shows the actual response of this PLL to the size $2\pi/3$ worst-case ramp, together with the response of the PLL with $B_N T = 0.28$. The narrower loop would swing outside the 1.2-rad specification for perhaps four symbol intervals. Altogether, the worst-case ramp disturbs the design PLL for about 5 symbols and disturbs the narrower loop for about 10.

Second Solution. As now designed, the synchronizer may not quite meet the error rate specification. The jitter in its output will drive the carrier phase error in the BPSK demodulator during worst-case ramps to relatively large values, which, combined with the poor channel SNR, will lead to a detection error rate poorer than 10^{-5}. One way or another, a better E_s/N_0 is needed in the channel. For example, increasing E_s/N_0 by the factor 1.57 (2.0 dB) leads to $B_N T = 0.44$ in the jitter equation (4.5-5), so that a PLL with bandwidth 0.44 now satisfies conditions (4.5-4). At the same time, the demodulator will now have a lower error rate when by chance the conditions (4.5-4) are not satisfied.

It is interesting to observe that the E_s/N_0 required by a BPSK-coded modem with a perfect carrier reference was 0.628. Without one, in the real link E_s/N_0 may need to be as high as 2.5. Here the cost of synchronization is 10 log 2.5/0.628 = 6 dB.

A Third Solution. Another, more subtle solution exists. Spacecraft electronics can often be reprogrammed from the ground. Suppose we replace the rate $\frac{1}{4}$ encoder with a rate $\frac{1}{2}$ one, like the one in Example 4.4-1. (A look at Chapter 6 will show that rate $\frac{1}{2}$ convolutional codes exist with 6-dB gain and about the same complexity as our rate $\frac{1}{4}$ code.) So that the data rate remains 300 b/s, the BPSK modem now runs at $1/T_s = 600$ symbols/s. Now E_s and E_b satisfy the relationship $E_s = \frac{1}{2}E_b$, which is to say that E_s is 3 dB higher throughout for the same E_b, compared to the rate $\frac{1}{4}$ transmission. The synchronizer feeds on more energy and it should work better.

[8]More likely, the 4-GHz signal is mixed down to a convenient intermediate frequency, and the VCO centers on twice that.

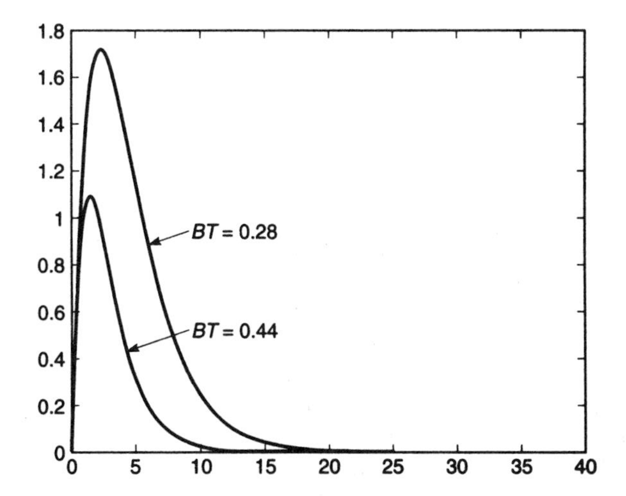

Figure 4.23 The VCO ramp phase responses to 2.09 ramp(t) for two double-integral PLLs having $B_N T = 0.44$ and 0.28 Hz-s/symbol. The axes are radians versus time in symbols. Synchronizer outputs after the frequency is divided by 2 are half those shown.

Unfortunately, a repeat of the first solution attempt will not yet give the desired outcome. Repeating (4.5-5) with the doubled symbol energy gives $B_N T = 0.56$, a PLL with twice the normalized bandwidth. But repeating (4.5-6), in which the ramp slope per symbol is now doubled, leads to $B_N T = 0.88$, which is still in conflict with (4.5-5). The root of the difficulty can be found by thinking about the physical reality seen by the synchronizer. Once it has squared the BPSK signal, it is simply a PLL observing an 8-GHz sine wave. The wave drifts in frequency and has background noise. Its power is the same in either case, since with either code the energy arriving per second is 300 E_b. In fact, the solution of the form (4.5-7) for the physical loop leads to the same PLL, because although $B_N T$ is now 0.88, T is doubled and B_N has the same numerical value that it has in (4.5-6).

At least one way out exists, however. We could retain the 1200 symbols/s BPSK rate, keep E_b the same, and transmit for half the time, in bursts perhaps. Now the physical transmission power truly is twice as large while a burst lasts, but long term, the power remains unchanged and the data bit rate is still 300 b/s. If the spacecraft allows this, (4.5-5) then gives $B_N T = 0.56$, (4.5-6) gives 0.44, and both are in harmony.

Conclusions. This example demonstrates the problems encountered when channel SNR is very low. The points raised are subtle and worth reiterating:

- The final level of E_s/N_0 was set by the synchronizer and not by the demodulator/decoder; as much as 6 dB more was needed.
- The synchronizer's hunger for energy would be worse with a more complicated modulation, such as QPSK, because a more accurate phase reference would be required.

- Low-rate parity-check codes may not be as useful as they appear.
- Simple phase transients may not be the dominant factor in a design.

Example 4.5-3: Cellular Radio Symbol Timing. In North America, the IS-54 digital cellular radio standard specifies a 48-kb/s data stream transmitted by a kind of QPSK with an 850-MHz carrier. Actually, three separate users alternate sending 162 symbol bursts at this speed in time division, so that each sees an average 16-kb/s data rate, but we need not concern ourselves with this now. (Further details are in Example 4.8-4.) The chief impairment in the channel is Rayleigh fading. Symbol timing is obtained by observing zero crossings in the in-phase baseband signal $I(t)$, forming a pulse train from these, and applying that to a PLL whose VCO is centered at the symbol rate, 24 kHz. The PLL is implemented in software and functions only during the data burst; otherwise, it holds its system state and "sleeps." Symbol timing is considered in detail in Section 4.7. For now, it is enough to think of the usual PLL, trying to track a 24-kHz oscillation with noisy phase. What is a good PLL design?

The disturbances to this transmission are very different from those in the preceding examples in this chapter, and so the design philosophy differs as well. Whereas the previous channels contained noise, phase transients, and data noise, the dominant problem with the channel here is that it fades at random and can even disappear entirely. The statistics of the phase change caused by these dropouts can be measured. At a vehicle speed of 80 km/h with an 850-MHz carrier, the bandwidth of the random fading disturbance to the symbol timing phase is about 70 Hz.[9] Alongside of fading, the other channel disturbances hardly matter. The zero crossings in $I(t)$ depend a little on the data, so there is data noise from this source. There may be oscillator drift, but this is compensated separately.

In this problem, then, the physical bandwidth of the PLL needs to be about 70 Hz—first, so that it tracks only real phase reference shifts and, second, so that its inertia carries it through deep fades. This means that $B_N T = 70(1/24000) = 0.0029$, a very small value. A startup transient will clear away faster with a two-pole loop than with a DI loop [compare Figs. 4.13b and 4.16c], and from Fig. 4.13b, $\zeta = 1$ seems to be a reasonable damping factor. With the aid of Table 4.1, $\omega_n = 8\zeta B_N = 560$ rad/s; the analog transfer function is

$$H(s) = \frac{(560)^2}{s^2 + 2(560)s + (560)^2} \tag{4.5-8}$$

Equivalently, this is a loop with gain $K = \omega_n/2\zeta = 280$ and loop filter $F(s) = 280/(1 + 0.00089s)$. For a discrete-time implementation, $H(s)$ will need to be converted to a z-transform, as in Section 4.6. Figure 4.24 shows the error response of (4.5-8) to a unit step.

---

[9]A complete discussion of random fading models appears in Section 5.4.

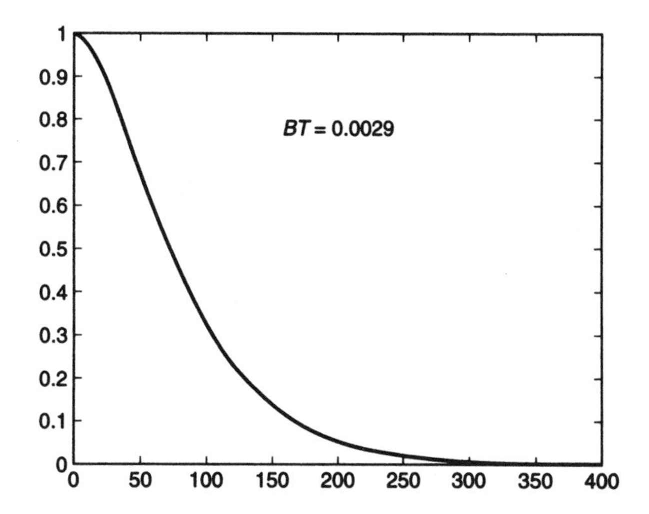

Figure 4.24 Unit step error response of solution to Example 4.5-3 (two-pole loop).

4.6 DIGITAL PHASE-LOCK LOOPS

Implementing a PLL in discrete time can have many advantages. The many kinds of digital hardware are often cheaper, more versatile, and more reliable than analog multiplier, VCO, and filter components. If the PLL can be implemented as *software,* then the hardware reality is a standard microprocessor of some kind, and the development process is simpler yet. The only major reason not to use a discrete-time implementation is speed: Digital hardware is slower than analog, when both are made from similar technology, and software running on the same technology in a processor will be considerably slower.

In addition to the cost advantage, digital PLLs can be more versatile. Consider the design examples in Section 4.5. In Example 4.5-1, the PLL bandwidth $B_N T$ needed to be very narrow in order to average out the data noise. Example 4.5-2 was a deep-space link that suffered from poor E_b/N_0, but space channels are otherwise very stable. So the phase acquisition could be made narrowband in order to make up for the poor SNR. Example 4.5-3 was a mobile radio channel that needed a very narrowband loop because of signal fading. With all these cases, the PLL structure can run at a much narrower bandwidth than the signal itself. The digital PLL, which can easily run at a submultiple of the symbol rate, offers a wider scope for design.

We will study digital PLLs by considering first the straight out conversion of the standard analog circuits of Sections 4.2–4.3. Next we will look at PLLs that differ but still resemble an analog loop at least somewhat. These loops can run at a submultiple of the sampling rate. Finally, we will look at digital PLLs that are quite different from the standard analog loop. These are closely related to symbol timing, and so we will discuss them in Section 4.7. Before beginning any of this, it will be useful to review the z-transform, which is the main analytical tool for digital PLLs.

4.6.1 Review of z-Transforms

Many excellent texts, such as Refs. 12 and 13, treat the z-transform. The bilateral z-transform of a sequence $\ldots, h_{-1}, h_0, h_1, h_2, \ldots$ is defined to be

$$H(z) = \sum_{n=-\infty}^{\infty} h_n z^{-n} \tag{4.6-1}$$

We will always take the sequence to be the samples $\ldots, h(-T), h(0), h(T), h(2T), \ldots$ at rate $f_s = 1/T$ of a function of time $h(t)$. With this assumption, Table 4.3 lists the most commonly used z-transforms together with their sample sequence, the $h(t)$, and the Laplace transform of $h(t)$. Since linear superposition applies to both transforms, the entries in the table can be combined linearly to give many more z-transforms and their corresponding Laplace transforms.

If $\{h(kT)\}$ is the isolated-pulse response of a linear time-invariant system at times kT, then for an input $\{x(kT)\}$ with z-transform $X(z)$, the transform of the output of the system is

$$Y(z) = H(z)X(z) \tag{4.6-2}$$

The *frequency response* of a discrete-time system $H(z)$ at ω radians per second is defined to be

$$|H(z)|\big|_{z=e^{j\omega T}} \tag{4.6-3}$$

When the Laplace transform $H(s)$ with corresponding $h(t)$ satisfies $H(j\omega) \approx 0$, $|\omega| > \pi/T$, then

TABLE 4.3 Useful z-Transforms and Related Laplace Transforms that Have Matching Time Response at Sample Times

z-Transform		Laplace Transform		
$h(kT), k = 0, 1, \ldots$	$H(z)$	Function	$h(t), t \geqslant 0$	$H(s)$
$A, 0, 0, \ldots$	A	Impulse	$A\delta(t)$	A
$A, A, \ldots$	$\dfrac{Az}{z-1}$	Step	$Au(t)$	A/s
AkT	$\dfrac{ATz}{(z-1)^2}$	Ramp	AT	A/s^2
$Ae^{-aTk} \sin \omega_0 Tk$	$\dfrac{Aze^{-aT} \sin \omega_0 T}{z^2 - 2ze^{-aT} \cos \omega_0 T + e^{-2aT}}$	Pole pair	$Ae^{-aT} \sin \omega_0 t$	$\dfrac{A\omega_0}{(s+a)^2 + \omega_0^2}$
$Ae^{-aTk} \cos \omega_0 Tk$	$\dfrac{A[z^2 - e^{-aT}(\cos \omega_0 T)z]}{z^2 - 2ze^{-aT} \cos \omega_0 T + e^{-2aT}}$	Pole pair	$Ae^{-aT} \cos \omega_0 t$	$\dfrac{A(s+a)}{(s+a)^2 + \omega_0^2}$
Ae^{pTk}	$\dfrac{Az}{z - e^{pT}}$	Simple pole	$Ae^{pt}u(t)$	$\dfrac{A}{s-p}$

$$H(j\omega) \approx H(z)|_{z=e^{j\omega T}} \tag{4.6-4}$$

and the two transforms have about the same frequency response. Otherwise, they do not, and the response of $H(z)$ is said to be *aliased*.

We will want to construct discrete-time circuits whose response is similar to analog circuits. The rows of the table illustrate a natural way of going between transforms, namely by matching up their discrete and analog time responses. The formal name of this idea is the *impulse invariant transformation*. For a particular system $H(z)$, we can imagine matching up other time responses, such as the step response or, in the case of a PLL, an error response. In these cases, the method is the same, but applied to the new response in time. In designing discrete-time filters, the most common method is quite a different one, the *bilinear transformation*; this one works by making the substitution

$$s = \frac{2}{T}\,\frac{z-1}{z+1} \tag{4.6-5}$$

into $H(s)$ in order to form $H(z)$. The bilinear transform, and its allied technique of frequency warping, generally give a better match between the frequency responses of $H(s)$ and $H(z)$, especially when there is some aliasing.

Sequences can often be described by a difference equation, which in turn can be written as a line of software code. In this case, the line of code can be said to have a z-transform. Suppose the general difference equation form is

$$\sum_{k=0}^{P} a_k\, y(n-k) = \sum_{k=0}^{Q} b_k x(n-k) \tag{4.6-6}$$

Then the transforms $Y(z)$ and $X(z)$ are related by $Y(z) = H(z)X(z)$, where

$$H(z) = \frac{b_0 + b_1 z^{-1} + \cdots + b_Q z^{-Q}}{a_0 + a_1 z^{-1} + \cdots + a_P z^{-P}} \tag{4.6-7}$$

A useful example of this technique is the discrete-time accumulator, or integrator, a device with difference equation $y(n) = y(n-1) + x(n-1)$. This is (4.6-6) with coefficients $a_0 = 1$, $a_1 = -1$, $b_0 = 0$, $b_1 = 1$, which implies by (4.6-7) that the summing junction has transform

$$H(z) = \frac{z^{-1}}{1 - z^{-1}} = \frac{1}{z-1} \tag{4.6-8}$$

4.6.2 Simple Conversion of Analog PLLs

By one of the s-to-z transformations just given, any of the analog loops in 4.2–4.3 can be converted to a z-transform, and then realized in software steps or in digital hardware. Software that performs the conversion is part of many commercial packages. MATLAB, for example, executes the bilinear transformation in (4.6-5), which will produce a discrete transform $H_D(z)$ that has the same order as a target analog transform $H_A(s)$.

The setup for simple conversion of a linear loop is in Fig. 4.25. Only *phase* signals in the transform domain are shown here, in the same spirit as Fig. 4.1b. In the analog version, the loop tries to follow the phase of a possibly noisy repetitive signal whose phase offset is $\psi_0(t)$. Although a VCO and loop filter are shown, only the total transfer function of the dashed box, $H_A(s)$, is of interest to us now. The PLL output is the VCO phase $\theta(t)$, the "controlled" signal, with transform $C(s)$. In the discrete-time version, the reference phase signal encounters first a prealias filter, which reduces its bandwidth to $1/2T_D$ hertz; then an analog-to-digital converter changes the signal into a sample sequence at rate $1/T_D$, whose transform is $R(z)$. The dashed box comprises the digital PLL, and we have shown an open-loop transfer function, although here again, only the total transfer function $H_D(z)$

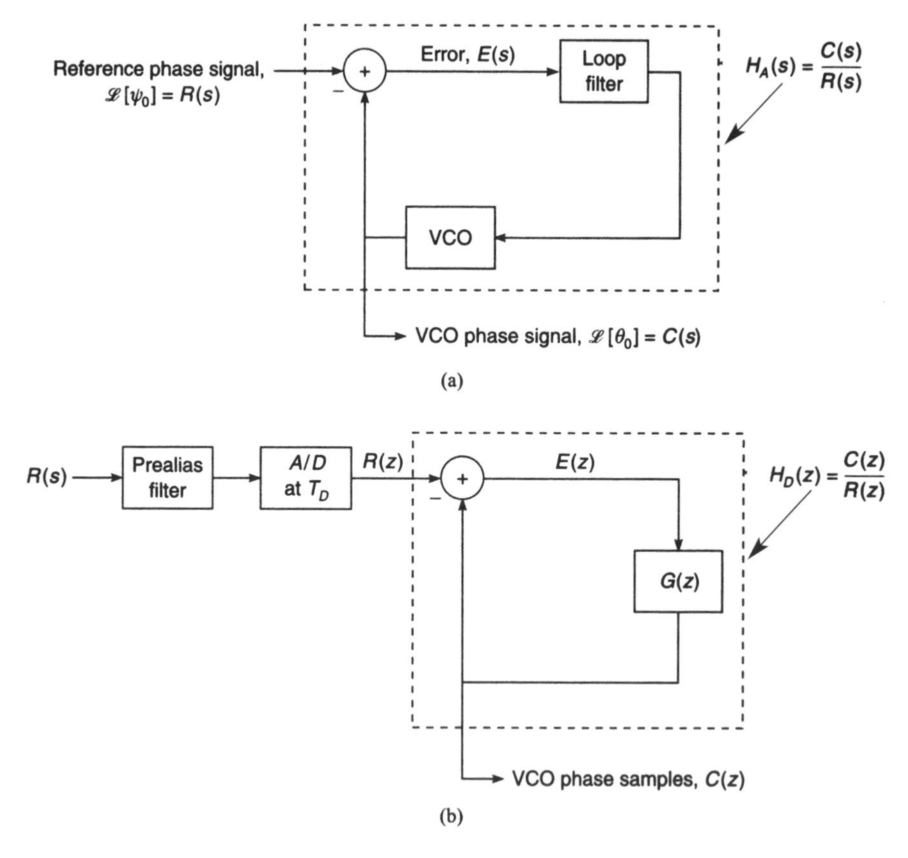

Figure 4.25 Transformation of **(a)** analog PLL to **(b)** equivalent discrete-time loop. Both loops attempt to lock to reference phase signal $\psi_0(t)$. All signals here are phases in transform domain.

$= G(z)/[1 + G(z)]$ matters. The PLL output is the VCO phase sample sequence with transform $C(z)$.

We aim now to match $H_A(z)$ to $H_D(s)$. This will, in general, be possible only if the sampling rate $1/T_D$ exceeds both twice the bandwidth of $\psi_0(t)$ and of the analog loop. In the example that follows, which revisits the QPSK synchronizer of Example 4.5-1, this is easily satisfied because the bandwidths in question are much less than the modulation symbol rate $1/T$. A word of caution, however. Any delay in the circuits of Fig. 4.25 is a delay in the response of the overall receiver system to a change in signal phase. In particular, the prealias filter in the discrete-time version is a new, additional delay. From circuit theory, the delay of any low-pass linear system is $\approx 1/f_c$, where f_c is the cutoff frequency of the system.[10] A PLL of bandwidth B, then, necessarily has delay about $1/B$. If the prealias filter in the discrete-time version is designed to the same cutoff, it too has delay $1/B$, and the delay in the whole synchronizer rises to about $2/B$. In receivers that cannot tolerate the double delay, the sampling rate can be raised somewhat, and the synchronizer response will be dominated, as it was in the analog version, by the PLL bandwidth.

A second word of caution about discrete-time loops generally is that they can be unstable even when the analog loops that have more or less inspired them are stable. One cause is aliasing during conversion of an analog $H_A(s)$ to an $H_D(z)$.

Example 4.6-1: QPSK Timing with a Narrowband Loop—Example 4.5-1 Revisited. In Example 4.5-1, a quadrupling loop (Fig. 4.3) achieves carrier phase synchronization to an RC pulse QPSK signal. The quaternary symbol rate is 0.772 MHz, the carrier frequency is 1.0 MHz, and the only significant challenge to the synchronizer is the data noise from the RC pulses. This noise implies that the 4-MHz PLL inside the quadrupler circuit needs to have a normalized bandwidth of only $B_N T = 0.086$ Hz-s/symbol. The analog design reached in Example 4.5-1 specifies Eq. (4.5-2), which is

$$H(s) = \frac{\omega_n^2}{s^2 + 2\zeta\omega_n s + \omega_n^2} \tag{4.6-9}$$

with

$$\omega_n = 500{,}000 \text{ rad/s}, \qquad \zeta = 1$$

What digital PLLs will have similar performance? Program 4.6-1 that follows converts $H_A(s)$ to $H_D(z)$ and plots the frequency responses of each. Figure 4.26a shows the original $H_A(s)$ frequency response and the responses of two discrete-time $H_D(z)$, one taken at sampling frequency 772 kHz and one at 193 kHz, which are the modulation symbol rate $1/T$ and a quarter of the rate, $1/4T$, respectively. At $1/4T$, the discrete loop comes out to be

[10]More precisely, Butterworth responses have delay close to $0.125n/f_c$ where n is the order; a verification of this rule can be seen in Fig. 2.17.

$$H_D(z) = \frac{0.319z^2 + 0.637z + 0.319}{z^2 + 0.257z + 0.017} \tag{4.6-10}$$

Program 4.6-2 that follows finds the error response to a step input for discrete and analog transfer functions. Figure 4.26b compares the analog step error response for $H_A(s)$ to the discrete $H_D(z)$ step error response at $T_D = 4T$. The discrete response is reasonably close and is quite acceptable, despite a sampling rate much lower than the symbol rate. Observe that the discrete response that results from the bilinear transformation oscillates a little, while the overdamped analog loop with its $\zeta = 1$ does not. The discrete loop is evidently stable; this can be verified by checking the pole positions in $H_D(z)$.

Program 4.6-1: Analog to Discrete-Time Transformation. The following MATLAB program transforms an $H_A(s)$ to an $H_D(z)$ via the bilinear transformation and plots the spectrum of both. The variables `num` and `den` are the coefficients of the numerator and denominator of $H_A(s)$; `numz` and `denz` are these for $H_D(z)$; `fs` is the sampling frequency $1/T_D$.

```
%    Find bilinear transformation
[numz,denz] = bilinear(num,den,fs)
%    Find discrete frequency response
```

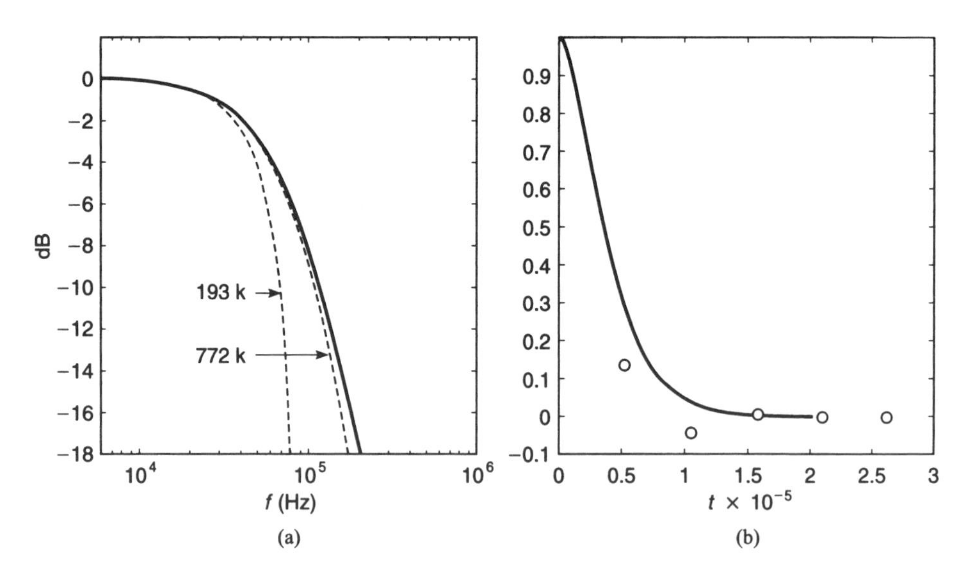

Figure 4.26 **(a)** Frequency responses in Example 4.6.1: analog closed-loop transfer function (4.6-9) (solid line) and discrete function (4.6-10) (dashed lines) with 193 and 772 kHz sampling rates. **(b)** Error responses to step input for analog and discrete 193-kHz loops in (a); circles indicate discrete response.

```
[hz,wz] = freqz(numz,denz,101);    %101 points on complex response
logfz = log10(wz*fs/(2*pi));
dbmagz = 20*log10(abs(hz));
plot(logfz,dbmagz);                %dB versus log freq plot
%   Find analog frequency response
wplot = [0:.01:1]*fs*2*pi;         %101 freq points up to fs
ha = freqs(num,den,wplot);         %101 points on complex response
logfa = log10(wplot/(2*pi));
dbmaga = 20*log10(abs(ha));
plot(logfa,dbmaga);                %dB versus log freq plot
```

In the calculation of Example 4.6-1, Eq. 4.6-1 has numerator and denominator `num = [0 0 500000^2]` and `den = [1 1000000 500000^2]`. The bilinear transformation with `fs = 193000` gives

```
numz = [0.319 0.637 0.319]
denz = [1 0.257 0.017]
```

which is the system (4.6-10). The analog and discrete frequency responses appear in Fig. 4.26a.

Program 4.6-2: Discrete-Time Step Error Response. The following MATLAB program finds the error response of a discrete-time $H_D(z)$ to a unit step input. The variables `num, den, numz, denz`, and `fs` are as in the previous program. Recall from (4.3-2) that the error response function for a system with transfer function H is $1 - H$:

```
%   Find numerator polynomial of 1-H; denominator is unchanged
nez = denz-numz;
%   Find discrete step error response
stp = ones(1,101);                    %Unit step of length 101
yz = filter(nez,denz,stp);            %Response
plot([0:1:100]/fs,yz);                %Plot 101 points
```

For completeness, here is the same for the analog system:

```
%   Find numerator polynomial
ne = den-num;
%   Find analog step error response
t = [0:1:100]/fs;                     %Same time points as above
[y,x,t] = step(ne,den,t);             %Step error at points
plot (t,y);                           %Plot 101 points
```

In Example 4.6-1, the error response polynomial `num-den` for function (4.6-9) is `[1 1000000 0]`. The analog and discrete step errors appear in Fig. 4.26b.

4.6.3 Second-Order Discrete PLLs Based on an Accumulator

Up to now we have designed discrete-time PLLs by simply converting an analog transfer function that is already known to work. A more fundamental approach is to design directly a discrete system $H(z)$ of some order. Many circuit design procedures exist that create an $H(z)$ with specified properties. We can also add new elements to the loop, elements that are overtly digital. One such element is the accumulator, defined by the difference equation $y(n) = y(n-1) + x(n-1)$, which is equivalent to

$$y(n) = \sum_{k=0}^{n-1} x(k)$$

This element is the discrete equivalent of integration, which, as was shown in Section 4.2, is precisely the circuit function of the VCO. Replacing the VCO with an accumulator inspires the digital PLL structure in Fig. 4.27. The accumulator is shown as its z-transform (4.6-8). Here, $F(z)$ is the loop filter.

The only free element in Fig. 4.27 is the loop filter. By restricting $F(z)$ to the general first-order form

$$F(z) = f_1 + \frac{f_2}{1 - f_3 z^{-1}} \tag{4.6-11}$$

we generate a class of second-order digital PLLs with closed-loop transfer function

$$\begin{aligned} H(z) \triangleq \frac{C(z)}{R(z)} &= \frac{F(z)[1/(z-1)]}{1 + F(z)[1/(z-1)]} \\ &= \frac{F(z)}{z - 1 + F(z)} \end{aligned} \tag{4.6-12}$$

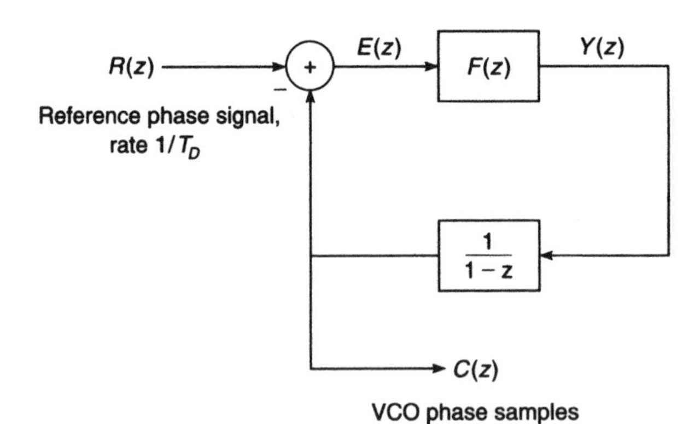

Figure 4.27 Discrete-time PLL based on accumulator. All signals are transforms of phase signals.

Some further manipulation shows that $H(z)$ directly in terms of f_1, f_2, f_3 works out to be

$$H(z) = \frac{(f_1 + f_2)z - f_1 f_3}{z^2 + (f_1 + f_2 - f_3 - 1)z + f_3(1 - f_1)} \tag{4.6-13}$$

We proceed just as we did after (4.3-1); the error transfer function is

$$\frac{E(z)}{R(z)} = 1 - H(z) = \frac{z - 1}{z - 1 + F(z)} \tag{4.6-14}$$

With the aid of (4.6-6), we can realize this loop equivalently in the following software steps. The phase detector junction in Fig. 4.27 performs

$$e(n) = r(n) - c(n) \tag{4.6-15}$$

where $r(0), r(1), \ldots$ is the loop's input. The difference equation

$$y(n) = f_3 y(n-1) + (f_1 + f_2)e(n) - f_1 f_3 e(n-1) \tag{4.6-16}$$

represents $F(z)$. The accumulator generates the loop output from

$$c(n) = c(n-1) + y(n-1) \tag{4.6-17}$$

All three of these can be combined into a single difference equation in $r(n)$ and $c(n)$ that implements the whole loop:

$$\begin{aligned} c(n) = {} & (-f_1 - f_2 + f_3 + 1)\, c(n-1) - f_3(1 - f_1)\, c(n-2) \\ & + (f_1 + f_2)\, r(n-1) - f_1 f_3 r(n-2) \end{aligned} \tag{4.6-18}$$

The z-transform of this relationship is (4.6-13).

In the rest of this section, we will explore the possibilities of these accumulator loops. This provides examples of how a new element creates new PLL possibilities, and how one can go about analyzing them. It needs to be reiterated that, unlike the analog case, not all second-order loops are stable. For example, when $F(z)$ in (4.6-12) is set to a simple gain factor K, there will be a z-plane pole of $H(z)$ at $z = 1 - K$; $H(z)$ is unstable whenever the pole lies outside the unit circle, which occurs unless $0 < K < 2$.

It will be convenient to work with the error response to a step. Proceeding from (4.6-13), we find its z-transform to be

$$\frac{z}{z-1}[1 - H(z)] = \frac{z^2 - f_3 z}{z^2 + (f_1 + f_2 - f_3 - 1)z + f_3(1 - f_1)} \tag{4.6-19}$$

How should the parameters f_i be set? A great many strategies can be imagined, but one with intuitive appeal is to match the step error response here to the step error response of an analog PLL. A reasonable candidate for a match-up is the analog two-pole loop, whose step error response is (see Table 4.1)

$$\frac{1}{s}\frac{s^2 + 2\zeta\omega_n s}{s^2 + 2\zeta\omega_n s + \omega_n^2} = \frac{s + 2\zeta\omega_n}{s^2 + 2\zeta\omega_n s + \omega_n^2} \tag{4.6-20}$$

We will assume an underdamped response with $\zeta < 1$.

The next step is to attempt to find f_i's that cause the time responses of (4.6-19) and (4.6-20) to match. This matches the step error responses. In terms of the standard transforms in Table 4.3, (4.6-20) is a sum of the fourth and fifth entries:

$$\frac{s + 2\zeta\omega_n}{s^2 + 2\zeta\omega_n s + \omega_n^2} = \frac{s + \zeta\omega_n}{s^2 + 2\zeta\omega_n s + \omega_n^2} + \frac{\zeta\omega_n}{s^2 + 2\zeta\omega_n s + \omega_n^2}$$

which equals

$$\frac{A(s+a)}{(s+a)^2 + \omega_0^2} + \frac{B\omega_0}{(s+a)^2 + \omega_0^2} \tag{4.6-21}$$

in which $a = \zeta\omega_n$, $\omega_0 = \omega_n\sqrt{1-\zeta^2}$, $A = 1$ and $B = a/\omega_0$.

From the table, the z-transform that matches this is

$$\frac{B[ze^{-aT}\sin\omega_0 T] + A[z^2 - ze^{-aT}\cos\omega_0 T]}{z^2 - 2ze^{-aT}\cos\omega_0 T + e^{-2aT}}$$

which is

$$\frac{z^2 + ze^{-aT}[(a/\omega_0)\sin\omega_0 T - \cos\omega_0 T]}{z^2 - 2ze^{-aT}\cos\omega_0 T + e^{-2aT}}$$

with the A and B above. Comparing this to (4.6-19) implies the following relationship for f_1, f_2, f_3:

$$\begin{gathered} e^{-aT}\left(\frac{a}{\omega_0}\sin\omega_0 T - \cos\omega_0 T\right) = -f_3 \\ e^{-2aT} = f_3(1 - f_1) \\ 2e^{-aT}\cos\omega_0 T = 1 + f_3 - f_1 - f_2 \end{gathered} \tag{4.6-22}$$

These can be solved in turn for f_3, f_1 and f_2.

An example calculation is shown in Fig. 4.28b. The solid line there is the step error response of an analog loop with $\omega_n = 490000$ rad/s and $\zeta = 0.9$. This is essentially the loop in Example 4.6-1; the parameters are changed a little in order to meet the

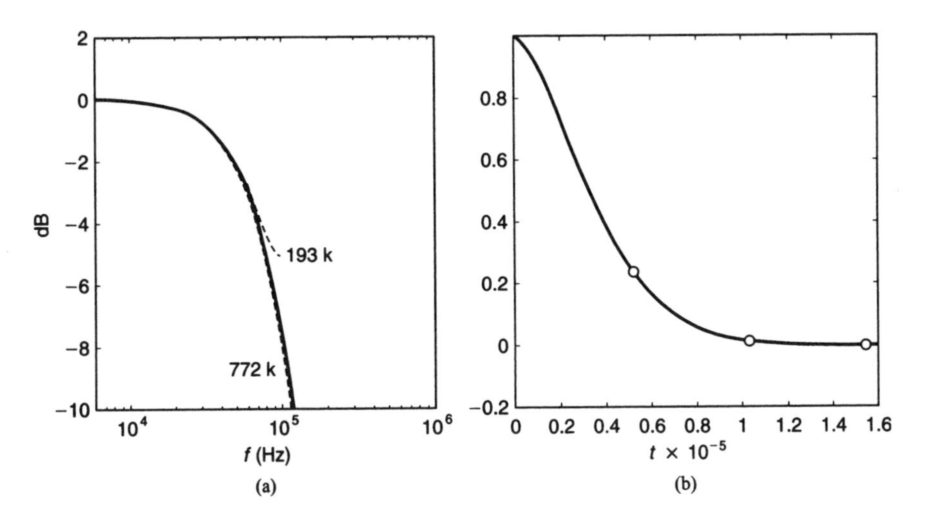

Figure 4.28 **(a)** Frequency responses in step-error-matched accumulator loop: analog two-pole closed-loop transfer function $H(s)$ with $\zeta = 0.9$ and $\omega_n = 490$ krad/s (solid line) and function $H(z)$ from (4.6-23) with 193 and 772 kHz sampling rate (dashed). **(b)** Matched error responses to step input for analog and discrete 193-kHz loop in (a); circles indicate discrete response.

condition $\zeta < 1$. Also shown there is the discrete loop step error response when the sampling rate is $1/4T = 193$ kHz. In this case,

$$\begin{aligned} &a = 441{,}000, \qquad \omega_0 = 214{,}000, \qquad T_D = 5.18 \times 10^{-6} \\ &f_1 = 1.073, \qquad f_2 = -0.306, \qquad f_3 = -0.142 \\ &H(z) = \frac{0.767z + 0.153}{z^2 - 0.0907z + 0.0104} \end{aligned} \tag{4.6-23}$$

The responses here are found by Program 4.6-2, working on $H(s)$ and $H(z)$. The discrete-time response lies on the analog response, as it must, and varies little from Example 4.6-1. Figure 4.28a compares the analog $H(s)$ frequency response to the discrete one when T_D is 1/193,000 and 1/772,000. The slower sampling rate exhibits a minor variation because the folding frequency of the response is only 96.5 kHz; the faster rate response is almost identical to the analog.

The discussion here has shown how a digital PLL based on an accumulator can be adapted to perform some of the phase synchronizing tasks that came earlier in the chapter. Discrete-time loops can be based on many other devices. Two of these are the charge-pump phase detector and the numerically controlled oscillator, and because they relate more closely to symbol timing, we will take them up in the next section. An idea of the wide variety of digital PLL circuits that exist can be found in published surveys. Two of these are the survey by Lindsey and Chie [5] and the papers collected by Razavi [14].

4.7 SYMBOL TIMING RECOVERY

Symbol timing is the sequence of sampling instants at the receive filter in a detector, the times nT at which the switch closes in the sampling receiver in Fig. 2.3, the linear receiver in Fig. 2.7, or the matched filter receiver in Fig. 2.32. The sampling receiver needs to know the exact instant at which one pulse takes on the transmission symbol value and all the others pass through zero. The matched-filter and linear receivers need to know when the exact correlation of pulse and signal appears at their filter output.

Timing recovery is relatively independent of carrier phase recovery, and it is usually obtained from the demodulated baseband signal. All of the points raised in Section 4.1 apply. The product of the recovery operation is a series of clock ticks, most likely a digital waveform, in which a repeating feature, or "edge," marks the clock times for the samples. The recovery circuit needs to adjust to changes in phase and frequency, which, in the timing case, are often relatively minor; it also must average out jitter and noise, which can be major. Consequently, timing recovery loops generally are quite narrowband.

Timing can be obtained from the data-bearing signal itself, or from a master clock, or from a pilot tone. The last two are common in communication networks. In the telephone network, for example, pilot signals can mark the timing instants for a whole bank of channels. In this way, the power and bandwidth penalty for the pilot signal is distributed among many channels. The separate functions of acquisition and tracking are required in timing recovery, although it is often known quite accurately in advance what the timing rate is.

We will look at timing recovery in two ways. The first is a generic method, called the early–late gate synchronizer, a method that applies also in such problems as frame timing and radar time delay extraction. The second way is the discrete-time PLL recovery circuit. The digital PLL approach is particularly well matched to timing recovery.

4.7.1 Early–Late Gate Synchronizer

The idea of this circuit is to sample the baseband signal twice, once before the supposed proper time and once after. If the signal waveform is symmetric near the correct sample time and the timing recovery is at present correct, the two samples will be the same on the average. A PLL checks for this.

This simple idea needs only a little development in order to work with BPSK, QPSK, and a wide variety of linear modulations of the QAM type. For concreteness, we will assume that the rest of the receiver is the standard quadrature receiver in Fig. 3.9 and that the transmission is a superposition of orthogonal pulses, specifically the *root* RC train that is illustrated in Fig. 2.8. The final early–late circuit is shown in Fig. 4.29. At the left is the receiver filter, which is matched to the single root RC pulse; it converts the root RC pulse train to an RC pulse train, an illustration of which is in Fig. 2.5. This waveform is the target for the early–late gate in the rest of the figure.

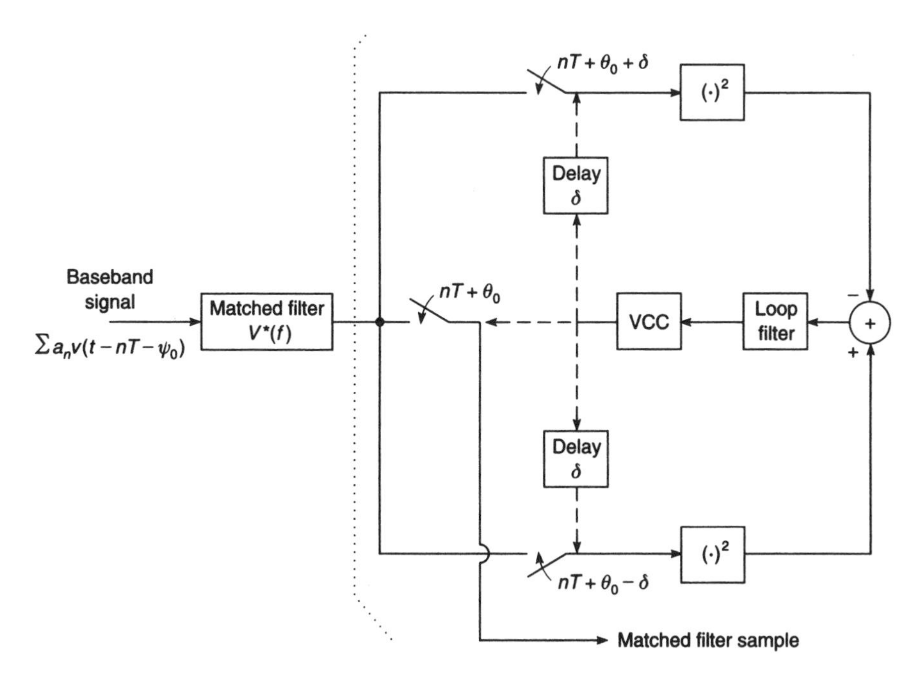

Figure 4.29 Early–late timing synchronizer with linear modulation signals: $v(t)$ is symmetric orthogonal pulse and dashed lines indicate timing signals. Correct timing is $nT + \psi_0$; loop is shown locked at $\theta_0 = \psi_0$.

A look at the Fig. 2.5 waveform will show that while each RC pulse is symmetric about its sampling time, the pulse *train* is certainly not. On the average, however, the train is symmetric in the neighborhood of its sample points. Some feeling for the average waveform can be obtained from an eye pattern such as Fig. 2.13a. Figure 2.13c shows that the averaging will work for noisy signals, too.

A formal proof that the average signal is acceptable goes as follows, for $v(t)$ an orthogonal pulse and a circuit with a sample squarer as in Fig. 4.29. Let $\{a_n\}$ be a sequence of IID data for which $\mathcal{E}[a_n] = 0$; let there be N pulses in the train. The receive filter sample squared at delay δ later than the correct instant has the value (let $\psi_0 = 0$ and neglect noise)

$$\left[\sum_n a_n v(\delta - nT)\right]^2$$

The expectation of this is

$$\mathcal{E}\left[\left(\sum_n a_n v(\delta - nT)\right)\left(\sum_m a_m v(\delta - mT)\right)\right]$$

$$= \mathcal{E}\left[\sum_n a_n^2 v^2(\delta - nT)\right] + 2\left[\sum_{n \neq m} \mathcal{E}[a_n a_m] v(\delta - nT) v(\delta - mT)\right]$$

The second set of terms are all zero, leaving just

$$\mathcal{E}[|a_n|^2]\sum_n v^2(\delta - nT) \tag{4.7-1}$$

Figure 4.30 is a plot of $\Sigma v^2(\delta - nT)$ for a 30% and 100% RC pulse [whose formula is Eq. (2.2-6)]. It is clear that the average of many observations should show the desired symmetry about $\delta = 0$.[11] The symmetry can be shown for the NRZ pulse as well (see the Problems).

We can now complete the design details in Fig. 4.29. The PLL there consists of a loop filter, a voltage-controlled clock (VCC), and a feedback through the summing junction at the right. The VCC replaces the VCO, and it should be thought of as putting out a clock pulse, or edge, that enables the receiver sampler. The loop needs to be narrowband compared to $1/T$, since the averaging in (4.7-1) is long term. Here, θ_0 is the timing phase extracted by the circuit. Some method is needed to remove the sign of the input signal; squarers are shown in the figure, but these can as well be absolute value blocks. A sampling time that is consistently late produces on the average a positive control voltage, which speeds up the VCC and advances the timing.

It is also possible to advance and retard the matched filter rather than the timing, and the loop can work at RF rather than baseband. Some of these details are discussed in Proakis [6]. He also discusses the fact that the early–late gate approximates an ML estimator for the offset ψ_0.

4.7.2 Timing Based on Zero Crossings

Timing recovery provides the clock signal for what is likely to be a digital circuit, and so there is a close relation between recovery and discrete-time processing. Timing marks in a digital circuit, for their part, often manifest themselves as zero crossings. Baseband signals of PSK and QAM contain zero crossings that are related to their symbol timing. The combining of all these ideas leads to a class of digital timing recovery circuits based on zero crossings, to which we turn now. We again focus on the linear receiver and orthogonal pulses, with the signal taken after the receive filter.

First, consider the receive-filtered baseband signal and its zero crossings. The easiest way to see these is with an eye pattern, such as the ones in Section 2.4. The desired zero crossings are near $\pm 0.5/T$ for pulses that are symmetrical about the time origin. When the pulse is smoothed, unfortunately, the crossings only cross at $\pm 0.5/T$ on the average; Fig. 2.13a illustrates this for binary 30% RC signaling. Consequently, we again need a narrowband tracking loop, as was the case in the early–late gate loop. Actually, the zero-crossing situation is worse than it appears, because many zero crossings are completely missing. A look at the RC pulse trains in Figs. 2.5 and 2.20 shows that zero crossings in simple linear modulations tend to be missing between transmission symbols with the same sign.

[11]The plot for the sinc pulse is a straight line; the circuit of Fig. 4.29 will not work!

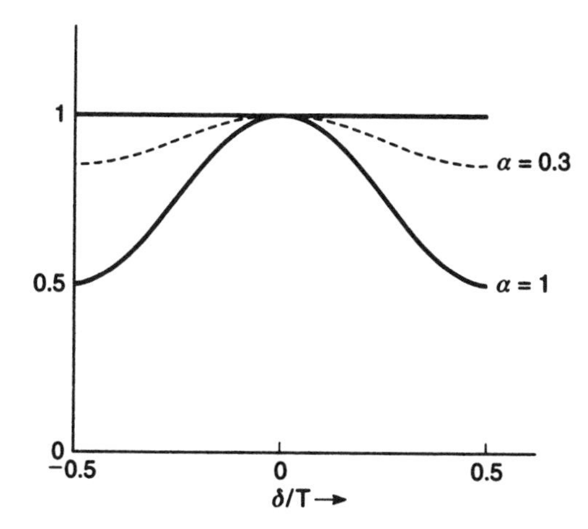

Figure 4.30 Expected square value of RC pulse train with random data observed δ before and after correct sampling time nT. Excess bandwidth factors 0.3 and 1. Binary data.

Figure 4.31 sketches a timing diagram of these signal zero crossings and the leading edge of a VCC pulse. The left part of the figure converts the zero crossings of the data signal into clean digital clock pulses; a standard way to do this is the chain shown there, a limiting amplifier plus differentiator/rectifier circuit. The rest of the clock recovery circuit is in Fig. 4.32. The leading edges of the VCC clock and the pulses from the differentiator/rectifier drive a three-state phase de-

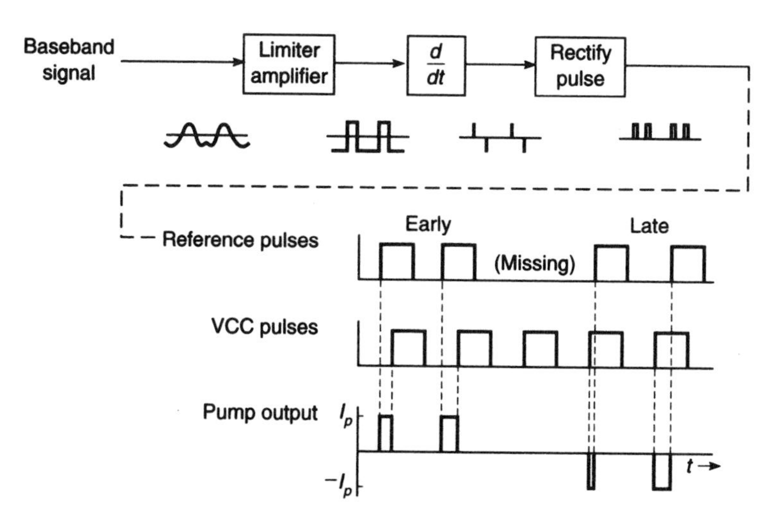

Figure 4.31 Preparing input to charge pump timing recovery circuit in Fig. 4.32, beginning at matched-filter output and ending with input to three-state phase detector. Timing diagram compares this reference input to VCC and charge pump pulses.

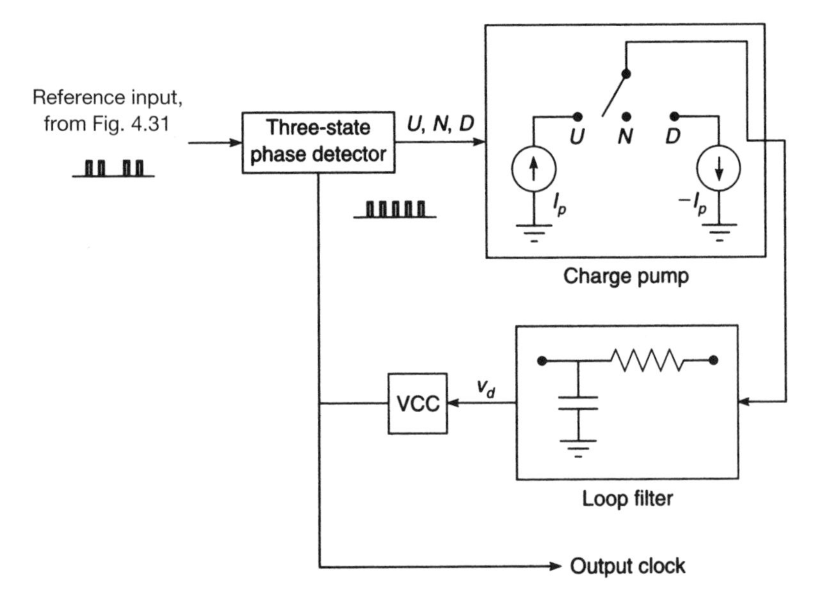

Figure 4.32 Charge pump timing recovery circuit. Three-state phase detector controls switch of charge pump; narrowband loop filter smoothes current into VCC control voltage v_d.

tector. This device puts out a positive pulse (state *U*) when the PLL input leads the VCC and a negative pulse (state *D*) when it lags; when there is no input pulse, the phase detector output is zero (state *N*).[12] The charge pump responds to these states: *U* produces a current $+I_p$, *D* produces $-I_p$, and no current flows at state *N*. The rest of the loop is a narrowband filter that smoothes the current pulses and the VCC. The overall effect is that the VCC advances or retards according to the average lead or lag of the reference input and missing reference pulses are simply ignored.

Gardner [15] analyzed the behavior of the charge pump loop for the case of the simple RC loop filter illustrated in Fig. 4.32. He found that the discrete-time transfer function of the loop for small phase excursions is of the form

$$H(z) = \frac{C(z)}{R(z)} = \frac{A_1 z}{(z-1)^2 + (z-1)A_2 + A_3} \tag{4.7-2}$$

Here $R(z)$ is the z-transform of the sequence of VCC phases at each clock pulse and $C(z)$ is the transform of the reference phases. There is always some ripple in the clock phase because of the pulsed VCC control voltage, but with symbol time especially, ripple can usually be tolerated. The loop will be unstable if its gain (i.e., the size of I_p) is too high.

[12]This circuit is easily realized from flip flops.

Another digital circuit element that is often a part of discrete-time loops is the *numerically controlled oscillator* (NCO). It could, for instance, replace the VCC in the charge-pump loop. A counter is the heart of the NCO. Figure 4.33 shows how it implements a VCC. A control voltage value enters at left and is converted into a binary number N by means of some decision logic (the logic can implement the loop filter as well). Beginning at binary zero, the counter counts up to this number, resets, and counts up again. The result is a clock edge at a factor N slower than the counter clock f_0; the counter is divided by N. The new clock rate f_0/N needs to be large enough so that it divides the reference phase into small enough increments. If this smallest desired increment is $\Delta\phi$ radians, then it is clear that N will be about

$$N \approx \frac{f_0}{f_c} = \frac{2\pi}{\Delta\phi} \tag{4.7-3}$$

A discrete-time analysis of loops that incorporate an NCO appears in Ref. 5.

To summarize, we have looked at several timing recovery circuits that are based on special digital circuit elements and that mostly sense zero crossings. For applications, we have concentrated on QAM/linear modulation signals. For other kinds of signals the same principles often apply, although in many cases, such as CPFSK, quite special circuits have been invented. Finally, it is worth reemphasizing that the standard analysis given in Sections 4.6–4.7 is for tracking and for small phase errors, and it needs to be modified for large errors and for the acquisition mode. In particular, our analysis cannot reliably predict stability in these modes.

4.8 FRAME SYNCHRONIZATION AND NETWORKING

There is a natural progression in synchronization from the small to the large. If the signal is carrier-modulated, the receiver must start with phase or frequency lock, without which it probably cannot obtain symbol timing. After symbol timing, the

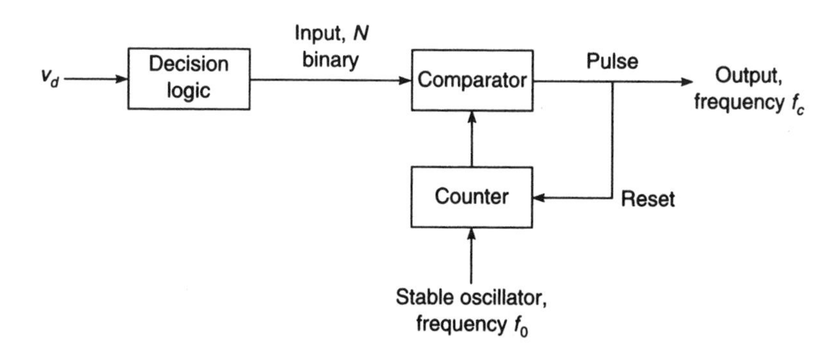

Figure 4.33 Block diagram of numerically controlled oscillator implementation of VCC at frequency f_c.

receiver moves on to locate words and other structures in the symbol stream. This last level, *frame synchronization,* is important in this section.

Until now in the chapter we have imagined that a sender and receiver wish to stay in synchronization with each other, but it often happens that many stations wish to stay mutually in synchronization. This is the related *network synchronization* problem. It brings with it some new challenges, since, for example, a station can be caught between two other stations that insist on slightly different symbol rates. Since networks often provide phase and symbol time references, and these are not in doubt, the means of resolving conflicts in flow rates generally involves manipulating the frame structure. There is thus a natural relation between frame and network synchronization, and so we take them together.

A discussion of network synchronization leads directly to the field of communication networks, a rapidly growing field that has become an engineering specialty in its own right. We must leave most of that field to other textbooks.

Frame synchronization depends on some properties of sequences, and we will review some of these first. Next come synchronization procedures and, finally, a review of large networks and their synchronization.

4.8.1 Sequences and Their Boundaries

How can the beginning of a frame be marked? This is our first question. We will start with the simpler problem of how to design a marker that simply repeats. Then we will turn to markers that appear in the pattern marker, data, marker, data, ...

To start, consider the simple marker pattern shown in Fig. 4.34, consisting of a repeating m-symbol marker $\mu(t)$. The symbols have length T_c and are called chips; if we think of chips as indivisible, T_c is the resolution of the marker. The object now is to locate the chip that begins the marker sequence, and so to obtain a clock that is aligned with the markers. This is the classical *epoch detection* problem.

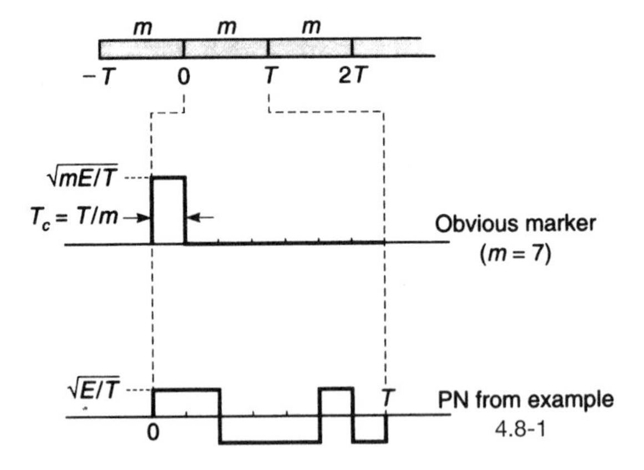

Figure 4.34 Examples of repeating marker sequence having resolution $T_c = T/m$ showing obvious solution consisting of single pulse and PN solution in Example 4.8-1.

Epoch detection can be approached in several ways. A detector can watch for a certain feature to go by, such as a peak. A more effective method in general is *sliding correlation*. In a succession of positions down the pattern, a correlation is performed between a clean replica of the marker and the repeating incoming pattern. When an incoming marker is exactly aligned with the clean replica, the correlation will be highest, at least in the absence of noise, and the detector then knows the most likely beginning symbol of the marker. To make the correlator idea more precise, define the periodic autocorrelation of the marker sequence with itself,

$$R_{\text{per}}(\tau) = \frac{1}{E_\mu} \int \mu(t)\mu_{\text{rp}}(t+\tau)\, dt \tag{4.8-1}$$

in which

$$E_\mu = \int \mu(t)^2\, dt$$

is the marker energy and $\mu_{\text{rp}}(t)$ denotes the repeating marker. For now, we will assume that t and τ are multiples of the chip time T_c; in this case, $R_{\text{per}}(jT_c)$ is really the correlation of $\mu(t)$ with itself cyclically shifted j chips left.[13] Some basic facts about epoch detection are as follows; the proofs of these can be found in advanced books, such as the classic by Stiffler [16].

1. The search time to find the beginning of the marker in noise is inversely proportional to the marker energy.
2. The search time to locate the marker beginning is a monotone function of the value $\max_j R_{per}(jT_c), j \neq 0$; since $R_{\text{per}}(0) = 1$ and $R_{\text{per}}(jT_c) < 1$, this max is the second-highest value of $R_{\text{per}}(jT_c)$.
3. For any marker sequence,

$$\max_{j\neq 0} R_{\text{per}}(jT_c) \geq \frac{-1}{m-1} \tag{4.8-2}$$

An easy solution to the marker design problem is shown in Fig. 4.34. A total of T seconds is devoted to the marker, and the chip time T_c is T/m. It is a single pulse of height $\sqrt{mE/T_c}$, followed by zero for $(m-1)T_c$ seconds. Its energy is E and its second-highest autocorrelation is 0, which is almost as good as (4.8-2). The marker, however, has a severe practical drawback. Its peak amplitude grows with $\sqrt{m}$ and will eventually strain the system electronics as the resolution becomes finer.

Pseudonoise Sequences. Fortunately for the electronics, an easily generated and detected sequence is available that essentially satisfies 3, yet consists of constant-energy chips: the pseudonoise (PN) sequence. The PN sequences are generat-

[13]A cyclic shift left means that each chip moves one left and the leftmost becomes the new rightmost.

ed by the binary linear feedback shift-register circuit, henceforth called an LFSR, of the type shown in Fig. 4.35. The circuit works with the binary symbols {0, 1} and modulo-2 multiplication and addition. It begins with at least one nonzero symbol in the register, say at X. At each clock time, the register contents move right, the tap calculation from the previous contents moves into the first cell, and the content of the last cell becomes the next LFSR output. An output sequence of an LFSR is called a *shift-register sequence,* and a PN sequence is a special case, which we will define after the next example.

Example 4.8-1: A Length-7 PN Sequence. Take a length-3 shift-register with $a_1 = a_3 = 1$ and $a_2 = 0$, as shown in Fig. 4.35, and start the process with a single 1 at X. This LFSR process generates a length-7 repeating output, which happens to be a PN sequence. The sequence of shifts in Fig. 4.35b occurs. Observe that after seven shifts, the circuit has returned to the initial state $XYZ = 100$. It will, in fact, endlessly pass through the sequence of states and outputs that appear above the dashed line, with the output being an endless repetition of 0011101. . . . Had the register been initialized in another state, instead of 100, the circuit would enter the output cycle at another point, but it would move through the same 7-bit cycle.

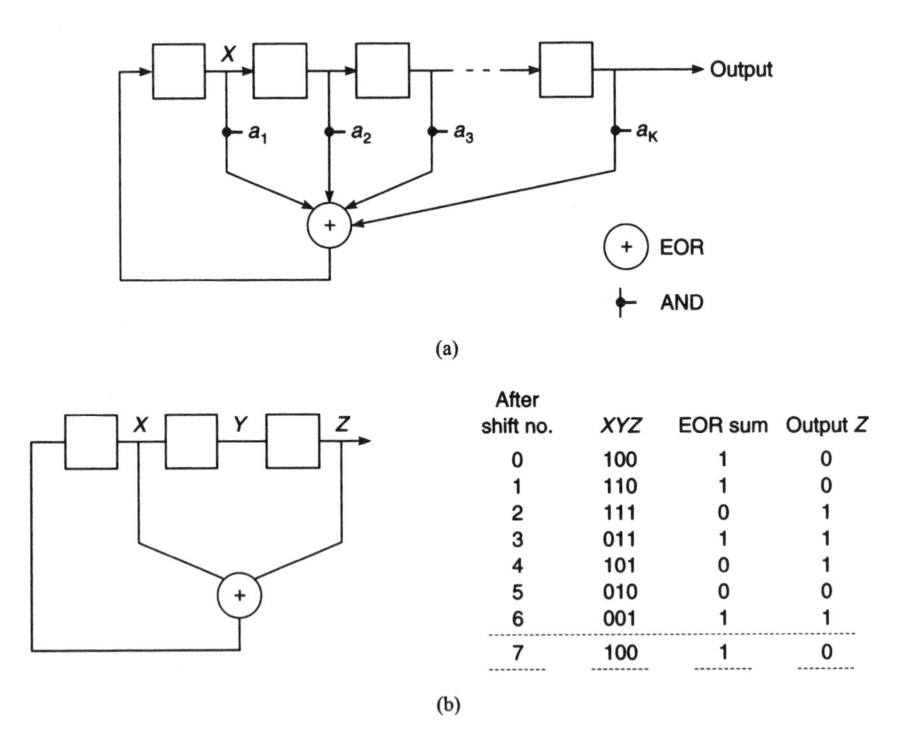

After shift no.	XYZ	EOR sum	Output Z
0	100	1	0
1	110	1	0
2	111	0	1
3	011	1	1
4	101	0	1
5	010	0	0
6	001	1	1
7	100	1	0

Figure 4.35 Linear feedback shift-register circuits: **(a)** general circuit and **(b)** Example 4.8-1. All arithmetic is binary; EOR stands for Exclusive-OR.

Here are some facts about shift-register sequences, whose proofs can be found in Refs. 16 and 19 or in the classic book by Golomb [20] on shift-register sequences.

1. There exist tap sets $\{a_k\}$ such that the output repeats only every $2^K - 1$ bits. These are called *maximal-length* shift register sequences.[14] Other tap sets lead to shorter cycles, of length $2^K/2$ or less.
2. Suppose the LFSR output is maximal length, and consider any two cyclic shifts. The bitwise exclusive-OR (EOR) sum of these shifts gives a third shift.
3. Any maximal length output sequence has $2^K/2$ 1's and $2^K/2 - 1$ 0's.

The term PN sequence refers to any complete output cycle of a maximal-length LFSR. The PN sequences thus have length $2^K - 1$ when the shift register is of length K. Obviously, any shift of a PN sequence is another PN sequence. In Example 4.8-1, K is 3, the output cycle is maximal since $2^K - 1 = 7$, and the output is thus a PN sequence. We can easily demonstrate 2 and 3. Let the output be $\boldsymbol{c}^{(0)} = 0011101$ and let $\boldsymbol{c}^{(1)}$ denote the one-step cyclic shift to the right, 1001110. The bitwise EOR sum $\boldsymbol{c}^{(0)} + \boldsymbol{c}^{(1)}$ is

$$\begin{array}{c} 0011101 \\ \underline{1001110} \\ 1010011 \end{array}$$

which is $\boldsymbol{c}^{(3)}$, the shift by 3. Finally, since there are 4 1's and 3 0's in any of the shifts, property 3 is verified.

The PN sequences are attractive for epoch detection because they have virtually the best possible autocorrelation given by (4.8-2). It is easiest to express this fact by associating the binaries $\{0, 1\}$ with the real numbers $\{+1, -1\}$; we need to make such an association in order to transmit the abstract symbols, and this is the one that was made in Chapter 3. With this done, the PN sequence as a time signal becomes

$$\mu(t) = \sum_{i=0}^{m-1} c_i p(t - iT_c)$$

where the c_i are ± 1 and the chip pulses $p(t)$ are assumed to be orthogonal to T_c shifts (cf. Definition 2.2-2). Consequently, the periodic autocorrelation (4.8-1) becomes

$$R_{\text{per}}(jT_c) = \frac{1}{m} \sum_{i=0}^{m-1} c_i \, c_{(i+j) \bmod m} \tag{4.8-3}$$

A way exists to evaluate this sum by using PN properties 2 and 3. Each element in the sum emulates the EOR sum of two bits in the binary representation of the PN

[14]No cycle can be longer than $2^K - 1$, since this is the number of nonzero states; an LFSR in the zero state will remain there forever.

sequence: If the binaries were 1 and 1 or 0 and 0, then the real number $c_i\, c_{(i+j)\text{mod}\, m}$ is 1; otherwise, it is –1. Furthermore, the sequence $c_{(i+j)\text{mod}\, m}$, $i = 0, \ldots, m-1$, is the cyclic right shift by j of sequence $c_0, \ldots, c_{m-1}$, and consequently, $c_i\, c_{(i+j)\text{mod}\, m}$, $i = 0, \ldots, m-1$, is the sum of the PN sequence and one of its shifts, with the binary-to-real mapping. By 2 any such sum is itself a PN sequence, and by 3 it contains one more –1 than +1. Thus the sum of elements in (4.8-3) must be –1 and the periodic autocorrelation $-1/m$ when $j \neq 0$. In Example 4.8-1, for instance, (4.8-3) evaluates to $\frac{1}{7}[4(-1) + 3(+1)] = -\frac{1}{7}$, unless $j = 0$, in which case (4.8-3) is unity.

The same argument works for any PN sequence and this shows that long PN sequences virtually achieve property 3 for epoch markers. Thus PN sequences are essentially the best markers at their length and energy for use in repeating-marker epoch detection.

Applications of PN Sequences. We can distinguish several epoch detection problems in digital transmission. With a known PN sequence inserted periodically into a data frame structure, and the chip time equal the data symbol time, the PN sequence marks a certain bit position, from which the rest of the frame structure can be found. Algorithms that find this position are discussed in Section 4.8.2. Actually, one can go further than simply tagging one symbol position. If the correlation variable τ in $R_{\text{per}}(\tau)$ is varied continuously, rather than in steps of T_c, $R_{\text{per}}(\tau)$ varies continuously, reaching a peak only at $\tau = 0$. Figure 4.36 shows an example of this when the chip pulse $p(t)$ is a root RC pulse. The precise location of the peak gives the symbol timing. Thus a finely varied sliding correlator can mark the symbol as well as the frame boundary. This method of symbol time acquisition is used in some digital cellular telephone systems (see Example 4.8-4).

The PN sequences are also important in many related communication engineering fields. It sometimes happens that the arrival or round-trip time of a burst needs to be known with great precision. A way to do this is to make up the burst out of one or more repetitions of a PN sequence. By a sliding correlation, the arrival time of the burst can be learned to a resolution T_c. This is the radar problem. Sonar and distance measuring are similar applications. In all of these, by lengthening m, the energy and hence the noise resistance in the sequence can be increased, while keeping the resolution and transmitter power fixed. Alternately, resolution can be improved at a fixed noise resistance.

We have mentioned PN property 3, that the appearance of 1's and 0's in the binary sequence is as close as it can be to 50% in an odd-length sequence. Actually, a similar statement applies to the occurrence of bit pairs 00, 01, 10, 11 and for the eight 3-tuples, and so on, when the sequence is longer. (See proofs in Ref. 19 and especially Ref. 20.) The name "pseudorandom" originates from this perfected randomness.[15] Because of it, PN sequences are used as test sequences when it is important that a system such as a communication link be driven through all its system states.

[15]In one point of view, such ironclad randomness is no randomness at all; truly random sequences virtually never have perfect distribution of 1,2,3, . . . tuples. True randomness is a difficult philosophical question.

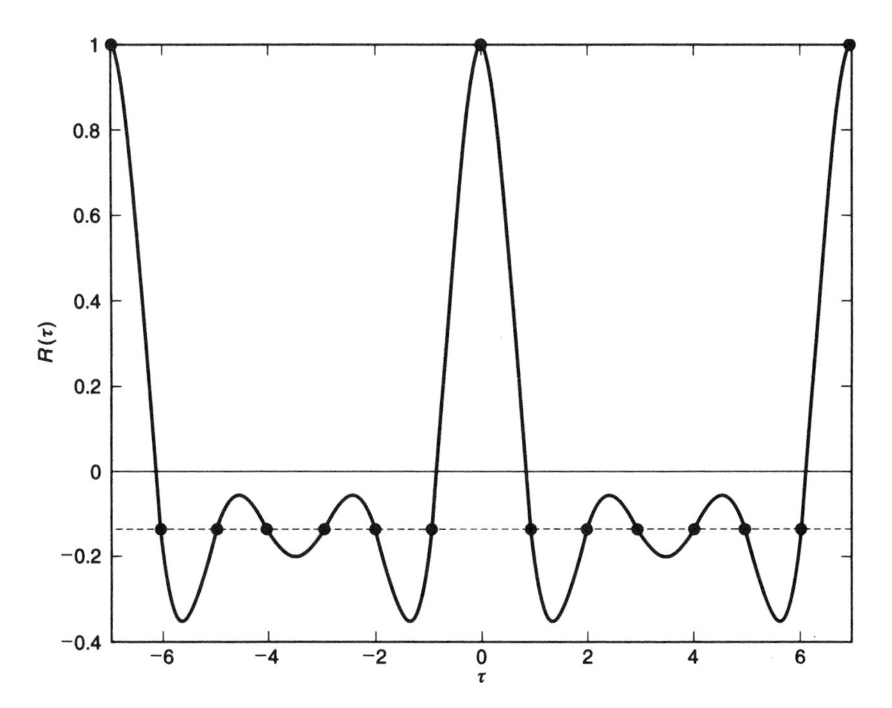

Figure 4.36 Correlation of PN sequence of Example 4.8-1 versus three repetitions of same marker when τ is taken as continuous variable and chip pulse is 30% root RC pulse. Black dots are periodic autocorrelation (4.8-3) for discrete case, where $\tau = jT_c$.

Pseudonoise randomness is the key to other applications as well. In direct sequence spread spectrum, individual data symbols are mapped to whole PN sequences, with the outcome that the modulated spread sequence, being pseudorandom, has a nearly perfect flat frequency spectrum. The receiver can be the sliding correlator that we have discussed. Spread spectrum techniques are discussed in Chapter 7. In multiple-access communication, several simultaneous transmissions of this type, with data symbols mapped to different whole PN sequences, will all look like background noise to each other. The m–sequences and Gold sequences (see Ref. 20) are subclasses of the PN sequences that find use in multiple-access signaling; these sequences have low correlation with each other as well as with shifts of themselves.

Barker Sequences. Until now, we have ignored the effect of bits transmitted on either side of a PN marker. In the absence of channel errors, the sliding correlator will still see a peak when it is perfectly aligned, but when its viewing window is offset, the correlation will be disturbed from its theoretical value of $-1/m$. Subsidiary peaks may lead to false lock, in case noise disturbs the value of the main peak. As well, a perfect replica of the PN marker could conceivably appear in the midst of the data. The solution of these difficulties is, first, to observe the putative marker position through several frames and, second, to design a marker that has minimal sensitivity to the symbols at the sides.

The solution to the second problem is called a *Barker sequence.* We will assume that the nonmarker symbols are random data. We can define the problem more precisely with the help of Fig. 4.37. Here the correlator attempts to match its marker, which is $m - i$ later, with a length-m part of the incoming sequence. The right $m - i$ bits of the correlator marker will correlate against data, and the left i bits will correlate against the right i bits of the incoming marker. We cannot control the interaction with the data, but we can look for some clever design of the interacting i-symbol marker pieces. The design problem boils down to making the i-bit prefixes of the marker as little as possible like the i-bit suffixes.

Once again, making the association $\{0, 1\} \leftrightarrow \{+1, -1\}$, we compute the m-symbol real-number correlation $R(i)$ in Fig. 4.37. On the average, it will be

$$\mathcal{E}[R(m-i)] = \rho_i + 0 \tag{4.8-4}$$

where ρ denotes the marker's i-suffix/i-prefix correlation and 0 is the average data/$(m - i)$ prefix correlation. The ρ_i can vary from $-i/m$ to $+i/m$. We want to minimize Eq. (4.8-4) simultaneously for all i, if possible. Barker [21] proved the following result for the set of ρ_i:

Suppose $\rho_i \leq 0$, *all* i. *If a marker with this set of* ρ_i *exists, it must satisfy*

$$\rho_i = \begin{cases} 0, & i \text{ even}, i = 0 \\ -1/m, & i \text{ odd} \end{cases} \tag{4.8-5}$$

This is an indirect approach, to be sure, but Barker went on to show that there are just three sequences that satisfy (4.8-5), namely,[16]

$$\begin{array}{l} 110 \\ 1110010 \\ 11100010010 \end{array} \tag{4.8-6}$$

When the correlation constraint is tightened to $\rho_i \leq -1/m$, Barker found that no sequences exist. When it is loosened to $\rho_i \leq +1/m$, many exist.

Longer sequences are, in general, better markers than shorter ones, but for a set length we should always choose one that satisfies a tight Barker condition, and for $m = 3, 7, 11$ the sequences (4.8-6) are surely the answers.

Example 4.8-2: A Naive Marker. It is interesting to compare to these Barker sequences a marker made from a simple run of 1's. This marker lacks all internal symbol structure and might occur to one who first thinks about the problem. Figure 4.38 plots the average correlation observed at various offsets for both the naive marker

[16]We omit complements and reversals, which work equally well; the reversal of 1110010 is a length-7 PN sequence, since two cyclic shifts of it give the outcome in Example 4.8-1.

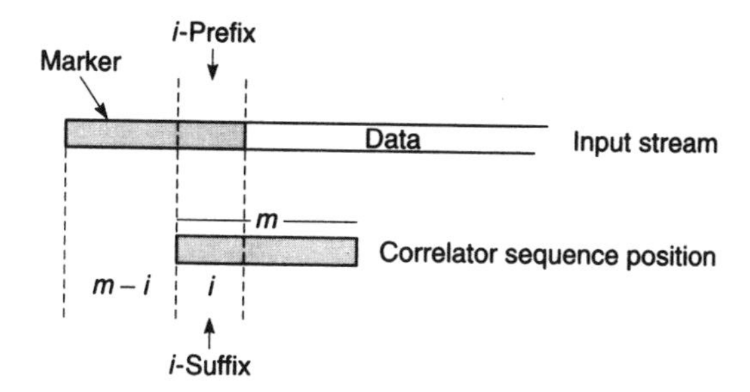

Figure 4.37 Illustration of Barker's problem: correlation of marker delayed $m - i$ versus marker and data.

with $m = 7$ and the same-length Barker sequence. On the average, the Barker is far better. Another view of the problem is obtained by assuming values for adjacent data bits and computing the correlation. With a data 1 to the right of the naive marker, there is a run of 8 1's and it is completely ambiguous which of two runs of 7 1's is really the marker. With a 1 to the right of the Barker marker, it is still clear where the marker is. In fact, some experiment shows that in the absence of channel errors, it is impossible to mistake the marker in any view with shift $-6 \leq j \leq 6$. A misalignment is possible only when the data randomly take the form of the marker, an event with probability 1/128. Two occurrences in succession have probability $(1/128)^2$.

4.8.2 Finding a Frame Marker

Now that we have discussed what is a good frame marker, it remains to find one in the transmission when the channel is noisy and the random data themselves can

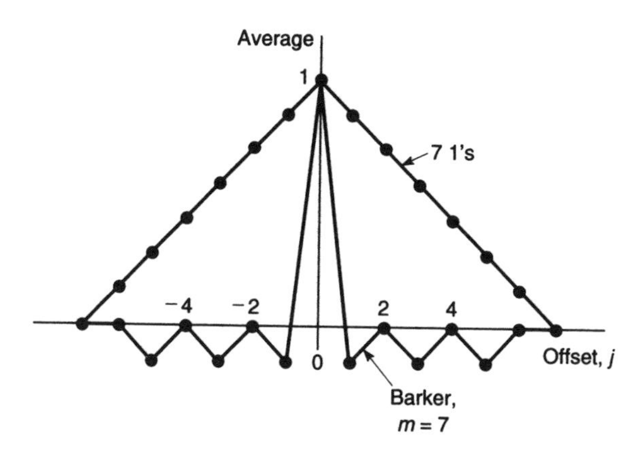

Figure 4.38 Average value of sliding correlator output for naive marker 1111111 and Barker sequence 1110010, both with random data on either side of marker. Horizontal axis is correlator offset to the right j.

take the form of the marker. This is the *framing* problem. We can state the general problem as follows: For each possible marker position in the frame structure, look for the marker repeatedly. Stop looking in this position when either Pr[Marker present | Observations] or Pr[Marker not present | Observations] is high enough.

The study of problems of this type is called *sequential analysis,* a study that traces back to a book [17] of the same name by Wald, a good reference on the subject that is still worth reading. Wald devised the sequential probability ratio test (SPRT), a test that decides whether the sequence of observations at slot position j should continue. First, let $A > 0$ and $B < 0$ be constants. After the nth observation, form

$$z_n = \log \frac{\Pr[\text{Marker present} \mid n \text{ observations}]}{\Pr[\text{Marker not present} \mid n \text{ observations}]} \tag{4.8-7}$$

There is a z_n for each observation at slot j. Then sum the z_n so far and perform the following:

$$\begin{gathered}\text{If} \sum_{k=1}^{n} z_k > A, \text{ decide marker in slot } j \\ \text{If} \sum_{k=1}^{n} z_k < B, \text{ decide marker not in slot } j \\ \text{Otherwise, make another observation.}\end{gathered} \tag{4.8-8}$$

The sequence of observations continues as long as $B < \sum z_k < A$. The choice of A and B is set by the performance level desired: the probability that a marker is decided where none exists, and that a slot is rejected when the marker really starts there. The details of the choice are in Ref. 17 or another text on the subject.

A flow chart of a procedure to acquire the marker position is shown in Fig. 4.39. The central decision is the SPRT, or a simplification of it, that looks at the frame symbols in a given slot and updates the judgment about them.

The foregoing is the solution of the frame lock problem in principle. It goes without saying that practical systems make assumptions and modifications. First, the frame pattern is assumed to repeat endlessly. If there is no customer data to transmit, an *idling sequence* takes the place of the data. It is important that the sequence not unduly disturb the framing procedure (e.g., the sequence should not contain the marker). The sequence should also not disturb the phase and symbol time synchronizers; for example, a run of 0s with simple pulse transmission will starve the symbol timing synchronizer for zero crossings. Second, the framing procedure continues to run after acquisition in order to watch for loss of framing. Third, markers are as often short simple sequences as they are the longer sophisticated sequences that we have discussed. Finally, the m marker bits can be distributed about in some rather odd places, instead of gathered together in one burst of m. *Added-digit framing* refers to adding just a single marker symbol to each data word, so that m words are needed to send a whole marker. In this case, the "View another frame" line in Fig. 4.39 becomes "View another set of m frames."

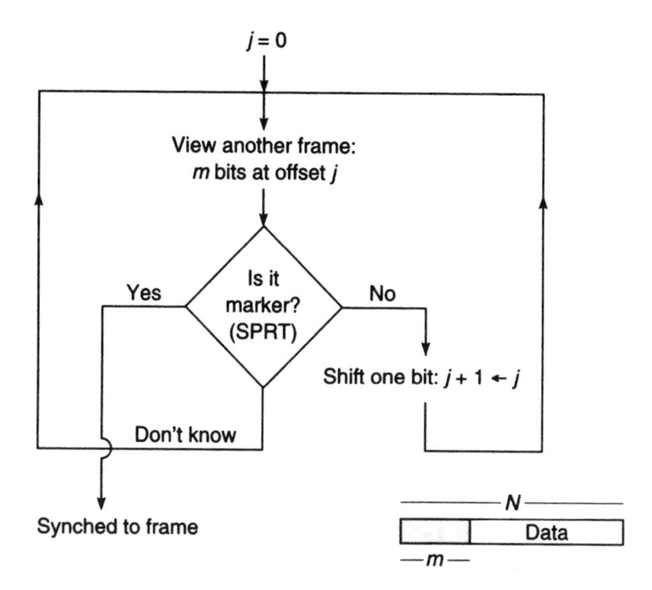

Figure 4.39 Procedure to find marker position in a frame.

With all these assumptions in place, the expected time until acquisition for added-digit framing is found ideally as follows, where T_s is the symbol time and there are $N - 1$ data symbols and a single marker symbol per frame. The time to view a full marker is the product $(T_s)(mN)$, and the correlation must shift through $\frac{1}{2}$ N slots on the average until it confronts the marker. So,

$$\mathcal{E}[\text{Lock time}] = (mN)(\tfrac{1}{2}N)(T_s) = \tfrac{1}{2}mN^2T_s \tag{4.8-9}$$

This assumes no noise and no false lock to data. Observe that when the observed slot is the added framing digit, the marker appears to repeat endlessly, in the manner of our discussion of PN sequences, and it makes sense to compute a periodic correlation like (4.8-1). When the slot is not the framing digit, the correlation is with the data and is on the average zero.

Example 4.8-3: T1 Frame Acquisition. The symbol rate in T1, DS1, and similar binary first-level standard systems in a transmission hierarchy is 1.544 Mbits/s, with added-digit framing employed. We can assume that the system is noise free. Suppose a length-4 frame marker is used (see Example 4.8-5). If the marker is avoided in the idling sequence, the expected lock time will be, from (4.8-9), $4N^2/2(1.544 \times 10^6)$ seconds. In the T1 standard, a frame is 24 8-bit data words plus the framing bit, for a total $N = 193$ bits. This yields an ideal expected lock time of 48 ms. A strategy to deal with false lock to data and noise might be to require several pattern repetitions before lock is declared and to ignore any single failure in the pattern.

The framing strategies we have discussed so far all involve adding marker bits to the data stream, either one by one or in a group of m. Other methods are employed but are less common. In robbed-digit framing, a data bit is "stolen" and its symbol slot is used for a marker bit. In speech or image transmission, for example, the noise caused by this bogus bit can often be tolerated. Robbed-digit framing avoids the throughput loss of added-digit framing. Another strategy is to look for unusual bits in the data themselves. A classic example is given by the bits in 8-bit standard PCM-digitized speech. The first bit of each 8, the sign bit, has probability 0.5 of being 1 or 0. The second bit, the most significant bit, is 1 with probability 0.95, since speech peaks are rare. The PCM structure can thus be recognized by its statistics, without a separate frame marker.

For a final example, we choose a cellular radio system that transmits its markers as m bits all in one place. The example highlights several other engineering issues that arise in cellular systems.

Example 4.8-4: American TDMA Digital Cellular Telephone Frame. This U.S./Canadian time-division cellular radio system, loosely referred to as IS-54 (Interim Standard 54), corresponds roughly to the GSM system used elsewhere in the world. It carries digitized telephone in 40-ms frames, each of which consists of six subframes of length 6.67 ms. A one-way call occupies two of these, either subframes 1 and 4, 2 and 5, or 3 and 6. In this way, three half calls are carried in time division over each radio carrier. One band of these carriers transmits from mobile to base station and another from base to mobile. Figure 4.40 shows the breakdown of one mobile-to-base transmission subframe (for a number of technical reasons, the base-to-mobile subframe varies somewhat). The modulation runs at the rate 24.3 ksymbol/s and employs a kind of differential root RC QPSK modulation. The subframe contains a marker of length 14 quaternary symbols, which appears amid 130 quaternary data symbols as shown. The marker performs several functions other than subframe alignment: There are actually six differ-

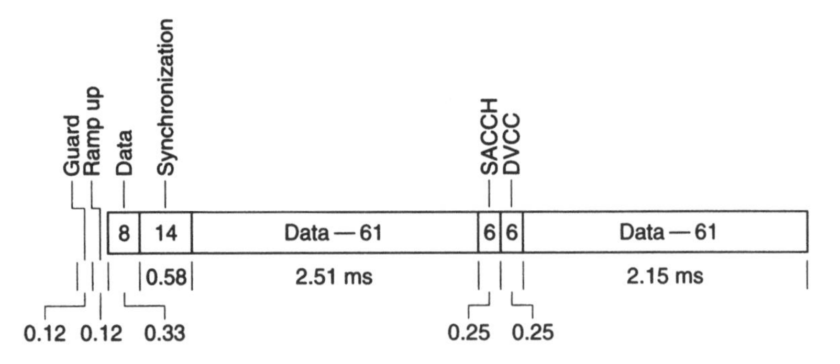

Figure 4.40 Mobile-to-base subframe structure in IS-54 digital cellular radio, showing quaternary symbols in and time devoted to each part (SACCH, Slow Associated Control Channel; DVCC, Digital Verification Color Code; these are control signals).

ent markers that identify the six subframes, and a sliding correlator acquires symbol timing with the marker in addition to subframe alignment (see Fig. 4.36). Several other points about the subframe are of interest. The SACCH and DVCC symbols, six each, are used for control. Three symbols worth of time, 0.123 ms, is devoted to guard time at the beginning of the subframe; this is needed because the mobile cannot time its transmission to arrive at the base precisely in the allocated time slot. Another three symbols are allocated to ramping up the transmitter power before the data symbols proper begin. Aside from electronics, there is a more subtle reason for allowing this time. The precursors of the first few root RC modulation pulses (see Fig. 2.2) appear in this ramp-up position; if these precursors were simply truncated, the bandwidth of the first few pulses would not be that of the root RC pulses, and unacceptable out-of-band interference would appear.

4.8.3 Hierarchical Networks and Their Synchronization

Large-scale networks are built-up hierarchies of streams that have agreed-upon bit rates. Streams are combined and broken down according to a set of rules. For example, in the commonly occurring DS hierarchy, four DS1 lines each at 1.544 Mb/s are combined to form a DS2 stream at 6.312 Mb/s. In North America this hierarchy extends two more levels: Seven DS2 streams combine to form a DS3 stream at total rate 44.74 Mb/s, and six DS3 streams form a DS4 stream at 274 Mb/s. Equipment tends to be specified for these rates and for the analog spectra that correspond. Other hierarchies in North America that coexist tend to match certain of these rates, especially the first one; see Examples 2.2-1 and 4.8-3.

An alternate hierarchy that is common in Europe, the CEPT, or E hierarchy, starts with a basic 2.048-Mb/s stream (e.g., thirty 64 kb/s PCM channels). Four of these combine to a second-level stream at 8.448 Mb/s; four of those make a third-level stream at 34.37 Mb/s; four third-level streams make a fourth level at 139.3 Mb/s.

The heart of the coordination of such networks is the multiplexer. A multiplexer is a device that combines many slow symbol streams into one faster one. A demultiplexer does the reverse. The notation M12 denotes a multiplexer across the first step in a hierarchy; similarly, D21 denotes a demultiplexer in the reverse direction. In the discussion of multiplexing that follows, we will illustrate with the DS system. Figure 4.41a shows a block diagram of the components in the M12 and D21 devices. In addition to combining and splitting streams, "mux" and "demux" equipment, as they are called, must also insert and remove various control bits, must maintain synchronization (maintain "frame"), and must recover from loss of synchronization with the input stream, a process called reframing.

Since it cannot be guaranteed that the incoming streams to a multiplexer have identical timing, some way of rationalizing the stream rates must be provided. The reverse is true at the demultiplexer: The different users at the output side sometimes cannot accept bits in perfect synchrony, and so some elasticity must be built into the

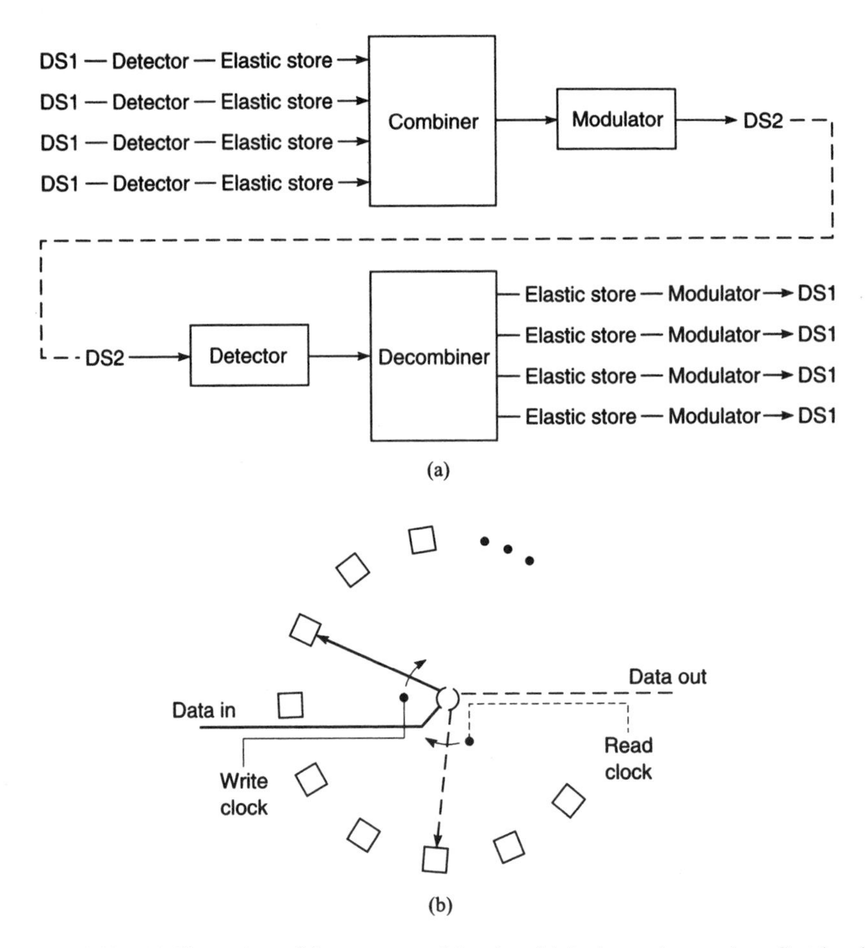

Figure 4.41 **(a)** Illustration of first- to second-level multiplexing and second- to first-level demultiplexing. **(b)** Detail of elastic store block; L storage cells are arranged in a circle, with independent read and write arms.

dispensing of bits. The critical element is called an *elastic store*. There is one for each incoming mux channel and one for each outgoing demux channel. An elastic store can be thought of as a circular buffer of length L, whose input, or write, clock is controlled by the incoming data clock and whose output, or read, clock is controlled by an outgoing data clock (in Fig.4.41b, this is the combiner clock in the mux and the T1 data sink clock in the demux). If the read clock is faster, it eventually overtakes the write clock, and if the overspeed continues, eventually L symbols are put out twice. If the read clock is slower, the write clock overtakes it, and if the underspeed continues, L symbols are overwritten and lost. Ideally, the two clock rates are the same on the average, and if L is long enough, symbols are virtually never duplicated or lost.

It may seem obvious that the way to control a network of multiplexers and demultiplexers is with one master clock. But even with perfect clock distribution to all network nodes, it is still not possible to guarantee exactly synchronous streams. It can be shown, for example, that a 1000-km coaxial cable that carries 300 Mb/s delivers about 200 extra symbols as a result of warming up by 1°C, simply because the cable dielectric warms; the symbol rate at the sending end can be kept the same, yet a receiver that accepts this clock must somehow dispose of 200 loose symbols. Here is a network that is threatened by the break of day! Diverse rate-matching problems like this exist. In any event, one master clock is a bad idea because if it fails, the whole network is lost: Too much control is given to one point in the network.

Several other network clocking methods have been proposed [18], but the one that is by far the most used is *pulse stuffing*. This is the adding of extra bits to slower arriving streams at announced places, in order to bring all incoming streams at a mux up to the same average rate. The corresponding demultiplexer reads the announcement and strips away the stuffed bits.

Figure 4.42 demonstrates in simplified form how stuffing works. The mux output rate is set higher than the rate expected at any input stream. Three streams come in at rates that are four-fifths, three-fourths, and two-thirds of the output rate. The stuff bits, shown by *X*'s, raise all three streams to the same rate, and they can now be synchronously combined into one triple-rate output stream.

The next example shows how a particular multiplexer works, the M12 conversion from four DS1 streams to one DS2. The example is meant to be typical and not to describe the exact details of one manufacturer or one international standard. As well as a stuffing implementation, the example demonstrates several common design techniques. The incoming streams are not combined in the simplest row/column way. Framing marker bits are scattered throughout the frame, rather than being collected in one place. The example does not show bits devoted to parity checking and system control that appear in higher muxes, such as a DS2-to-DS3 conversion.

Example 4.8-5: DS1-TO-DS2 Multiplexer. A DS2 stream, the combiner output in Fig. 4.41b, runs at 6.312 Mb/s and consists of four 1.544-Mb/s DS1 streams. These are combined as shown in Fig. 4.43. Each "[48]" in the figure consists of 12 bits from each of the four DS1 streams, digit by digit interleaved in the stream pattern 12341234. . . . The overall DS2 frame is a block of four rows of these [48]

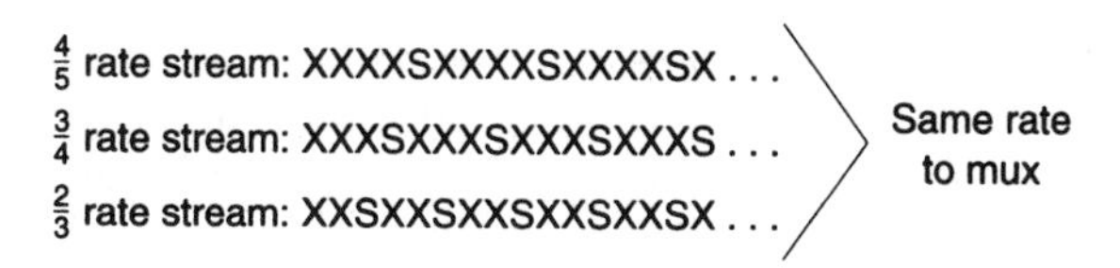

Figure 4.42 Illustration of pulse stuffing: X denotes data; S denotes stuffed symbol.

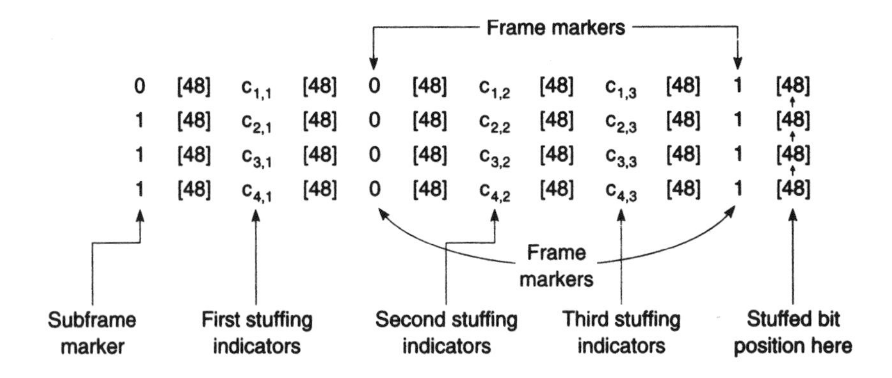

Figure 4.43 A DS2 frame structure showing data bits (in 48-bit groups denoted [48]), stuffing indicators (denoted C), and frame marker bits. Total bits: 1176; 5367 frames/s. Stuffed bits, if any, are in rightmost [48] in each row. Four DS1 streams are interleaved here, as explained in text.

units, interspersed with system bits and transmitted row by row. Note that although there are four incoming streams and four rows, the rows do not correspond to the streams. The system bits are of three kinds, frame markers, subframe (row) markers, and stuffing indicators. The frame bits are dispersed around the frame as shown, and, taken in order of time, they are 01010101. The framing synchronizer looks for this marker, but, once found, the marker does not indicate which row is the top one in the array. That is done by the subframe marker, which is 0111, taken in order. (Alternately, the 12 bits could be taken as the one marker 001 101 101 101, in order of transmission, with the spaces inserted for clarity; the M12 mux here instead acquires frame synchronization in two stages, column followed by row synchronization.) One stuffed bit position exists in each row, and whether or not this is a stuffed bit or a true data bit is shown by the third kind of system bits, the stuffing indicators. Consider the stuffing position for DS1 stream 1. If $c_{1,1} = c_{1,2} = c_{1,3} = 1$ in the first row, then an extra bit is stuffed into stream 1; the position is the first stream 1 position in the last [48] group in the first row. If the three indicators are 0, the position is a true data bit. The three indicators constitute a single-error-correcting code, a protection that is needed because of the seriousness of a stuff position error. Indicator bits $c_{2,1}$, $c_{2,2}$, $c_{2,3}$ appear in the second row and apply to DS2 stream 2; the stuff position is the first stream 2 position in the last [48] group in row 2—that is, the second bit in this [48] group. And so on for the last two rows and streams. The degree to which symbol rate mismatch can be tolerated is estimated as follows. Suppose the DS1 incoming stream arrives at the nominal rate 1.544 Mb/s or lower; that is, the arrival rate lies in the range $(1.544M - f_1, 1.544M)$, with the lower rate corresponding to full stuffing and the higher rate meaning that all stuff positions are data. Each frame contains $12 \times 6 \times 4 = 288$ data positions for each DS1 stream, counting the single stuff position. If this position never carries a data bit, the corresponding lower rate is

$$(1.544 \text{ Mb/s})(287/288) = 1.544 \text{ Mb/s} - 5.4 \text{ kb/s}$$

So the width of the arrival rate range is $f_1 = 5.4$ kb/s. This is a frequency stability of 3500 parts per million, far wider than a crystal oscillator, which achieves perhaps 10 ppm. However, the network design should allow for chains of many oscillators in tandem; this and other details may be found in books such as Ref. 18.

Network switching and synchronization are complex subjects, and we have been able to touch only some main points here. References 22 and 23 provide the next level of detail.

Optical Networking. Optical fibers are steadily replacing other media as the main means of transmission in large-scale networks. A fiber is not necessarily cheaper per bit for low-speed transmission than a wire or cable, because the handling and maintenance of fibers has a certain difficulty associated with it. But fibers are unexcelled at rates of 10–100 Mb/s and higher.

Although fibers carry lightwave signals, it is generally electrical signals that arrive at and leave the fiber medium. This means that network signals are specified in two forms, electrical and lightwave. The two most common hierarchies are given in Table 4.4. They are SONET (for synchronous optical network), which has evolved primarily in North America, and SDH (for synchronous digital hierarchy), which is a world CCITT standard. SONET uses the STS electrical signal hierarchy and SDH uses the STM, and both interface to the same OC optical signal hierarchy.[17] Both STM and STS are similar in principle to the hierarchies earlier in this section, and their lower levels are meant to interface to the higher levels of the earlier DS, E, and so on. The higher optical levels run at much faster data speeds than do earlier electrical systems, and several very high speed levels have been left off the table.

Compared with electrical techniques such as coherent modulation and time-division multiplexing, lightwave techniques are still crude. Lightwave modulation is mostly on–off signaling, and the signals are more difficult to switch. SONET and SDH work like a conveyor belt that carries data packages without accurate timing. Special pointers and well-marked data package boundaries allow streams to be added and removed easily, without precise harmonizing of symbol rates. It is also relatively easy to replace a failed lightwave path with another one.

These are all attractive features, of course, but accessing a stream inside one of the light packages is still relatively difficult. To do it, one must demodulate from light to electrical form, then demultiplex and remove the desired stream, then laser-modulate once more to optical form. The lowest optical levels are still rather complex electrical signals, with bit rates of 50 Mb/s and above. The intention in an optical network is that these, but not slower, streams can be switched in and out easily.

[17]These acronyms stand for synchronous transport signal, synchronous transport module, and optical channel, respectively.

TABLE 4.4 Part of the SONET and SDH Optical Hierarchies

Electrical signal				
SONET	SDH	Optical Signal	Rate (Mb/s)	Notes
STS-1	—	OC-1	52	Matches DS-3
STS-3	STM-1	OC-3	156	Matches E-4
STS-9	STM-3	OC-9	467	
STS-18	STM-6	OC-18	933	
STS-24	STM-8	OC-24	1244	
STS-48	STM-16	OC-48	2488	Optical backbone

A readable introduction to optical networking appears in the text by Leon-Garcia and Widjaja [24].

PROBLEMS

Most of the problems here are best solved with the aid of software tools such as MATLAB and the Programs in the text.

Phase-Lock Loops

4-1. Consider a first-order PLL whose gain constant is $K = 1/100T$. Let $T = 1/2400$ s, a number typical of telephone modems.

(a) Give the PLL closed-loop transfer function $H(s)$.

(b) Give the error transfer function.

(c) Give the unit error step response. How long does it take for the error to fall to 0.1 radians?

(d) What is the steady state error for a unit ramp input?

(e) Give the noise-equivalent bandwidth B_N for this PLL.

4-2. *PLL impulse response.* A sudden reference phase shift followed by a return to the original phase can be modeled by an impulse $\delta(t)$ at the PLL input. Suppose a low $B_N T$ of 0.01 is needed.

(a) Find the response of a first-order loop with this $B_N T$ to $\delta(t)$.

(b) Find the two-pole loop response when $\zeta = 1/\sqrt{2}$.

4-3. *PLL pulse response.* Plot the unit pulse error response for the two-pole loops in Fig. 4.13; that is, find the error response to the input $u(t) - u(t - T)$, where $u(t)$ is the unit step.

4-4. Consider a first-order PLL.

(a) Find the loop gain K that will give a step response that achieves 90% of the step input before 10 symbol intervals have elapsed. For simplicity, set the symbol time T to 1.

(b) Find, or look up, the 3-dB point of the loop frequency response $H(j\omega)$. Compare this to the noise-equivalent bandwidth.

(c) Find the unit ramp response and give the steady-state error.

4-5. Show that the noise-equivalent bandwidth B_N defined in (4.3-15) is $\omega_n/8\zeta$ for a two-pole loop with the standard pole locations (4.3-12).

4-6. A transmission system carries data at a symbol rate $1/T = 1$ MHz, with a 10-MHz carrier. Phase synchronization is obtained by a PLL that tracks a steady pilot tone near 10 MHz. There is background AWGN whose power spectral density is $N_0/2$ W/Hz. Let E be the energy devoted to the pilot per time period T. It is desired that the system stay reliably in synchronization even when E/N_0 for the pilot falls to 2, and consequently $B_N T$ is set to 0.01.

(a) Suppose the VCO frequency error in the PLL suddenly becomes 1 kHz. Give the steady-state phase error for first order, two-pole, and DI loops. (Ignore the noise here.)

(b) Now return the VCO frequency error to 0 Hz. Give the rms phase jitter due to noise for these three when $E/N_0 = 2$. If need be, assume that $\zeta = 1/\sqrt{2}$.

4-7. Suppose that a BPSK carrier recovery circuit is such that it adapts to a sudden reference phase jump of 45° at time zero with error

$$e(t) = 45\, e^{-0.2t/T}$$

for $t > 0$. It is desired that the error symbol rate be 0.000001.

(a) What E_b/N_0 is required for detection if there is perfect phase tracking?

(b) How many intervals will go by until the error probability returns to within 10% of the specification if the receiver phase has the error $e(t)$? Assume that the initial phase in an interval is the one that applies to the whole interval.

4-8. A special kind of PLL is based on a *lead-lag* loop filter, whose transfer function is $K(1 + as)/(1 + bs)$. Here $K > 0$ is the total PLL loop gain.

(a) Give the closed-loop transfer function $H(s)$ for the standard PLL with this filter. Express it in the standard second-order form for which the denominator is $D(s) = s^2 + 2\zeta\omega_n s + \omega_n^2$.

(b) Give an expression for the damping coneefficient ζ in terms of a, b, and K. What does the fact that ζ is real and positive imply about a and b?

(c) Which a and b lead to a stable $H(s)$?

(d) Let $K = 10$ and $\zeta = 1/\sqrt{2}$. Give two pairs a, b that lead to this. (This shows that the PLL bandwidth can be adjusted independently of K and ζ, which is important in some applications.)

Synchronizer Designs

4-9. Consider a QPSK digital mobile radio system, running at 2.4 kb/s with carrier 400 MHz. Because of motion-induced Doppler shifts, the carrier can vary by up to ±100 Hz. Phase synchronization is derived by a fourth-power loop. Explore the design of a second-order PLL inside the synchronizer. Define the PLL by its damping factor and its noise equivalent bandwidth $B_N T$.

(a) If the Doppler shift is interpreted as a ramp phase excitation to the PLL, what is the maximum phase slope of the ramp in radians per symbol?

(b) Assume that the synchronizer phase error, after the frequency divide by 4, should never exceed 0.3 radians. Find the two-pole PLL with the least $B_N T$ that meets this specification. Pay heed to the damping factor. This problem is best done by cut-and-try, with one of the Programs.

4-10. Repeat Problem 4-9, but find the *double-integral* PLL with least $B_N T$.

4-11. A deep space probe transmits back to earth over a NRZ-BPSK channel that incorporates a powerful error-correcting code, so that the E_s/N_0 in the channel is 0 dB, where E_s is the energy in a channel symbol. The value of the data E_b/N_0 could, for instance, be 5–10 dB. A squaring loop that incorporates a two-pole PLL is used for phase synchronization.

(a) Find the $B_N T$ needed in the PLL so that the jitter variance in the synchronizer output (after the frequency is divided by 2) is 0.01 rad^2; that is, the standard deviation is 0.1. Allow for a squaring loss of 3 dB.

(b) A slow drift of 0.01 radians/symbol occurs in the transmit or receive oscillator. What synchronizer output phase error will this cause in the steady state?

(c) In the worst case, the synchronizer phase error will be the sum of (a) and (b). Is there a design such that the sum of these seldom exceeds 0.1 radian?

4-12. Consider rectangular 16QAM with NRZ pulses. A fourth-power device quadruples the phase of this signal. A limiter then reduces all the amplitude factors to the same value A but leaves the phases untouched. Relative to the 4× carrier, what phases are present, and with which probabilities, assuming equally likely data? Design a PLL for the phase synchronizer, by considering the noise from these phases only; ignore other disturbances to the PLL.

4-13. BPSK is transmitted with pulse shape $v(t) = \sin(\pi t/T)$, $0 < t \leq T$. That is, the signal is $\Sigma a_n v(t - nT) \cos \omega_0 t$, with $a_n = \pm 1$. A bandpass limiter could be employed to obtain phase synchronization, but instead a squaring loop is used (Fig. 4.2, but note that the amplitude factor A is not constant).

(a) Find the SNR at the squarer/BPF output due solely to data noise.

(b) Without actually computing a PSD, estimate the loop bandwidth needed to reduce VCO phase jitter to 0.1 radians.

4-14. Establish the accurate relationship between the symbol error probability P_s of QPSK and synchronizer phase error $\Delta\phi$. Use Eq. (4.2-5) with its approximation $1 - [1 - Q(x_1)][1 - Q(x_2)] \approx Q(x_1) + Q(x_2)$ and proceed as follows. (*Note:* You may find it convenient to add routines to find $Q(\cdot)$ and $Q^{-1}(\cdot)$ to your MATLAB toolbox, if they are not already present.)

(a) Plot $\log_{10} P_s$ against E_b/N_0 in dB (a standard waterfall curve), for $\Delta\phi = 0, 0.1, 0.2$ radians.

(b) Plot $\log_{10} P_s$ against $\Delta\phi$ over $[0, \pi/4]$ radians for E_b/N_0 fixed at 9.59 dB, the value giving $P_s = 10^{-5}$ at $\Delta\phi = 0$.

Discrete-Time PLLs

4-15. Find the pulse response of a digital PLL as follows.

(a) Give the z-transform of a sequence that consists of two unit samples and is otherwise zero.

(b) Derive the error response to a pulse for a PLL with z-transform $H(z)$; assume the pulse is as in (a).

(c) Plot the pulse error response for the 193-kHz case in Example 4.6-1 and compare to the step error response in Fig. 4.26b.

4-16. A very narrowband analog PLL having bandwidth $B_N T = 0.01$ is to be realized in discrete-time form with sampling time $20T$. Let the analog loop be the two-pole one in Example 4.6-1, Eq. (4.6-9). By means of the bilinear transformation, find an equivalent discrete PLL. Plot the frequency and step error response and compare to Fig. 4.26.

4-17. Demonstrate that the single z-transform (4.6-13) carries out the entire accumulator digital PLL by proving the following.

(a) Show that the series of steps (4.6-15)–(4.6-17) leads to the difference equation (4.6-18).

(b) Demonstrate that the z-transform of (4.6-18) is (4.6-13).

Timing Recovery

4-18. Show that with NRZ pulses the average waveform viewed by an early–late gate synchronizer in the vicinity of the correct sampling instant is symmetric and reaches a peak at the correct time, in the way pictured in Fig. 4.30.

4-19. Making reasonable assumptions, propose a design for the symbol time recovery for the 4800-b/s V.32 QPSK modem standard (see Table 5.4). Assume a perfect channel and 30% root-RC pulses, and concentrate attention on the signal zero crossings. *Note:* There is no single approach to this design; one way to begin is to simulate the zero crossings of the QPSK signal and measure their statistics.

4-20. Making reasonable assumptions, propose a design for the symbol time recovery in the deep-space link of Example 4.5-2.

4-21. *Synchronizer for deBuda's Fast FSK.* This modulation method is the FSK scheme depicted Fig. 3.21 when $h = \frac{1}{2}$. The signal is

$$s(t) = A\cos(\omega_0 t + a_n \pi t/2T + \psi_n)$$

where $a_n = \pm 1$ and ψ_n is selected each symbol time to make the phase continuous. The scheme is equivalent to MSK (see Example 3.6-3). deBuda [25] proposed a phase and symbol synchronizer that first squared the signal $s(t)$. This produces exactly the same Fig. 3.21 signal but with $h = 1$. It can be shown that the squared signal has a continuous spectrum together with two spectral lines at the frequencies $\omega_0 \pm \pi/T$. Two narrowband PLLs extract these lines, producing the signals $\cos(\omega_0 t +$

$\pi t/T$) and $\cos(\omega_0 t - \pi t/T)$. Design the rest of a system whose output is the carrier $\cos \omega_0 t$ and the symbol timing signal $\cos 2\pi/T$.

Network Synchronization

4-22. A length-7 sequence is to be used in the acquisition of a timing epoch. We are to locate the epoch by correlating what is received with a copy of the length-7 sequence at the receiver. The transmitted symbols are the antipodal values ± 1, henceforth denoted by + and –. Immediately before and after the sequence, nothing is transmitted, which the receiver interprets as the value 0.

(a) Find the correlation that will be observed if the sequence + – + – + – + is employed, and compare your answer to what would happen if the PN sequence + + + – + – – in Example 4.8-1 were used instead.

(b) Now suppose the first and seventh digits are received in error (i.e., inverted). Compare the two correlations.

4-23. *Extension of Example 4.8-3.* The example shows that the expected lock time for a standard T1 frame is 48 ms over a noise-free channel, if the frame marker does not appear in the data. Let the channel be a simple binary symmetric channel (see Section 6.1) that inverts bits with probability p. The marker is 1100. During acquisition of lock, the data is the idling sequence 111 . . . 1. A frame lock system works as follows: During the acquisition phase, *frame lock* is declared if the marker is observed twice in succession; during the tracking phase, *frame loss* is declared if an acquired marker twice fails to appear where it is expected. In what follows, make simplifying assumptions as needed and let $p = 0.01$. In parts (a) and (b) the system is trying to acquire the marker.

(a) Estimate the probability of false lock—that is, a lock that includes one or more symbols in the idling sequence.

(b) How long does it take to reach lock on the average, whether true or false?

(c) In the next two parts, assume the system is in lock to the true marker at some time t_0. Estimate the probability that declaration of loss occurs two marker repetitions later.

(d) What is the mean time to frame loss, at any time later than t_0?

4-24. It can be difficult to distinguish frame markers from ordinary data. Let the sequence + + – (a short Barker sequence) be used to mark the beginnings of frames. The sequence appears in the pattern . . . + + – *DDDDD* + + – *DDDDD* + + – *DDDDD* . . . , where *D* denotes a random binary data symbol and the data appear in the 5-bit blocks shown. Suppose the channel error probability is 0.1 and that an error complements the bit.

(a) What is the probability that some run of 3 data bits in a 5-bit data word looks exactly like the marker?

(b) Consider one complete frame + + – *DDDDD* in the midst of the transmission. What is the probability of seeing the marker beginning at an offset of 0, 1, . . . , 7 positions to the right of the correct position? Remember that there is an error probability.

(c) Repeat for a receiver that looks for two markers in succession. (The answers will be much improved.)

4-25. A length-7 marker is used to mark the frame boundary in an error-free channel; a single complete frame looks like *MMMMMMMDDDD . . . DD*, with *M* a marker bit and *D* a data bit, and this pattern repeats endlessly. The data are random and the data blocks *DDDD . . . . DD* contain many more than 7 bits.

(a) Plot the probability of observing the marker $+ + + - - + -$ in any position.

(b) Repeat for the naive marker $+ + + + + + +$.

4-26. *Extension of Example 4.8-5.* The DS3 digital signal carries 7 DS2 streams. In one version, the M23 multiplexer combines 672 bits from each of the DS2 streams, plus 8 housekeeping bits each stream, into one $(672 + 8) \times 7 =$ 4760-bit DS3 superframe. Included within each 672-bit group is one stuff position. When carrying PCM telephone calls, a DS3 carrier is sometimes called a Mastergroup.

(a) How many calls are there in a Mastergroup (assume 64-kb/s PCM)?

(b) Take the nominal 44.736-Mb/s DS3 bit rate as the highest allowed multiplexer output rate. With the single stuff position, what are the highest and lowest allowed DS2 input rates?

(c) How many total housekeeping bits are there per DS3 superframe, counting levels DS1–DS3? Do not count stuffing bits.

REFERENCES[18]

1. A. J. Macdonald, Characterization and development of ML-based synchronization structures for trellis-coded modulation, Ph.D. Dissertation, ECSE Department, Rensselaer Polytechnic Institute, Troy, NY, May 1991.
2. J. P. Costas, Synchronous communication, *Proc. IRE,* vol. 14, December 1956, pp. 1713–1718.
3. R. E. Ziemer and C. R. Ryan, Minimum-shift keyed modem implementations for high data rates, *IEEE Commun. Mag.,* vol. 21, October 1983, pp. 28–37.
4. *R. E. Ziemer and R. L. Peterson, *Digital Communications and Spread Spectrum Systems,* Macmillan, New York, 1985.
5. *W. C. Lindsey and C. M. Chie, A survey of digital phase-locked loops, *Proc. IEEE,* vol. 69, April 1981, pp. 410–432.
6. *J. G. Proakis, *Digital Communications,* 3rd ed., McGraw-Hill, New York, 1995.
7. W. F. Egan, *Frequency Synthesis by Phase Lock,* John Wiley & Sons, New York, 1981.
8. H. L. van Trees, *Detection, Estimation, and Modulation Theory (Part I),* John Wiley & Sons, New York, 1968.
9. G. F. Franklin, J. D. Powell, and A. Emami-Naeini, *Feedback Control of Dynamic Systems,* 3rd ed., Addison-Wesley, Reading, MA, 1990.
10. *F. M. Gardner, *Phaselock Techniques,* 2nd ed., John Wiley & Sons, New York, 1979.

[18]References marked with an asterisk are recommended as supplementary reading.

11. H. Meyr and G. Ascheid, *Synchronization in Digital Communications,* John Wiley & Sons, New York, 1990.

12. *J. A. Cadzow, *Foundations of Digital Signal Processing and Data Analysis,* Macmillan, New York, 1987.

13. *L. B. Jackson, *Signals, Systems, and Transforms,* Addison-Wesley, Reading, MA, 1991.

14. B. Razavi, ed., *Monolithic Phase-Lock Loops and Clock Recovery Circuits,* IEEE Press, New York, 1996.

15. F. M. Gardner, Charge-pump phase-lock loops, *IEEE Trans. Commun.,* vol. COM-28, November 1980, pp. 1849–1858.

16. J. J. Stiffler, *Theory of Synchronous Communications,* Prentice-Hall, Englewood Cliffs, NJ, 1971.

17. A. Wald, *Sequential Analysis,* John Wiley & Sons, New York, 1947.

18. Members of Bell Laboratories Staff, *Transmission Systems for Communications,* various editions, ATT Inc., New York.

19. E. A. Lee and D. G. Messerschmitt, *Digital Communication,* 2nd ed., Kluwer, Boston, 1994.

20. S. W. Golomb, *Shift Register Sequences,* Holden-Day, San Francisco, 1967; rev. ed., Agean Park Press, Laguna Hills, CA, 1982.

21. R. H. Barker, Group Synchronizing Of Binary Digital Systems, in *Communication Theory,* Butterworths, London, 1953, pp. 273–287.

22. J. C. McDonald, *Fundamentals of Digital Switching,* Plenum, New York, 1983.

23. J.W. Pan, Synchronizing and multiplexing in a digital communications network, *Proc. IEEE,* vol. 60, 1972, pp. 594–601.

24. *A. Leon-Garcia and I. Widjaja, *Communication Networks,* McGraw-Hill, New York, 2000.

25. R. deBuda, Coherent demodulation of frequency-shift keying with low deviation ratio, *IEEE Trans. Commun.,* vol. COM-20, June 1972, pp. 429–435.

APPENDIX 4A: DATA NOISE IN QPSK

In this Appendix we analyze the data noise properties of QPSK with root RC pulses. What follows is typical of what might be employed with other modulations such as QAM. The method is to measure the noise produced in the basic phase error signal (4.4-8) by a sequence of random data.

In order to analyze (4.4-8), we make the assumption that θ_0 is quasi-static, as we did in the case of channel noise. This is a good assumption whenever the PLL bandwidth is much narrower than the bandwidths of I and Q. With a quasi-static θ_0 and ψ_0, $\sin 4(\psi_0 - \theta_0)$ and $\cos 4(\psi_0 - \theta_0)$ act as constants and the factors with $\tan^{-1}(\cdot)$ inside act as noise. For example, at quasi-static error $\psi_0 - \theta_0 = 0.1$ radian, the error signal (4.4-8) becomes

$$0.389 \cos\left[4 \tan^{-1}\left(\frac{Q(t)}{I(t)}\right)\right] + 0.921 \sin\left[4 \tan^{-1}\left(\frac{Q(t)}{I(t)}\right)\right] \qquad \text{(4A-1)}$$

We can equate this to the form

$$\mu \sin 4(\psi_0 - \theta_0) + \lambda(t) \tag{4A-2}$$

in which the first term is the long-term average value of the noisy signal and $\lambda(t)$ is a zero-mean noise. Eventually, the mean value will drive the PLL to lock, while $\lambda(t)$ creates a variance in the VCO phase.

Often, the easiest way to analyze a signal like (4A-2) is to create pulse trains and simply measure it. Figure 4.44 traces the actual signal (4A-1) for the 0.1-radian error when the RC pulse trains $I(t)$ and $Q(t)$ each carry 20 random ±1 data symbols. In

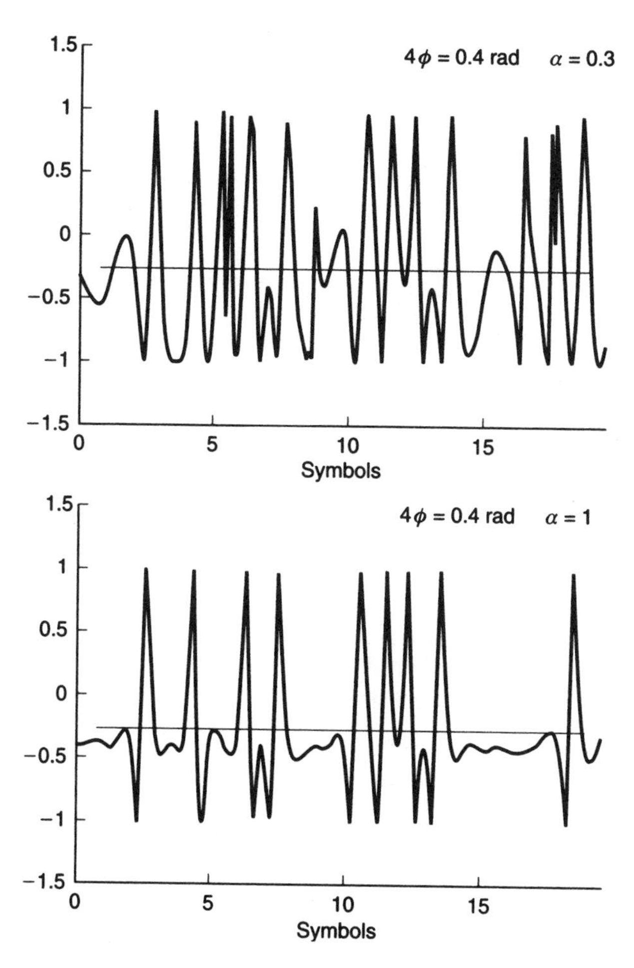

Figure 4.44 A QPSK quadrupler loop phase error signal (4.A-1) for 20 random I and Q data symbols; RC pulses with excess bandwidth $\alpha = 0.3$, 1 and signal phase error $\phi = \psi_0 - \theta_0 = 0.1$. Mean values shown by line; in both cases value is $4(\psi_0 - \theta_0)\mu \approx 0.28$; variances about mean are 0.32 and 0.18.

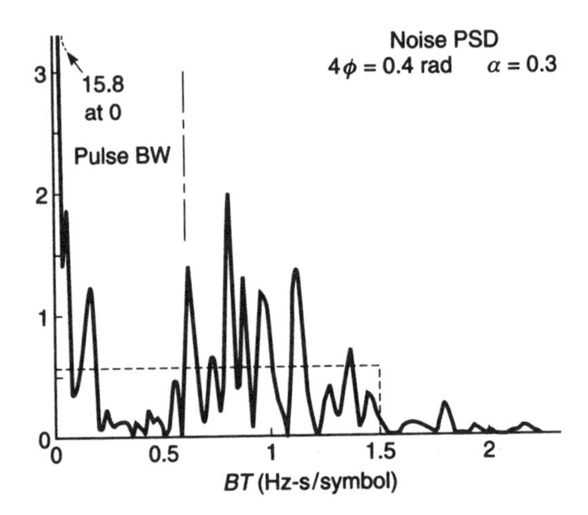

Figure 4.45 Plot of $(1/N)|Y(n)|^2$, where $Y(n)$, $n = 0, \ldots, 511$ is 512-point FFT of Fig. 4.44 for $\alpha = 0.3$ case. Frequency axis is normalized to symbol time. Because of padding zeros, true PSD is 2.4 times values shown.

one plot, the RC excess bandwidth factor α is 0.3, while in the other it is 1, which creates wide-bandwidth pulses that are closer to the NRZ shape. In either case, the average signal value is about the same, but the variance of the noise component $\lambda(t)$ in the $\alpha = 1$ case is only half as much.[19] The noise manifests itself as spikes when $Q(t)/I(t)$ changes sign. Figure 4.45 is an FFT that approximates the power spectral density of signal (4A-1) for the $\alpha = 0.3$ case.[20] Note how the bandwidth of the pulses (0.65 Hz-s) is about tripled by the fourth-power nonlinearity.

By imagining that the data noise is flat Gaussian noise, we can gain a rough but useful insight into the data noise's effect on the VCO phase. Before beginning, we need to take the perhaps subtle step of scaling the noise by $1/\mu$. The reason for this is that the total PLL response will be the superposition of two responses, a rapid jitter response to the noise and a long, slow response to the mean value $\mu \sin 4(\psi_0 - \theta_0)$. The data noise has the effect of inserting a μ below 1—in our example it lies near 0.6—and for a fair comparison to the NRZ pulse case, we need to scale up the "drive" of the PLL to $\sin 4(\psi_0 - \theta_0)$. The separate PLL responses to the scaled noise and the mean value are thus modeled and superposed as in Fig. 4.46.

Now we replace the data noise with a surrogate Gaussian source. For Fig. 4.45, this could be done as shown by the dashed lines: The FFT is replaced by a new one

[19]Note that the average value sign is opposite to the phase error in both cases. This sign must be reversed at the PLL input, as it is in Fig. 4.3.

[20]For K samples $y(1), \ldots, y(K)$ of a random process $y(t)$ taken at frequency f_s hertz, the PSD is approximated by $P(n) = (1/K)|Y(n)|^2$, where $Y(0), \ldots, Y(N-1)$ is the N-point FFT of these samples with $N - K$ padding zeroes. The frequency f corresponding to FFT point n is $f_s n/N$ hertz. Let I_0 be the integral $\int P(f)\, df$ over the range $[0, f_s)$, or alternately, let I_0 be $(f_s/N) \sum P(n)$. Then the power in $y(t)$ is approximately I_0. The estimate to the second moment of a single observation is I_0/f_s.

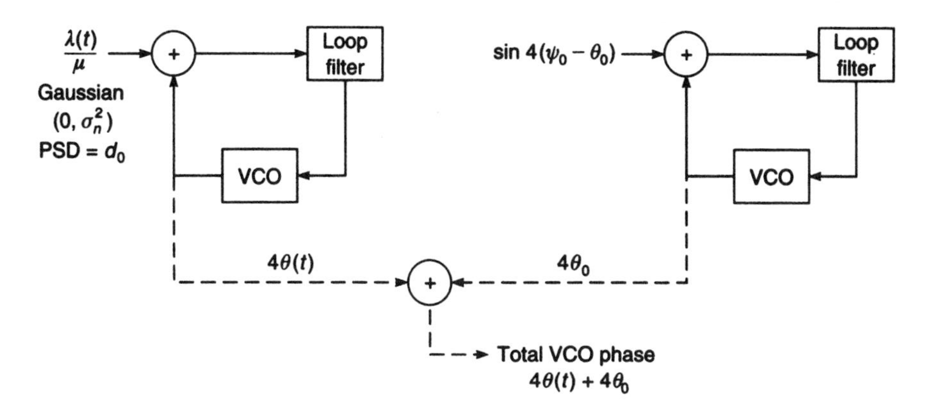

Figure 4.46 Gaussian equivalent noise PLL model for data noise response. Jitter response $\theta(t)$ **(left)** is superposed on much slower response to reference phase offset $4\psi_0$ **(right)**.

of constant height 0.55 and effective single-sided bandwidth $B_{\text{eff}} = 1.5/T$, dimensions that are chosen to give the same power as the original PSD. Because of padding zeroes, the height 0.55 corresponds to a power spectral density of $d_0 = 2.4(0.55) = 1.33$ W/Hz. Assuming now that the data noise really is Gaussian, we can repeat the argument that led to Eq. (4.4-5). For a density d_0, the power of the jitter is

$$\sigma_\theta^2 = \int_{-\infty}^{\infty} d_0 |H(j2\pi f)|^2 \, df$$
$$= 2\, d_0 B_N$$

where B_N, the noise-equivalent bandwidth, is taken to be less than B_{eff}. From its definition, $d_0 = (\sigma_n/\mu)^2/2B_{\text{eff}}$, where σ_n^2 is the data noise power. So, the jitter power is, with two T factors inserted,

$$\sigma_\theta^2 = \left(\frac{\sigma_n}{\mu}\right)^2 \frac{B_N T}{B_{\text{eff}} T} \tag{4A-3}$$

This states that the power in $\lambda(t)/\mu$ is reduced by the symbol-normalized ratio of the loop to the noise bandwidths. The principle contained here, that of reducing data noise by narrowing the loop bandwidth, applies widely. In our QPSK example, the second moment of a noise observation scaled by $1/\mu$ is close to unity, and reducing the VCO jitter power to, say, 0.05–0.10 will require a loop $B_N T$ product close to these same numbers. Example 4.5-1 discusses such noise calculations further.

CHAPTER 5

CHANNELS

Earlier chapters have focused on generating and detecting signals, as well as on the laws that govern transmission over the Gaussian channel. We have mostly ignored how signals pass from one place to another in the real world or how they pass from one time to another in a storage medium. These communication channels are a rich source of engineering challenges. Channels are where most signal distortion occurs, and both nature and man tend to work in ways that frustrate analytical tools.

In information theory, a channel is conceived of as a simple conditional probability that a symbol is received given that one was sent. Most of the time, successive uses of the channel are taken as independent. This conception is too simple for us now. In reality, a channel works upon continuous-time signals and inserts dependencies across frequency and time. How it does this depends on the channel's physical nature.

Among the real-life channels there is a natural progression from easy to hard, and this forms the organization of the chapter. The easiest channels are the various kinds of wires and cables that work in a shielded, self-contained medium that is free or almost free from the outside world. To a first approximation they distort signals but do not add noise. We call these *guided channels*; they are the subject of Section 5.1. A fundamentally different channel is the simple radio link in Section 5.2. Now the signals are carried by propagating radiation that weakens and spreads, without protection from the outside world. With some oversimplification, we can think of these as adding noise but not otherwise distorting the signal. This is still a simple channel if the propagation path is free of obstacles, a classic example being the space channel. Terrestrial radio channels, in Section 5.3, add a major source of distortion, namely Earth, and are much more difficult to work with. Mobile terrestrial channels are more difficult still, because the terminals are moving. Combined with nearby obstacles and reflectors, motion leads to severe difficulties, which consume the last section of the chapter.

Digital Transmission Engineering, Second Edition. By John B. Anderson

Noise plays a role in most channels. Were is not for noise, the simple weakening of signals in a channel would be of no consequence, because they could just be amplified back to size. From the noise point of view, channels can be subdivided into three types: front-end, background, and man-made noise. Front-end noise refers to the thermal and electronic Gaussian noise that is always present in electronic components. Since it usually dominates in radio channels and in guided channels where noise is a factor, we will devote Section 5.2.2 to it. With especially quiet circuitry, background radiation on Earth or in space may dominate instead. It also is Gaussian. Man-made noise covers a multitude of sources, from adjacent-channel transmissions (see Section 3.8) to noise from electrical machinery, to radio junk that runs loose in the ionosphere.

An object of this chapter is to present enough information to allow at least a first cut at a transmission link design, for most situations that may arise. Toward this end, many tables of physical channel data are given. These are drawn from many sources and represent, for example, a generic 22-gauge wire pair or a typical 3 meter antenna, rather than the exact data for particular equipment. After a first design cut based on this chapter, a next step might be to include some finer details of the link and consult data for particular components. Some standard handbooks that give details are Refs. 2, 3, and 6.

Another goal in this chapter is to understand fading channels. Fading comes about because of dispersion of radio waves combined with motion. There are not only periodic losses of signal but also smearing of the signal in both time and frequency. Fading is the critical element in cellular and personal radio systems, as well as local area radio networks.

5.1 GUIDED CHANNELS

A guided channel is one that contains its signals in some sort of conduit, protected at least mostly from the outside world. Everyday examples are coaxial cable, in which an outer conductor shields an inner one, and waveguide, which behaves in an electrically similar way but without the inner conductor. Another everyday guided channel at lower frequencies is the wire pair, which is what it says, a pair of wires.[1] The pair, singly or in a bundle with others, can be enclosed in a shield as well. A final example is the optical fiber.

While these channels work best at different frequencies and are based on different electrophysics, they share the property that they distort a signal by acting as a distributed-parameter bandpass filter. Signal energy also dissipates as it flows down the conduit, through losses to the dielectric (cables, fibers) or in the conducting wires and walls (pairs, guides, cables). For a cable or pair of a given length, the total effect is that of a *linear filter* with some transfer function $H(f)$ and impulse response $h(t)$, an impairment that we have already discussed in Section 2.4. As long

[1]Another name for a wire pair is a twisted pair; the wires are often twisted together to cancel common-mode interference.

as the signal does not fall near the ambient noise level, then $h(t)$—and its remedy—will dominate the problem.

A major analytical tool is the eye pattern. The receiver can in principle be a matched filter receiver, as in Section 2.6, matched to the response $h(t)$, but with linear modulation it is more likely to be the linear receiver (Fig. 2.7), in the hope that the ISI is not too bad. An improvement on this is the equalization technique, which is explored in Chapter 7. A standard approach in long guided lines is *repeaters*: Break the total length into links short enough so that the distortion is not severe, and detect and regenerate a clean signal after each link.

Waveguides and microstrip lines behave at microwave frequencies in a similar way. Since they find use chiefly in radio systems, we will discuss them in Section 5.2. A glass fiber band-limits the light that passes through it, but the effect occurs at such wide bandwidth as to have little effect on modulation bandwidths in the hundreds of megahertz. Fibers chiefly dissipate rather than filter, and the mechanism is sufficiently different that we will discuss it separately at the end.

5.1.1 Characteristics of Wires and Cables

To design links with wire pairs and coaxial cables, we need to gain some appreciation of their bandwidth and attenuation. The latter is particularly easy to deal with, since it follows an exponential law and can be characterized in decibels per unit length.

According to standard transmission line theory, a line can be thought of as a succession of many small sections of the kind shown in Fig. 5.1. The inductance and capacitance of the line section are given as L and C per unit length, and the line dissipation losses are R ohms per unit length down the line and G mhos per length across the line; all these are shown in the figure times a differential piece of length dy. When many small sections are taken in succession, it can be shown that the ratio of voltage to current on the line is the *characteristic impedance*

$$Z_0 = \sqrt{\frac{R + j\omega L}{G + j\omega C}}, \qquad \omega = 2\pi f \tag{5.1-1}$$

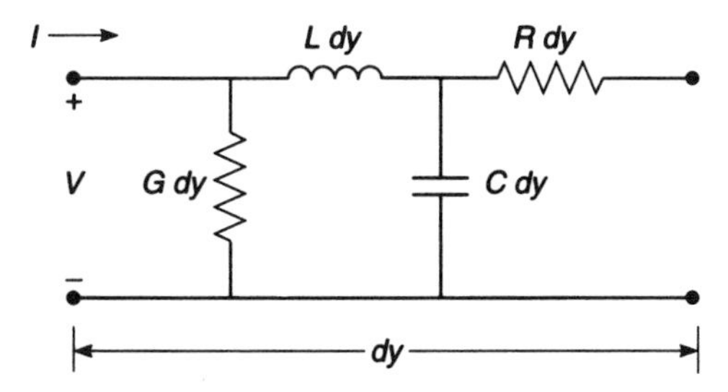

Figure 5.1 Transmission line segment of length dy modeled as lumped-parameter circuit.

Another important line parameter is the *propagation constant,*

$$\gamma = \sqrt{(R + j\omega L)(G + j\omega C)} \tag{5.1-2}$$

If $V(j\omega)e^{j\omega t}$ is a voltage (or current) of radian frequency ω entering the line, it can be shown that the voltage at time t and length y down the line is

$$v(y, t) = V(j\omega)\exp(-\gamma y)e^{j\omega t} \tag{5.1-3}$$

The entering phasor $V(j\omega)$ decays along the line as $V(j\omega)\, e^{(-\gamma y)}$; that is, its phase shifts by $\angle\{e^{-\gamma y}\}$ and its amplitude decays as $e^{-\alpha y}$. Here α, the real part of γ, is called the attenuation constant, and it is more convenient to express the loss in decibels, as

$$20\log_{10} e^{-\alpha y} = 20\alpha y \log_{10} e \qquad \text{(dB)} \tag{5.1-4}$$

The wave velocity along the line is $\omega/\mathrm{Im}\{\gamma\}$.

It needs to be stressed that all these parameters depend on frequency. In particular, R is proportional to $\sqrt{f}$ because of the skin effect in conductors. Aside from frequency, R, G, L, C depend on the conductor geometry and the dielectric. Design formulas are available in handbooks that calculate line parameters from these characteristics. For a line without dissipative losses, (5.1-1) and (5.1-2) show that $Z_0 = \sqrt{L/C}$ and $\gamma = j\omega\sqrt{LC}$. When the loss is reasonably small, $\sqrt{L/C}$ can still be taken as the characteristic impedance, although Z_0 becomes more capacitive at low frequencies. When $R \ll \omega L$—that is, at high frequencies—the real part of (5.1-2), α, is approximately

$$\alpha \approx \frac{R}{2}\sqrt{\frac{C}{L}} \tag{5.1-5}$$

Recall that R will vary as $\sqrt{f}$ here. In whatever case, if the line is terminated by impedance Z_0, the input impedance will also be Z_0 and the loss factor will be α. Otherwise, there will be reflections from the termination and the input Z needs to be calculated from transmission line theory. A simple explanation of this is given in Ref. 2.

For the link designer, what is important about guided links is their attenuation and their overall frequency transfer function, assuming that all line terminations are roughly matched. The attenuations for selected wire pairs and coaxial cables are shown for different frequencies in Table 5.1, together with some numbers for waveguides (these are discussed in Section 5.2.3). The central fact in the table is that as frequency grows, one must move progressively from pairs to cables to waveguides.

Table 5.2 summarizes the complete parameters for some typical wires and cables. The approximate behavior of others can be inferred from this data. What follows is an example featuring the details of one standard wire pair.

TABLE 5.1 Attenuation for Selected Wire Pairs, Coaxial Cables, and Waveguides

	dB/km			dB/100 m							
	100 Hz	10 kHz	100 kHz	1 MHz	10 MHz	100 MHz	1 GHz	4 GHz	8 GHz	12 GHz	14 GHz
Wire Pairs (see Table 5.2 for details)											
1. 22 AWG, Example 5.1-1	0.36	3.0	4.2								
2. 19 AWG toll pair	0.21	1.5	3.0								
3. Copper pair, spaced 20 cm	0.026	0.058	0.16								
Coaxial Cables (see Table 5.2)											
1. RG-58C/U					4.6	18	66				
2. RG-213/U, RG-8A/U, RG-11A/U, RG-12A/U				0.72	2.2	7.2	29				
Waveguides (see Table 5.3)											
1. RG-48/U							*	2.5			
2. RG-50/U								*	7.6		
3. RG-52/U								*	21	14	
4. RG-53/U										*	54

Note: Waveguides are in $TE_{1,0}$ mode. An asterisk denotes a value below cutoff. Typical data come from various sources.

TABLE 5.2 Measured Characteristics of Selected Wires and Cables

Wire Pairs	f	R (Ω/km)	L (mH/km)	C (μF/km)	α (dB/km)	Z_0 (Ω/km, °/km)
1. 22 AWG, Example 5.1-1 pair	100 Hz	107	0.60	0.05	0.36*	1850 at –45°*
	10 kHz	—	0.60	0.05	3.0*	190 at –35°*
	100 kHz	—	0.60	0.05	4.2*	110 at –8°*
2. 19 AWG toll pair, no loading coils	100 Hz	52	0.68	0.04	0.21	1440 at –45°*
	10 kHz	52	0.68	0.04	1.5	160 at –25°*
	100 kHz	85	0.68	0.04	3.0	130 at –6°*
3. Copper wire pair, spaced 20.3 cm, wire diameter 2.6 mm; 33 insulators/km	100 Hz	6.4	2.1	0.0055	0.026	1380 at –39°*
	10 kHz	8.0	2.1	0.0055	0.058	620 at –2°*
	100 kHz	22	2.1	0.0055	0.16	620 at –0.5°*

Coaxial Cables	Diameter (mm) Inner Conductor	Diameter (mm) Dielectric	C (pF/m)	α (dB/100 m)	Z_0 (Ω, real)
1. RG-58 C/U	0.9	2.9	94	4.6	50
2. RG-213U, RG-8A/U	2.3	7.2	96	2.2	50
3. RG-11A/U, RG-12A/U	1.2	7.2	67	2.2	75

Notes on cables: Measurements at 10 MHz (otherwise see Table 5.1); polyethylene dielectric; inner conductor is stranded bundle. An asterisk denotes a value calculated via Eq. (5.1-1). Typical data come from various sources.

Example 5.1-1: 22 AWG Wire Pair Parameters. Telephone local loops often consist of wire pairs, often combined by the dozens or hundreds into cables. The parameters vary somewhat, but generalizations are possible for audio frequencies. With reference to Fig. 5.1:

$$C \approx 0.05\ \mu\text{F/km (does not vary at audio frequencies)}$$

$$L \approx 0.6\ \text{mH/km (drops above 100 kHz)}$$

$$R \approx 107\ \Omega\text{/km (grows as } \sqrt{f} \text{ above 10 kHz)}$$

$$G \approx 0\ \text{(no shunt dissipation)}$$

The following are calculated from (5.1-1) and (5.1-2) at 1000 kHz, with some actual measurements shown in brackets:

$$Z_0 = 584\ \angle 316°\ \Omega\ [640\ \angle 315°\ \Omega]$$

$$\gamma = 0.183\ \angle 46°\ \text{km}^{-1}, \qquad \alpha = 0.127\ \text{km}^{-1}$$

$$20\alpha \log_{10} e \approx 1.1\ \text{dB/km}\ [0.94\ \text{dB/km}]$$

For example, a $y = 10$ km line will have total $C = 0.5$ μF, $L = 6$ mH, DC resistance 1070 Ω, and $Z_0 = 584$ Ω and attenuation $10 \times 1.1 = 11$ dB, all at 1 kHz. A voltage phasor V entering the line will appear as phasor $V\exp(-10\gamma) = V\exp(-10(0.183\angle 46°) = V(0.070 + 0.272j)$ at the other end, if the load is Z_0. It is interesting to look further at the behavior at 3.3 kHz, the nominal band edge of a telephone channel. At 10 km, $\gamma = 0.222 + 0.250j$ and the line transfer function is

$$H(j2\pi\,3300) = e^{-10\gamma} = -0.0865 - 0.0652j = 0.1084\angle -143°$$

That is, the response to an input $\cos 2\pi 3300t$ is

$$\mathrm{Re}\{0.1084\, e^{-j2\pi 143/360}\, e^{j2\pi 3300t}\} = 0.1084\cos(2\pi 3300t - 143°)$$

This is a loss of $20\log_{10} 0.1084 = 19.3$ dB. At 30 km, the loss is 58 dB. The speed of propagation at 3.3 kHz in this wire pair is $2\pi\,3300/\mathrm{Im}\{\gamma\} = 83{,}100$ km/s. Figure 5.2 shows the loss in dB at 10 and 30 km as a function of frequency f, and below that the group delay versus f (this is $-d\phi(\omega)/d\omega$, where $\phi(\omega)$ is the angle of $H(j\omega)$, $\omega = 2\pi f$). The group delay is reasonably flat across the telephone bandwidth, which means that all signal frequency components will arrive after about the same delay; in particular, a 3.3-kHz signal arrives after a delay of 10 km/83,100 km/s ≈ 0.12 ms. But the component amplitudes are badly distorted. Thirty kilometers is considered to be the longest practical length for a compensated wire pair that carries an analog telephone signal; we can see that without compensation voice signals will be nearly unintelligible.

5.1.2 The Telephone Channel

The telephone channel is a ubiquitous reality that everyone knows, yet it is frustrating to define because it has no single physical reality. Rather than a wire or cable, the telephone channel is any analog medium with a certain character—in rough terms, a linear channel with passband 300–3300 Hz. In reality, the telephone channel could be one frequency-division multiplexed subchannel in a wideband radio channel, or it could be realized as a 64-kb/s digital transmission over, for example, an optical fiber.

There are as many actual, physical telephone channels as there are telephone connections in the world. In fact, it is necessary to define the telephone channel *statistically,* since no fixed definition is practical. Extensive studies of the telephone network have been made in different parts of the world. Reference 4 is an authoritative study of the U.S. analog toll connections as they existed in 1971, and Table 5.3 summarizes some of its main conclusions. After measuring many physical connections, the study concluded that 99% of connections in the United States will be better than these specifications. The figures are for dialed long-distance ("toll") connections that pass through a number of repeaters, filters, digitizers, and other processors. Another way to define the telephone channel is in terms of its typical frequency transfer function, and Fig. 5.3 shows some of these, again for U.S. toll

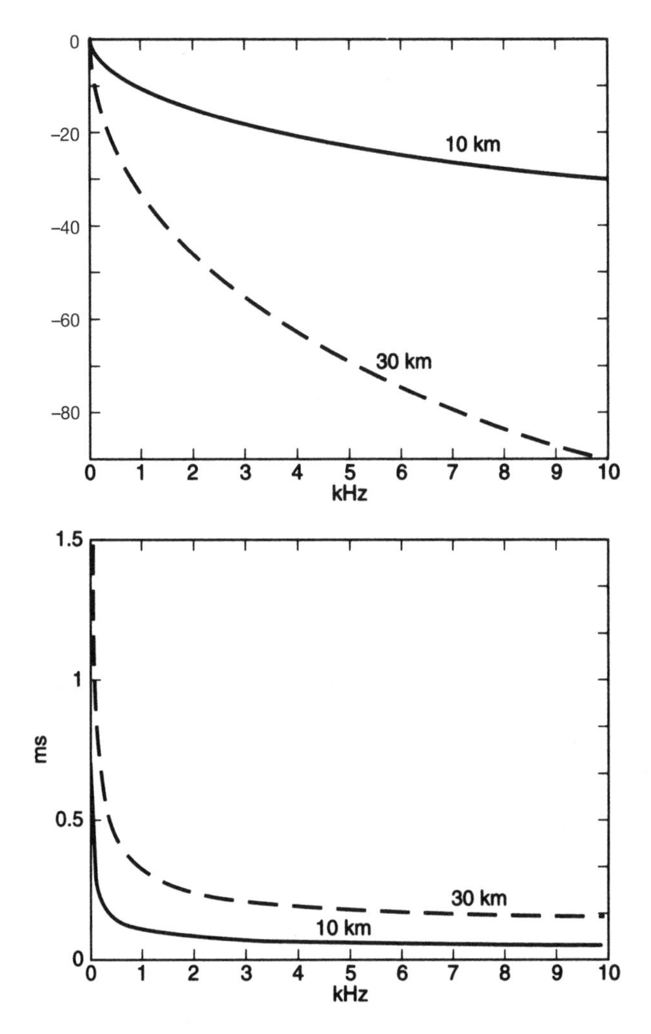

Figure 5.2 Transfer function magnitude (in dB) and group delay (ms) for 22 AWG wire pair (Example 5.1-1) at 10 km (solid) and 30 km (dashed).

TABLE 5.3 Worst-Case Measurements for Telephone Channels

Attenuation, end to end, at 1 kHz	27 dB
SNR (with special weighting)	20 dB
Frequency offset	3 Hz
Peak-to-peak phase jitter (20–300 Hz)	13°
Phase jumps greater than 20°	1 per minute
Noise impulses, 4 dB below mean signal or higher	4 per minute
Delay (terrestrial)	50 ms

Note: Selected from Ref 4; 99% of U.S. telephone channels performed better at the time of the study.

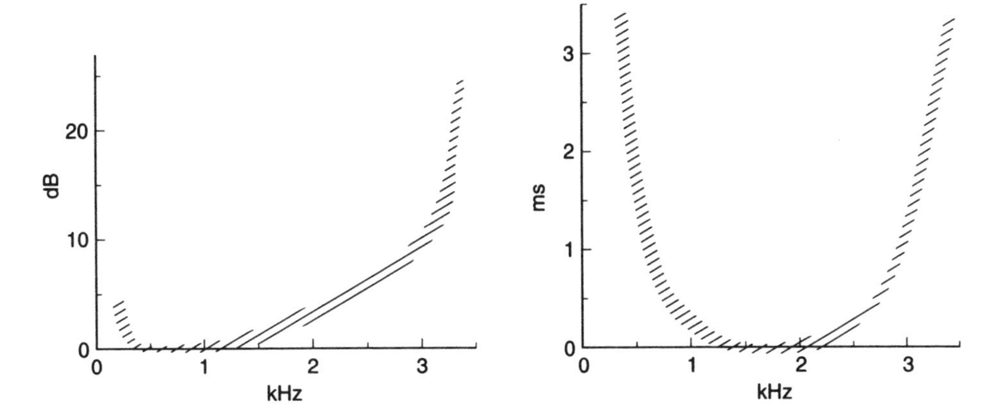

Figure 5.3 Approximate range of attentuation (**left**, in dB) and delay (**right**, in ms) for analog U.S. toll lines in 1984. Loss and delay are relative to their minimum values, which occur at 1 and 1.7 kHz, respectively. Data adapted from Ref. 5.

lines and this time in 1984 [5]. The plots show a relative amplitude response range in decibels and a range of group delays. Today, most, but not all, such toll lines in developed countries have been replaced by 64-kbit/s digital lines, and the effects discussed here are in that case limited to local analog wire pairs and A-to-D conversion steps.

The term *leased line* refers to a connection that is permanently allocated to a customer, rather than dialed anew at each use.[2] The behavior of leased lines falls in a narrower range than that of dialed lines, and they are thus easier to design for.

A connection that lies entirely within a local switching area (a "central office") is called a *local loop connection,* and it has much better behavior than a toll line. These loops are sometimes simply a 1- to 3-km wire pair and can then have quite a wide bandwidth, which can be calculated as in Example 5.1-1. In other cases, a local loop passes through a line concentrator, a device that stacks up subchannels on a single line by frequency division or digitizing; in this case, bandwidth, noise, and phase jitter will be worse but still better than a toll line. Some local loops are poor because they are very long wire pairs. Others have inscrutable impairments that simply have to be measured.

The sources of noise in the telephone channel are digital quantization noise, thermal noise in detectors, crosstalk between adjacent lines, impulses, and 50/60-Hz induction from nearby power lines. The first is the dominant one in recent years, because digitally modulated glass fibers have taken over toll transmission. It typically lies 35 dB below the average signal power. Both thermal and quantization noise can be viewed as an equivalent Gaussian noise. Other major impairments that occur on telephone lines include echoes from unmatched line terminations and phase jumps

[2]Dialed lines can also be "hot switched"—that is re-routed during use.

and jitter. The last two stem principally from frequency conversions along the connection path. Some feeling for phase impairments is given in Table 5.3.

Increasingly, the telephone channel is used to transmit digital data, via a telephone line modem. We have encountered some standard modem modulations in the examples of Section 3.5. Table 5.4 summarizes some present modem standards that are designed for less-than-perfect channels with $\approx$ 3400-Hz bandwidth. The higher speed standards require a better channel, and they must revert to a lower standard if the channel degrades. It may seem odd that a digital fiber channel is sometimes the reality of an analog telephone channel, which in turn carries digital modem signals, but the utility of one standard communication network that reaches everywhere cannot be overemphasized. This is provided by the telephone channel.

5.1.3 Glass Fibers

Optical fibers have huge advantages over wires and cables and have been rapidly replacing them since 1980. Indeed, it can be said that the important physical channels today are the fiber, when both ends are fixed, and the radio medium, when one end moves or when the network is sparse.

Present-day fibers consist of a central core with a high refraction index and a surrounding tube that has a low index. The geometry of core and tube is such that light rays in the core hit the refraction index boundary at less than the critical angle θ (critical angles are defined later in Section 5.3). Consequently, all the light is reflected internally and stays in the core. At larger core sizes, a number of θ are possible, and these fibers are called multimode. If the core is small enough, only one θ is possible, and the fiber is single-mode. Glass has various physical properties. The to-

TABLE 5.4 Selected CCITT International Telephone Line Modem Standards

Bit Rate	Symbol Rate	Modulation[a]	CCITT[b] Standard
330	300	2FSK	V.21
1,200	600	QPSK[c]	V.22
2,400	600	16QAM[c]	V.22bis
1,200	1,200	2FSK	V.23
2,400	1,200	QPSK	V.26
4,800	1,600	8PSK	V.27
9,600	2,400	Fig. 3.13a	V.29
4,800	2,400	QPSK	V.32
9,600	2,400	16QAM	V.32ALT
14,400	2,400	128QAM, TCM	V.32bis
28,800	3,429	1024QAM, TCM	V.fast(V.34)

Note: Many more details than these appear in the standards.

[a]TCM stands for TCM-coded modulation.

[b]Comité Consultatif International de Téléphonie et Télégraphie (an international standard setting body based in Geneva).

[c]Two transmissions share the channel in frequency division multiplex.

tal-reflection concept works better at shorter wavelengths, but inhomogeneities and impurities affect the light less at longer wavelengths. The best compromise lies at 1.3–1.6 μm, in the infrared band. The signal bandwidth available in this band exceeds 100,000 GHz, and present-day attenuation constants are 0.2 dB/km or better.

The effect of the fiber on its signal can be characterized by an impulse response or a frequency transfer function, just as we have done earlier with wires and cables. Time dispersion occurs within a multimode fiber because different rays can take different reflection angles, which imply differing paths, which means varying arrival times at the far end. Even in single-mode fibers, a small dispersion occurs because of Rayleigh scattering and a speed versus wavelength relationship in the glass dielectric. All these effects grow with distance, because the ray arrival times are subject to more variety. In consequence, there is a trade-off between sustainable bit rate and repeater spacing. Short fibers can sustain a pulse rate of perhaps 10 gigapulses/s. After 10 km, dispersion reduces this limit; by 100 km, the fiber is limited to 1 gigapulse/s. Beyond 100–150 km, attenuation and its aftereffects limit the pulse rate.

The light sources in fiber transmission are light-emitting diodes, which are incoherent and must be amplitude-modulated, and injection lasers, which can be coherently modulated. Powers are limited to 1–10 mW in order to avoid nonlinear optical effects.

Light detectors at present are either PIN (positive/intrinsic/negative layer) diodes, or avalanche photo diodes (APDs). Both devices convert incident light photons into current flow. If $h = 6.6 \times 10^{-34}$ J-s is Planck's constant, P is the light power (watts), and f is the frequency (Hz), we have

$$\text{Incident photons/s} = \frac{P}{hf} \tag{5.1-6}$$

Diodes convert with an efficiency of η charge carriers/photon, $\eta < 1$; consequently, the signal current flow is $\eta qP/hf$ amperes, where $q = 1.6 \times 10^{-19}$ coulombs is the charge of an electron. For example, a diode with $\eta = 1$ at 1.3-μm wavelength, working with a 1-mW source and a 100-km fiber at 0.2-dB/km attenuation, will produce a current of

$$\frac{(1\text{ mW})(0.01)(1.6 \times 10^{-19}\text{ C})}{(6.6 \times 10^{-34}\text{ J-s})(2.3 \times 10^{14}\text{ Hz})} = 11\ \mu\text{A}$$

The PIN diodes have efficiency near unity when they operate in a narrowband mode, but low efficiency when they are wideband. The APDs can have high efficiency at wide bandwidth, but they are noisy because of the avalanche principle of their operation. Another distinction between the two diodes is that PINs work at baseband and often at low output, so that the detector output is characterized by the ordinary thermal "KT" noise in a following preamplifier. The APDs work at light frequency, are themselves inherently noisy, have the higher quantum noise of Eq. (5.2-7), and need no preamplifier. These noise types are explained in Section 5.2.2.

The discussion here only touches on some of the many design factors in a fiber link. Further information appears in brief form in Ref. 1, Chapter 5, and Ref. 13, Chapter 12; a full text is Personick [14].

5.2 THE SIMPLE RADIO LINK

The simplest radio propagation channel is one in which the signal energy simply propagates through space from sender to receiver without reflections or scattering. To design such a link, we need to quantify first the behavior of antennas and propagation; then comes noise, which in the simple link case is almost always white Gaussian; finally come the hardware losses that occur in a practical system. The totaling of all the gains and losses in a channel is called a *link budget*. Just as with a household budget, a healthy system requires a positive balance, an amount called the margin in a link budget. The margin is a measure of protection against change and the unexpected; the larger the margin, the more reliable the link.

5.2.1 Antennas and Propagation

Once transmitted, radio waves in free space propagate without limit and spread out. It is useful for the budget analysis to imagine that radio waves depart from a single point called an *isoradiator*. This imaginary antenna has no favored direction, and radio waves leave it uniformly in all directions. If P_t watts leave the isoradiator, then the density of power at distance d meters is

$$\frac{P_t}{4\pi d^2} \qquad (\mathrm{W/m^2}) \tag{5.2-1}$$

This is because the power passes uniformly through a sphere of surface area $4\pi d^2$ at this distance. Equation (5.2-1) is the familiar inverse-square law of radio propagation in free space.

A real antenna either directs or collects energy in a preferred direction. It is easiest to think about this process in terms of dish antennas as shown in Fig. 5.4. We can think of a transmit antenna as focusing energy in one direction and a receive antenna as gathering energy in a manner proportional to its area. From wave physics, the gain in power density forward of a parabolic dish, compared to an isoradiator, can be shown to be

$$G \approx \frac{4\pi A}{\lambda^2} \tag{5.2-2}$$

This is the *antenna power gain,* where A is the dish area, λ is wavelength, and the measurement point is located at least several wavelengths exactly in front of the dish. The gain formula is approximate because of imperfections in the antenna and in the wave theory theory model, but the true gain of an antenna is in any case easily measured. Some measured gains for dishes are shown in Table 5.5. They are typ-

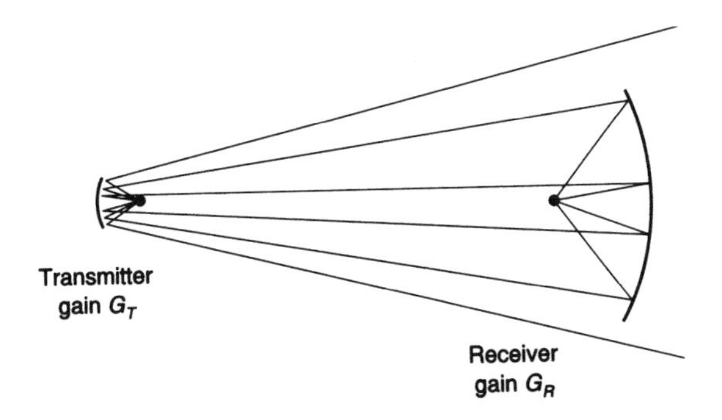

Figure 5.4 Transmit and receive dish antennas.

ically 54% of Eq. (5.2-2). It can be seen that the gains depend strongly on wavelength.

Other kinds of antennas, such as the dipole, are tuned radiators and do not have an explicit area. However, a gain may still be derived or measured. A strong influence on all antenna gains is the observation angle in 3-space of the transmitter by the receiver or vice versa. Gain plots like Fig. 5.5 are a convenient way to show

TABLE 5.5 Typical On-Axis Gains of Antennas

	Gain If Not Frequency-Dependent (dB)	Gain If Frequency-Dependent (dB)				Comments
		2GHz	4GHz	6GHz	11GHz	
Horn reflector (standard pyramidal, 2 × 3 m)		—	39	43	48	Common in terrestrial microwave
Dishes [~ $2.2\pi A/\lambda^2$]						
1 m		24	30	33	39	Space/Earth use
2 m		30	36	39	45	
10 m		44	50	53	59	Full Earth station
Open end of waveguide 2 of Table 5.7 (~ $10\ A/\lambda^2$)		—	1	5	10	
Parallel element broadside array	5–7[a]					SW-UHF
Three-element Yagi	4–8[a]					Rooftop TV antenna, SW-UHF
Half-wave dipole	2.15[b]					
Turnstile	0.65[b]					
Vertical whip	1.85[b]					VHF-UHF, if good ground plane

Note: Isotropic = 0 dB. Measured at point of maximum radiation, several wavelengths above ground.
[a]Depends on tuning and element spacing.
[b]From theory.

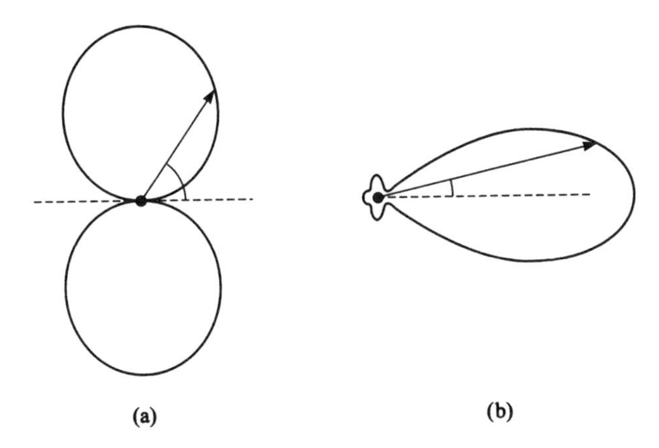

Figure 5.5 Some antenna patterns. **(a)** Approximate dipole pattern and **(b)** small dish, both in free space. Dashed lines are antenna axes. Line length from center point to pattern is relative absolute gain.

this. Note particularly that antennas may have virtually no gain in some directions, such as off the back of a dish or off the end of a dipole. In whatever patterns the energy leaves the antenna, it spreads at each azimuth according to the inverse-square law. A far observer sees no difference between P_t radiated by an antenna with gain G_t and G_tP_t watts radiated by an isoradiator. The product G_tP_t is called the EIRP ("effective radiated power referenced to isotropic") for the transmitter.

At the receiver, the dish antenna of area A_r acts to collect the power flux. The collected power P_r will be the product of the area and the density in the region, which will be

$$P_r = \frac{(P_tG_t)(A_r)}{4\pi d^2} \qquad \text{(W)} \tag{5.2-3}$$

By the reciprocity principle for antennas, we can equally well think backwards and use Eq. (5.2-2) to express this A_r in terms of an antenna with gain G_r at wavelength λ. The A_r is then $\lambda^2 G_r/4\pi$ (we have incorporated any imperfections in the antenna into a reduced effective area). A substitution into (5.2-3) gives

$$P_r = P_tG_tG_r\left(\frac{\lambda}{4\pi d}\right)^2 \tag{5.2-4}$$

The factor $(\lambda/4\pi d)^2$ here, a very small number, is called the *free-space loss*. It represents the loss of signal power in free space due to radio propagation or, more simply, distance. Free-space loss is denoted L_p. With it, the basic propagation equation (5.2-4) becomes

$$P_r = P_tL_pG_tG_r \qquad \text{(W)} \tag{5.2-5}$$

That is, received power is the product of transmitted power, the free-space loss, and the two antenna gains. Any of these may be traded for the others.

L_p grows worse as the square of carrier frequency, but the two antenna gains both improve at the same rate, if they are dishes. Here is an important point to observe in the propagation equation (5.2-5): For fixed L_p and dish antennas, the received power grows as the square of frequency. When distances are very long, we are forced to higher frequencies.

Further details about waves and antennas may be obtained from an electromagnetics text such as Ref. 16.

5.2.2 Receiver Noise

Now that we have the received power, it remains to find the noise power, in order to have an SNR to use in the probability of error equations in Chapters 2 and 3. Most often in radio links, noise at the input stage of the receiver dominates and is Gaussian. Other noises, such as shot noise (due to charge carriers), antenna noise, and PCM conversion noise, are well-modeled as Gaussian. Noise analysis is a complex specialty. Our goal here is to learn simple modeling techniques that enable link budgets to be set up and to learn some basics of low-noise systems.

The classic source of Gaussian noise is thermal noise, the white noise due to the vibration of molecules is a resistor. In experiments by Johnson [9] in 1928, he reported that the mean-square voltage across a warm resistor satisfied

$$v_{\text{rms}}^2 = 4kTRB \tag{5.2-6}$$

Here k is Boltzmann's constant, 1.38×10^{-23} J/K, T is Kelvin temperature, B is the observation bandwidth in hertz, and R is resistance in ohms. Johnson traced the dependence of (5.2-6) on R, T, and B and even made a crude estimate of Boltzmann's constant. In the same journal issue, Nyquist [8] derived (5.2-6) from the kinetic theory of heat. His model actually went beyond random motion of electrons and included, for example, crystal lattice vibration, ion motions, and oscillations (see, e.g., Ref. 10, Chapter 6).

The Johnson–Nyquist theory states that all resistive bodies produce Gaussian noise, whose power is zero only at $T = 0$, absolute zero. In reality, thermal noise is indeed white up to 100 GHz or more; after that, Planck's Law (or "quantum") noise dominates and gives a spectral density that rises with frequency. The full formula, incorporating both types of noise, says that the power delivered is

$$\frac{hf}{e^{hf/kT} - 1} \qquad \text{(W/Hz)} \tag{5.2-7}$$

in which h is Planck's constant (see Ref. 10). This formula must be used in optical fiber problems.

Returning now to radio frequencies, consider a receiver. Somewhere at the beginning is a resistance R across which the received signal develops a signal voltage.

Assume it is a pure resistor,[3] the one in Fig. 5.6a. The rms voltage in Eq. (5.2-6) is represented by an equivalent voltage source; this is the open-circuit voltage that will be seen by an infinite-impedance measuring instrument. How much of this is delivered to the receiver? Figure 5.6a shows as dashed lines a load R_L and a gain G that represent the rest of the receiver. The gain is whatever is needed for the particular application, and the dashed parts of the receiver are noise free; all noise is modeled as coming from one place, R. A basic theorem of electrical circuits says that the maximum power transfer from a source with resistance R occurs when $R_L = R$. Under these conditions, the noise power dissipated in the total of R and R_L is

$$\frac{v_{\text{rms}}^2}{R + R_L} = \frac{4KTRB}{2R} = 2kTB$$

Since the resistors split the power, the power delivered to R_L must be

$$P_{\text{avl}} = kTB \qquad \text{(W)} \tag{5.2-8}$$

This white-noise power, the maximum that can be transferred from this or any warm resistor, is called the *available noise power*. Observe that P_{avl} does not depend on R, but only on temperature and bandwidth.

Throughout this book we characterize white Gaussian noise power by a two-sided power spectral noise density $N_0/2$. The power in a bandwidth B, where B counts the positive frequencies only, is $2B(N_0/2) = BN_0$. From (5.2-8), then, the parameter N_0 at the front of the radio must be

$$N_0 = kT \qquad \text{(W/Hz)} \tag{5.2-9}$$

Example 5.2-1: A Room Temperature Resistor. In noise calculations, a special standard value 290 K, called "room temperature," T_0, often appears.[4] From (5.2-6), the rms open-circuit noise voltage across a 50-Ω resistor at this temperature is

$$v_{\text{rms}} = \sqrt{4kT_0RB} = (4 \times 1.38 \times 10^{-23} \times 290 \times 50)^{1/2}\sqrt{B} = 9 \times 10^{-10}\sqrt{B} \qquad \text{(V)}$$

In a 1-MHz bandwidth, v_{rms} is 0.9 μV, not an inconsequential value. The maximum power delivered to a load at this bandwidth is $P_{\text{avl}} = kT_0B = 4 \times 10^{-15}$ W, and the value of N_0 is 4×10^{-21} W/Hz.

In a general circuit, noise can arise in many places. With electronic devices and collections of components, as with a receiver, it is convenient to imagine an equivalent warm resistor that would produce the same available noise power. Thevenin's theorem says that a circuit with an output port can be replaced by an equivalent

[3]Most often, antenna circuits are 50 Ω.

[4]The room, it appears, lies in a chilly Northern European country, since T_0 is about 17°C (62.6°F).

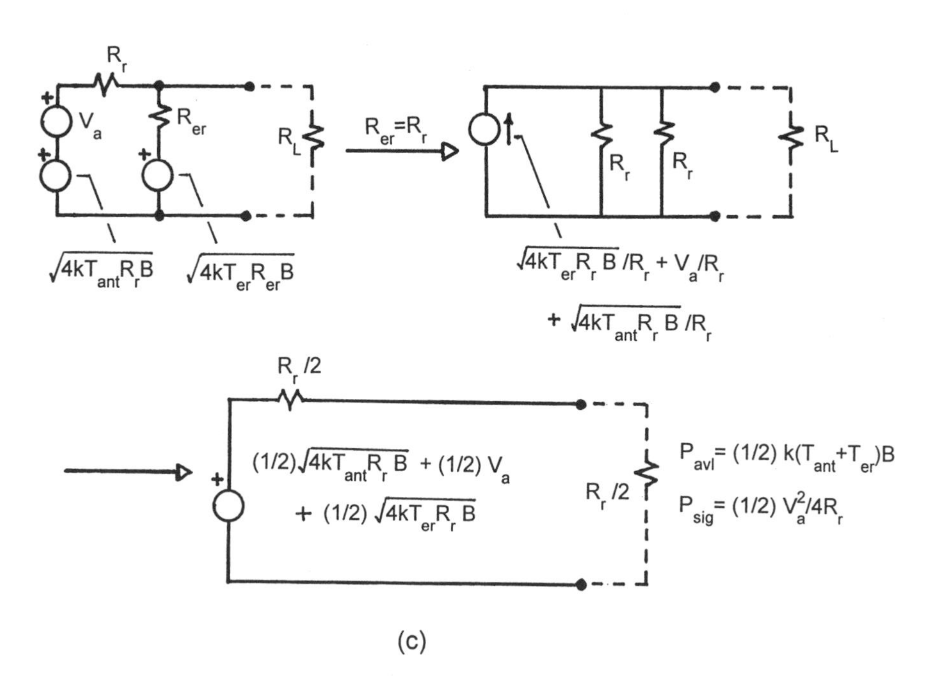

Figure 5.6 Noise-equivalent circuit modeling for (a) resistor, (b) antenna, and (c) antenna-receiver system. All voltages and currents are rms.

voltage source and a series resistance R_{th}. Similarly, we can work backward from a measured N_0 or an open circuit rms voltage to a resistor with an apparent Kelvin temperature T_e; this is called the *equivalent noise temperature* of a circuit. The resistor has value R_{th} and temperature T_e. The circuit is driven by two voltage sources, the Thevenin source and a Gaussian source with rms value $\sqrt{4kTR_{th}B}$. Both sources deliver power to the next circuit, and the noise taken alone delivers P_{avl} in the case of a matched load. For example, a power 5.52×10^{-21} W measured over 1 MHz means that the equivalent noise temperature of the circuit leading up to that point is $(5.52 \times 10^{-21})/(10^6 \times 1.38 \times 10^{-23}) = 400$ K. This is a typical value for a low-noise transistor amplifier.

Alternately, T_e can be derived mathematically. As an illustration, consider the case when the noise stems from two resistors at different temperatures (or two circuits equivalent to such) that are in series. The mean-square open circuit voltages add, since the noises are uncorrelated Gaussians, and so by the Johnson–Nyquist law the total mean-square voltage is

$$v_{rms}^2 = 4kT_1R_1B + 4kT_2R_2B = 4k(T_1R_1 + T_2R_2)B \qquad (5.2\text{-}10)$$

The combination can be taken as one resistor $R_1 + R_2$ (the Thevenin resistor) at a noise equivalent temperature of $(T_1R_1 + T_2R_2)/(R_1 + R_2)$, since this clearly leads to the same v_{rms}^2. For example, some electronics with $T_e = 400$ K and $R_{th} = 25\ \Omega$ in series with a room temperature resistor of 50 Ω makes a 75-Ω total resistance at equivalent temperature $(25 \times 400 + 50 \times 290)/(25 + 50) = 327$ K. The power delivered to a matched load is $P_{avl} = kT_eB = 1.38 \times 10^{-23} \times 327 \times B$, and the load needs to be 75 Ω.

The discussion so far has assumed nonreactive components and matched loads throughout. The Johnson–Nyquist formulas extend to reactive impedances Z, with the conditions that R in (5.2-6) needs to be replaced with $\text{Re}\{Z\}$ and the matched load leading to (5.2-8) needs to be $Z_L = Z^*$, the conjugate of Z. Note here that a pure L or C is noiseless ($T_e = 0$), but real components have some pure resistance, especially at high frequencies. See Refs. 2, 3, 10, and 11 for more. It is also possible to derive the noise behavior of complicated electronic circuits. References 3 and 12 introduce this challenging subject. It turns out for example that different designs optimize power gain and noise temperature.

Noise Factor. Another way to characterize the noise in a device or circuit is by its *noise factor*[5] F. The logic in the derivation of F is one of measuring. Consider the circuit with power gain G in Fig. 5.7. Its internal workings introduce a white noise power density N_{int} which gets amplified by G and appears as density GN_{int} at the output. The noise factor for the circuit is defined to be

$$F = \frac{\text{actual output power in } B \text{ Hertz}}{\text{output power in } B \text{ if no internal noise}}$$

[5]Noise factor is also called noise figure. Often the word figure connotes F in decibels. We will use the two names interchangeably.

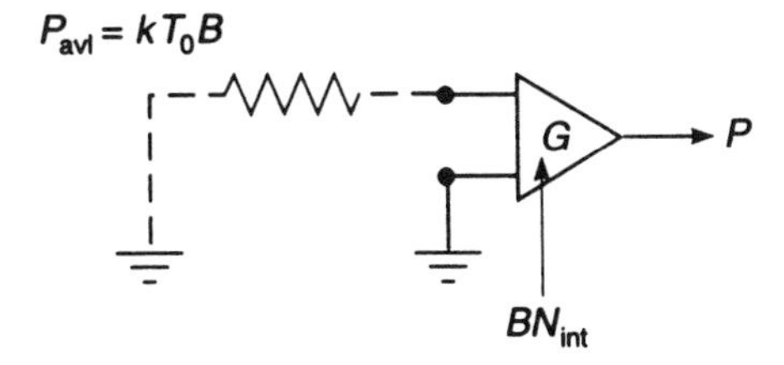

Figure 5.7 Modeling the noise factor calculation.

There is a problem here, because if no external noise applied, F is infinite. Furthermore, measurements with a different external input will lead to different F. The solution is an agreed-upon noise input. By convention this is a source at room temperature, $T_0 = 290$ K, matched to the input impedance. From (5.2-8), this transfers kT_0B watts into the circuit. F is thus defined as

$$F \triangleq \frac{N_{\text{int}}BG + kT_0BG}{kT_0BG} \tag{5.2-11}$$

F can be measured easily, by measuring the total power output, the input kT_0B, and G, all in the same bandwidth. Clearly, $F \geq 1$.

Another way to measure and define noise factor is to form the ratio of the input and output SNR at T_0:

$$F = \frac{\text{SNR}_i}{\text{SNR}_o} \tag{5.2-12}$$

A little arithmetic shows that this gives the same result as (5.2-11).

From F, one can compute a noise equivalent temperature and vice versa. We replace the internal noise $GN_{\text{int}}B$ with a warm resistor at the input which has the same effect. Its temperature will have to satisfy $kT_eBG = N_{\text{int}}BG$. Substituting this into (5.2-11) leads to the relation

$$F = 1 + T_e/T_0 \qquad \text{or} \qquad T_e = (F - 1)T_0 \tag{5.2-13}$$

Observe that room temperature plays an important role. Some noise factors and their equivalent temperatures are listed in Table 5.6 for a number of important circuits and situations. The usual convention is to express F in decibels, which is 10 $\log_{10} F$. It is obvious that some noisy devices have a noise temperature much higher than their physical temperature. The F measure, in decibels, is more convenient at high temperatures and is usually employed there, and the T_e measure is convenient in low-noise systems.

Example 5.2-2: Measuring *F*. One instrument for measuring the noise factor of a system is based on two matched resistors. One lies in a bath of liquid nitrogen at T_C

TABLE 5.6 Selected Equivalent Temperatures and Noise Figures for Amplifiers and Antennas Taken as Noise Sources

	Gain (dB)	T_e (K)	NF(dB)
Preamplifiers			
Parametric amplifier			
He cooled	10–20	15	0.22
Low-cost cooled (large Earth station)	10–20	35–50	0.5–0.7
GaAs transistor			
Cooled	10–20	70	0.94
No cooling (small Earth station)	10–20	100–150	1.3–1.8
Ordinary transistor low-noise amplifier	10	400	3.8
Ordinary IF amplifier (chip)	50	1150	7.0
Noise seen by antennas (microwave)			
In space, pointed at space		3	0.04
In space, pointed at Earth		290	3.0
Pointed at sun		10,000	15.5
On Earth, pointed up		20	0.3
On Earth, pointed near horizon		50	0.7

= 77.3 K, and the other lies in water at $T_H = 373$ K. Both liquids (gently) boil, which guarantees the two temperatures with great accuracy. The noise power at the output of the test system, N_C and N_H respectively, is measured carefully with both inputs. If we accept the Johnson–Nyquist formula (5.2-8), then definition (5.2-11) gives $N_H = kT_HBG + (F - 1)kT_0BG$ and $N_C = kT_CBG + (F - 1)kT_0BG$. The quotient of these two is

$$\frac{N_H}{N_C} = \frac{T_H + (F-1)T_0}{T_C + (F-1)T_0}$$

which does not depend on the measurement bandwidth B or the system gain G. We can substitute $T_H/T_0 = 1.286$ and $T_C/T_0 = 0.267$ and solve for F, to obtain

$$F = 1 + \frac{1.286 - 0.267(N_H/N_C)}{(N_H/N_C) - 1}$$

For example, an observed $N_H/N_C = 2$ leads to $F = 1.75$. A cheaper way to build the instrument would be to make T_C simply 290 K; then $N_H/N_C = 2$ would lead to $F = 2$.

Antenna Temperature. A special element in communication noise analysis is the antenna temperature. The antenna structure picks up a signal as well as background noise, which is Gaussian in the simple radio link case. The noise generally stems from background thermal radiation, either from the warm atmosphere or, in the case of an antenna in space, the 3 K background microwave radiation in the universe. The antenna need not be aimed at a noise source in order to pick it up; noise can enter through sidelobes in the antenna pattern. Noise thus needs to be measured at each desired antenna position. The antenna converts the signal electromagnetic fields to voltages and currents in a way that depends on its physical structure. The ratio of these is called the *radiation resistance,* R_r, of the antenna. We assume that R_r is real or, equivalently, that the antenna is matched by some means to the transmission line at the operating frequency. More on this can be found in Schwartz [10] or in a standard electromagnetic text such as Jordan [16].

Figure 5.6b models this situation with a resistor and two voltage sources, a signal source V_a and a noise source. The resistor plays two roles. Its value R_r gives the correct relation between the RF signal voltage and current, and its *temperature* is set to precisely that value T_{ant} which yields the right mean-square noise according to (5.2-6) in the receiver bandwidth B. The consequent noise is viewed as coming from the second voltage source.

In applications it is probably best to measure both the RF signal and the noise at the antenna terminals and perhaps adjust the structure, aiming, and matching so as to optimize their ratio. Table 5.6 lists some standard outcomes for dish antennas at microwave frequencies. It can happen that antenna currents flow through some physical resistance, as opposed to the radiation resistance, which has a temperature T and contributes an independent noise power kTB. Antennas are designed to minimize this, by for example incorporating a preamplifier directly into the RF feed structure. What dissipation that remains can be counted in with the losses in the line leading to the receiver. These are called *lead-in* losses.

The equivalent noise temperature of the whole system, including antenna and receiver, is called the *system noise temperature,* T_{sys}. Its modeling is shown in Fig. 5.6c. The antenna noise model of Fig. 5.6b at T_{ant} is in parallel with the receiver noise model of Fig. 5.6a at T_{er}. R_{er} is set to R_r. A Norton equivalent circuit is formed, then a Thevenin; the Thevenin resistor is $R_r/2$. The two noises are uncorrelated and their powers add; the available noise power delivered to a matched load $R_L = R_r/2$ is $k(T_{\text{er}} + T_{\text{ant}})B/2$, half what we might expect. However, the signal power is delivered half to R_{er} and half to R_L, and thus it is reduced in half as well. The system thus acts as if it has equivalent temperature $T_{\text{er}} + T_{\text{ant}}$ and we have

$$T_{\text{sys}} = T_{\text{er}} + T_{\text{ant}} \tag{5.2-14}$$

Attenuators. Another important figure in communication is the noise factor for an attenuator. A major reason is that waveguides and lead-ins for antennas function as attenuators. If T is the attenuator temperature and $G \leq 1$ is its gain, it turns out that F and T_e when $T = T_0$ are simply

$$F = \frac{1}{G}, \qquad T_e = T_0\left(\frac{1}{G} - 1\right) \tag{5.2-15}$$

The calculation of the result at any T is a little subtle but it makes a good exercise in noise analysis. Assume that the attenuator is resistive and matched at both ends;

these are good assumptions in practical systems. Place at the input a signal source delivering S watts in series with a matched resistor at temperature T_0. By the SNR definition of noise factor (5.2-12),

$$F = \frac{\mathrm{SNR}_i}{\mathrm{SNR}_o} = \frac{S/kT_0B}{GS/(kT_0BG + \gamma kTB)} = \frac{T_0G + \gamma T}{T_0G} \tag{5.2-16}$$

Here γ is a factor to account for how much of the noise generated by the attenuator itself appears at the output. Because some of this is attenuated, it is not obvious how much of it is delivered. But two things are clear. First, the noise delivered must be proportional to kTB. Second, when T is T_0, then *all* the noise must come from the attenuator; this is because the F calculation assumes that a matched resistor at T_0, and nothing else, is placed at the input. At $T = T_0$ the attenuator simply acts as a resistor at T_0 and delivers power kT_0B, independently of its gain. The last fact identifies γ: It must be that the output noise factor $kT_0BG + \gamma kTB$ in (5.2-16) satisfies $kT_0BG + \gamma kT_0B = kT_0B$. It follows that $\gamma = 1 - G$. The complete noise factor at T becomes

$$F = 1 + \frac{T}{T_0}\left(\frac{1}{G} - 1\right), \qquad T_e = T\left(\frac{1}{G} - 1\right) \tag{5.2-17}$$

Cascaded Circuits. Receivers are long cascades of circuits. Fortunately, only the first has a strong influence on the receiver noise performance as a whole. This will be clear from the formula for the noise factor for several circuits in cascade, which we derive now.

First consider the cascade of two circuits with gains and noise factors G_1, G_2 and F_1, F_2, as diagrammed in Fig. 5.8. The final available noise power P_2 here is the sum of $N_{\mathrm{int},2}G_2B$ and the first stage noise P_1 amplified by G_2. By definition (5.2-11) applied to circuit 2, $N_{\mathrm{int},2}G_2B = F_2(kT_0BG_2) - kT_0BG_2$. Thus

$$P_2 = P_1G_2 + F_2kT_0BG_2 - kT_0BG_2$$

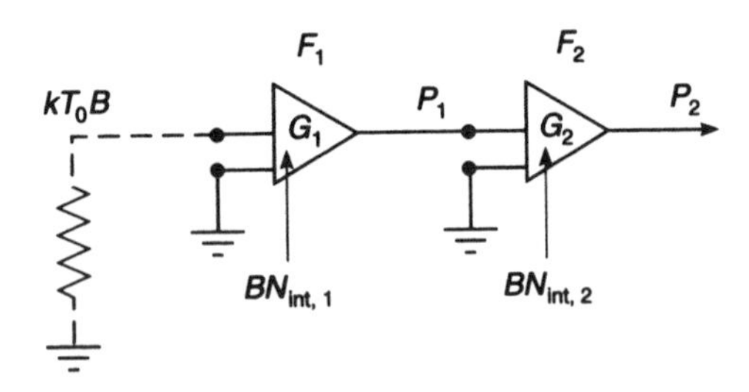

Figure 5.8 Noise power levels in cascade of two noisy circuits. Total noise factor is $P_2/kT_0B(G_1G_2)$.

By definition once more, $P_1 = F_1kT_0BG_1$, so that

$$P_2 = kT_0BG_1G_2F_1 + kT_0BG_2F_2 - kT_0BG_2$$

For the cascade, F_{tot} is defined to be $P_2/kT_0BG_1G_2$, and so

$$F_{\text{tot}} = \frac{G_1G_2F_1 + G_2F_2 - G_2}{G_1G_2} = F_1 + \frac{F_2 - 1}{G_1}$$

By iterating the argument, we can show for L circuits in cascade that

$$F_{\text{tot}} = F_1 + \frac{F_2 - 1}{G_1} + \frac{F_3 - 1}{G_1G_2} + \ldots + \frac{F_L - 1}{G_1 \ldots G_{L-1}} \quad (5.2\text{-}18)$$

By substitutions from (5.2-13) we obtain the equivalent noise temperature relation,

$$T_{\text{tot}} = T_1 + \frac{T_2}{G_1} + \frac{T_3}{G_1G_2} + \ldots + \frac{T_L}{G_1 \ldots G_{L-1}} \quad (5.2\text{-}19)$$

Equation (5.2-18) can incorporate the antenna by using the special values $F_1 = 1 + T_{\text{ant}}/T_0$ and $G_1 = 1$ for the first element in the cascade; these represent the antenna noise factor.

These results show that the first noise factor F_1 dominates the overall noise factor, particularly when it can be contrived that the first stage has significant gain. The receiver example that follows is typical and illustrates some important low-noise design principles:

1. Avoid a lossy first element in the cascade, such as a long transmission line.
2. The need for high signal levels can be met with a noisy high-gain amplifier, provided that it is preceded by a low-noise amplifier with moderate gain.
3. The same can be said for mixers, which tend to be noisy: Precede them by a low-noise amplifier.
4. Money spent to obtain a low antenna temperature is wasted if the receiver has a much higher noise temperature.

Example 5.2-3: Standard Receiver with Long Lead-In. The radio-frequency part of a standard superheterodyne receiver with antenna and lead-in is shown in Fig. 5.9. The lead-in is a matched transmission line at room temperature with attenuation 3 dB; the rest of the circuit is a good solid-state RF low-noise amplifier (LNA), a mixer that is rather noisy, and a standard IF amplifier chip, which is quite noisy. Using Eq. (5.2-17) for the attenuator and Eq. (5.2-18), we get the noise factor

$$F = 2 + \frac{0.69}{0.5} + \frac{2.07}{0.5(10)} + \frac{3.97}{0.5(10)5} = 3.95, \qquad \text{or } 856 \text{ K}$$

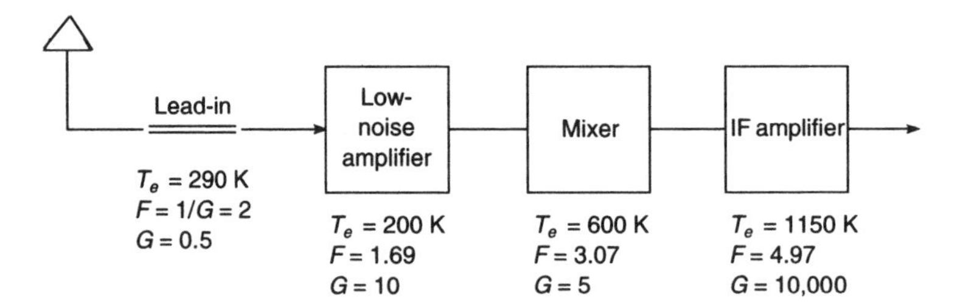

Figure 5.9 Radio-frequency sections of receiver in Example 5.2-3. Antenna noise figure is ignored.

The noisy IF amplifier hardly contributes to this figure. A much better noise figure is obtained by reversing the RF amplifier and the attenuator; physically, this is done by locating the LNA inside the antenna and running its output through the lead-in down to the mixer. The resulting noise figure is 2.36, or 395 K. This is a 3.4-dB improvement in the noise power spectral density at a subsequent detector. In either case, the IF amplifier contributes just 0.16 to F and its gain of 10,000 plays no role. As for the antenna, its noise temperature T_{ant} might be about 290 K, assuming that it is pointed at an earthly object. With the low-noise design, this would give a system noise temperature T_{sys} of about $290 + 395 \approx 700$ K. If the antenna is pointed at space, it appears that an expensive antenna design is not worth the cost unless the LNA is replaced with a cooled one having lower T_e. The system temperature is dominated by the LNA.

5.2.3 Detailed Link Budgets

Antennas and front-end noise are the major elements of a simple radio channel, but it remains to work out the smaller details of link budgets. There are a number of minor gains and losses to describe. When all are expressed in dB, the sum of all items in the budget, large and small, minus the noise power in dB is a final signal-to-noise power ratio at the detector, denoted C/N. For the receiver, we define a *receiver threshold* (RT), which is the signal-to-noise power ratio at the detector that gives the maximum acceptable error rate. C/N needs to be greater than the RT. The notation C/N distinguishes C/N, which is a power ratio, from the E_b/N_0 which we use in error calculations, which is a ratio of *energy* E_b to noise *density*. It is necessary sometimes to convert back and forth between C/N and E_b/N_0, but this is a detail for later.

Here is a list of losses that occur commonly in radio links, both in simple ones and in the terrestrial and mobile ones that follow later. The notations and units will be new, but they are the ones used in the radio engineering profession. For completeness, we list again the gains from Section 5.2.1.

The following count as gains:

1. The *transmit power* P_t, reckoned in units of dBW (decibels relative to 1 watt)

2. The *transmit antenna gain* G_t, in dBi (decibels relative to an isoradiator)
3. The *receive antenna gain* G_r, also in dBi

The following count as losses; all are dimensionless and expressed in dB. Not all appear in every link budget:

4. The *free-space loss* L_p is the loss due to radio propagation.
5. *Feed losses* L_f are losses in connecting cables, waveguides, lead-ins, and so on, and need to be estimated as in Section 5.1. For waveguides, representative data are given in Table 5.7. Waveguides can transmit radiation in a number of electromagnetic modes, and each mode cannot be sustained below its cutoff frequency. The table is based on the simplest mode, called $TE_{0,1}$.
6. *Branching losses* L_b occur in splitters and circulators; for example, a 2:1 splitter causes a 3-dB power loss in each branch.
7. *Circuit losses* L_c occur in the coupling between power amplifier and antenna at the transmitter, and between antenna and lead-in and first stage at the receiver. We will count the transmit circuit loss as 0 dB (that is, P_t is net power output).
8. *Pointing loss* L_θ occurs when either antenna is not directly pointed at the other. A formula for the loss in dB for a parabolic dish is [7]

$$L_\theta \approx 12.0\left(\frac{\phi_e}{\phi_{1/2}}\right)^2 \quad \text{(dB)} \tag{5.2-20}$$

in which ϕ_e is the angular error and $\phi_{1/2}$ is the half-power beamwidth of the dish. The beamwidth drops with rising frequency. Antennas have other losses, but we will include these in their net gain.

9. *Atmospheric loss* L_a is loss from rain, snow, and the like. It depends on frequency and will be discussed in Section 5.3.
10. *Fade margin FM,* also called link margin, is an extra power reserve in case of losses from multipath fading, solar cell degradation, and various other losses, expected and unexpected. Fade margin will be discussed in Section 5.3.

TABLE 5.7 Attenuations and Frequencies for Some Standard Inch-Metric Waveguides Operating in $TE_{0,1}$ Mode

Standard	Dimensions (mm)	Wall Thickness (mm)	$TE_{1,0}$ Range Cutoff (GHz)	Low end (GHz)	Attenuation (dB/100 m)	High end (GHz)	Attenuation (dB/100 m)
1. RG-48/U	76.2 × 38.1	2.0	2.08	2.8	3.6	3.9	2.5
2. RG-50/U	38.1 × 19.0	1.6	4.29	5.9	9.4	8.2	7.5
3. RG-52/U	25.4 × 12.7	1.3	6.56	8.2	21	12.4	15
4. RG-53/U	12.7 × 6.4	1.0	14.1	18	68	26.5	49

What is left after 1–10 is the power C that enters the receiver first stage. To obtain the power SNR ratio C/N, C is divided by the noise power entering the receiver, which in all likelihood is $N = kTB$ watts, as obtained in Section 5.2.2.

The interworking of all these budget items is easiest to absorb from examples, and so we continue with two standard ones, a geostationary satellite link and a deep-space link.

Example 5.2-4: UHF Uplink to a Geostationary Satellite. This example link is typical of a SARSAT (satellite search and rescue) system. Downed aircraft in remote areas activate a transponder that works at UHF with a crude antenna and a low data rate. A geostationary satellite hovers overhead and relays transponder signals back to a central rescue station on Earth. Take the following system specifications:

$f = 400$ Mhz
$d = 35800$ km (geostationary orbit distance, to Earth surface)
$P_t = 10$ dBw (i.e., 10 watts)
$G_t = 1$ dBi (crude dipole transmit antenna, including losses)
$G_r = 22$ dBi (4-m dish receive antenna at satellite; gain includes all front-end coupling losses)
$T = 100$ K (a good front end amplifier at the satellite)
Modulation: 2400 bit/s BPSK, 100% root-RC pulses

The free-space loss in this link is

$$L_p = 10 \log_{10}\left(\frac{\lambda}{4\pi d}\right)^2 = 20 \log\left(\frac{0.75 \text{ m}}{4\pi\, 3.58 \times 10^7 \text{ m}}\right) = 176 \text{ dB}$$

Take the pointing loss L_θ as 0 dB, and take the sum of the feed and circuit losses as 2 dB, split between transmitter and receiver. Finally, include the atmospheric loss in the fade margin and take both as 3 dB. Then the link budget goes as follows:

Transmitter power P_t	+10 dBW
Transmitter antenna gain G_t	+1 dBi
Transmitter losses L_f, L_c, L_θ	–1 dB
Link EIRP	+10 dBW
Receiver antenna gain G_r	+22 dBi
Receiver losses L_f, L_b, L_c, L_θ	–1 dB
Free-space loss L_p	–176 dB
Receiver power P_r	–145 dBW
Fade margin, *FM*	–3 dB
Worst-case power C	–148 dBW

Now find the noise power. A reasonable assumption is that the receiver bandpass is the bandwidth of the 2400-bit/s BPSK signal, which from Section 3.2 is (2400 bit/s)(2 Hz/bit) = 4800 Hz. The 2 here follows from the 100% excess bandwidth in the BPSK. Since the receiver front end temperature is 100 K, the noise power delivered is

$$N = kTB = (1.38 \times 10^{-23}\ \mathrm{J/K})(100\ \mathrm{K})(4800\ \mathrm{Hz}) = 6.6 \times 10^{-18}\ \mathrm{W}$$

which is –171.8 dBW. Thus C/N is about 24 dB. This is the worst-case SNR planned for in the budget.

In the example, a strong influence on the final C/N is the low data rate. An increase to data rate 24 kbit/s opens the front end 10-fold, and it drives C/N down to 14 dB. But there is plenty of C/N in the link. We could consider reducing the transponder to a more comfortable power, say, 2 W (7 dB less). Alternately, we could widen the front-end bandwidth: The 4800-Hz bandwidth is very narrow in comparison to the UHF operating frequency, and a wider bandwidth would make the system more tolerant of oscillator drift.

Conversion from C/N to E_b/N_0. The C/N value is needed for circuit and synchronizer design, but in order to use the error probability analysis of Chapters 2 and 3, we need to express the receiver SNR in terms of E_b/N_0. Finding E_b is easy:

$$E_b = \frac{C}{R} \quad \text{(J)} \tag{5.2-21}$$

where R bits/symbol is the data bit rate. Under AWGN front-end conditions, $N_0 = kT_e$, from (5.2-9). Thus

$$\frac{E_b}{N_0} = \frac{C}{kT_eR} \tag{5.2-22}$$

It could be asked where the bandwidth restriction on the noise lies here, since without one the noise power will be very high. The answer is that the matched filter assumed in the ML error probability derivation is a baseband lowpass filter. Any other filter leads to a poorer error rate, either by damaging the signal or by letting in too much noise. In a practical receiver the matched filter may come late in the receiver chain, and in this case the noise should be limited in the front end by a filter with bandwidth wide enough not to interfere with the matched filtering.

The next example illustrates many facts about deep-space communication.

Example 5.2-5: Earth-To-Mars Link. It is desired to transmit a high data rate over this very long distance. A large terrestrial antenna is available, and an X-band frequency of 8.4 GHz will be used. The specifications are:

$f = 8.4$ GHz

$d = 4 \times 10^8$ km, worst case

$P_t = 23$ dBW (a 200-W final amplifier)

$G_t = 51$ dBi (a 5.4-m dish at the satellite; gain includes circuit, feed, and pointing losses)

$G_r = 73$ dBi (a very large and very costly 70-m steerable dish; all losses included)

$T = 20$ K (assumes a maser front end integrated with the antenna, combined with 3 K space background and 6 K atmospheric noise temperatures)

Modulation: 10 Mbit/s QPSK, 80% root-RC pulses

As in Example 5.2-3, we will assume 2 dB for any further circuit losses and 3 dB for the fade margin. The latter stems largely from rain near the receive antenna. The link budget is as follows.

Transmitter power P_t	+23 dBW
Transmitter ant. gain G_t	+51 dBi
Transmitter losses	−1 dB
Link EIRP	+73 dBW
Receiver ant. gain G_r	+73 dBi
Receiver losses	−1 dB
Free-space loss L_p	−283 dB
Receiver power P_r	−138 dBW
Fade margin	−3 dB
Worst-case power C	−141 dBW

In order to find the receiver C/N, we must make an assumption about the receiver bandwidth. A reasonable estimate is 9 MHz, the theoretical bandwidth of 10 Mb/s QPSK with 80% excess bandwidth. Together with a front-end equivalent temperature of 20 K, this gives a noise power of

$$N = kTB = (1.38 \times 10^{-23}\ \text{J/K})(20\ \text{K})(9\ \text{MHz}) = 2.5 \times 10^{-15}\ \text{W}$$

or −146.0 dBW. C/N is about 5 dB, which is a barely adequate value. The E_b/N_0 is computed as follows:

$$N_0 = kT = 2.8 \times 10^{-22} \quad \text{W/Hz}$$

$$E_b = \frac{\text{antilog}(-141\ \text{dBW})}{10\ \text{Mb/s}} = 7.9 \times 10^{-22}\ \text{J/bit}$$

Thus E_b/N_0 is 2.9, or 4.6 dB. The difference from C/N stems from the physical bandwidth of the 80% QPSK.

Since the E_b/N_0 will yield only a 2% symbol error probability (see Fig. 3.17), we really should try to improve the link budget. Larger antennas, higher transmit power, or a quieter receiver front end will all help, but at least 3 dB is needed, and this is a doubling, for example, of either the antenna area or the power. A more attractive option is error-correction coding. A rate 1/2 convolutional code from Chapter 6 with 6-dB coding gain yields the following revised calculation of E_b/N_0, beginning at the worst-case received power of –141 dBW:

Worst-case received power C	7.9×10^{-15} W (–141 dBW)
Consequent E_b, no coding	7.9×10^{-22} W
E_b with 6 dB coding gain	3.1×10^{-21} W
Revised E_b/N_o	11.4 (10.6 dB)

This is an adequate E_b/N_0. It should be pointed out that the rate 1/2 coding doubles the physical bandwidth to 18 MHz, and consequently C/N decreases to 1.9 dB.

This example differs from the preceding one in an important way: Dish antennas are used at *both* ends. As a result, two factors of $1/\lambda^2$ are introduced into the budget through G_t and G_r, which counteract the λ^2 factor in the free space loss. Each doubling of carrier frequency adds 6 dB to the link budget. This fact is crucial to the design of space links. Were the 400 MHz of Example 5.2-4 employed in Example 5.2-5, the budget would have suffered by 26 dB! Higher frequencies than 8.4 GHz are even more attractive for deep-space links, but these are subject to rain attenuation at the Earth terminal. One solution would be a Mars to near-Earth link at a higher frequency, with a link down to Earth at 8.4 GHz. This is explored in Ref. 7, Section 1.3. An authoritative reference on the subject of space links is Yuen [26].

More on Deep-Space Channels. The successful design after 1960 of links that could reach other planets was of course a major feat of communication engineering. It is instructive to compare Example 5.2-5 to an early Mars link, the one used in the 1964 Mariner 4 project. The frequency was 2.3 GHz, the transmitter power used for data transmission was only 3 W, and the receiver noise temperature was 55 K. The spacecraft transmit antenna diameter was 0.79 m and the Earth receive diameter was 26 m. Combined with the 2.3 GHz frequency, these yield gains of just 23 and 53 dB, respectively, compared to 51 and 73 dB in the example. Adding these differences to the lower power, we get that the 1964 link had to work with 66 dB less received power! To this must be added the effect of the 55 K noise temperature, a degrading by another $10 \log_{10} 55/20 = 4.4$ dB. In consequence, Mariner 4 had a bit rate of only 8 bits/s. The 1974 Mariner 10 was able to raise this to 118 kbits/s.

Space missions to the far planets are even more challenging. The Voyager space missions are the longest link budgets to date. The antenna gains, wavelength, transmitter power, and noise temperature were not far from those in Example 5.2-5.[6] For

[6]They were, respectively, 48 dBi, 76 dBi, 8.4 GHz, 13 dBW, and 25 K. Details appear in Ref. 27; numbers given here include later corrections to that reference.

the 1986 Uranus encounter, at distance 3×10^9 km, 30 kbit/s was possible over this link. This was achieved only with an advanced convolutional Reed–Solomon coding system (see Chapter 6), which had a coding gain of 6–7 dB compared to the QPSK transmission in Example 5.2-5. The steady improvement in link performance over 25 years was the product of many technical advances: Coding, higher microwave frequencies, better power amplifiers, larger spacecraft, better Earth antennas, and tracking/navigation accurate enough for very narrow antenna beams.

5.3 TERRESTRIAL RADIO CHANNELS

The next radio channel must deal with reflections and bendings of its beam by nearby objects. Most channels of this sort are UHF or microwave terrestrial radio channels, although the channel need not of course be on Earth or in the microwave. Still, we will make the terrestrial UHF/microwave assumption in what follows and treat the other cases as exceptions. The fundamental problem in this section is that radio beams reach the receiver through several paths with differing delays. The constructive and destructive interference that results is called multipath interference. Other problems that occur are absorption and scattering of the beam.

5.3.1 Reflection, Refraction, Diffraction, and Absorption

Refraction is the bending of a beam as it passes through media with changing refraction index. *Diffraction* is the re-radiation of energy that occurs when a beam hits an edge or an object whose dimension is small compared to wavelength. *Reflection* needs no definition. All these produce extra beams or move a desired beam away from its target, and, what is worse, the effect can change with time. We will take these three in turn, starting with reflection. A fourth impairment to propagation is *absorption* of the beam. A fifth is *scattering*, the dispersion of waves by small (compared to wavelength) objects in the path. When this dominates, we have Rayleigh fading, which we will take up under mobile radio in Section 5.4. Most of what follows is based on electromagnetic wave theory. This is a deep subject, and we will hit only some high points that affect link budgets. Details can be found in electromagnetics texts such as Refs. 16 and 17; a recent treatment from a mobile communication point of view is Rappaport [29].

Radio reflections, like those of all waves, obey the Snell's law geometry shown in Fig. 5.10, in which the angles of incidence and reflection, θ, are the same. This is true regardless of the wave polarization. Reflections occur from objects that are smooth, large, and planar in comparison to the wavelength, such as the ground. Many reflections occur, but only the one with the geometry in the figure reaches the antenna and interferes with the direct ray. In addition to the reflected wave, a surface, or "ground," wave is initiated and may also reach the receive antenna. Above 100 MHz, we can assume that the reflection from a surface like ordinary ground or a large building is strong and the ground wave is small.

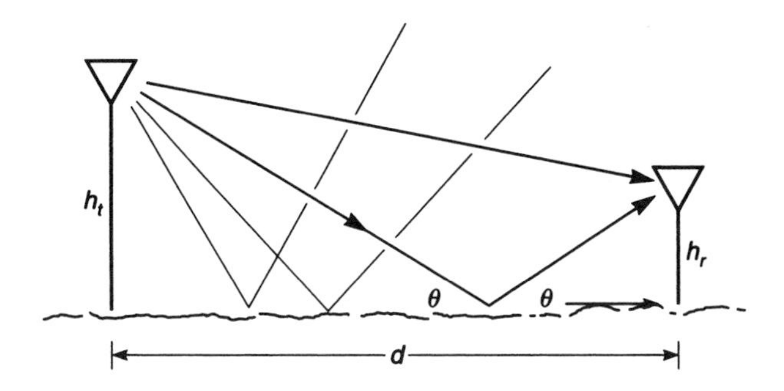

Figure 5.10 Snell's law reflection with ground wave.

The geometry means that the phase between the direct and reflected waves is

$$\begin{aligned} \phi &= \frac{2\pi d}{\lambda}\left[\sqrt{\left(\frac{h_t+h_r}{d}\right)^2+1}-\sqrt{\left(\frac{h_t-h_r}{d}\right)^2+1}\right] \quad \text{rads} \\ &\approx \frac{4\pi h_t h_r}{\lambda d} \quad \text{for large } d \end{aligned} \tag{5.3-1}$$

Here h_t and h_r are the heights of transmitter and receiver antennas d apart, and λ is the wavelength. With perfect reflection and small ϕ, the arriving power P_r compared to the direct ray power P_0 is

$$P_r = P_0|1-e^{j\phi}|^2 \approx P_0 \sin^2 \frac{4\pi h_t h_r}{\lambda d} \approx P_0\left(\frac{4\pi h_t h_r}{\lambda d}\right)^2 \tag{5.3-2}$$

(Note that reflection reverses phase.) Were reflections not present, P_r would be $P_t G_t G_r(\lambda/4\pi d)^2$, as in Eq. (5.2-4); now instead it will be

$$\begin{aligned} P_r &= P_t G_t G_r\left(\frac{4\pi h_t h_r}{\lambda d}\right)^2\left(\frac{\lambda}{4\pi d}\right)^2 \\ &= P_t G_t G_r\left(\frac{h_t^2 h_r^2}{d^4}\right) \end{aligned} \tag{5.3-3}$$

The factor replacing the free-space loss in (5.2-4) is $h_t^2 h_r^2/d^4$. We can see here some interesting facts about this terrestrial channel. Its propagation loss factor is independent of wavelength, and antenna heights have a strong effect. Most interesting, P_r obeys a fourth-power law in distance.

Efficient reflection such as this occurs only when a surface is smooth. A measure of smoothness is the Rayleigh criterion

$$C = \frac{4\pi\sigma\theta}{\lambda} \tag{5.3-4}$$

in which the incidence angle θ (radians) and λ are as before and σ is the rms size of the surface irregularities. Here $C < 0.1$ gives a near-perfect reflection, while $C > 10$ gives virtually none. In microwave links C is typically 0.2–0.4 [15].

If the reflector is a perfect conductor, then all the wave energy is reflected; otherwise some enters the reflector. The amount depends on polarization, θ, λ, and the conductivity and dielectric parameters of the reflector.

We turn next to refraction. Bending occurs when a beam passes through a change in the refraction index. The bending is governed by Descartes' law, which is illustrated in Fig. 5.11. The figure is drawn with and without the curvature of the Earth, which often plays a role in refraction. Without the curvature, Descartes' law simply states that the departure angles α and α_0 in media with refraction index n and n_0 obey $n \cos \alpha = n_0 \cos \alpha_0$. With curved index boundaries in a planetary atmosphere, the angles are relative to tangents and the law depends on the planet radius ρ (nominally 6370 km for Earth) and the height of the index shift, h. We can sum up both cases as:

$$\frac{n_0 \cos \alpha_0}{n \cos \alpha} = \begin{cases} 1, & \text{no curvature} \\ \dfrac{\rho + h}{\rho}, & \text{curved Earth} \end{cases} \tag{5.3-5}$$

For refraction in the atmosphere, the index n at the higher altitude is normally the smaller one, which implies that the beam is bent toward the Earth, as illustrated in the figure. This downward bending helps radio beams—and beams of the setting sun—to follow the curvature of the Earth.

In reality, the index of the atmosphere drops in a continuous manner from a nominal sea level value of 1.000301 to the free-space value of 1. For refraction on a planetary scale, both this and the curvature need to be taken into account. A good model for the drop in n is the linear law $n \approx n_0 - bh$, in which the gradient b is 3.92

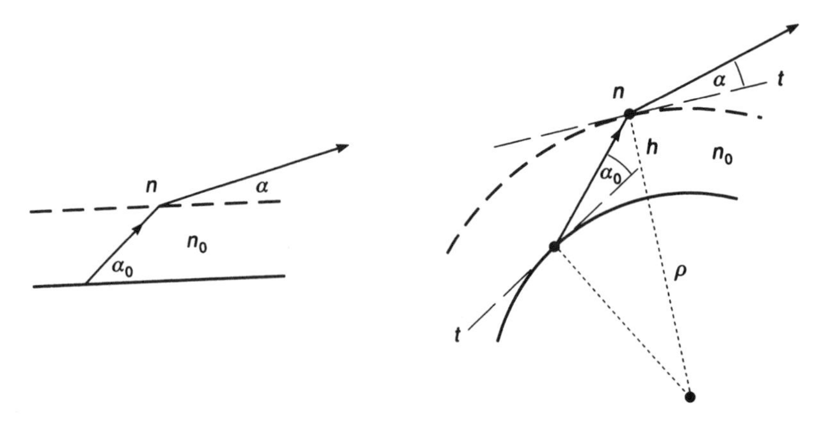

Figure 5.11 Descartes' law of refraction, with and without a curved Earth. Atmosphere depth at right is exaggerated.

$\times 10^{-5}$ km^{-1}, h runs up to 7.7 km, and n_0 is the sea level index. If Eq. (5.3-5) is applied repeatedly to many curved layers with linearly declining index, the result is the law

$$\frac{\cos \alpha_0}{\cos \alpha} \approx \frac{\rho + h(1 - b\rho)}{\rho} \qquad (5.3\text{-}6)$$

Note that there is no explicit index in this; the index is assumed to decline from $1 + hb$ to 1. α is now the angle with a tangent at the last thin layer. A useful way to think about this is to consider the case of a revised Earth with no atmosphere and the new radius $\rho' = (4/3)\rho$. Substituting the new ρ' and any $h < 7.7$ km into (5.3-5) and taking $n_0 = n = 1$ gives the same relation between α and α_0 as (5.3-6) gives. In other words, the linear drop in index can be modeled by a $\frac{4}{3}$–size Earth with no atmosphere. If the Earth were 4/3 bigger with the same atmosphere, refracted rays would simply follow its curvature.

The fine details of positioning antenna beams are beyond our scope, but a major issue is that high-gain antennas have narrow beams, and refraction can cause the beam to miss its intended target altogether. A second issue is that refraction can change, and much higher gradients than the nominal one can apply. A particularly damaging case occurs when a higher index overlies a lower index, since then the beam is bent up, away from terrestrial receivers. Overlying humid air, for example, has a higher index than dry air. The following example illustrates another common refraction problem.

Example 5.3-1: Low-Angle Earth-Satellite Link. When the space end of an Earth-space link appears to be on the horizon as seen from Earth, it is actually somewhat below. We can estimate the true angle by using (5.3-6). That equation assumes that the refraction index drops linearly from its nominal sea level value to its free space value at 7.7-km altitude, after which there is no further bending. Figure 5.12a is a $\frac{4}{3}$–Earth picture, on which the bending that occurs until 7.7 km appears as a straight line of length D. From the Pythagorean theorem, $D \approx \sqrt{2Rh} = 362$ km. Figure 5.12b shows the normal-Earth view, on which the propagation path is bent; we will assume that both it and the straight line connecting the observer and the 7.7-km point have length D. The satellite from the observers position is actually at angle $\alpha - \theta$, even though it appears to be on the horizon. From (5.3-6) with $\alpha_0 = 0$, $h = 7.7$ km, and $\rho = 6370$ km, we get that α is 2.44°. The geometry of Fig. 5.12b shows that

$$k^2 + \ell^2 = D^2, \qquad (6370 - k)^2 + \ell^2 = (6370 + 7.7)^2$$

Solving these gives $k = 2.6$ km and $\ell = 362$ km. The angle θ is $\arcsin(\ell/6377.7)$, which is 3.25°. In conclusion, the true angle at the observer location is 0.81° below the horizon. Since high-gain antennas for space use can have beam width on the order of 1°, it is necessary to account for curved-Earth refraction during aiming.

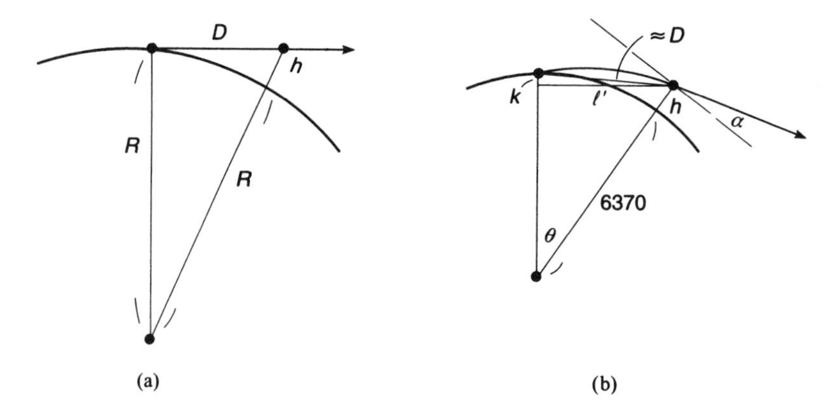

Figure 5.12 Calculation of apparent observation angle $\alpha - \theta$ in Example 5.3-1, starting from **(a)** $\frac{4}{3}$ Earth and **(b)** progressing to normal Earth. Here, R is $\frac{4}{3}$ (6370), where 6370 km is Earth radius.

Whereas refraction causes path loss by bending the beam away, diffraction scatters the beam, by a process of re-radiation from encountered objects. Objects such as trees and buildings, whose size is on the order of a wavelength, will scatter a beam (larger objects will tend to reflect it). Re-radiation also occurs from an encountered edge, such as a building or even a hill crest. The reason is easily explained by Huygens' principle, which says that each point on a propagating wavefront acts as a new point source. Further propagation is the sum of all these contributions. With a plane wave in free space the result is another plane wave, but a partial obstruction changes things dramatically. Figure 5.13 sketches two cases of an idealized thin obstruction called a "knife edge." Diffraction from such an object is well understood in electromagnetics (see Refs. 15–17). In Fig. 5.13a the knife edge lies somewhat above the line of sight. Even so, radiation works around the edge and reaches the receiver. Huygens' principle, illustrated at the left, shows that point radiators near the knife point make the wavefront appear to bend. It can be shown that when $\ell, \ell_e \gg h$ and $h/\sqrt{\ell\lambda} \ll 1$, the ratio of received power to that hitting the knife edge, P_r/P_0, is [15]

$$\frac{P_r}{P_0} \approx \frac{1}{h^2 4\pi^2}\left[\frac{\lambda \ell_e(\ell - \ell_e)}{\ell}\right] \tag{5.3-7}$$

Here ℓ_e is the knife-to-receiver distance, ℓ is the total path length, and h is the knife height above line of sight.

Diffraction is not always friendly. Figure 5.13b shows what happens when the knife edge is below the line of sight, but not by much. There is again re-radiation at the knife point, but now it acts to reduce the signal at the receiver. By Huygens, the point radiators blocked by the knife cannot properly reinforce the plane wavefront at the receiver, and the knife casts a shadow even though it does not block the line of sight. The figure shows a line of sight that misses the knife by Δ. Theory and practice [3] show that path loss will be small when

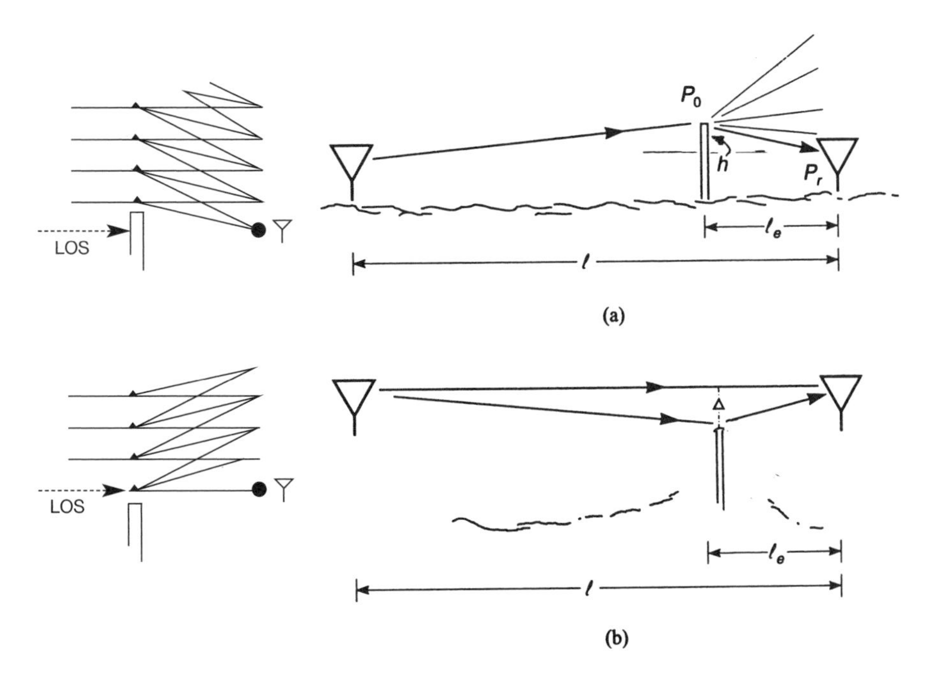

Figure 5.13 Huygens' principle and two cases of diffraction: **(a)** from an obstructing knife edge, and **(b)** from a knife edge near the LOS path. At left is Huygens' principle in action; at right is text notation.

$$\Delta^2 \geq \frac{\lambda \ell_e(\ell - \ell_e)}{\ell} \qquad (5.3\text{-}8)$$

The right-hand side here is the bracket term in (5.3-7).[7]

The following example treats the standard terrestrial microwave link, a 50-km link that will be explored in more detail in Example 5.3-3.

Example 5.3-2: Diffraction in a 50-km Link. A standard design length for terrestrial microwave links is ℓ = 50 km, and a common frequency is 4 GHz. Suppose a 10-m-high sharp-edged apartment block lies halfway down the line-of-sight path (as in Fig. 5.13a). In (5.3-7), the bracket term is 938 m^2; the equation then yields

$$P_r/P_0 \approx \frac{1}{(10 \text{ m})^2 \, 4\pi^2} (938 \text{ m}^2) = 0.237 \qquad (-6.2 \text{ dB})$$

[7]The square root of the bracket in (5.3-7) is the radius of the first Fresnel zone of the tranmission. A rule of thumb states that transmission paths should be unobstructed over their first Fresnel zone (see Ref. 29, Chapter 3).

(Before accepting this we check that $h/\sqrt{\ell\lambda} = 0.16 \ll 1$.) The free-space loss up to the knife edge, from (5.2-4), is

$$\left[\frac{\lambda}{4\pi(\ell/2)}\right]^2 = 5.7 \times 10^{-14} \qquad (-132.4 \text{ dB})$$

So the total path loss is 138.6 dB. Without the obstruction, the path loss would be $(\lambda/4\pi\ell)^2$, which is a loss of 138.4 dB. One needs the decimal tenths—which are meaningless in this example—to see the effect of the obstruction. If the building is replaced by a 30-m object, the calculation of P_r/P_0 gives

$$P_r/P_0 \approx \frac{1}{(30 \text{ m})^2 \, 4\pi^2} (938 \text{ m}^2) = 0.0264 \qquad (-15.8 \text{ dB})$$

and a total path loss of 148.2 dB. (But $h/\sqrt{\ell\lambda}$ is now about 0.5). This is considerably worse that the 138-dB loss without the obstruction. If we wish to move the line-of-sight path away from the obstruction and avoid diffraction entirely, it will need to miss by Δ in (5.3-8), which is $\sqrt{938}$ m $\approx$ 31 m.

A last major source of terrestrial path loss is absorption in the atmosphere. Three major absorptions need to be considered. Water vapor has a broad molecular absorption peak at 24 GHz; near this frequency, losses of $\approx$0.2 dB/km can be encountered. Molecular oxygen has a sharp absorption peak at 60 GHz, with loss 10 dB/km. Regions near both these frequencies are avoided in terrestrial links.

Absorption from precipitation is proportional to frequency and is particularly troublesome above 10 GHz or so. At 10 GHz, absorption ranges from 0.01 dB/km for a light mist up to 3 dB/km for very heavy rain. At 100 GHz, these losses are 1 dB/km and 40 dB/km. Rain loss is considerably tempered by the fact that cells of heavy rain are small and, in the case of space links, by the fact that high-angle paths encounter little atmosphere.

With this, we turn away from a theory-based view of terrestrial propagation. There are other influences, such as polarization effects, propagation inside and around structures, and reception via antenna side lobes, to name a few. The interaction among all these factors is complex and rates a book in itself. Some classic references for those wishing to take another step are Jakes [15] and the various Bell Laboratories manuals, such as Ref. 3.

5.3.2 Medium-Wave and Shortwave Propagation

For completeness, we detour briefly to the medium-wave (300 kHz–3 MHz) and shortwave bands (3–30 MHz). Propagation in these bands is in large measure the interplay between ground wave propagation, which hugs the Earth, and sky wave propagation, in which several layers in the ionosphere act in varying degrees as reflectors and absorbers. Ground wave propagation declines with frequency, since higher frequencies are more absorbed by the conducting Earth. Propagation is thousands of kilometers at 300 kHz but declines to perhaps a hundred at 30 MHz. Ionospheric behavior depends critically on the time of day and year, since ionization is proportional to the sun intensity.

The charged layers of the ionosphere can be thought of as having two conflicting properties, namely, reflection and absorption. Absorption increases with charge density and is inversely proportional to frequency. In the lower medium-wave range, energy is absorbed by the highly charged daytime ionosphere; at night the charge density declines, these frequencies are reflected, and they propagate further than the ground wave. Medium-wave propagation is thus better at night. At a given charge density there is a highest frequency that can be reflected, with higher frequencies passing through into space. This frequency grows with density. Higher frequencies are also less absorbed. At daytime densities, frequencies in the range 12–30 MHz are strongly reflected and lower ones tend to be absorbed. At lower night densities, these frequencies pass into space; frequencies in the range 4–12 MHz are still reflected and are no longer absorbed.

Shortwave channels are characterized by fading, time dispersion, and Doppler effects, all of which trace back to the constantly moving and changing ionospheric layers. Noise is generally not Gaussian, for at least two reasons. First, natural and man-made interferences are much stronger that the Gaussian background below 30 MHz, and with the changing ionosphere, rogue signals from far away can appear in unwanted places. Second, these interferences are usually stronger than the receiver front-end noise.

5.3.3 Statistical View of Terrestrial Channels

What we have described so far are the physical reasons behind loss of signal in a terrestrial radio link. However beautiful the physics, it applies imprecisely to a real channel, and never in the same way for very long. Most of the time, it is more productive to take a statistical view. In what follows, we concentrate once more on the UHF and microwave.

A great many measurements have been made, and fortunately they obey some fairly clear empirical rules. The following rule is in common use and stems from the work of Vignants [18] and Barnett [19]. It is therefore called the Barnett–Vignants formula:

$$Pr\left\{\frac{P}{P_0} < u\right\} \approx 6 \times 10^{-7}\, Cf\ell^3 u \tag{5.3-9}$$

where

P = power received in terrestrial channel

P_0 = power that would be received in free space

f = frequency, GHz

ℓ = path length, km

C = terrain factor

The formula estimates the probability that a given observation of the received power will be faded by a ratio u or more. This can be thought of as an *outage probability,* relative to the threshold u. The terrain factor C depends on the smoothness of the path and its humidity, the two factors that most promote fading. For average humidity and terrain $C = 1$. For a very smooth and damp path, C might be 4–6; for propagation over a rough or mountainous desert, it might be 0.2 or lower. Since the formula is only an empirical rule, C need not be set with great precision. The formula does assume that a line-of-sight path exists in the absence of refraction (which is included in the formula). The loss from knife-edge diffraction, Eq. (5.3-7), needs to be added in separately.

Formula (5.3-9) for the first time brings probability into our discussion of link design. Now we will think as follows. Nothing in life is certain, least of all fading radio links. A sensible strategy is to set a limit on P/P_0 and design for a set level of *reliability,* as measured by the probability that this ratio lies above a limit. A reliability of .9999, which means the outage probability in (5.3-9) is .0001, is a failure time of 53 min/year; similarly, .99999 is 5.3 min/year and .999999 is 32 s/year.[8] Another version of the Barnett–Vignants formula sets the reliability, $\mathcal{R}$, equal to $1 - Pr\{P/P_0 < u\}$, and takes u as FM, the fade margin term in Section 5.2.3. When all quantities are converted to decibels, Eq. (5.3-9) becomes (base 10 logs)

$$\mathrm{FM_{dB}} \approx 30 \log \ell + 10 \log(6fC) - 10 \log(1 - \mathcal{R}) - 70 \qquad (5.3\text{-}10)$$

with f in gigahertz and ℓ in kilometers. The fade margin has been taken as a positive decibel number, which means reversed signs on the right.

Example 5.3-3: Standard Terrestrial Link. Microwave links on Earth average about 50 km long. The most common frequency is 4 GHz. If we add to this an average terrain ($C = 1$) and a reliability of 5.3 min/year (.99999), we can call the result a standard link. Putting all these in (5.3-10) gives

$$\begin{aligned} \mathrm{FM} &= 30 \log 50 + 10 \log(6 \times 4 \times 1) - 10 \log(10^{-5}) - 70 \\ &= 51 + 14 + 50 - 70 = 45 \text{ dB} \end{aligned}$$

This fade margin is only an estimate. As well, there is always the possibility that the link will work perfectly for 10 years and then fail for 53 minutes. A standard rule of thumb for terrestrial microwave links says take FM as 40 dB.

A second empirical rule exists for the *length* of fades, as opposed to the depth of fade. Many observations have shown that under average conditions at microwave,

$$\text{Fade duration} \approx 400\sqrt{u} \qquad (5.3\text{-}11)$$

[8]In link design parlance, .9999 is called "4 9s," .99999 "5 9s," and so on.

where u is the fade depth as before and duration refers to the time the fade remains below u. For example, a fade that falls below 20 dB will last about $400 \times \sqrt{0.01} = 40$ s. During some of this time the fade may drop below 30 dB; if so, the periods below 30 dB will last about 13 s each. The power of u in (5.3-11) will be a little smaller than 0.5 (longer fades) for transmission over water, and it will be a little larger than 0.5 over deserts.

5.4 FADING CHANNELS

Now we come to the most difficult case of radio propagation, when one or both of the terminals are moving. A moving antenna probably passes through a succession of poor channels of the type in Section 5.3, and this alone leads to a time-varying channel. Absorption, reflection, and multipath change with position, and so also do the intersymbol interference (ISI) and band-limitation discussed in earlier chapters. But motion leads to new phenomena, including Doppler effects and signal strengths that fluctuate rapidly, perhaps as fast as the signal modulation. Moreover, these effects vary with terrain and differ for vehicles, ships and aircraft.

This section describes these new phenomena. The general name for motion-induced signal variation is *fading*. The subject is complex, and we begin with some general ideas about fading, many of which are everyday experience. The subject of what we can *do* about fading forms part of Chapter 7.

5.4.1 Basic Ideas of Fading

What is meant by motion? One or both of the transmitter and receiver can be moving, but in most cases just one terminal moves. This one is called the "mobile," and other one is called the "base." It simplifies the exposition to assume that only the mobile moves. Nothing is actually lost; by the reciprocity principle of antennas, the terminals see the same channel whether the mobile or the base is transmitting. A more subtle point is that the channel itself is probably in motion. The physical parts of it—myriad reflectors and pieces of atmosphere—are probably all in independent motion. Even with transmitter and receiver fixed, fading due to motion occurs because the medium is in motion. Thus we can expect fading even with fixed terminals. Such fading is easily observed with a shortwave radio.

A second point about mobile and everyday cellular and local area network channels is that they generally lack a simple line-of-site connection between transmitter and receiver. Propagation is said to be *diffuse*. The terminals are in a radio half light, with energy reaching them from many directions, from a multitude of reflections and diffractions. In its pure form, this kind of radio propagation is called *Rayleigh fading,* and we will take that up in Section 5.4.2. Diffuse propagation and motion are separate phenomena but they tend to occur together in mobile radio.

A third point that we have all observed is that signal strength is correlated at closely spaced times or positions. Fades last a certain time. The dependence on position is easily observed with an FM car radio: Cars in a traffic jam can encounter long fades that remain until the traffic moves to a new place. A car radio further demonstrates that fades come and go more rapidly as the car gains speed. The variation of signals with a gross change of position is called *shadowing*. The terminals move in and out of weak signal zones—darker shadows—along the path of motion. Shadowing stems from motion of the terminals, not of the medium. Here again, shadowing is a separate phenomenon, but it occurs together with other fading phenomena. Many phenomena vary the signal amplitude, and if an amplitude signal modulation varies at about the same speed, it will be difficult to distinguish the data from the fading.

Finally, we should expect to see some Doppler phenomena. These make it problematic to carry information in the simple value of phase or frequency. Which shifts are motion and which are modulation? We have already seen that rapid fading can make amplitude modulation difficult to use. Doppler effects mean that frequency and phase modulations may need to be differential ones or ones with wide enough frequency shifts so that the Doppler shifts are insignificant. Most analog mobile radio systems use wideband FM; these have a wide bandwidth, which was the original reason that mobile radio was placed in the VHF to microwave bands.

Even with the best physical intuition, fading is still a subtle phenomenon. One way to conceptualize mobile radio fading is to think of it as stemming from two facts: The physical environment changes from place to place, and diffuse reception combines with motion. Each induces, more or less, its own signal variation. A useful way to think about signal loss in fading is the equation (in dB)

$$\text{Losses} = \text{Average path loss} + \text{Shadowing loss} + \text{Rayleigh loss} \qquad (5.4\text{-}1)$$

In free space, the path loss would be the factor $L_p = (\lambda/4\pi d)^2$ in Eq. (5.2-4), with λ the wavelength and d distance. This is inverse square law propagation. In a messy mobile scenario, it seldom applies and it is best simply to measure the average signal strength over the envisioned service area. Rural and city cellular and indoor radio communication all behave differently, with indoor radio through walls having as bad as inverse sixth power propagation. A good all-around assumption is an inverse cube law—that is signal propagation according to $1/d^3$. The shadowing loss term adds to this and is the allowance for radio shadows that are more severe than average; we will see that 6–10 dB is sufficient for this extra loss. The Rayleigh term is the allowance for diffuse reception; it will need to be 30–40 dB. Fading thus costs typically 40 dB in signal power, plus the effect of the poorer propagation law.

Another conceptual tool is Figure 5.14. It shows the myriad rays that arrive at a receiver as a result of random scatterings and reflections. These are organized into delay shells, loci of paths that have the same delay. Three are shown. Each consists of many independent paths that together make up a diffuse signal arrival at delay τ_i. In reality, some rays bounce more than once, some are refracted, and the shells exist in three dimensions, not just the plane of the paper. However, nothing is lost by simplifying the model. The important point is that each shell is an independent diffuse arrival process, which has the Rayleigh statistics in the next section. It is reasonable to claim that the signals from each shell are independent, because a small τ corresponds to a significant distance, and at VHF frequencies and above, it corresponds to many multiples of 2π in the carrier phase.

There is a different Fig. 5.14 at each locality in the mobile system. The figure gives an important insight: We can divide fading at this locality into two components, *time dispersion,* or spreading, and *time variation*. The first occurs because a transmission arrives spread over delays, and the second occurs because reflectors within a shell change with time. Another name for dispersion is multipath distortion, although this often refers to spreading with several distinct arrivals rather than a continuum. Dispersion is discussed in Section 5.4.3. The standard concept of fading is built on these two contributions. The amount of each, dispersion and variation, is not easily modeled and is best found by measurements. We will take them up in Section 5.4.3.

The third element of fading, shadowing, is the subject of Section 5.4.4.

Time and frequency effects are intermingled in fading. One leads to the other, and the insights of basic Fourier theory are important in understanding the interaction. Theory tells us that aside from a pure sinewave, a signal cannot flunctuate in time without spreading in frequency; conversely, changing a signal's spectrum modifies its time evolution. This is time–frequency duality. Time and frequency distortion in fading are thus tied together. Sometimes it is more natural to talk about an effect in one domain or the other. For example, Doppler effects at first glance are about frequencies, and multipath effects occur in time. But all fading effects can be described in either time or frequency, and we will do so it what follows.

Two full-length discussions of fading in the literature are Jakes [15] and the third part of the Schwartz et al. text [22], both classics that have been reissued. More recent shorter treatments that focus on cellular radio are Rappaport [29], Steele [30], and Sklar [24].

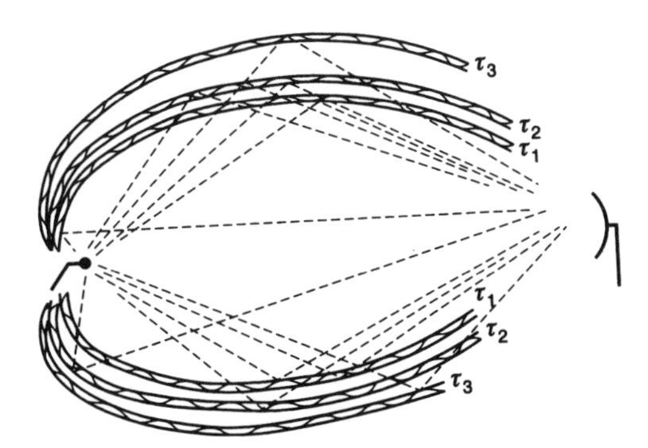

Figure 5.14 Rays arriving from equal-delay reflecting shells for shells at delays $\tau_1 < \tau_2 < \tau_3$. Plane of paper is horizontal plane. Other rays due to refraction and diffraction arrive as well.

5.4.2 The Rayleigh Fading Model

We have seen that in a fading channel energy usually reaches the receiver in independent packets from a multitude of directions. We show now that the total received intensity at a given moment obeys a Rayleigh probability distribution. Sometimes the mobile receives both a Rayleigh component and a line-of-sight (LOS) component; this is also called the specular component. It can be a single strong reflection. A LOS component is preferred, of course, but it is still possible for the Rayleigh components to cancel the LOS ray. This milder kind of fading is called *Rician fading,* because its strength follows the Rician distribution.

We will limit the analysis to the most common and simplest case, a vertical mobile antenna and vertically polarized waves. This antenna will respond to the E-field only and is omnidirectional in the horizontal plane. If it is moving, its speed is v. A pure sinewave is transmitted and we consider one shell in Fig. 5.14. Random-sized rays reach the antenna uniformly from all directions and within a time interval that is small compared to the envisioned modulation symbol time, and each has phase uniformly distributed in $[0, 2\pi)$. The model here is often called the dense scatterer model. Figure 5.15 diagrams a few rays. Each is affected by a Doppler shift, and for any arriving ray, the Doppler effect states that the apparent arrival frequency is shifted by

$$\Delta f = \frac{v}{\lambda} \qquad \text{Hz} \tag{5.4-2}$$

with λ the transmitted wavelength and the signed variable v the velocity of approach. For example, in the all-too-common case of a 10-GHz police radar, a return from an oncoming driver at 100 km/h (27.8 m/s) is shifted by +926 Hz. In Fig. 5.15, waves arrive at random angles ϕ_n, and each sees an approach velocity of $v \cos \phi_n$; the shifts are thus $\Delta f_n = (v/\lambda) \cos \phi_n$.

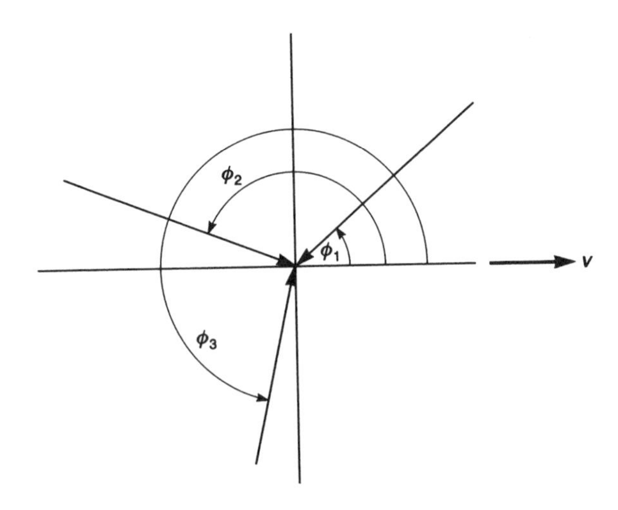

Figure 5.15 Waves arriving from random angles to antenna moving at velocity v.

If the arriving waves have amplitudes B_n and shifts Δf_n and are vertically polarized, the total E-field seen by the antenna is

$$E_r = \sum_{n=1}^{N} B_n \cos(\omega_0 t + \theta_n), \qquad \theta_n = \frac{v}{\lambda}\, 2\pi t \cos\phi_n + \psi_n \tag{5.4-3}$$

The ψ_n here are uniformly distributed in $[0, 2\pi)$, which arises from the random distances the arriving waves travel. We next rewrite (5.4-3) by the standard I/Q transformation (3.1-3) to get

$$\begin{gathered} E_r = I(t)\cos\omega_0 t - Q(t)\sin\omega_0 t \\ I(t) = \sum_{n=1}^{N} B_n \cos\theta_n, \qquad Q(t) = \sum_{n=1}^{N} B_n \sin\theta_n \end{gathered} \tag{5.4-4}$$

Now comes the critical step: Since the elements at time t in the $I(t)$ sum are independent, the observation $I(t)$ must be approximately Gaussian by the Central Limit Theorem. The same applies to $Q(t)$, and the observations $I(t)$ and $Q(t)$ are uncorrelated by the orthogonality of the sine and cosine of θ_n. A calculation shows that both have mean zero and variance $\mathcal{E}\{|B_n|^2\}/2$, where the average is over ϕ_n and ψ_n, both of which are uniformly distributed.

In fact $I(t)$ and $Q(t)$ are low-pass Gaussian processes, but for now we are concerned only with a single observation at t. Received field (5.4-4) has envelope $A(t) = \sqrt{I(t)^2 + Q(t)^2}$ and phase $\arctan[Q(t)/I(t)]$. Basic probability theory tells that such a phase is uniformly distributed and that $A(t)$ is Rayleigh distributed, with density function

$$p(a) = \begin{cases} (a/\sigma^2)e^{-a^2/2\sigma^2}, & a < 0 \\ 0, & \text{otherwise} \end{cases} \tag{5.4-5}$$

with

$$\sigma^2 = \tfrac{1}{2}\mathcal{E}\{|B_n|^2\}$$

The distribution has mean $\sigma\sqrt{\pi/2}$ (the average envelope value) and second moment $2\sigma^2$ (the sum of the powers in I and Q).

Thus any isolated observation is Rayleigh. We can pause to think about what this means. Many rays reach the antenna from all directions uniformly, and at uniformly distributed phase shifts. The ray amplitudes in the derivation differ, but the fact is that the Central Limit argument works even if they are all the same. One can say that the universe began at a certain time, unfolded to a certain present state at t, and (5.4-5) is the distribution of a at t. One could as well say that the universe was set up according to the random model and has not changed until t, at which time we first observe it. Either way, it is interesting that we did not use the velocity v in arriving at (5.4-5), and in fact the velocity makes no difference with this single, isolated observation. The same distribution applies for any other isolated observation, say at t', given that the propagation model holds. For v to matter, there must in fact be two observations of the signal.

Observations of real radio waves from VHF to microwave all verify the Rayleigh result (see, e.g., Ref. 15, Chapter 1), so long as there is no LOS path or dominant reflection. In fact the superposition of as few as a half dozen rays in a practical channel leads to a strongly Rayleigh signal. This is what justifies the model. With careful consideration of actual E and H fields, a complete analysis leads to many useful results. Spectra and cross-correlations of E and H and the signal envelope are available; correlations between different spatial points can be computed; useful tricks with wave polarization and antenna directivity can be devised. The wider analysis is summarized in Jakes [15], and much of the pioneering mathematics is due to Rice [20].

Spectral Spreading Due to Motion. The power spectral density of a stationary random process implicitly assumes multiple observations, since it is derived from the process autocorrelation function. In the foregoing, a simple sine wave cos $\omega_0 t$ was transmitted. Upon arrival, it arrives as uniformly diffused rays that appear to a moving receiver to have a range of frequencies. The ray arrangement may not have changed since the beginning of time, but this is hardly natural, and now is a good time to introduce a new element of randomness and make the evolution of the diffuse rays a stationary random process. This means that the phases and amplitudes of the rays evolve in a stationary, random way, and consequently so does E_r in (5.4-3). Let $S(f)$ be the power spectral density of the bandpass E_r process.

The following is a simple derivation of $S(f)$ given by Jakes [15]. Since the power near $f = f_0 + (v/\lambda)\cos\phi$ can only arrive from angles $\pm\phi$, the power $S(f)df$ in a small band df is, with an omnidirectional antenna,

$$S(f)\,df \approx 2G_r P_0 \frac{1}{2\pi}\,d\phi$$

Here $(1/2\pi)\,d\phi$ is the power that arrives from a small range of angle $d\phi$, G_r is the antenna gain[9] and P_0 is the power received by an isotropic antenna. We can let ϕ range over $[0, \pi]$ only and double the result. From the f-versus-ϕ constraint,

$$df = \frac{v}{\lambda} - (\sin\phi)\,d\phi = \frac{v}{\lambda}\sqrt{1-\left(\frac{f-f_0}{v/\lambda}\right)^2}\,d\phi$$

Substitution of this yields

$$S(f) = \frac{G_r P_0/\pi}{(v/\lambda)\sqrt{1-\left(\dfrac{f-f_0}{v/\lambda}\right)^2}} \qquad (5.4\text{-}6)$$

when $|f| < f_0$. The integral of this PSD is $G_r P_0$. Figure 5.16 plots it for $f_0 = 850$ MHz, which is typical of cellular phones, and a speed of 100 km/h. The maximum frequency deviation due to motion is

[9]From Table 5.2-1, G_r would be 1.85 for a vertical antenna.

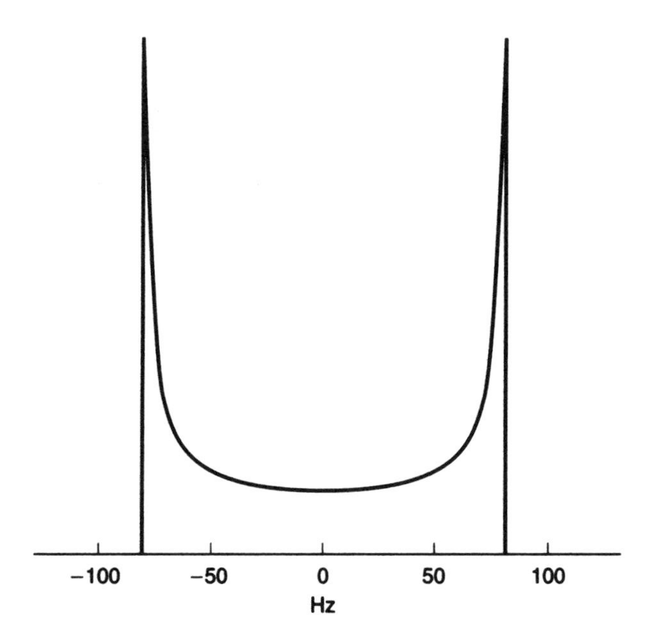

Figure 5.16 Theoretical Doppler power spectral density for 850-MHz transmitted carrier in Rayleigh channel at speed 100 km/h.

$$\pm\frac{v}{\lambda} = \pm\frac{100\text{ km/h}}{(3600\text{ s/h})0.353\text{ m}} = \pm 79\text{ Hz}$$

Although the mathematics is tedious, it is possible to calculate power spectral densities for the signal phase, its instantaneous frequency, its envelope, and other quantities as well (Refs. 15, Sections 1.3–1.4). In particular, further analysis of the envelope process shows that its autocorrelation function narrows as speed grows, which means that envelope values at nearby times become less correlated and the fading becomes more rapid and violent. As the mobile slows, fading slows.[10] An actual record of a Rayleigh fading envelope when the mobile is a car is shown at the lower right in Fig. 5.17. Above it are two examples of simulated Rayleigh fading based directly on the mathematical model. Different fading rates are evident in the figure.

Analysis of Fading Rate. Fading rate is the most important aspect of Rayleigh fading, and we will give now a relatively nonanalytical view of it that is used in applications. By fading rate we mean the comparatively slow (10–1000 Hz) variation with time that would occur in the RF envelope if a pure sine wave were transmitted. The variation stems from one delay shell; it is distinct from variations due to other

[10]Fading rate is technically defined as the number of times per second that the envelope crosses a set level, such as the median. Recall that a fading rate can exist even with fixed terminals, since the medium can be in motion.

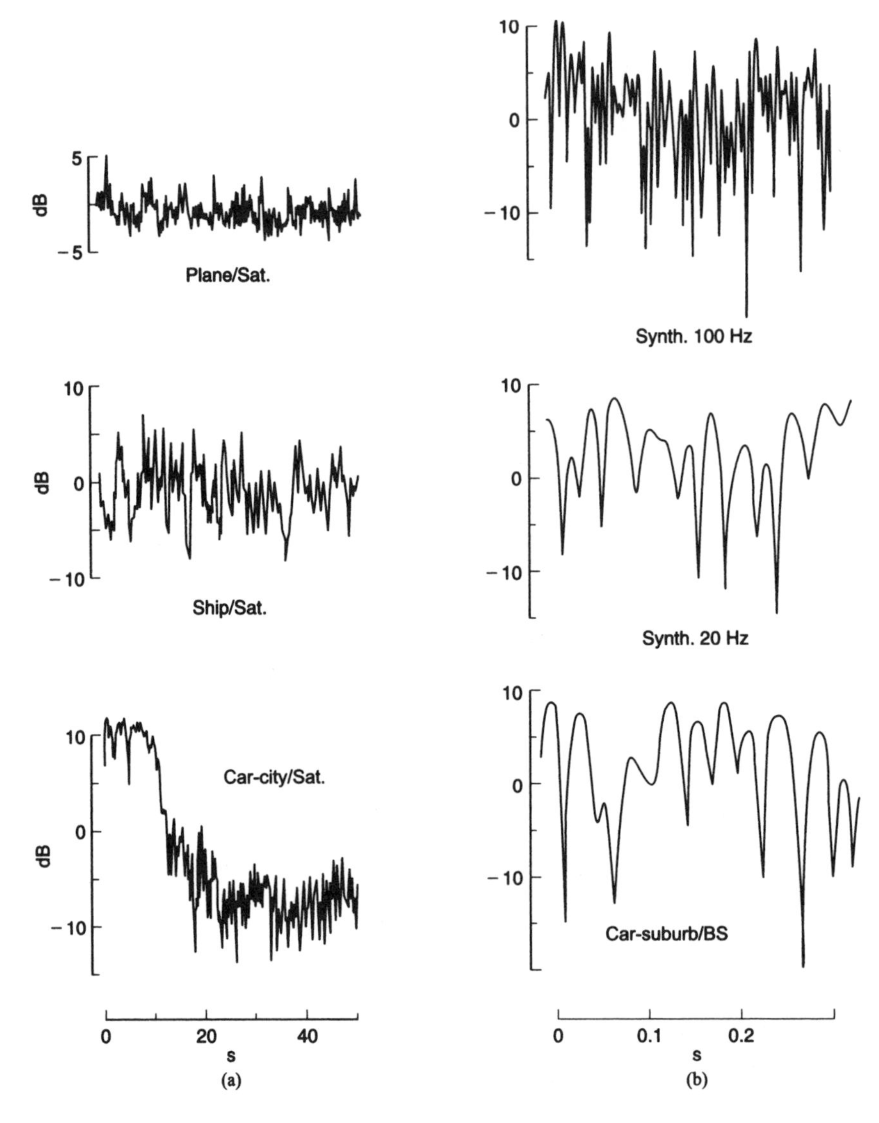

Figure 5.17 Rician and Rayleigh fading examples, plotted in dB versus time. All examples have the same vertical scale. **(a)** *Rician examples*. All are data at 1.54 GHz; antenna gains 3–5 dB. The Rayleigh character grows from top to bottom. *Top:* Plane to satellite, speed 600–800 km/h, satellite elevation 5°. *Middle:* Ship to satellite, speed 20 km/h, elevation 4°. *Bottom:* Car to satellite, speed 10–20 km/h, elevation 26°; large drop is from obstruction to line of site. (Adapted from data of E. Lutz et al., German Aerospace Research, Munich.) **(b)** *Rayleigh examples*. Note much wider swings. *Top:* Synthetic Rayleigh fading, baseband bandwidth 100 Hz; first method of Appendix 5A used. *Middle:* Synthetic Rayleigh as above, but 20-Hz bandwidth. At 25 km/h and 836 MHz v/λ is $\approx$ 20 Hz. *Bottom:* Measured Rayleigh fading, car to base station, suburban area at 25 km/h and 836 MHz. Compare to middle example. (Adapted from Jakes [15, p. 14]. Copyright ATT IMP Corp. Used by permission.)

shells (as in Section 5.4.3) or to modulation. We made the assumption earlier that the shell envelope process is stationary, and so an autocorrelation $R(\tau)$ may be defined for it, where τ is the time separation between two process samples. The Fourier transform of $R(\tau)$,

$$F(\nu) = \int R(\tau)e^{-j2\pi\nu\tau}d(\tau) \tag{5.4-7}$$

is called the *Doppler spectrum* of the fading, and with perfect single-shell dense scattering it will be the power spectral density in Fig. 5.16. Otherwise, a test RF carrier can be transmitted and the spectrum measured at the receiver. The support of $F(\nu)$ is called the *Doppler spread,* denoted B_{dopp}. It is usually enough to work with a rough estimate of this, say, the width at which $F(\nu)$ drops to half of its peak.

A quantity closely related to B_{dopp} is the coherence time, T_{coh}. Formally, this is the support of the autocorrelation $R(\tau)$. It is the approximate time that the response to a test sinewave, both phase and amplitude, remains constant. From Fourier transform properties, it must be that

$$B_{\text{dopp}} \approx \frac{1}{T_{\text{coh}}} \tag{5.4-8}$$

Coherence time—or, equivalently, inverse Doppler spread—measures the time the fading effect holds constant. *Slow fading* refers to long T_{coh} or narrow B_{dopp}, and fast fading is the opposite case. Naturally, the construction of a detector is far easier when the coherence time exceeds the detector processing time. Otherwise, the received pulse is distorted in phase or amplitude, and it is hard to obtain a phase reference or design a matched filter.

Figure 5.18b shows two car-to-satellite measured Doppler spectra, one at 4 km/h and one at 40 km/h. There is some LOS component here, which creates a peak at zero Doppler deviation. At the slower speed, a classic Doppler "horn" is visible at 8 Hz; the theoretical position, from (5.4-2), is $v/\lambda = (1.2 \text{ m/s})/19 \text{ cm} \approx 6$ Hz. The coherence time is about $1/B_{\text{dopp}} \approx 0.15$ s. No visible horn was measured at 40 km/h, but we can say roughly that B_{dopp} is 10–20 Hz. When a modulated signal is transmitted instead of a tone, a distorted spectrum will appear at the antenna terminals unless the modulated spectrum is much wider than B_{dopp} or, alternately, if the modulation symbol time is much less than $1/B_{\text{dopp}}$. In both cases in the figure, so long as the time is shorter than 0.05–0.1 s, there will be no variation in the channel during a symbol time.

Figure 5.18b corresponds to the fading record in Fig. 5.17a at the lower left. It is difficult to count fades precisely, but the rate appears to lie in the range 0.1–1 per second. In the pure Rayleigh fading in Fig. 5.17b, lower right, the fading rate is clearly 20–30 Hz. Here the speed and wavelength are 25 km/h and 36 cm, and the predicted Doppler spread is $v/\lambda = (7 \text{ m/s})/36 \text{ cm} \approx 20$ Hz.

If instead of time, fading intensity is plotted over distance traveled in wavelengths, the rate of fading depends only on speed. This can be derived as follows.

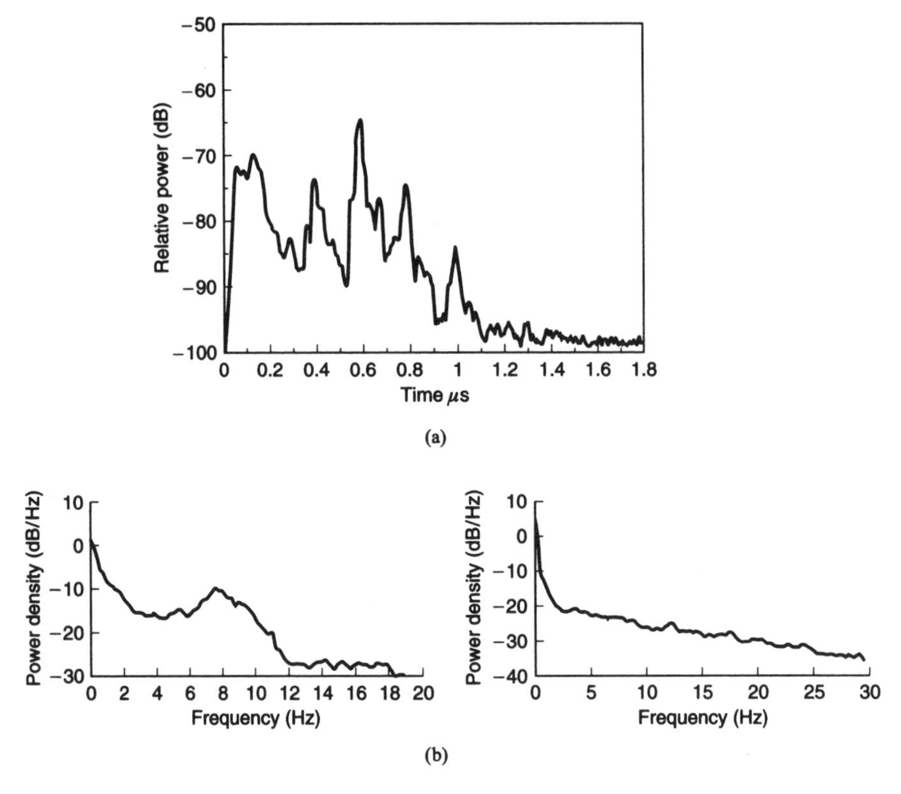

Figure 5.18 **(a)** Delay power profile measured inside a building at 850 MHz. (From Chuang [28]. Copyright IEEE. Used by permission.) **(b)** Measured Doppler spectra for car to satellite 1.54-GHz Rician channels. Satellite elevation 24°. City, speed 4 km/h (*left*); highway, speed 40 km/h (*right*). Note that Δf in Eq. (5.4-2) is 6 Hz at 4 km/h and 57 Hz at 40 km/h. Highway channel has strong LOS component. (From Lutz et al. [21]. Copyright IEEE. Used by permission.)

Replace the Fourier relation in (5.4-8) with the formula $T_{\text{coh}} = 1/4B_{\text{dopp}}$. T_{coh} should be the approximate time over which signal amplitudes are first unrelated. The distance between nulls should then be close to $2vT_{\text{coh}} = v/2B_{\text{dopp}}$. Replacing B_{dopp} with the theoretical maximum Doppler shift v/λ, we get a distance $\lambda/2$ meters. This is what is in fact commonly observed: Rayleigh fades occur every half-wavelength. The constants in the argument have been chosen to give this answer. The half-wavelength span gives the relation

$$\text{Fade rate} \approx \frac{2v}{\lambda} \qquad \text{fades/s} \tag{5.4-9}$$

Rician Fading. When a nonfading LOS ray superposes on a Rayleigh fading reception, Rician fading occurs. If $A\cos\omega_0 t$ is the LOS ray, the new envelope value at time t is $a = \sqrt{(I(t) + A)^2 + Q(t)^2}$. Define the ratio $K = A^2/2\sigma^2$, the ratio of LOS to Rayleigh power. The LOS component has power $A^2/2$; when $A = 0$ we have the Rayleigh fading in (5.4-3), and $\sigma^2 = \mathcal{E}\{|B_n|^2\}/2$ is the power in this case. It can be shown that the distribution of a in the general case is the Rician distribution

$$p(a) = \frac{a}{\sigma^2} e^{-(K+a/2\sigma^2)} I_0\left(\frac{\sqrt{2K}a}{\sigma}\right) \tag{5.4-10}$$

Here $I_0(\cdot)$ is the modified Bessel function of order zero. It can be seen that (5.4-10) reverts to (5.4-5) when $K = 0$ [use $I_0(0) = 1$]. The received phase is no longer uniformly distributed.

Rician fading is present in varying degrees in the fading examples at the left in Fig. 5.17. The transmitter is a satellite and consequently is visible to the plane, ship, and car. The fading is 10–20 dB less in these Rician examples, compared to the Rayleigh ones.

Several other distributions are used in more specialized fading models [23, 29].

5.4.3 Time Dispersion in Fading Channels

We have seen that Rayleigh fading widens the bandwidth of a sinusoidal tone and that this appears in the time domain as a relatively slow time variation in the signal intensity. Fading also introduces a *dispersion over time,* because returns from different reflectors and scatterers take different times. The chief impairment is to the modulation, and so we now modulate the sine wave in the previous section with an information-bearing signal. The modulation symbol time is T_s.[11] The delay shells in Fig. 5.14 will spread out and distort the pulses.

Figure 5.18a is a plot of the average impulse response energy against time for an 850-MHz indoor radio channel.[12] $h(\tau)$ can represent the response from the shell with delay τ. But we need to be more careful here because the value of $h(\tau)$ is the outcome of a stochastic process, since the response of the shells is free to evolve in time. The shell response at τ should be written as $h(\tau|t)$, a response parameterized in time t. We will call t the clock time; the whole fading medium evolves with t. Finally, we assume that the evolution is stationary. Then we can define an autocorrelation

$$R_{hh}(\tau) = \mathcal{E}_t\{|h(\tau|t)|^2\} \tag{5.4-11}$$

the time average of $|h(\tau|t)|^2$ over t. A plot of measured $|h(\tau|t)|^2$ versus τ, called a delay power profile, is a common way to measure the multipath part of fading. The dispersion in Fig. 5.18a is about 1 μs. This value (or the rms width or some similar

[11]Here it is easiest to think about linear (pulse) modulation. With nonlinear or constant envelope modulation, we take T_s as the modulation response time to a transmitted symbol.

[12]The LOS transmission time is the time-axis zero and is subtracted from all times in the discussion to follow.

quantity) is called the *multipath spread,* or delay spread, of the channel, T_{mult}. Profiles like the figure are measured by transmitting a very short burst of carrier, observing the response, and then averaging over many trials.

If T_{mult} is considerably less than T_s, then time dispersion has little effect on detection. This condition is called *flat fading*. Another view is to take a Fourier transform of $R_{hh}(\tau)$. What results is a kind of typical power spectrum of the h response (the precise spectrum at each clock time t would be the transform of $h(\tau|t)$ at that t). The support of $\mathcal{F}\{R_{hh}(\tau)\}$ (or the rms width or similar) is B_{coh}, the *coherence bandwidth* of the fading. From Fourier symmetries we know that

$$B_{\text{coh}} \approx \frac{1}{T_{\text{mult}}} \tag{5.4-12}$$

Flat fading, then, is the case where B_{coh} considerably exceeds $1/T_s$, which we can take as the approximate modulation bandwidth. The dispersion acts as a "flat" filter that has little effect on the transmission.

When the condition $B_{\text{coh}} < 1/T_s$ holds, the fading is said to be *selective*. Now the dispersion, viewed as a filter, can strongly modify the modulation spectrum. If the dispersion evolves slowly and we can identify $h(\tau|t)$, we can hope to correct the filter effect at the receiver. Circuits that do this are called equalizers and are the subject of Sections 7.1–7.3.

Different coherence bandwidths apply in practical fading channels. We saw that indoor radio can have a bandwidth around $1/T_{\text{mult}} = 1/1\ \mu\text{s} = 1$ MHz. The bandwidth B_{coh} for a 900 MHz cellular channel lies usually in the 0.3- to 1-Mhz range, but can be much less in the presence of distant reflectors like mountains. Shortwave radio B_{coh} moves about in the 0.5- to 10-kHz range; this is often audible in voice transmissions, which are typically 8 kHz wide. Fading in the shortwave bands stems from the ionosphere, whose troublesome nature presents a special challenge. Multipath spreads are longer, and fading tends to be selective. Schwartz et al. [22], Section 9.3, gives a particularly good survey of the shortwave channel. Other challenging fading channels are the underwater channel, which has a very long dispersion, and the tropospheric-scatter channel. In each case the modulation bandwidth should lie, if possible, inside the coherence bandwidth.

Summary. We can sum up these deliberations now and place them in the context of the fast and slow fading that arose in Section 5.4.2. Figure 5.19 shows the independent time variation and time dispersion dimensions of fading as a slow/fast axis and a flat/selective axis. The picture next repeats in the frequency domain. These are equivalent; the first is driven by T_{coh} and T_{mult}, respectively, and the other by B_{dopp} and B_{coh}. Fading, for example, can be flat and slow, meaning that the fades occur slowly and apply equally to a wide frequency band. The worst case is fast selective fading: Signal amplitude and phase change rapidly and change differently in each part of the tranmission band.

What constitutes "fast" or "selective" depends on the symbol time T_s. It needs to be longer than the multipath (so the dispersion is insignificant) but shorter than the

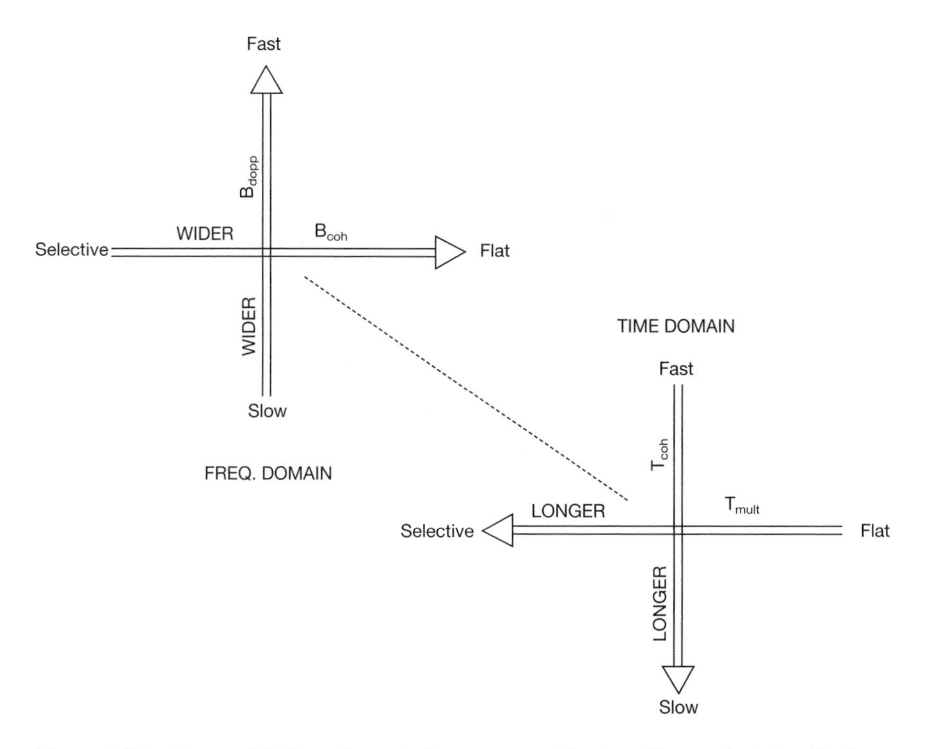

Figure 5.19 Types of fading, shown in frequency and in time. The symbol time T_s places the signal somewhere in these two planes.

signal variation (so the receiver can adapt to each local reality). This relationship is expressed by

$$T_{\text{mult}} < T_s < T_{\text{coh}} \tag{5.4-13}$$

it can as well be expressed as

$$B_{\text{coh}} > \frac{1}{T_s} > B_{\text{dopp}} \tag{5.4-14}$$

This says that the modulation bandwidth must be less than the coherence bandwidth but greater than the Doppler.

An interesting way to appreciate (5.4-13)–(5.4-14) is to plot fading signal intensity as a function of time and frequency. This is done for synthetic Rayleigh fading in Fig. 5.20. The picture shows the amplitude (log scale) that applies to signals at range of relative frequencies and times (an example might be carrier plus 0–10 kHz and 0–0.1 s). In this picture, T_{coh} is about 10 time units and B_{coh} is about 25 frequency units; that is, there are about 10 major changes in time and 4 major changes

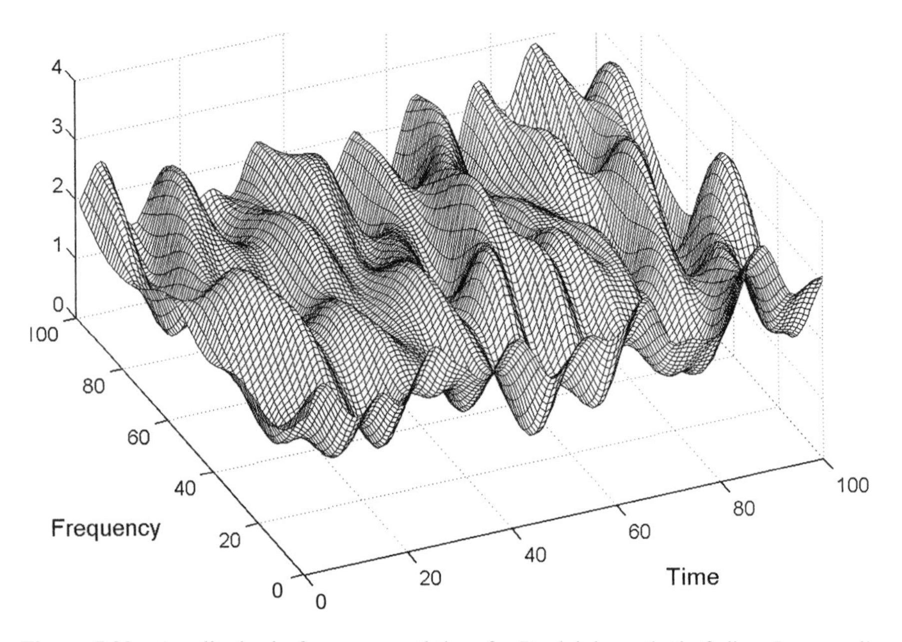

Figure 5.20 Amplitudes in frequency and time for Rayleigh synthetic fading. Log amplitude scale and arbitrary frequency and time scales. Variation in time (fading speed) is 4 times faster than variation in frequency (fading selectivity).

in frequency. One symbol's worth of a transmission should occupy less than a hilltop or a valley floor in the picture.

Example 5.4-1: Fading in a GSM Mobile Telephone. In the GSM second generation digital telephone system, the symbol rate $1/T_s$ is 271 ksymbols/s, which consumes a total RF bandwidth of about 250 kHz (see also Example 3.7-1). Data bits (about 140) are organized into slots of length 0.577 ms, eight of which form a frame of 4.62 ms. A typical rms delay spread is 2–4 μs, the lower figure applying to urban fading, the higher to rural. The Doppler formula (5.4-2) gives (28 m/s)/(33 cm) = 84 Hz for a 900-MHz carrier and 100-km/h (28-m/s) speed. Applying (5.4-14), we get that

$$250\text{–}500 \text{ kHz} > 1/T_s > 84 \text{ Hz}$$

The 271-ksymbol/s rate of GSM easily exceeds the right-hand Doppler limit but barely avoids the left-hand coherence limit. In reality, GSM receivers do encounter intersymbol interference due to multipath, and they include an equalizer to mitigate it (see Chapter 7). Another important question is how often the GSM channel medium undergoes a major change. This is the shadowing question, which we will take up in the next section. The GSM receiver identifies and adapts to the ISI and the signal phase and amplitude once each 0.577-ms slot. The channel should therefore not change much during a slot. Changes stem from two sources, namely, shadowing and standard Rayleigh fading. The second we have taken as 84 Hz, but it is more like 200 Hz at train and daredevil car speeds. GSM is also used in a higher 1900-MHz band, which doubles this again to 400 Hz. Shadowing changes are probably slower. To conclude, GSM barely meets the multipath spread limit in (5.4–14); and its system is such that its slot rate (1733 Hz), not its symbol rate, must fall below the Doppler limit, which it does by a good margin.

Before leaving this discussion, we should emphasize that a fading signal sometimes simply disappears. Equations (5.4-13)–(5.4-14) show how to avoid multipath and Rayleigh time variability, but when the signal disappears in a deep fade, communication is not possible. There is no channel! Because of deep fades, fading error probability is inversely proportional to SNR, just $1/(E_b/N_0)$ instead of the exponential dependence in Chapters 2 and 3. In Section 7.4 we will return to this error probability and its mitigation.

5.4.4 Long-Term Model Variation: Shadowing

In fading, the environment changes as the mobile moves from place to place. The medium can shift as well and, especially in the case of the ionosphere, be in constant motion. These changes shift the dispersion and especially the phase and amplitude of the mechanisms in the previous two subsections.

The mean value of the Rayleigh distribution (5.4-5) is $\sigma\sqrt{\pi/2}$. For brevity we denote this as γ. The quantity γ is an average over the effects of random reflection and scattering, as seen in one place. This is to be distinguished from gross variations in local signal level, such as occur when the mobile moves out from behind a building or simply gains distance from the transmitter. These variations can look like Rayleigh fading, but they are slower and fundamentally different in origin.

A consistent observation in real mobile channels is that the deviations in the local mean value γ are *lognormally* distributed. This means that log γ has a normal distribution as the mobile moves from place to place. It is convenient to express this log in dB, as $\gamma_{\text{dB}} = 20 \log_{10} \gamma$, where γ_{dB} is distributed as

$$p(\gamma_{\text{dB}}) = \frac{1}{\sqrt{2\pi}\,\sigma_{\text{dB}}} e^{-\gamma_{\text{dB}} - m_{\text{dB}})^2/2\sigma_{\text{dB}}^2} \qquad (5.4\text{-}15)$$

The lognormal distribution describes the so-called "excess path loss," the cubic or other propagation loss minus the actual local average value. The variance σ_{dB}^2 does not depend much on frequency, distance, or antenna configuration, although the mean m_{dB} does. σ_{dB}^2 is typically in the range 6–10, meaning that the standard deviation of γ in dB is 6–10 dB. The mean m_{dB}, expressed as an excess loss over basic propagation, is perhaps 35 dB (rural areas) to 45 dB (cities).

The lognormal distribution of the local mean still applies when the local fading is Rician. In either case the envelope value is distributed as

$$p(a) = \int p(a|\gamma) \frac{20}{\gamma\sqrt{2\pi}\,\sigma_{\text{dB}}\ \ln 10}\, e^{-(\log_{10}\gamma - m_{\text{dB}})^2/2\sigma_{\text{dB}}^2}\, d\gamma \qquad (5.4\text{-}16)$$

Here $p(a|\gamma)$ is either (5.4-5) or (5.4-10), and the rest comes about from distribution (5.4-15) with a change of variable from γ_{dB} to γ. Some measurements of lognormal parameters are found in Refs. 15 and 25. These statistics, with (5.4-16), allow a signal at or above a given level to be guaranteed with a specified probability.

Whatever the source of change, fading channel parameters continually shift. It is hoped that they will not shift faster than the receiver can adapt. Figure 5.21 illustrates the transmission of a simple stairstep waveform through a changing channel. The changes in the figure are rapid, to provide a better example. The stairstep is in fact ruined.

Here is a step-by-step explanation of how all the contributions in the figure add up. To make the exposition as concrete as possible, we consider only the baseband I component rather than the whole passband signal. Time passes in small intervals δt. The model is governed by a clock k that marks the integer multiples of δt. Once each k, the channel impulse response changes, and the succession of these responses is plotted in the middle of the figure. The response sequences are denoted . . . $\beta(y|-1)$, $\beta(y|0)$, $\beta(y|+1)$, . . . , with the second digit denoting the clock time. Each sequence $\beta(0|k)$, $\beta(1|k)$, . . . is the response to a pulse at clock time k. Each element is a Gaussian random variable. It is the nature of fading that these variables can be correlated in either the y or the k dimension, or in both.

The figure breaks the transmitted $I(t)$ down into four pulses and shows at the right the response to each one. In a physical channel, the first ray to arrive is the LOS one.[13] Without harm to the model, we have simplified the time keeping by subtracting the LOS propagation delay at the receiver. The responses add at the receiver as shown, to produce the received $I'(t)$. Because the channel time dispersion is of the same order as the duration of $I(t)$, the received signal is strongly distorted, both in time and amplitude.

The responses in the top part of Fig. 5.21 are from the point of view of the sender: An impulse at clock time k produces the response shown, which evolves in future time; a response beginning at $k + 1$ evolves though most of this time, too, but it need not take the same evolution. It is often more convenient to view the time-varying evolutions from the *receiver* perspective. At time k at the receiver, the channel response appears as $\beta(0|k)$ times the $I(t)$ pulse at time k, plus $\beta(1|k-1)$ times the I pulse at time $k-1$, plus $\beta(2|k-2)$ times the pulse at time $k-2$, and so on. This is because of the spread of channel delays. We can define a response from the receiver point of view at clock time k as

$$h(z|k) = \beta(z|k-z), \qquad z = 0, 1, 2, \ldots, \qquad (5.4\text{-}17)$$

[13]Note that in a medium with changing refraction index, the LOS path may be curved.

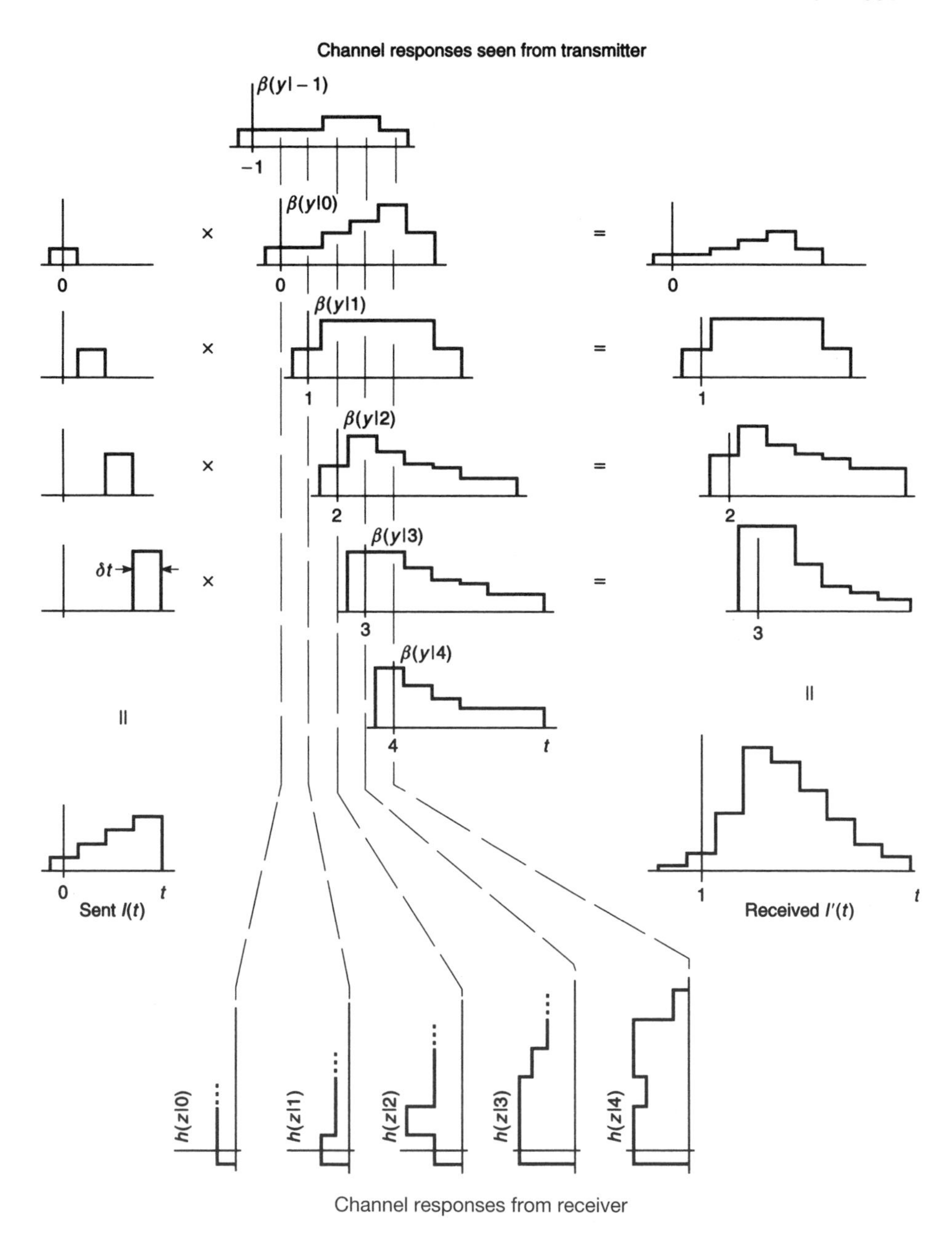

Figure 5.21 Illustration of transmission of four-sample waveform through time-varying channel; each sample pulse produces different output shape at right. In the middle is a set of channel impulse responses as seen from transmitter; across the bottom are responses as seen at receiver.

where z is time evolution at the receiver. When the β's are side-slipped as in the figure, the h response simply runs up the columns in the pattern. The outcome is another set of responses, this time from the receiver perspective, in which the channel clock time and dispersion in the transmission time have been transposed.

In terms of h responses, the received signal is the convolution

$$I'(k) = \sum_{\tau=0}^{3} h(k-\tau|k)I(\tau)$$

For example, $I'(3) = h(3|3)I(0) + h(2|3)I(1) + h(1|3)I(2) + h(0|3)I(3)$. At this point we can pass to the integral and write the convolution as

$$I'(t) = \int_{-\infty}^{\infty} h(t-\tau|t)I(\tau)\, d\tau \qquad (5.4\text{-}18)$$

Except for the fact that the response can change each t, this is the usual form of a linear filter response, where the response of the filter is $h(\tau|t)$. We can take the Fourier transform for clock time t as well and obtain

$$H(f|t) = \int h(\tau|t)e^{-j2\pi f\tau}\, d\tau \qquad (5.4\text{-}19)$$

As with any linear filter, $I'(f|t) = H(f|t)V(f)$, and an inverse transform gives back $I'(t)$. The difference now is that we must change the $H(f|t)$, potentially, at each t.

Conclusion. These thoughts can be continued, and a full stochastic process model of fading can be built up. The model is complex, and a standard simplification is to assume at least stationarity, both in the evolution of the dispersion and the Rayleigh amplitude/phase. The stationary model is introduced in Appendix 5B. Most of the time, however, it is enough to characterize fading in terms of the coherence and Doppler parameters and then base a design on the limits expressed in Eqs. (5.4-13)–(5.4-14). This will evade the worst effects of fading. It remains to combat the occasional disappearance of the signal and the ISI from time dispersion. This is done by diversity transmission and equalization, which are covered in Sections 7.1–7.4.

PROBLEMS

Wire and Space Channels

5-1. (a) Using the data for R, L, and C for case 3 in Table 5.2, find theoretical values for the attenuation of that copper wire pair and compare to the measured values in the table. (This heavy-gauge spaced wire pair is meant to model two separate wires strung on poles.)

(b) Plot transfer function absolute value and group delay up to 0.1 MHz for this pair at 10 km, and compare to Fig. 5.2. (Interpolate R values in the table.)

(c) Find the speed of propagation at 3300 Hz.

5-2. It is desired to send 10 Mb/s for a short distance, and the only convenient channel is a 22 AWG wire pair. Making reasonable design assumptions concerning modulation, signal levels, and receiver noise, estimate the longest distance over which this transmission can be made. At what characteristic impedance will your system run?

5-3. Find the free-space loss in decibels for a geostationary-orbit satellite channel at 4 GHz. Repeat for 12 GHz and 30 GHz.

5-4. A standard frequency pair for commercial satellite use is 4 and 6 GHz, one for the up link and one for the down link. Assume that a 4 m dish antenna is installed on the satellite.

(a) Find the antenna gains at these two frequencies. Which one would you use for the down link?

(b) Repeat for the next standard pair, which is 12 and 14 GHz.

Noise

5-5. Suppose an RF amplifier in a receiver has a noise temperature of 200 K and a power gain of 10. The remainder of the receiver has a bandwidth of 10 kHz, and so this is the bandwidth of interest. The receiver has a 50-Ω input resistance, which presents a matched load to the antenna. When no RF signal is present, an open-circuit rms voltage of 3.7×10^{-8} is observed at the antenna's terminals when the observation bandwidth is 10 kHz.

(a) What is the equivalent noise temperature T_{ant} of the antenna?

(b) What is the system noise temperature of the antenna-RF amplifier combination?

(c) The antenna is connected to the receiver, an RF signal is picked up, and 10 μV rms is observed at the RF amplifier input. What signal-to-noise power ratio will be observed at the RF amplifier output? (Assume here that the rest of the receiver is connected and that it presents a matched 50 Ω load while the measurement is being made.)

5-6. An antenna is connected to a preamplifier with noise figure $F = 1.6$ dB, gain 20 dB, and bandwidth 10 MHz. The antenna is pointed at the sky in such a way that it has a $T_{\text{ant}} = 30$ K.

(a) Find the effective noise temperature of the combination.

(b) Find the available noise power at the preamplifier's output.

5-7. A good shortwave receiver can achieve an output signal-to-noise ratio of 10 dB with an RF input voltage of less than 0.5 μV. For example, the ICOM Corporation R71 series receivers claim that during single sideband operation, 0.15 μV at the 50-Ω antenna terminal will produce an SNR of 9.5 dB. Estimate the noise factor and equivalent temperature of this receiver. Some assumptions are needed. The half-power bandwidth of the receiver is close to 2 kHz under SSB operation. There are a number of filters and protection circuits that come between the antenna terminal and the first RF amplifier; assume that these act as an attenuator and have a 2-dB loss. Assume that 0.15 μV refers to the peak value of an input test sinusoid; for a power ratio you thus need to compute the rms value. Assume that all stages are matched at 50 Ω.

5-8. A low-noise satellite microwave receiver is made up of the following elements, connected in the order listed: (1) an antenna pointed at space, with noise temperature T_{ant} = 15 K; (2) a short piece of waveguide at room temperature with loss 0.25 dB, which connects the antenna focus to (3) a maser amplifier with noise temperature 4 K and gain 20 dB; this leads to (4) a cheaper amplifier with noise factor 6 dB and gain 20 dB, which connects to (5) a down-converter (mixer) with noise factor 12 dB and gain 0 dB. Find the system noise temperature of this receiver; that is, include the antenna.

5-9. Consider a simple circuit consisting of a perfect capacitor C in parallel with a resistor R at T Kelvin. Find the rms open circuit noise voltage across C in terms of k, T, R, C. What voltage appears at 290 K?

5-10. Three amplifier units are to be connected in sequence. Units 1, 2, and 3 have equivalent noise temperatures 100 K, 150 K, and 200 K, respectively, and power gains 2, 10, and 50.

(a) What is the best arrangement of the units? Give the noise equivalent temperature.

(b) The total gain here is 1000. If instead of gains 2, 10, and 50, you are free to assign gains to the units in any way such that the total is 1000, what is the best assignment?

5-11. A cheap receiver has a noise factor F = 15 dB. It is desired to upgrade the receiver to 10 dB, by placing a preamplifier before it. Preamplifiers with a 6-dB noise factor are available. What gain must the preamplifier have?

Reflection and Diffraction

5-12. A recent contest among radio amateurs involved setting distance records for communication at 10 GHz over a terrestrial path without line of sight. The propagation works because a mountain lies in between and there is diffraction from that. Consider a 100-km path with a 1000-m mountain halfway down the path; take the receiver and transmitter as being at sea level. Using the theory of knife-edge diffraction in Section 5.3, compute the loss over free space in this path. Be careful here: Diffraction can occur from the mountain or from any object on its surface. What size object is of interest in this diffraction? How is path loss affected by the position of the mountain? Where is the best spot for the mountain?

5-13. Prove Eq. (5.3-2), that is, show that under simple reflection interference in a terrestrial link, the ratio of the received power to the direct ray power alone, P_r/P_0, is approximately $(4\pi h_t h_r/\lambda d)^2$. Give the conditions on the heights h_t, h_r, the distance d, and the wavelength λ under which the approximation is valid.

Link Budgets

5-14. A microwave transmitter has an output of 0.1 W at 2 GHz. Assume that it is used in a link where the transmit and receive antennas are 1.3-m-diameter dishes. The receiver acts as if its equivalent noise temperature is $T_e = 400$ K.

(a) Find the gain of these antennas, assuming that it falls short of theory by the factor 0.54.

(b) Find the EIRP.

(c) If the antennas are 24 km apart, find the received power P_r under free-space conditions.

(d) Allow a 40-dB fade margin and assume that all other losses are 0 dB. Let the digital modulation be 16QAM working at a symbol error rate of 10^{-5}. Find the maximum data rate in bits/s that can be supported under these conditions.

(e) How will these answers change if the path length is 48 km?

5-15. In Problem 5-18 a short television link with terrestrial impairments is analyzed. As a comparison for the answers there, assume that the transmission is instead line of sight in free space. Find the required signal power C. In addition to the information given in Problem 5-18, make the following assumptions:

- The receiver T_e is 400 K.
- The television signal bandwidth is 6 MHz.
- The C/N required in this bandwidth is 20 dB, where N is the noise power in 6 MHz.

5-16. This problem calculates the link budget for a satellite system. Some of the information given may not be needed, and some may be missing so that assumptions are required. Some assumptions that you will probably make are: (*i*) because of high power on the ground, up-link transmission is perfect; (ii) noise enters the system only in the front end of the Earth receiver; (iii) there are no losses from channel filtering; (iv) the ratio of total signal power to total noise power (C/N) is numerically equal to the ratio E_b/N_o. With these provisos, the satellite system is to have 4-m antennas, both satellite and Earth; it runs at 4/6 GHz and is geostationary, that is, it is the system in Problems 5.3 and 5.4. System losses are measured at 5 dB. Front-end receiver noise measures –155 dBW. The bit rate is 10 Mb/s with BPSK modulation, and the error rate specification is 10^{-6}. Give a link budget for the down link that meets the error rate specification. If you cannot meet the specification in a reasonable way, propose a revised system that meets the specification.

5-17. Some of the longest terrestrial microwave links occur over deserts that are surrounded by mountains. Perform such a link budget analysis for a 200-km link operating at 4 GHz. Assume to start that the reliability is 0.999999 and that antennas are both 2 m in diameter. Use the Barnett–Vignant formula to estimate the fade margin needed. Calculate the needed transmitter power. Redo with other specifications if you feel the power required is not acceptable (a suggested range is 1 mW–10 W). Propose a reasonable bit rate for your system.

5-18. A local television station sends out a mobile unit that transmits live video from special events, accidents, and so on. The transmit antenna is a 0.5-m dish that sits 10 m high on a truck. The receive antenna is a 1 m dish and is also 10 m high. Assume 50-km distance, flat terrain, and a temperate climate like that of Northern Europe. The frequency is 4 GHz. Transmission is line of sight, but fading occurs due to reflection and rain. Calculate a fade margin for this link in two ways:

(a) Use the Barnett–Vignant formula.

(b) Use Eq. (5.3-3) to compute a fade margin. This gives the power needed to combat solely reflection, compared to free space.

5-19. A 50-W mobile radio base station is to be set up at 1800 MHz. As a first attemp at a design, the link is to be designed on the basis of simple line of sight transmission. Assume the following:

Transmit dish antenna diameter 1 m
Receive antenna gain 0 dB
Distance less than 10 km
Rain, pointing, and circuit losses are 20 dB total
QPSK modulation; $P_e \approx 10^{-5}$

Each mobile channel requires 16 kbit/s. Estimate how many such channels can be supported. (The solution depends on the noise factor F, or alternately the N_0, of the mobile receiver. Make two assumptions for this, say, $F = 2$ and 8 dB.

Mobile Radio

5-20. Equation (5.4-5) gives the Rayleigh density function $p(\text{a})$ for parameter σ. Here a is the amplitude of a Rayleigh faded signal.

(a) Show that the cumulative distribution function $Pr\{A < a\}$ is

$$1 - e^{-a^2/2\sigma^2}, \qquad a \leq 0$$

(b) Now find the probability that an observation of a Rayleigh variable is below its mean.

(c) Find the probability that it is 10 dB below the mean.

(d) Suppose there are two independently fading transmissions of the same information; this is called diversity transmission in Section 7.4. What is the probability that *both* are 10 dB below the mean.

5-21. Consider transmission at 120 MHz (an air traffic control frequency) to an airliner moving at 1000 km/h. The airline does not have good line of sight to the ground station, and consequently the channel obeys the Rayleigh fading model. For a transmitted sine wave tone at 120 MHz, calculate and plot the theoretical observed Doppler spectrum.

5-22. In the previous problem, assume that the spectrum there is the power spectral density of a Rayleigh fading stochastic process. By inverse Fourier transformation (possibly numerical), find the delay power profile of the process and the rms delay spread T_{mult}. Express these for a general frequency–speed product and for the 120-MHz, 1000-km/h combination in particular. (What you have found here is the delay spread of the standard dense scatterer fading model.)

5-23. Derive an expression for the distribution of phase in a Rician fading signal. [The amplitude distribution appears in Eq. (5.4-10)].

5-24. Using two-dimensional transforms, show that the autocorrelation (5B-3) and the coherence transform (5B-5) are a Fourier transform pair.

5-25. By means of MATLAB or a similar package, generate and plot the outcomes of a Rayleigh fading envelope process. Do this by generating the I and Q components as Gaussian processes like those in Eq. (5.4-4). Set the bandwidth of the Rayleigh process by controlling the bandwidth of the Gaussians, and repeat for bandwidths 10, 100, and 500 Hz.

5-26. A communication system for high-speed trains is to be designed. The operating frequency is 2 GHz and the multipath spread is $\leq 20\ \mu$s. The fading is assumed to be Rayleigh. Specify reasonable symbol and slot rates for a GSM-like system (that is, 8 user slots share a frame in time division, a frame contains about 150 symbols, and there should be little change in the channel during a slot time). Justify your answers.

5-27. Cellular communication systems are designed for a compromise scenario that does not hold in extreme situations. Consider the GSM system in Example 5.4-1. Take the 1900 MHz operating frequency. How would you modify the design for the following extreme scenarios?

(a) The multipath spread is 50 μs. (This can occur in mountainous areas such as Zürich or San Francisco.)

(b) The speed is 330 km/h. (This occurs on high-speed trains.)

(c) The mobile is in a long tunnel.

REFERENCES[14]

1. *E. A. and D. G. Messerschmitt, *Digital Communication,* 2nd ed., Kluwer, Boston, 1994.
2. *Reference Data for Radio Engineers,* various editions, H. W. Sams, New York.
3. *Members of Bell Laboratories Staff, *Transmission Systems for Communications,* various editions, ATT, Inc.

[14]References marked with an asterisk are recommended as supplementary reading.

4. F. P. Duffy and T. W. Thatcher, Jr., 1969–70 Connection survey: Analog transmission performance on the switched telecommunications network, *Bell Syst. Tech. J.,* vol. 50, April 1971, pp. 1311–1347.
5. M. B. Carey, H.-T. Chen, A. Descloux, J. F. Ingle, and K. I. Park, 1982/83 End Office Connection Study: Analog Voice and Voiceband Data Transmission Performance Characterization of the Public Switched Network, *ATT Bell Lab. Tech. J.,* vol. 63, Nov. 1984, pp. 2059–2119.
6. W. D. Reeve, *Subscriber Loop Signaling and Transmission Handbook,* IEEE Press, New York, 1995.
7. M. K. Simon, S. M. Hinedi, and W.C. Lindsey, *Digital Communication Techniques,* Prentice-Hall, Englewood Cliffs, NJ, 1995.
8. H. Nyquist, Thermal agitation of electric charge in conductors, *Phys. Rev.,* vol. 32, 1928, pp. 110–113.
9. J. B. Johnson, Thermal agitation of electricity in conductors, *Phys. Rev.,* vol. 32, 1928, pp. 97–109.
10. *M. Schwartz, *Information Transmission, Modulation, and Noise,* 4th ed., McGraw-Hill, New York, 1990.
11. W. R. Bennett, *Introduction to Signal Transmission,* McGraw-Hill, New York, 1970.
12. R. S. Carson, *Radio Communications Concepts: Analog,* John Wiley & Sons, New York, 1990.
13. H. Stark, F. B. Tuteur, and J. B. Anderson, *Modern Electrical Communication,* 2nd ed., Prentice-Hall, Englewood Cliffs, NJ, 1988.
14. S. D. Personick, *Fiber Optics Technology and Applications,* Plenum, 1985.
15. *W. C. Jakes, Jr., ed., *Microwave Mobile Communications,* John Wiley & Sons, New York, 1974; reissued by IEEE Press, New York, 1995.
16. E. C. Jordan, *Electromagnetic Waves and Radiating Systems,* Prentice-Hall, Englewood Cliffs, NJ, 1950.
17. S. Ramo, J. R. Whinnery and T. Van Duzer, *Fields and Waves in Communication Electronics,* John Wiley & Sons, New York, 1965.
18. A. Vignants, Space diversity engineering, *Bell Syst. Tech. J.,* vol. 54, January 1975, pp. 103–142.
19. W. T. Barnett, Multipath propagation at 4, 6 and 11 GHz, *Bell Syst. Tech. J.,* vol. 51, February 1972, pp. 321–361.
20. S. O. Rice, Mathematical analysis of random noise, *Bell Syst. Tech. J.,* vol. 23, July 1944, pp. 282–332; vol. 24, January 1945, pp. 46–156. Also see Statistical properties of a sine wave plus random noise, *Bell Syst. Tech. J.,* vol. 27, January 1948, pp. 109–157.
21. E. Lutz et al., The land mobile satellite communication channel—recording, statistics, and channel model, *IEEE Trans. Vehicular Technology,* vol. VT-40, pp. 375–386, May 1991.
22. *M. Schwartz, W. R. Bennett, and S. Stein, *Communications Systems and Techniques,* McGraw-Hill, New York, 1966; reissued by IEEE Press, New York, 1995.
23. *J. Proakis, *Digital Communications,* 3rd ed., McGraw-Hill, New York, 1995.
24. *B. Sklar, *Digital Communications: Fundamentals and Applications,* Prentice-Hall, Englewood Cliffs, NJ, 1988; *see also* Rayleigh fading channels in mobile digital communication systems, Parts 1 and 2, *IEEE Commun. Mag.,* vol. 35, no. 9, September 1997, pp. 136–155.

25. S. H. Jamali and T. Le-Ngoc, *Coded-Modulation Techniques for Fading Channels,* Kluwer, Boston, 1994.
26. J. H. Yuen, ed., *Deep Space Telecommunications Systems Engineering,* Plenum, New York, 1983.
27. E. C. Posner and R. Stevens, Deep space communications—Past, present, and future," *IEEE Commun. Mag.,* vol. 22, no. 5, May 1984, pp. 8–21.
28. J. C.-I. Chuang, The effects of time delay spread on portable radio communications channels with digital modulation, *IEEE J. Sel. Areas Commun.,* vol. SAC-5, June 1987, pp. 879–889.
29. *T. S. Rappaport, *Wireless Communications,* Prentice-Hall PTR, Upper Saddle River, NJ, 1996.
30. R. Steele, ed., *Mobile Radio Communications,* Pentech, London and IEEE Press, New York, 1992.

APPENDIX 5A: SIMULATION OF FADING

It might seem that the easiest way to observe fading is simply to go out and measure it, but in reality there is not much literature that reports such measurements. This is because the measurement of fading channels at VHF and higher is quite expensive. A GSM-type channel, for example, has a Doppler spread of perhaps 100 Hz and a carrier near 1 GHz. In order to measure frequency deviations accurately, a reference oscillator is needed with stability one part in 10^8–10^9 and a correspondingly narrow phase noise bandwidth (see Section 4.5). Such a frequency standard is not cheap. Consequently, most research and development with fading relies on fading channel simulation.

From the foregoing exposition, several simulator designs present themselves. Simple Rayleigh fading of the signal envelope can be obtained through generating a sequence of IID zero-mean Gaussian amplitude factors for the I and Q of the transmitted signals. The factors simulate Eq. (5.4-4) for a pure sine-wave transmission; multiplying them by the actual transmitted signal I and Q will simulate flat fading of a modulated transmission. By low-pass filtering the Gaussians, it is possible to simulate correlated Rayleigh fading.[15] This approximates the effect of vehicle speed, although multipath spread is more difficult to simulate this way. The simulated fading in Fig. 5.17 is obtained by this approach. Rician fading can be simulated by adding a constant to the I-channel attenuation factor. In either case, the simulated received signal at RF is created by

$$r(t) = I'(t) \cos \omega_0 t - Q'(t) \sin \omega_0 t$$

where I' and Q' are I and Q multiplied by the attenuation factors that have been generated.

[15]Note that the *Gaussians,* not the Rayleigh variates, should be filtered. Filtering the latter destroys their Rayleigh character.

The filtered Gaussian method is simple and often good enough, and so we give a MATLAB function to produce a sequence of correlated Rayleigh amplitudes.

Program 5A-1: Correlated Rayleigh Amplitudes. The function `y = raysim (L,ctf,fs)` returns a length L vector y of correlated Rayleigh amplitudes. The variable *ctf* is the bandwidth of the variables in hertz, and *fs* is the sample rate in hertz. The sequence thus occupies L/fs seconds. The program generates I and Q Gaussians and filters them with a 6-pole Butterworth filter whose 3-dB bandwidth is *ctf* Hz. The filter order and type can be changed in the second line. Applying a square root to $I^2 + Q^2$ produces the Rayleigh variates. Since the filter has a startup transient, the first 100 are ignored.

```
function y = raysim(L,ctf,fs)
%  Set up filter
  wn = 2*ctf/fs;                    %Normalized cutoff
  [b,a] = butter(6,wn);             %6-p Butterworth polynomials
%  Form sum of 2 filtered and squared Gaussians; then take sqrt
  y = filter(b,a,randn(1,L+100)).^2 + filter(b,a,randn(1,L+100)).^2;
  y = sqrt(y(101:(L+100)));         %Ignore first 100 & take sqrt
```

As an example, set L = 600, *ctf* = 50 Hz, and *fs* = 2000 Hz. Compute `y = raysim(L,ctf,fs)` and plot in decibels versus time with `plot (1/fs:1/fs:L/fs, 20*log10(y))`. The result will extend over 0.3 s and look like one of the two Rayleighs in Fig. 7.17.

Another simulation method is based on recreating the single-tone spectrum in Fig. 5.16. Alternately, a given Doppler spectrum can be approximated. Jakes [15] suggests simulating the desired spectrum with a bank of $2N - 1$ low-frequency offset oscillators as follows, each with a phase shift θ:

$$
\begin{array}{cc}
2 \sin \theta_1 \cos \omega_1 t & 2 \cos \theta_1 \cos \omega_1 t \\
2 \sin \theta_2 \cos \omega_2 t & 2 \cos \theta_2 \cos \omega_2 t \\
\cdots & \cdots \\
2 \sin \theta_{N-1} \cos \omega_N t & 2 \cos \theta_{N-1} \cos \omega_{N-1} t
\end{array}
$$

$$\cos \omega_N t$$

These are quadrature modulated to form the Rayleigh-faded passband signal

$$
\begin{aligned}
r(t) = 2\left[\sum_{n=1}^{N-1} \cos \theta_n t \cos \omega_n\right] \cos \omega_0 t - 2\left[\sum_{n=1}^{N-1} \sin \omega_n t \sin \theta_n\right] \sin \omega_0 t \\
+ \cos \omega_N t[\cos \omega_0 t - \sin \omega_0 t]
\end{aligned}
\tag{5A-1}
$$

Two sets of constants need to be chosen: The set of offset frequencies $\omega_1, \ldots, \omega_N$ are chosen such that the final output has the desired Doppler spectrum, and the set of phases $\theta_1, \ldots, \theta_{N-1}$ are selected so that the output has uniformly distributed phase. There are analytical ways to do this, or the selections can be made by trial and error. Only 5–10 sets of oscillators are needed. If transmission of a modulated signal is to be simulated, rather than a simple carrier, the brackets in (5A-1) respectively multiply the signal's I and Q components. The Jakes method is perhaps the most successful one for simulating motion-induced Doppler, but it still does not directly address multipath spread.

A third approach focuses directly on the time dispersion issue. As shown in Fig. 5.22, sampled I and Q signal components pass down a delay line. After each delay, Gaussian random variables that represent the in-phase and quadrature fading amplitudes multiply the components. The result of all the delays is summed to give the I and Q components of the final faded signal. In cellular radio simulations, it has been found that just two rays—with a single delay of 0.5–1 μs—produce an acceptable fading model; this has come to be called the *two-ray model.*

Ray models do not directly model speed-induced Doppler. This can be added to the model by filtering the Gaussians or, better, by replacing the Gaussians with a Jake's model generator at each delay. It also does not adequately simulate long dispersion channels, such as the underwater and ionospheric channels. These require special techniques.

APPENDIX 5B: STOCHASTIC PROCESS ANALYSIS OF FADING

As was introduced at the end of Section 5.4.4, we define the fade stochastic process by the impulse response $h(\tau|t)$; this is the dispersion over τ that applies to an impulse transmitted at system clock time t. Let $h^I(\tau|t)$ and $h^Q(\tau|t)$ apply to the I and Q components, respectively. Furthermore, both processes are uncorrelated and have the same statistics. The total bandpass fading process is

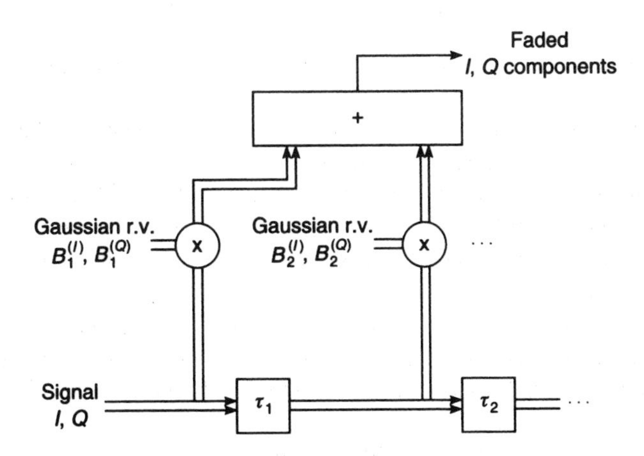

Figure 5.22 Simulation of fading by multipath ray model (r.v. = random variable).

$$\eta = h^I(\tau|t) \cos \omega_0 t - h^Q(\tau|t) \sin \omega_0 t \tag{5B-1}$$

(Bandpass processes of this kind are discussed in Appendix 3B.) At this point the standard analysis makes two simplifying assumptions: (i) The process η is stationary, both with respect to the clock time t and the delay parameter τ, and (ii) the components h^I and h^Q are wide-sense stationary. Fading that is stationary with respect to τ is said to be wide-sense stationary uncorrelated scattering (WSSUS). It holds in most fading, and it simply states that the scatterers in each delay shell work independently from those in other shells.

In general the autocorrelation of η is $R(\tau_1, \tau_2|t_1, t_2)$, a function of two delays and two clock times. From Theorem 3B-1 and the assumptions about h^I and h^Q, we get that

$$R(\tau_1, \tau_2|t_1, t_2) = \mathcal{E}[h^I(\tau_1|t_1)h^I(\tau_2|t_2)] \cos \omega_0(t_1 - t_2)$$

From the stationarity with respect to the clock times t_1, t_2, this can be expressed as

$$R(\tau_1, \tau_2|\Delta t) = \mathcal{E}[h^I(\tau_1|t)h^I(\tau_2|t + \Delta t)] \cos \omega_0 \Delta t \tag{5B-2}$$

Out of this, we are interested in just the correlation of the baseband h^I, or equivalently h^Q, which is

$$R_{hh}(\tau_1, \tau_2|\Delta t) = \mathcal{E}[h(\tau_1|t)h[\tau_2|t + \Delta t)] \tag{5B-3}$$

Here, h without the superscript denotes h^I or h^Q.

Because of the WSSUS assumption, we can define a new autocorrelation $R_{hh}(\tau|\Delta t)$, which is a function of just one τ. Here $R_{hh}(\tau|\Delta t)$ is the autocorrelation within delay shell τ between h at times t and $t + \Delta t$. Expression (5B-2) is related by

$$R_{hh}(\tau, \tau_1|\Delta t) = R_{hh}(\tau|\Delta t)\delta(\tau - \tau_1) \tag{5B-4}$$

We can now derive the parameters in Section 5.4.4 that measure fading.

- *Delay Power Profile.* Let the two clock times be equal, so that $\Delta t = 0$. Then $R_{hh}(\tau|\Delta t)$ is the expected value over t of $|h(\tau|t)|^2$, which is the average power in the channel response from delay shell τ. A plot of $R_{hh}(\tau|0)$ versus τ is the delay power profile. Figure 5.18a is a sample outcome of one; it was collected over nonzero time, but this is assumed to be ≈ 0 as far as R_{hh} is concerned. The width of $R_{hh}(\tau|0)$ is the multipath spread T_{mult}.
- *Coherence Bandwidth.* Another measurement tool results when the autocorrelation of the transform $H(f|t)$ of $h(\tau|t)$ is taken. We take formal transforms, and then as in the derivation leading to (5B-3), we end up with

$$R_{HH}(f_1, f_2|\Delta t) = \mathcal{E}[H(f_1|t)H^*(f_2|t+\Delta t)] \tag{5B-5}$$

This measures the correlation between the frequency response at f_1 for clock t with the response at f_2 for clock $t + \Delta t$ (t should make no difference if the fading is stationary). It can be shown that (5B-3) and (5B-5) are a two-dimensional transform pair. Furthermore, within the WSSUS assumptions, (5B-5) depends only on the difference $f_1 - f_2$. Therefore it makes sense to write $R_{HH}(\Delta f|\Delta t)$ instead of (5B-5). The coherence bandwidth, B_{coh}, refers to the width of Δf over which $R_{HH}(\Delta f|0)$ is roughly constant and nonzero. At a given clock time t, $H(f_1|t)$ and $H(f_2|t)$ within this Δf will be closely correlated, and signal components at f_1 and f_2 should be affected equally. The fading is flat over this range.

- *Doppler Spectrum.* The Fourier transform of $R_{HH}(\Delta f|\Delta t)$ when $\Delta f = 0$ is the mathematical Doppler spectrum. Formally, this is

$$F(\nu) = \int R_{HH}(0|\Delta t)e^{-j2\pi\nu\Delta t}\, d(\Delta t) \tag{5B-6}$$

To explore this, imagine the carrier f_0 alone. The baseband transfer function of the channel carrying f_0 is $H(0|t)$ (f_0 corresponds to 0 Hz at baseband). The received magnitudes are sample values along a random process and $R_{HH}(0|\Delta t)$ measures the correlation of the received tone magnitude at two times spread by Δt (its phase is uniform-random). Transform (5B-6) is the PSD of the process. The support of $F(\nu)$ is B_{dopp}.

- *Coherence Time.* Formally, T_{coh} is the support of the function $R_{HH}(0|\Delta t)$ of Δt. It is the approximate time that the baseband response to a sine tone remains constant. Slow fading means $R_{HH}(0|\Delta t)$ is slowly varying.

The modeling of fading as a multivariable random process is challenging to understand. Perhaps the best full treatment of it is given by Schwartz et al. [22].

CHAPTER 6

ERROR CORRECTION CODING

Earlier chapters have developed the concept of digital modulation, which is the sending of symbols by waveforms, and introduced the common transmission media. Channels are imperfect, and sometimes the demodulator makes an incorrect decision about a data symbol. Error-correction coding is for this situation: Given that the remaining symbols are mostly correct, error correction tries to repair incorrect decisions. This chapter introduces the encoders and decoders that make error correction possible.

The earlier chapters included much of the engineering reality of channels and transmission schemes, but the concepts of data symbol and channel that are used in error correction coding are by comparison rather limited and antiseptic. A data source puts out a sequence of independent, abstract *symbols* in this chapter; most of the time, these are not thought of as voltages or real-number values. A channel converts each symbol independently to a symbol in the same or different alphabet according to a probability distribution. It is this distribution that will define the channel; the distribution may stem from some reality of nature, but the distribution first defines the channel, not nature.

Section 6.1 describes several of these channel models, and it also reviews Shannon's information theory. The Shannon theory is a free-standing body of probability limit theorems that can be proven when information and channels are defined by probability distributions. It is often not easy to draw hard, engineering conclusions from these theorems, but the theory is a useful way to think about communication. We can touch on it only briefly in this book.

Sections 6.2 and 6.3 introduce parity-check codes. These work by adding extra symbols to the end of a transmitted block according to some mathematical rule; if errors occur, they can be corrected by working backward from the extra symbols. Section 6.4 introduces trellis codes and the important class of parity-check trellis codes, the convolutional codes. It also introduces coded modulation, a kind of

Digital Transmission Engineering, Second Edition. By John B. Anderson

coding that brings the modulator into the design. Section 6.5 looks at decoding schemes.

6.1 CHANNEL MODELS AND INFORMATION THEORY

The aim in this section is to derive basic channel and source models for some of the simpler transmission situations that occur in Chapters 2, 3, and 5. As part of this, we introduce some information theory that bears on the models, and we also work out the maximum-likelihood (ML) receivers where appropriate. The error-correcting codes that follow in the chapter are based on these models.

6.1.1 Basic Channel Models

The simplest of all channel models is the *binary symmetric channel* (BSC). The input and output to the channel is the binary symbol set $\{0, 1\}$, and ordinarily it is assumed that the inputs are independent. The BSC carries a symbol correctly with probability $1-p$ and erroneously with probability p, where p is called the *crossover probability*. The operation of the BSC can be sketched as shown in Fig. 6.1a, in which transitions from input to output symbols are labeled with the conditional probabilities that an output is received, given that some input is the case. A sketch

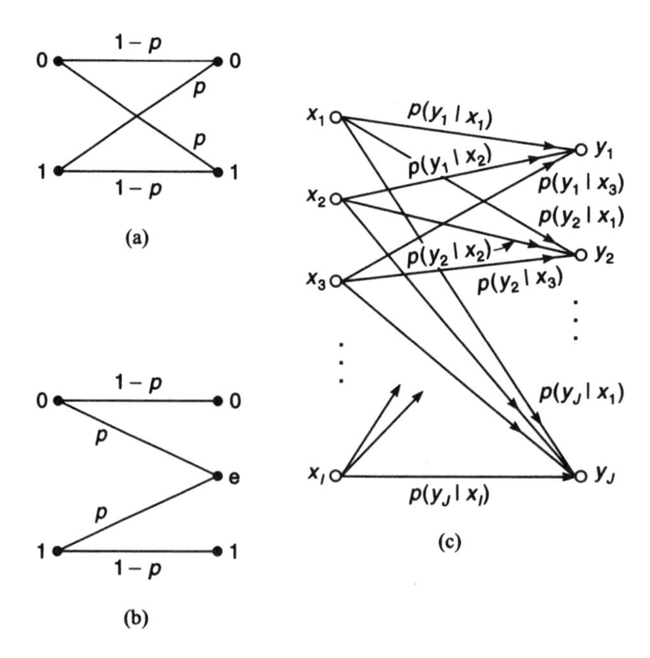

Figure 6.1 Discrete memoryless channel transition diagrams: **(a)** BSC; **(b)** BEC; **(c)** general case.

like this is called a channel transition diagram. The lines show the several ways that an input can be converted by the channel.

For the general case of a channel with a discrete set of I inputs and J outputs, we adopt a notation in which X is a channel input and Y is an output. We define the transition probability $p(y_j|x_i)$, the probability that the jth channel output y_j occurs, given that input x_i is the case. All these probabilities can be collected in a matrix called the channel transition matrix, which looks like

$$\begin{bmatrix} p(y_1|x_1) & p(y_2|x_1) & \cdots & p(y_J|x_1) \\ p(y_1|x_2) & p(y_2|x_2) & \cdots & p(y_J|x_2) \\ \cdots & \cdots & & \cdots \\ p(y_1|x_I) & p(y_2|x_I) & \cdots & p(y_J|x_I) \end{bmatrix} \tag{6.1-1}$$

The channel transition diagram for the general case is sketched in Fig. 6.1c. Since for every input, some output must occur, each row in the matrix sums to 1. If in addition the channel operation is independent from use to use, we have defined a *discrete memoryless channel* (DMC).

In coding and information theory, the input to a DMC is usually taken as an IID random variable X, for which a distribution $p(x_i)$, $i = 1 \ldots I$ is defined.[1] A distribution $p(y_j)$ can now be defined as

$$p(y_j) = \sum_{i=1}^{I} p(y_j|x_i)p(x_i) \tag{6.1-2}$$

The discussion here illustrates how coding and information theory view inputs and outputs as random variables. When the DMC transmits a block of N independent inputs, denoted by the vector $\boldsymbol{x}^{(N)}$, its transition probability is the product form

$$p(\boldsymbol{y}^{(N)}|\boldsymbol{x}^{(N)}) = \prod_{n=1}^{N} p_n(y|x) \tag{6.1-3}$$

where the subscript n now signifies the nth channel use.

The BSC makes severe assumptions about the channel operation, and it is an appropriate model only when nothing is known about the physical channel except its error rate. If the actual channel is BPSK transmitted over the additive white Gaussian noise channel (see Section 3.2), and the channel is truly an undegraded AWGN, then we could use a BSC model with $p = Q(\sqrt{2E_s/N_0})$. For example, $E_s/N_0 = 4.33$ dB yields $p = 0.01$ (see Table 2.1 for more). As discussed in Section 3.2, the BPSK transmission symbol +1 is identified with BSC symbol 0 and –1 with BSC symbol 1. This is an example of a *hard output* channel: Even though the AWGN channel puts out a real-number value, the BSC model acts as if the value has been quantized to one of the two hard symbols, which may as well be 0 or 1. Whatever further information was in the real-number value is lost.

[1]In practical communication, X is generally not independent and its distribution changes with time; for efficient transmission of speech and images, the true character of the input must be taken into account.

Section 3.2 points out that QPSK can be thought of as two independent BPSKs, and so a similar model for QPSK is two independent BSCs acting in parallel. Note that this modeling assumes perfect synchronization.

A one-step-more-complicated model than the BSC is the *binary erasure* channel (BEC), shown in Fig. 6.1b. The input is still binary but the output now consists of 0, 1 and the new symbol e, which occurs with probability p. Observe that if either 1 or 0 is received, it is certain to be correct; if e is received, it can equally well have come from either input. The BEC is one model for the fading channel in Chapter 5. When the channel fades, it "erases" the input and when the channel is strong, there are virtually no detector errors. A receiver can easily be designed to sense this 'e' condition. It is important to make use of a BEC model if it is appropriate, since it can be shown that a receiver/decoder that takes account of the e condition can correct twice as many errors as one just based on the BSC.

By modeling a channel as a DMC with more and more output symbols, the channel model can be made progressively "softer." The ultimate soft output channel is the *discrete-input AWGN* channel, first given in Fig. 2.33 and shown more abstractly in Fig. 6.2. Because the output Y takes infinitely many values, it no longer makes sense to portray the channel as in Fig. 6.1. The channel is a good model for the orthogonal modulations in Chapter 3, provided that the Gaussian channel in undistorted and the synchronization and bit timing are perfect. For BPSK, X takes the values $\{+\sqrt{E}, -\sqrt{E}\}$, which are sometimes replaced by the binary symbols $\{0, 1\}$; for QPSK, the model is two Fig. 6.2s in parallel; for nonbinary PAM, X takes more than two values, and for QAM there are again two models in parallel.

6.1.2 Information Theory in Brief

Information theory has three main divisions, channel transmission, information in sources, and rate-distortion theory. The last is about source reproduction when a certain degree of distortion is allowed. We are mainly interested now in the first part, the theory of information transmission through channels.

The Shannon theory views data sources and channels as probability distributions, and it proves some limit theorems about sequences of variables with these distributions. It does not manipulate symbols, noise, or signal spaces distances per

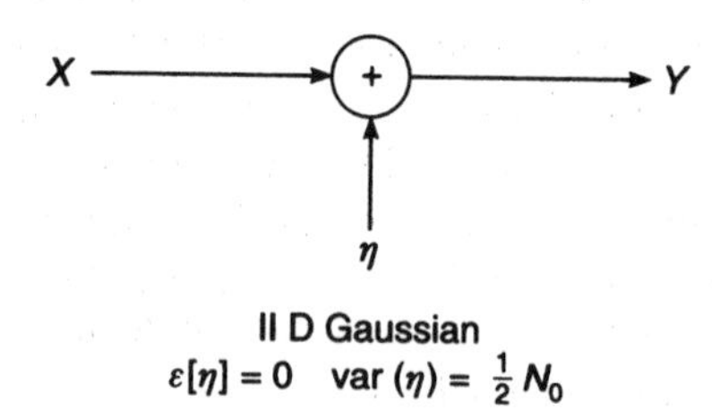

Figure 6.2 Discrete input AWGN channel. The BSC is a hard-output version of this soft-output channel.

se, nor does it study actual decoders; all these are the domain of coding theory, which follows in the later sections. We will now review some basic results. A complete introduction appears in the classic text of Gallager [1] or in the newer text by Cover and Thomas [2]. The classic 1948 and 1949 papers of Shannon [3, 4], which introduced source and channel information theory, are models of clarity and still eminently worth reading.

As we have mentioned, a data source to an information theorist is a random variable X with a probability distribution $p(x_i)$. The source can be a whole sequence of variables $X_1, \ldots, X_K$, represented by the vector $\boldsymbol{X}^{(K)}$ with distribution $p(\boldsymbol{x}^{(K)})$. The information in a particular scalar outcome x is defined to be $\log_2 p(x)$, and the measure of information in the source X as a whole is its *entropy,* defined for discrete variables to be

$$H(X) = -\sum_{i=1}^{I} p(x_i) \log_2 p(x_i) \qquad \text{(bits/outcome)} \tag{6.1-4}$$

The set of x_i is called the alphabet of the source. If X is a continuous variable, the sum sign in (6.1-4) becomes an integration. The term $H(X)$ can be interpreted as the expected information in X. It may also be viewed as the uncertainty in X; for example, $H(X)$ takes its largest value, $\log I$, when the I outcomes x_i are equiprobable, and X is maximally uncertain. When the vector variable $\boldsymbol{X}$ represents an IID sequence $X_1, \ldots, X_K$, it is easy to show that $H(\boldsymbol{X}) = KH(X_1)$. This simply states that the entropy in independent outcomes adds.

The unit of entropy and of information is the bit. This is a real-number measure like the meter, and it is important not to confuse this meaning of bit with the one in the rest of the book, where bit means a binary symbol. An 8-symbol binary message, for example, can be said to comprise 8 bits, and its entropy is 8.000 bits if the symbol outcomes X are equiprobable; but if the probability of symbol 0 is 0.05, and the symbol outcomes are IID, then (6.1-4) shows that the entropy in the message is $8H(X) = 8(-0.05 \log_2 0.05 - 0.95 \log_2 0.95) = 2.29$ bits. If the outcomes are dependent, this figure is reduced still further, but we will assume independence throughout the chapter.

The entropy of a binary data source when one outcome has probability p is given by the *binary entropy function*

$$h_B(p) = -p \log_2 p - (1-p) \log_2 (1-p) \tag{6.1-5}$$

The function is given a special name because it occurs often in communication and elsewhere. We will encounter it several more times.

Shannon proved a fundamental theorem, the *Noiseless Coding Theorem,* which is stated as follows: A source code exists that converts the source X outcome stream to a binary symbol stream, whose symbols occur at the rate $H(X)$ symbols per source output.

When there are two sources of data—call them X and Y—it is possible that they are correlated and that each thus says something about the other. The important example for us now is the case when X is the input and Y is the output of a channel; we certainly hope now that X and Y are closely correlated. For a particular outcome y, the *conditional entropy* of X is defined to be

$$H(X|y) = -\sum_{x} p(x|y) \log p(x|y) \qquad \text{(bits/outcome)} \tag{6.1-6}$$

(For economy of notation we will henceforth drop the i and j subscripts on x and y; furthermore, logarithms are base 2 unless otherwise noted.) Here $p(X|Y)$ is a conditional probability distribution, which expresses the dependence between X and Y. The quantity $H(X|y)$ is the entropy in the variable X if the outcome y is known. A further expectation of $H(X|y)$ can be taken, this time over y, to obtain

$$H(X|Y) = \sum_{y} H(X|y)\, p(y) \qquad \text{(bits/outcome)} \tag{6.1-7}$$

This is called the *equivocation* between X and Y. The equivocation is simply $H(X)$ if X and Y are independent. It can be viewed as the remaining uncertainty in X, on the average, if the outcome of Y is known.

Yet another measure of information between X and Y is the *mutual information,* which is defined to be

$$I(X; Y) = H(X) - H(X|Y) \tag{6.1-8}$$

This can be viewed as the reduction in the uncertainty in X, on the average, if Y is known. If X and Y are independent, $I(X; Y) = 0$ and there is no reduction; if X and Y are tightly correlated, as they hopefully are in a channel, $H(X|Y)$ is close to zero and $I(X; Y)$ is close to $H(X)$. Some manipulation shows that

$$I(X; Y) = \sum_{x} \sum_{y} p(x, y) \log\left[\frac{p(x, y)}{p(x)p(y)}\right] \tag{6.1-9}$$

and furthermore that $I(X; Y) = I(Y; X)$. The last is an expression of the reciprocity that exists between correlated variables.

The striking accomplishment of Shannon's 1948 article [3] was its assertion of channel capacity. Previously, it was thought that noise could not be removed from a transmission, but Shannon asserted that there was a largest measure of information that a given noisy channel can carry with vanishingly small error probability, and this he called its *capacity, C*. He defined C in terms of the probabilistic measures of information that we have just discussed. Formally, his definition is

$$C = \max_{p(X)} I(X; Y) \qquad \text{(bits/channel use)} \tag{6.1-10}$$

that is, C is the largest possible mutual information in the X–Y relationship that defines the channel, when we are free to maximize over the distribution of the source

X. The parameter C is the largest measure of information that can be learned about X through this channel. Shannon's *channel coding theorem* states that a channel code exists that carries source symbols across the channel at an information rate arbitrarily close to C bits/channel use.

The theorem does not, however, hint at how to construct such a code. It also does not give a precise result for the symbol error probability after decoding, other than to say that it tends to zero as the code's length grows. These, and the fact that the source is viewed probabilistically, are examples of how the Shannon theory lies at one remove from engineering reality. Nonetheless, the theory gives suggestive benchmarks, and we will look now at examples of capacity that figure later in the chapter.

Example 6.1-1: BSC Channel Capacity. When the data source is IID binary and the channel is the BSC, the capacity (6.1-10) is particularly easy to compute. Since the channel is symmetric, the $p(X)$ that maximizes in (6.1-10) must be the uniform one. From (6.1-8), we can write the capacity as $I(Y; X) = H(Y) - H(Y|X)$ with this $p(x)$. Working out the calculation, we get

$$I(Y; X) = H(Y) + \sum_x p(x) \sum_y p(y|x) \log p(y|x)$$

The conditional probabilities are those given in Fig. 6.1, and X and Y both have alphabet $\{0, 1\}$. The result is that

$$\begin{aligned} I(Y; X) &= H(Y) + \sum_x p(x)[p \log p + (1-p) \log (1-p)] \\ &= H(Y) + p \log p + (1-p) \log(1-p) \end{aligned}$$

Since X is uniform and the BSC is symmetric, Y must be uniform too, and consequently $H(Y) = 1$. Therefore the capacity is

$$C = 1 - h_B(p) \tag{6.1-11}$$

This is plotted in Fig. 6.3 as a function of the crossover probability. For example, $C = 0.5$ when $p = 0.11003$, which is an 11% error rate.

Example 6.1-2: BEC Channel Capacity. For the erasure channel in Fig. 6.1b, the capacity can be computed from the conditional probabilities, as in the previous example, again using symmetry. The result is that

$$C = 1 - p \tag{6.1-12}$$

This is also shown in Fig. 6.3.

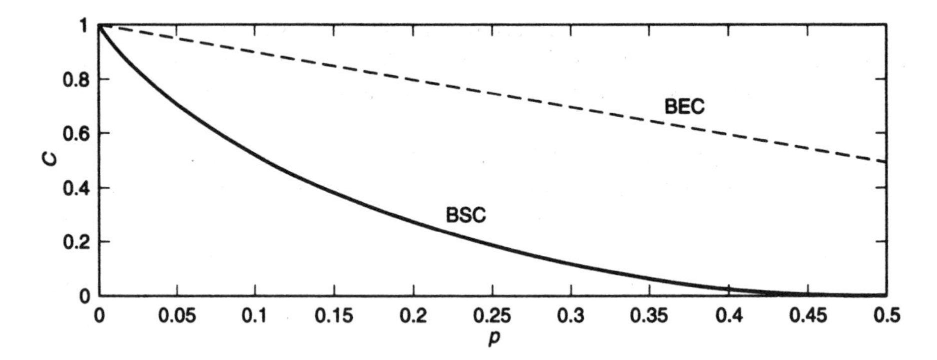

Figure 6.3 Capacity versus parameter p for binary symmetric and binary erasure channels.

Example 6.1-3: The Binary-Input AWGN Channel. This is the channel of Fig. 6.2, when X takes the antipodal values $+\sqrt{E}$ and $-\sqrt{E}$. Since the channel output Y is real-valued, the y-summations in (6.1-9)–(6.1-10) become integrals. Once again, the maximizing distribution of X must be the uniform one. We will use (6.1-9) and the fact that $p(y|x)p(x) = p(x, y)$. The result is that

$$C = \sum_x \int p(y|x)p(x) \log\left[\frac{p(y|x)p(x)}{p(x)p(y)}\right] dy$$

$$= \int \frac{1}{2}\, p(y| + \sqrt{E}) \log\left[\frac{p(y| + \sqrt{E})}{p(y)}\right] dy$$

$$+ \int \frac{1}{2}\, p(y| - \sqrt{E}) \log\left[\frac{p(y| - \sqrt{E})}{p(y)}\right] dy$$

where

$$p(y| + \sqrt{E}) = \frac{1}{\sqrt{\pi N_0}} \exp\left[\frac{(y - \sqrt{E})^2}{N_0}\right]$$

$$p(y| - \sqrt{E}) = \frac{1}{\sqrt{\pi N_0}} \exp\left[\frac{(y + \sqrt{E})^2}{N_0}\right] \qquad (6.1\text{-}13)$$

$$p(y) = \tfrac{1}{2}p(y| + \sqrt{E}) + \tfrac{1}{2}p(y| - \sqrt{E})$$

The capacity here is a function only of the ratio E/N_0. Figure 6.4 plots it (the circles) against this ratio; the plot axis is E_s/N_0 because the energy here is analogous to symbol energy in the earlier chapters. It is interesting to compare the new capacity to the BSC capacity (6.1-11) under the assumption that the BSC is derived from antipodal modulation with the same E_s/N_0. Capacity for the BSC is then $1 - h_B(p)$, where $p = Q(\sqrt{2E_s/N_0})$. It can be seen that the new capacity is only a lit-

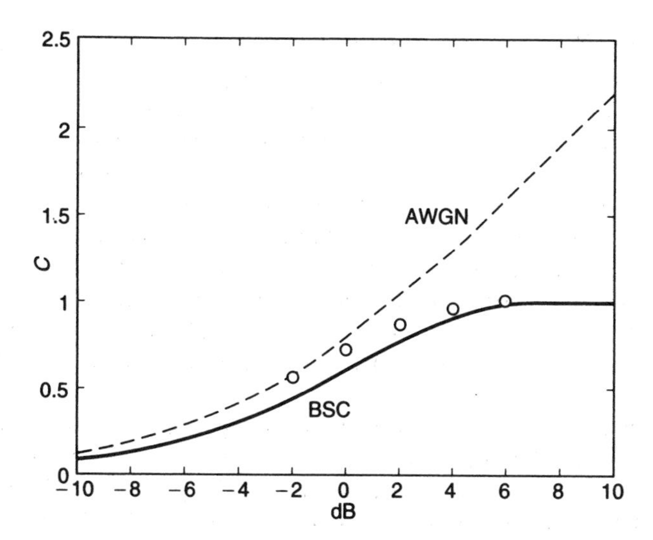

Figure 6.4 Capacity versus E_s/N_0 (in dB) for BSC, two-input AWGN channel (circles), and AWGN channel with unconstrained input. E_s is the energy per channel symbol.

tle larger than the BSC one. The idea of comparing capacities is an information theoretic one, and in this instance it is a little misleading: We will see that the error rate of actual codes is *strongly* affected by the switch from the BSC to the AWGN model.

In all these examples, the input to the channel is taken as a binary data source, either $\{1, 0\}$ or $\{\pm\sqrt{E_s}\}$. If the input data represented a larger set of real values, such that the energy of X is still E_s, then the capacity is much larger at higher E_s/N_0. With some effort, it can be shown that capacity is

$$C = \tfrac{1}{2}\log_2[1 + 2E_s/N_0] \qquad \text{(bits/channel use)} \qquad (6.1\text{-}14)$$

for the continuous-input–continuous-output case. This also is drawn in Fig. 6.4. The capacity calculation seems to suggest that we should abandon binary transmission at higher E_s/N_0, and this time practical experience says the same thing.

6.2 BASICS OF PARITY-CHECK CODES

In this section we will discuss error correction codes that work by adding extra symbols called parity-check symbols to the data. These symbols are mathematically related to the data symbols, and the relations allow data errors to be detected and corrected. The concept of a parity-check code is an extension of the idea of adding a single parity bit to a computer memory word in order to detect error, and from this comes the name. Parity-check codes do not have to work with binary symbols, but they usually do and we will assume binary alphabets, except when noted, throughout Sections 6.2–6.4. Parity-check coding is in fact closely wedded to the idea of the binary symmetric channel. A little thought will show that with a BSC, no other way exists to make an error-correcting code, other than adding extra symbols to the data. This is not true when a "softer" channel model is used.

Error-correction coding is a lengthy and complex subject which we can only summarize. Advanced parity-check codes are based on the theory of finite algebraic fields, a major subject for which there is no room here. Luckily, it is usually possible to engineer a communication system without knowing too many of these details. For those in need of a deeper treatment, we can recommend the texts [1, 5–8].

We will consider binary codes that encode a sequence $\boldsymbol{u}^{(K)}$ of K data bits into a channel code word $\boldsymbol{x}^{(N)}$ whose length is N bits. The $N - K$ parity-check bits are added to the data. The channel makes errors and changes $\boldsymbol{x}$ to the length – N received word $\boldsymbol{y}$. The entire transmission model is shown in Fig. 6.5. For completeness, source encoder and source decoder blocks are shown by dashed boxes, since these are part of a full coding theory view of the transmission problem. The boxes are important, but we will not discuss them further in this book. The notation $\boldsymbol{u}$, $\boldsymbol{x}$, and $\boldsymbol{y}$ is chosen because it is often used in coding theory and because we need to distinguish the problem from the modulation transmission model of Chapters 2–5. The entire modulation problem is contained in the channel block—that is, in the transition probabilities there. Sometimes the channel input and output x and y equate to the modulator transmission symbols a and $\hat{a}$ in, for example, Figs. 2.7 and 2.33. Other times, binary symbols x and y combine somehow to form quite different transmission symbols.

The set of all code words $\boldsymbol{x}$ makes up the code book, or dictionary, of the code. All possible data words $\boldsymbol{u}$ map uniquely to a code word. The codes we will discuss are called *block* codes because all code words are of the same length N. The *rate* of a binary block code, in data bits per channel use, is the ratio K/N. It is equally possible to have words of different length in the code book, and this is called a variable length code. An everyday example is the encoding of telephone numbers in Europe. Variable length codes and related properties are discussed in any standard coding text.

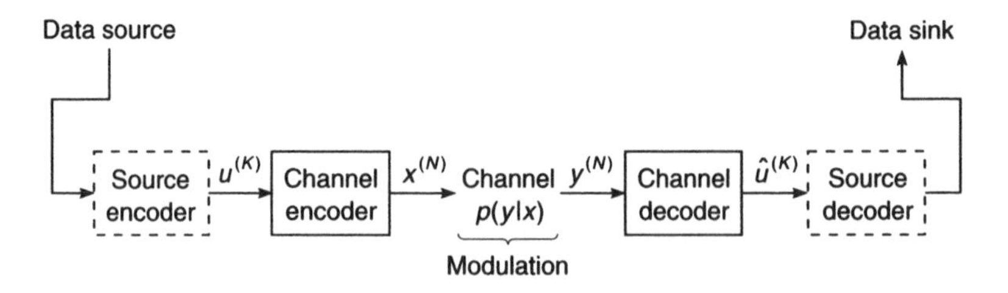

Figure 6.5 System model when source and channel coding are employed. Modulation process is contained in transition matrix probabilities.

6.2.1 Maximum Likelihood Reception with Block Codes and the BSC

Except when indicated, we shall assume the binary symmetric channel. When a code word passes through a BSC, the channel inverts some of the bits of an $\boldsymbol{x}$ to form $\boldsymbol{y}$. We can write this as $\boldsymbol{y} = \boldsymbol{x} + \boldsymbol{e}$, where the addition between vectors is bitwise and modulo-2 and $\boldsymbol{e}$ is a length $-$ N binary sequence called the *error sequence*.[2] An error in the nth position means $e_n = 1$; if there is no error, $e_n = 0$. These errors occur independently with probability p.

In analogy to Section 2.5, we can derive a maximum likelihood receiver for the BSC with a binary block code word input. As developed in the argument leading to Eq. (2.5-4), the ML receiver should find the maximum over ℓ of the conditional probability $p(\boldsymbol{y}|\boldsymbol{x}_\ell)$. The probability $p(\boldsymbol{y}|\boldsymbol{x}_\ell)$ is the same as $P[\boldsymbol{e} = \boldsymbol{y} - \boldsymbol{x}]$, the probability that the error sequence is $\boldsymbol{y} - \boldsymbol{x}$. With modulo-2 addition, $\boldsymbol{y}+\boldsymbol{x} = \boldsymbol{y} - \boldsymbol{x}$. Thus the ML receiver becomes

$$\text{Find } \ell \text{ that achieves} \qquad \max P[\boldsymbol{e} = \boldsymbol{y} + \boldsymbol{x}] \tag{6.2-1}$$

In a BSC channel with p, the probability of this outcome is

$$(1-p)^{N-d_\ell} p^{d_\ell}, \qquad p < 1/2 \tag{6.2-2}$$

where d_ℓ is the number of bit positions in $\boldsymbol{x}_\ell$ and $\boldsymbol{y}$ that differ. Just as was done before Eq. (2.5-18), we can take the logarithm of (6.2-2) without affecting the outcome of the max function. This yields

$$\text{Find } \ell \text{ that achieves} \qquad \max\, (N - d_\ell)\log(1-p) + d_\ell \log p$$

and after deleting the constant factor $N \log(1-p)$ and changing sign, we get

$$\text{Find } \ell \text{ that achieves} \qquad \min_\ell d_\ell \log[(1-p)/p]$$

The logarithm factor may be deleted, and the ML receiver that remains is

$$\text{Find } \ell \text{ that achieves} \qquad \min_\ell d_\ell \tag{6.2-3}$$

In other words, the ML receiver looks for the code word that differs in the fewest positions from $\boldsymbol{y}$, an entirely reasonable outcome.

The number of bits that differ between two binary vectors $\boldsymbol{x}$ and $\boldsymbol{y}$ is called the *Hamming distance* between them, denoted $h_d(\boldsymbol{x}, \boldsymbol{y})$. The distance h_d is the number of 1s in $\boldsymbol{x} + \boldsymbol{y}$. This number is also called the *weight* of $\boldsymbol{x} + \boldsymbol{y}$. It can be shown that Hamming distance has all the required properties of a distance measure, as did Euclidean distance in Chapters 2 and 3. The ML receiver finds the closest code word in Hamming distance to the received vector $\boldsymbol{y}$.

[2]In mod-2 arithmetic, $1 + 1 = 0 + 0 = 0$ and $1 + 0 = 1$; this is the exclusive-OR operation. If $\boldsymbol{x} = 1010$ and $\boldsymbol{y} = 1100$, for example, $\boldsymbol{x} + \boldsymbol{y} = 0110$.

6.2.2 Parity-Check Code Definition; Syndrome Decoding

The direct way to define a parity-check code is by its *generator matrix,* the $K \times N$ matrix that maps a data bit word $\boldsymbol{u}$ to a code word $\boldsymbol{x}$ by

$$\boldsymbol{x} = \boldsymbol{u}\boldsymbol{G} \tag{6.2-4}$$

All operations are bitwise modulo-2. From the form here, it is obvious that each row of $\boldsymbol{G}$ is a code word. Also every bitwise modulo-2 sum of rows is a code word; for example, the data word $\boldsymbol{u}$ = 01011000 with a 1 in positions 2, 4, and 5, and 0s otherwise, maps to the code word that is the sum of rows 2, 4, and 5 of $\boldsymbol{G}$. Finally, $\boldsymbol{0}^{(K)}$ maps to the all-zero code word. In fact, the words of this code form an algebraic group, since (*i*) the sum of any two words is another code word, (ii) there is a zero element (the all-zero word) and each word has an inverse (itself), and (iii) a commutative law applies. Such a code is called a *linear* code, or in former times, a group code.

A code for which the data bits explicitly appear in the code word bits is called a *systematic* code. We may as well take these as the first K bits in the code word, since rearranging the bit positions in the code words does not essentially change the code. Then $\boldsymbol{G}$ for the systematic code will have the form

$$\boldsymbol{G} = [\boldsymbol{I}|\boldsymbol{P}] \tag{6.2-5}$$

in which $\boldsymbol{I}$ is the $K \times K$ identity matrix and $\boldsymbol{P}$ is a $K \times N - K$ matrix that actually defines the code.

By analogy to the minimum distance defined in Section 2.6, we can define the minimum distance of an error-correcting code as the least Hamming distance between any pair of code words. For a linear code, the minimum distance is the weight of the least-weight code word other than the all-zero one. This can be shown as follows. For the all-zero word, the closest word is obviously the minimum weight one. Call this one $\boldsymbol{w}$ and let its weight be d. Now consider any other code word $\boldsymbol{x}$. Since $h_d(\boldsymbol{x} + \boldsymbol{w}, \boldsymbol{x})$ equals the weight of $(\boldsymbol{x} + \boldsymbol{w}) + \boldsymbol{x} = \boldsymbol{w}$, the distance from $\boldsymbol{x}$ to $\boldsymbol{x} + \boldsymbol{w}$ is again d. Moreover, no word $\boldsymbol{z}$ can lie closer to $\boldsymbol{x}$, because if it did, the word $\boldsymbol{z} + \boldsymbol{x}$, which must exist in the code, would lie closer than d to the word $\boldsymbol{0}$. Thus d is truly the minimum distance. The argument here can be extended to show that all words in a linear code have an identical constellation of neighbors. It is only necessary, therefore, to study the neighbor structure of the all-zero word.

From the triangle inequality, it follows that a linear code corrects all weight-δ error patterns if and only if $\delta < d_{\min}/2$; that is, a code with minimum distance 7 or 8 can correct up to 3 errors. A standard designation for block codes is (N, K), or alternately (N, K, d), meaning that there are K data bits, the block length is N and d is the minimum distance.

The simplest parity-check code is the repetition code, a code that carries one data bit per code word by simply repeating that bit N times. The code rate is $1/N$. Taking a majority decision on the received bits will correct up to $\lfloor N/2 \rfloor$ errors in the channel, where the brackets mean integer part. The opposite extreme is the single parity-check code, which carries $N-1$ data bits per word, with the Nth being a parity bit chosen to give the whole word even parity. Some thought will show that this code cannot correct errors, but it can detect that any odd number has occurred. For example, it can detect the most common case of error in a good BSC, that a single error has occurred. The rate is $(N-1)/N$.

An alternate way to define a parity-check code is by defining the code word set to be the null space of a matrix $\boldsymbol{H}$. This is the *parity-check matrix*. Row vector $\boldsymbol{x}$ is a code word if and only if

$$\boldsymbol{xH} = \boldsymbol{0} \tag{6.2-6}$$

Here, $\boldsymbol{H}$ is a binary $N \times N-K$ matrix, $\boldsymbol{0}$ is the all-zero row vector of length $N-K$, and, as usual, all arithmetic is modulo-2. The single parity-check code has $\boldsymbol{H} = (11\ldots1)'$. For a systematic code defined as in (6.2-5), $\boldsymbol{H}$ takes the form

$$\boldsymbol{H} = \left[\frac{\boldsymbol{P}}{\boldsymbol{I}}\right] \tag{6.2-7}$$

where $\boldsymbol{I}$ is the size $N-K$ identity and $\boldsymbol{P}$ is the same $K \times N-K$ matrix as in (6.2-5). Equation (6.2-6) collects in matrix form a set of $N-K$ parity-check equations, and all must be satisfied simultaneously if $\boldsymbol{x}$ is a code word.

When a channel error occurs and $\boldsymbol{y} = \boldsymbol{x} + \boldsymbol{e}$, substitution of $\boldsymbol{y}$ into (6.2-6) gives

$$\boldsymbol{yH} = (\boldsymbol{x} + \boldsymbol{e})\boldsymbol{H} = \boldsymbol{0} + \boldsymbol{eH} \tag{6.2-8}$$

The last term, $\boldsymbol{eH}$, is called the *syndrome* of $\boldsymbol{y}$. If $\boldsymbol{e}$ contains a single 1 in the nth place, the nth row of $\boldsymbol{H}$ is the syndrome; if $\boldsymbol{e}$ shows errors in several places, $\boldsymbol{eH}$ is the sum of the respective rows; the syndrome $\boldsymbol{0}$ indicates no errors. The word syndrome is taken from medicine, where it is the collection of symptoms that indicates a disease. Here the illness is channel error.

Example 6.2-1: Hamming Codes. These are a class of codes that correct all single errors. By definition, the rows of $\boldsymbol{H}$ for an (N, K) Hamming code consist of all the length $N-K$ words not equal to $\boldsymbol{0}$. Since there are $2^{N-K}-1$ of these, and one corresponds to each single error, the length of a code word is $N = 2^{N-K}-1$. The first Hamming code has $N-K=2$, which leads to $N = 2^2 - 1 = 3$; K is then 1, and the code is in fact the (3,1) repetition code. The next Hamming code has $N-K=3$, which creates the $\boldsymbol{H}$ and the code shown in Fig. 6.6. The code there has $N=7$, $K=4$, and rate $\frac{4}{7}$ and is in systematic form; this (7, 4) code has $2^4 = 16$ words. A syndrome $\boldsymbol{eH} = 111$ can occur, for example, when $\mathbf{e} = 0010000$, which indicates an error in the third place. In the list of code words, rows 2–5 make up $\boldsymbol{G}$. Rearranging the order of the rows of $\boldsymbol{H}$ leads to a different version of the same code. There are

$$H = \begin{bmatrix} 1 & 1 & 0 \\ 0 & 1 & 1 \\ 1 & 1 & 1 \\ 1 & 0 & 1 \\ 1 & 0 & 0 \\ 0 & 1 & 0 \\ 0 & 0 & 1 \end{bmatrix}$$

0000 000	0101 110
1000 110	0011 010
0100 011	1110 010
0010 111	1101 000
0001 101	1011 100
1100 101	0111 001
1010 001	1111 111
0110 100	
1001 011	

Figure 6.6 Hamming (7, 4) code: Parity-check matrix and dictionary of code words.

several least-weight words, all of weight 3, and $d_{\min}$ is thus 3, confirming that the code does correct single errors. The next code in the series has $N - K = 4$, $N = 15$, $K = 11$, and rate $\frac{11}{15}$. All Hamming codes have minimum distance 3. A code that corrects every t or fewer combination and no other error combinations in this way is called *perfect;* the only perfect binary codes are the Hamming codes and a (23, 12, 7) code called the Golay code.

If every combination of t or fewer rows of $\boldsymbol{H}$ is distinct, then any combination of t or fewer errors out of N can in principle be recognized since there is a syndrome for each one. A decoder that computes the syndrome $\boldsymbol{yH}$, maps to a set of error positions, and then corrects them is called a *syndrome decoder*. The idea here is clarified by a table called the standard array, an example of which is shown for a (6,3) code in Fig. 6.7. The code has $\boldsymbol{H}$ matrix and $\boldsymbol{G}$ matrix:

$$\boldsymbol{H} = \begin{bmatrix} 1 & 1 & 1 \\ 1 & 0 & 1 \\ 0 & 1 & 1 \\ 1 & 0 & 0 \\ 0 & 1 & 0 \\ 0 & 0 & 1 \end{bmatrix}, \qquad \boldsymbol{G} = \begin{bmatrix} 1 & 0 & 0 & 1 & 1 & 1 \\ 0 & 1 & 0 & 1 & 0 & 1 \\ 0 & 0 & 1 & 0 & 1 & 1 \end{bmatrix}$$

There are $2^3 = 8$ code words and $2^{6-3} = 8$ possible syndromes. Across the top of the standard array are the eight code words; three of these are the rows of $\boldsymbol{G}$, and the rest are all the sums of rows. Since the lightest nonzero word has weight 3, the minimum distance is 3. Down the left side are all the error patterns $\boldsymbol{e}$ that will be corrected, and down the right side are the eight syndromes, $\boldsymbol{eH}$. A syndrome applies to all sequences in its row. The error pattern shown against each syndrome is the lightest one having that syndrome; this pattern is chosen because of (6.2-3), which states that for the BSC with $p < \frac{1}{2}$ the ML decoding is the closest word in Hamming distance to $\mathbf{y}$. The rest of the standard array is the sum $\boldsymbol{x}+\boldsymbol{e}$,

				x					
	000000	001011	010101	011110	100111	101100	110010	111001	000
	000001	001010	010100	011111	100110	101101	110011	111000	001
	000010	001001	010111	011100	100101	101110	110000	111011	010
	000100	001111	010001	011010	100011	101000	110110	111101	100
y	001000	000011	011101	010110	101111	100100	111010	110001	011
	010000	011011	000101	001110	110111	111100	100010	101001	101
	100000	101011	110101	111110	000111	001100	010010	011001	111
	100001	101010	110100	111111	000110	001101	010011	011000	110

Figure 6.7 Standard array for (6, 3) code. Correctible error patterns are at left, code words run across top, and syndromes are at right. All 64 6-bit sequences appear somewhere.

for each pattern $\boldsymbol{e}$ and code word $\boldsymbol{x}$. All 64 words of length $N = 6$ appear somewhere, since $2^K 2^{N-K} = 2^N$.

To implement a syndrome decoder with the standard array, one finds the column in which lies the received word; the word at the head of the column is the closest code word. The code in Fig. 6.7 corrects all single errors, as did the (7, 4) Hamming code, and it corrects one double-error pattern, namely 100001. Any other error pattern leads to decoder error; for example, if word $\boldsymbol{x} = 111001$ is sent and two errors occur in the pattern $\boldsymbol{e} = 000011$, then $\boldsymbol{y} = 111010$ is received, and leads to $\boldsymbol{yH} = 011$. This syndrome connects to code word 110010, which is wrong in three positions.

The syndrome decoder is guaranteed to correct all combinations of up to $t = \lfloor \frac{1}{2}(d_{\min} - 1) \rfloor$ errors, where the brackets mean "integer part." It cannot guarantee more than t because no decoder based on minimum distance can do so. That the syndrome decoder corrects every error pattern up to t follows from the fact that $\boldsymbol{yH}$ cannot equal $\mathbf{0}$ for any $\boldsymbol{y}$ with weight less than $d_{\min}$ (otherwise the minimum weight cannot be $d_{\min}$). This means that no fewer than $d_{\min}$ rows of $\boldsymbol{H}$ sum to $\mathbf{0}$. Thus no combination of t or fewer rows can sum to any other sum of t or fewer. Consequently, $\boldsymbol{eH}$ is unique for every error pattern of weight up to t, and each of these can have its own syndrome.

This look-up procedure here consumes storage and is workable only for simple codes. Much research has been devoted to finding decoders that need small storage and computation. Most of these are based on the properties of the code's algebraic group.

In a sense, every parity-check matrix can be a generator matrix, and vice versa. Since $\boldsymbol{xH} = \mathbf{0}$ for any code word, it must be true that $\boldsymbol{uGH} = \mathbf{0}$ for any $\boldsymbol{u}$; that can only be true if $\boldsymbol{GH} = \mathbf{0}$, with $\mathbf{0}$ now the $K \times N - K$ all-zero matrix. It states the same thing to say that $(\boldsymbol{GH})^{\text{tr}} = \boldsymbol{H}^{\text{tr}}\boldsymbol{G}^{\text{tr}} = \mathbf{0}$. This says that $\boldsymbol{H}^{\text{tr}}$ can be a generator matrix and that $\boldsymbol{G}^{\text{tr}}$ is its parity-check matrix, for a length $-\ N$ code with $N - K$

data bits and the new rate $(N - K)/N$. These two codes are said to be *dual* codes to each other. The $(N, N - 1)$ single parity-check code and the $(N, 1)$ repetition code are such a dual-code pair. Several useful decoding algorithms are based on the duality concept.

6.2.3 Cyclic Codes

A large subclass of the linear codes are the *cyclic* codes. The words in these codes are all cyclic shifts of each other. One such code is the (7, 4) Hamming code in Example 6.2-1. Words number 2, 3, 7, 13, 8, 11, and 5 in the list form a set of right shifts; the words 1111111 and 0000000 are shifts of themselves; the remaining seven words form another set of shifts. For any cyclic code, it can be shown that there is a generator matrix whose rows are cyclic shifts, which generates the list of code words.

An alternate way to express the generation of code words is by a delay polynomial notation. This notation is particularly useful with cyclic codes. Consider the word $\boldsymbol{x} = [x(N-1), x(N-2), \ldots, x(0)]$, where bit $x(0)$ is the rightmost one in the word. The polynomial notation for this word is

$$x(D) = \sum_{i=0}^{N-1} x(i)D^i \tag{6.2-9}$$

where the variable D in the polynomial can be thought of as indicating a delay of one symbol position.

A way to generate the words of a cyclic code is by the polynomial multiplication $u(D)g(D)$, where $u(D)$ represents the data bits in the same way as (6.2-9), $g(D)$ is a certain generator polynomial, and all the coefficient arithmetic is mod-2. The operation may produce terms beyond power $N - 1$, and in this case, $x(D)$ is taken as the remainder when $u(D)g(D)$ is divided by $D^N - 1$. The complete operation may be written as

$$x(D) = u(D)g(D) \bmod (D^N - 1) \tag{6.2-10}$$

It can be shown that $g(D)$ generates a cyclic code if and only if $g(D)$ is a factor of $D^N - 1$.

As an example, we take the (7, 4) Hamming code again. It can be shown in general that $g(D)$ will generate a code via (6.2-10) if it is taken as the code word polynomial of least degree in the code list. For the (7, 4) code, this is $D^3 + D^2 + 1$, which is word 0001101 in Fig. 6.6. If the data word is $\boldsymbol{u} = 0010$, which is represented by the polynomial D, the product $u(D)g(D)$ is $D^4 + D^3 + D$, which is code word 0011010. In the Fig. 6.6 list, this word is produced by data 0011, which is not the $\boldsymbol{u}$ just taken, but it is still true that the (6.2-10) process generates the same total list if continued through all possible $u(D)$. Figure 6.6 and the new method are said to generate equivalent codes, even though the mapping between data and code words is not the same.

A rich theory of cyclic codes exists, a theory based on the properties of certain finite groups. The BCH[3] and Reed–Solomon (RS) codes are cyclic codes for which $g(D)$ consists of factors taken from a certain kind of polynomial. These factors are given in algebraic coding texts such as Refs. 5 and 6. In BCH coding, the polynomials and the $g(D)$ have binary coefficients, while in Reed–Solomon coding the coefficients are nonbinary. For each $m > 2$, an (N, K) BCH code exists that corrects t or fewer errors, with $N = 2^m - 1$ and $K \geq 2^m - 1 - mt$. The BCH are very powerful codes, and in fact the ones with $t = 2$ are the most powerful ones that can exist at their (N, K), in terms of what error patterns they correct. The double-error correcting (15, 7) BCH code, which has $t = 2$ and $m = 4$, is generated by $g(D) = D^8 + D^7 + D^6 + D^4 + 1$; it has $2^{15-7} = 256$ syndromes, which can be allocated to all 121 patterns of 0–2 errors, as well as 135 of the 3-error patterns. The (N, K) fixes the number of syndromes; for this number and a small crossover p, there is no better way to allocate them. Another BCH code is the Hamming (7, 4) code. A selection of BCH code parameters is given in Table 6.1.

Reed–Solomon codes have nonbinary coefficients in $x(D)$, $g(D)$, and $u(D)$. The coefficient alphabet ordinarily has size 2^ℓ. Instead of correcting up to t binary-symbol errors, RS codes correct up to t 2^ℓ–ary symbol errors. We can take as an example the (15, 9) RS code with hexadecimal symbols (i.e., $\ell = 4$) that corrects 3 symbol errors. Its generator is

$$g(D) = D^6 + 7D^5 + 9D^4 + 3D^3 + CD^2 + AD + A$$

with the usual hex alphabet $0, \ldots, 9, A, B, \ldots, F$. Addition and multiplication rules between symbols are specified, which we will not give. This RS encoder accepts 9 hexadecimals (36 bits) and codes them into 15 symbol (60 bit) words.

6.3 CODE PERFORMANCE

We look now at several ways to measure the performance of block codes. When the channel is well-defined—for example, a BSC or an AWGN channel—an estimate of decoder error probability can be calculated from the minimum distance of the code. Sometimes there is not enough information about the channel to define such a probability, but we can still guarantee that a certain number t of errors are corrected. Either way, we would like to know how the minimum distance of codes grows as a function of their length and rate.

6.3.1 Decoder Error Probability

Code performance to an engineer most often means decoder probability of error. For the simplest case, the BSC with crossover p, the error probability may be bounded from above by the following argument. Suppose that the decoder corrects

[3]BCH codes are named after their inventors, Bose, Ray-Chaudhuri, and Hocquenghem.

TABLE 6.1 Parameters of Certain High-Rate (N, K) BCH Codes

N	K	t	N	K	t	N	K	t
7	4	1	255	247	1	1023	1013	1
				239	2		1003	2
15	11	1		231	3		993	3
	7	2		215	5		973	5
	5	3		179	10		923	10
			511	502	1		873	15
31	26	1		493	2		828	20
	21	2		484	3		778	25
	16	3		466	5		728	30
				421	10		698	35
63	57	1		376	15			
	51	2		340	20			
	45	3						
	36	5						
127	120	1						
	113	2						
	106	3						
	92	5						

Note: t errors are corrected. Only $t = 1, \ldots, 5, 10, 15, \ldots$ are shown.

all combinations of t or fewer errors. For example, we found that the syndrome decoder corrects up to $t = \lfloor \frac{1}{2}(d_{\min} - 1) \rfloor$ errors. If there are more than t errors, the decoder may or may not decode incorrectly, so that the decoder probability is overbounded by the probability that more than t BSC channel errors occur; that is,

$$\begin{aligned} P_w &\leq \sum_{n=t+1}^{N} \binom{N}{n} p^n (1-p)^{N-n} \\ &= 1 - \sum_{n=0}^{t} \binom{N}{n} p^n (1-p)^{N-n} \end{aligned} \tag{6.3-1}$$

The subscript w in P_w is to stress that the decoder has put out an entire word, which is right or wrong. The data bits carried by the word cannot all be correct if the word is incorrect, but it is conceivable that only a few are wrong. The relation between data bit and code word error probabilities is a complex subject, discussed further in the references.

The tightness of the bound in (6.3-1) depends among other things on how many error patterns of weight more than t are corrected. The Hamming codes correct none, so (6.3-1) is tight for them; the convolutional codes in Section 6.4 correct many such patterns, and so the bound is looser. Another influence on the tightness of (6.3-1) is p. For fixed t and block length N, the bound (6.3-1) tends as $p \to 0$ to the single term

$$P_w \approx W_{t+1} p^{t+1} \tag{6.3-2}$$

in which W_{t+1} is the number of uncorrectable patterns of weight $t + 1$. The remaining terms in (6.3-1) are small by comparison. The only way to refine (6.3-1) further is to enumerate all the correctable patterns.

As an example, we can return to Example 6.2-1 and calculate the word error probability of a Hamming (7, 4) code when $p = 0.01$. From (6.3-1), the probability is

$$P_w = 1 - \sum_{n=0}^{1} \binom{7}{n} p^n (1-p)^{7-n} = 1 - (0.99)^7 - 7(0.99)^6(0.01) = 0.0020$$

The equality follows from the fact that the code corrects all single errors and no heavier patterns at all. If there were no code, the four data bits in each code word would be received correctly as a unit with probability $1 - (0.99)^4 = 0.039$. Thus the code reduces the error probability of a 4-bit data word by almost 20-fold. The "cost" of doing this is a little processor complexity and a $\frac{7}{4}$-fold expansion of bandwidth. As p becomes small, the limiting form in (6.3-2) becomes $P_w = 21p^2$, since $\binom{7}{2} = 21$. At $p = 0.01$, this differs by only 5% from the true value of 0.0020.

The discussion here can be extended to the AWGN channel model, but it will be more useful to do so in the framework of coding gain.

6.3.2 Coding Gain

More often than not, the BSC is a model of an underlying channel for which symbol energy and noise can be defined. If this is so, we can interpret the above savings in error probability as savings in transmitter energy. A standard assumption is that the channel behind the BSC is an AWGN channel with noise density N_0 and an antipodal modulator (such as BPSK) with channel bit energy E_s. As was discussed in Section 6.1.1, this has Fig. 6.2 as a model and $\pm\sqrt{E_s}$ as inputs. Therefore, the BSC crossover probability becomes $p = Q(\sqrt{2E_s/N_0})$. This BSC model reduces the physical channel output to one of two symbols, and it is thus a hard-output model for the underlying AWGN reality.

When a code of rate R bits/channel use is employed with this BSC, the code uses NE_s units of energy to carry K data bits. We need to define now two data bit energies, one if coding is used, called E_b^c, and another, E_b^{uc}, if no coding is used. Without coding, E_b^{uc} is simply the E_s above. With coding, the energy for each data bit is $E_b^c = NE_s/K = E_s/R$; a substitution for E_s then gives that the crossover p is $Q(\sqrt{2RE_b^c/N_0})$. At small p, the tight approximation to Q in Eq. (2.6-3) yields

$$p \approx \frac{1}{2} \exp\left(-\frac{RE_b^c}{N_0}\right)$$

By using the word probability formula (6.3-2) and taking logarithms, we get that

$$\log P_w \approx (t+1)\left(-\frac{RE_b^c}{N_0}\right)\log e + \log\left(\frac{1}{2}W_{t+1}\right) \tag{6.3-3}$$

Without coding, the word error probability in the BSC is $P_w = 1 - (1-p)^K$. This may be closely estimated by the following series of approximations:

$$\begin{aligned} P_w &= 1-(1-p)^K \approx Kp \\ &= KQ\left(\sqrt{\frac{2E_b^{uc}}{N_0}}\right) \approx \frac{K}{2}\exp\left(-\frac{E_b^{uc}}{N_0}\right) \end{aligned}$$

Again taking logarithms, we have

$$\log P_w = \left(-\frac{E_b^{uc}}{N_0}\right)\log e + \log\frac{K}{2} \tag{6.3-4}$$

We can get an interesting quality measure for codes by equating (6.3-3) and (6.3-4), so that both alternatives have the same word error probability, and by letting the two energies E_b^c and E_b^{uc} grow large. The result will be that

$$(t+1)\left(-\frac{RE_b^c}{N_0}\right) \approx \left(-\frac{E_b^{uc}}{N_0}\right)$$

since the other terms are small by comparison. Some rearranging gives a ratio $E_b^{uc}/E_b^c \approx R(t+1)$, called the *hard decision coding gain,* which expresses how much more energy per data bit is needed by the uncoded hard-decision AWGN transmission system, in order to achieve the same word error. Expressed in decibels, it is

$$G_h = 10\log_{10} R(t+1) \tag{6.3-5}$$

The coding gain applies strictly speaking only to word probabilities in a good channel, but it usually predicts real parity-check code behavior in a reasonable channel to within a fraction of a decibel.

A soft-output channel model is one that takes in transmission symbols but puts out symbols in some larger alphabet; in the AWGN channel model, the alphabet is the real numbers. We can carry out the coding gain calculation for the soft case by once again assuming that the code word bits are sent as $\pm\sqrt{E_s}$ by an antipodal modulation like BPSK. With coding, one unit of Hamming distance between two code word symbols becomes a piece of Euclidean square distance of size $4E_s$ in signal space (see the Problems); a Hamming minimum distance of size $d^h_{\min}$ in the code word set becomes square distance $d^h_{\min}\,4E_s$ in signal space. When this square distance is normalized to twice the data bit energy, it becomes

$$\frac{4E_s d^h_{\min}}{2E^c_b} = \frac{4E_s d^h_{\min}}{2E_s/R} = 2Rd^h_{\min} \tag{6.3-6}$$

The probability of deciding in favor of a word when its nearest neighbor was actually sent is a 2-signal detection problem from Section 2.6, with one dimension in signal space corresponding to each use of the antipodal modulation. The 2-signal probability is, from Eq. (2.6-4), $Q(D/\sqrt{2N_0})$, with D the Euclidean distance between the two words. We have

$$Q\left(\frac{D}{\sqrt{2N_0}}\right) = Q\left(\frac{\sqrt{d^h_{\min}4E_s}}{\sqrt{2N_0}}\right) = Q\left(\sqrt{\frac{d^h_{\min}2RE^c_b}{N_0}}\right)$$

in which E^c_b is the energy per data bit with coding. Without coding, the antipodal-signal error probability for each bit is, as usual, $Q(\sqrt{2E^{uc}_b/N_0})$. By following the same steps that led to (6.3-3) and (6.3-4), we get that the ratio of E^{uc}_b/E^c_b that equates the word error probability at large signal-to-noise ratio is $Rd^h_{\min}$. This is the *soft output coding gain*. Expressed in decibels, it is

$$G_s = 10 \log_{10} Rd^h_{\min} \approx 10 \log_{10} 2R(t + \tfrac{1}{2}) \tag{6.3-7}$$

Comparing (6.3-5) and (6.3-7), we see an important fact about coding: *In theory, the soft output channel has 3 dB more coding gain.* Technically, the coding gains G_h and G_s apply only with the antipodal/AWGN combination, but with other modulation and channel combinations, the gain from using soft channel outputs is of similar size. The design of decoders that use soft channel outputs is an important subject. With the antipodal/AWGN combination, these have typically 2 dB or more gain over BSC decoders at useful data bit error rates. It is interesting to compare this conclusion to Fig. 6.4, which shows capacities, not error rates, and seems to imply that there is little to be expected from soft channel outputs.

6.3.3 Behavior of the Minimum Distance

When channel probabilities cannot be defined, the quality measure of a parity-check code becomes distance alone. The trick in designing a good code of a given rate is to obtain the desired distance at as short a block length N as possible. Among codes of the same length, the achievable distance declines as the rate grows. Coding theorists all over the world search for new code designs, and for some years catalogs of these have been maintained; one appears in Ref. 5. Table 6.2 is a compilation of part of that catalog, together with some recent additions. The table shows the smallest N of any known (N, K, d) code, for each data size K and distance d; the code details can be found in Ref. 5 and other references. As examples of the use of this table, the Hamming codes appear down the first column (which is distance 3) at the positions $K = 4, 11, 26$. The repetition codes appear across the first row, where $K = 1$. Some very long best codes appear additionally in Table 6.1.

TABLE 6.2 Shortest Block Length N of Any Known (N, K, d) Code for Selected K and d

$K \backslash d$	3	4	5	6	7	8	9	10	11	12	13	14	15	16	17	18	19	20	21	22	23	24	25
1	3	4	5	6	7	8	9	10	11	12	13	14	15	16	17	18	19	20	21	22	23	24	25
2	5	6	8	9	11	12	14	15	17	18	20	21	23	24	26	27	29	30	32	33	35	36	38
3	6	7	—	—	13	14	—	—	20	21	24	25	27	28	31	32	34	35	38	39	41	42	45
4	7	8	—	—	14	15	—	—	—	—	—	—	29	30	—	—	—	—	—	—	44	45	48
5	—	—	12	13	15	16	—	—	—	—	—	—	30	31									
6	—	—	13	14	17	18	22	23	25	26	29	30	31	32									
7	—	—	14	15	18	19	23	24	26	27	31	32	35	36	40	41	43	44	46	47	49	50	55
8	12	13	15	16	19	20	25	26	28	29	33	34	37	38	41	42	44	45	47	48	50	51	57
9	13	14	17	18	20	21	26	27	29	30	34	35	38	39	44	45	47	48	52	53	54	55	58
10	14	15	18	19	21	22	28	29	30	31	36	37	39	40	46	47	48	49	53	54	55	56	59
11	15	16	19	20	22	23	29	30	31	32	37	38	41	42	48	49	50	51	—	—	57	58	60
12	—	—	21	22	23	24	30	31	34	35	40	41	43	44	49	50	51	52	—	—	58	59	61
13	—	—	22	23	26	27	—	—	35	36	41	42	44	45	52	53	54	55	—	—	59	60	65
14	—	—	23	24	27	28	—	—	36	37	42	43	46	47	53	54	55	56	58	59	60	61	66
15	—	—	25	26	29	30	—		37	38	43	44	47	48	—	—	57	58	59	60	61	62	69
16	—	—	26	27	30	31	—	—	38	39	47	48	49	50	56	57	58	59	60	61	62	63	70
17	—	—	27	28	32	33	37	38	39	40	48	49	50	51	58	59	59	60	61	62	63	64	71
18	—	—	28	29	33	34	38	39	40	41	49	50	52	53	59	60	61	62	63	64	66	67	75
19	24	25	29	30	34	35	39	40	41	42	50	51	53	54	60	61	63	64	69	70	74	75	80
20	25	26	30	31	35	36	40	41	42	43	51	52	54	55	61	62	65	66	72	73	77	78	81
21	26	27	31	32	37	38	41	42	43	44	52	53	55	56	62	63	66	67	74	75	78	79	84
22	27	28	32	33	38	39	43	44	44	45	53	54	57	58	64	65	67	68	75	76	79	80	85
23	28	29	34	35	39	40	44	45	45	46	54	55	58	59	65	66	68	69	76	77	80	81	86
24	29	30	35	36	40	41	45	46	46	47	55	56	59	60	66	67	69	70	77	78	81	82	93
25	30	31	36	37	41	42	47	48	47	48	58	59	60	61	68	69	70	71	80	81	82	83	96
26	31	32	37	38	42	43	—	—	53	54	59	60	61	62	69	70	71	72	81	82	83	84	97
27	—	—	38	39	43	44	50	51	54	55	60	61	62	63	70	71	72	73	82	83	84	85	98
28	—	—	39	40	44	45	51	52	55	56	61	62	63	64	71	72	78	79	83	84	85	86	99
29	—	—	40	41	45	46	52	53	56	57	62	63	65	66	73	74	79	80	84	85	86	87	101
30	—	—	41	42	46	47	54	55	57	58	63	64	66	67	74	75	80	81	85	86	87	88	104
31	—	—	42	43	47	48	55	56	58	59	65	66	67	68	75	76	81	82	86	87	93	94	105
32	—	—	43	44	48	49	56	57	59	60	66	67	68	69	76	77	82	83	92	93	94	95	106
33	—	—	44	45	49	50	57	58	60	61	67	68	69	70	77	78	83	84	93	94	95	96	107
34	—	—	45	46	50	51	58	59	61	62	68	69	70	71	78	79	85	86	94	95	—	—	108
35	—	—	46	47	51	52	59	60	62	63	69	70	71	72	79	80	86	87	95	96	—	—	109
36	—	—	47	48	52	53	60	61	63	64	70	71	72	73	80	81	87	88	—	—	102	103	111
37	—	—	48	49	53	54	61	62	65	66	72	73	76	77	81	82	88	89	—	—	103	104	112
38	—	—	49	50	54	55	62	63	66	67	73	74	77	78	82	83	89	90	—	—	104	105	113
39	—	—	50	51	55	56	63	64	67	68	76	77	78	79	83	84	90	91	—	—	106	107	114
40	—	—	51	52	56	57	64	65	70	71	77	78	79	80	84	85	91	92	102	103	107	108	115

Note: Most entries are taken from best code tables in Ref. 5. (Adaptation courtesy of P. Ståhl, Lund University. Used with permission.)

There are several bounds to the growth of d which are easy to prove, and these are given as the next two lemmas. The first is called the Hamming bound, and it gives an upper bound to the minimum distance that can be obtained from a (N, K) code.

Lemma 6.3-1 (*Hamming Bound*). Suppose a binary (N, K) block code with rate K/N corrects all combinations of t or fewer errors. The code need not be linear. It must satisfy

$$\sum_{i=0}^{t} \binom{N}{i} \le 2^{N(1-R)} \tag{6.3-8}$$

Proof. A simple proof works as follows. The number of error patterns $\boldsymbol{e}$ that must be corrected for each code word is the left side sum in (6.3-8). For each code word $\boldsymbol{x}$, the received words $\boldsymbol{y} = \boldsymbol{e} + \boldsymbol{x}$ are all different and must lie closer to $\boldsymbol{x}$ than to any other code word. Otherwise a minimum distance decoder will not correct them, and we know that the ML decoder is such a decoder. Consequently, there must be

$$2^K \sum_{i=0}^{t} \binom{N}{i}$$

such received words in all. The total cannot exceed 2^N, the number of sequences of length N. From this, (6.3-8) follows.

The next lemma is a lower bound. It says that there must exist a linear code with minimum distance equal to the largest d that satisfies (6.3-9).

Lemma 6.3-2 (*Gilbert–Varshamov Bound*). There exists a linear binary (N, K) code with Hamming minimum distance d, provided that

$$\sum_{i=0}^{d-2} \binom{N-1}{i} < 2^{N(1-R)} \tag{6.3-9}$$

Proof. We have already argued in Section 6.2 that the code will have minimum distance d if no $d - 1$ or fewer of the N rows of the $\boldsymbol{H}$ matrix sum to the $\boldsymbol{0}$ vector. Can such an $\boldsymbol{H}$ be constructed? To show that it can, choose any nonzero length $N - K$ vector as the first row, and proceed inductively. Suppose i rows have been chosen such that no combination of $d - 1$ or fewer sums to $\boldsymbol{0}$, and attempt to choose row $i + 1$. The row cannot be any sum of $d - 2$ or fewer out of the first i rows, since then $d - 1$ out of the new set of $i + 1$ rows will sum to $\boldsymbol{0}$. There are

$$\binom{i}{1} + \binom{i}{2} + \cdots + \binom{i}{d-2} \tag{6.3-10}$$

different such forbidden sums. At the same time, there are $2^{N-K} - 1$ nonzero vectors of length $N - K$; as long as the number of forbidden sums is less than this, a new row $i + 1$ can be selected. The selection of N rows will be a success if (6.3-10) is less than $2^{N-K} - 1$ when $i = N - 1$, since we will then have a candidate left over for row N. From this and the fact that $\binom{N-1}{0} = 1$, (6.3-9) follows. For a given N and K, the largest d that satisfies (6.3-9) is an underbound to the minimum distance possible with (N, K) codes, since we have constructed a code with this d.

The two bounds just given here are useful tools for judging the quality of a linear code. Figure 6.8 plots the outcome for both bounds as functions of the data length K, for rate $\frac{1}{2}$ codes; that is, the block length N is twice the K shown on the axis. The Hamming bound expression is converted from an upper bound on t to one on distance by using the relation $d^h_{\min} \le 2t + 2$. The circles on the plot are a selection of the best-known codes of rate $\frac{1}{2}$. For example, there exists a code with $K = 20$ data bits (hence $N = 40$ codeword bits) with $d_{\min} = 9$; this code corrects all patterns of 4 or fewer errors. At short K it can be seen that some codes in the figure actually hit the upper bound, and it is therefore a hopeless exercise to search there for better ones. The plot may be viewed conversely, which is to say that it gives the least K at which a distance may be achieved at rate $\frac{1}{2}$. For example, $d^h_{\min} = 16$ apparently requires $K = 40$ at least, which is a (80, 40) code.

Although it only just begins to appear in the plot, the growth in $d^h_{\min}$ versus K for the best codes matches the Gilbert–Varshamov bound slope when K is large. This bound has an interesting asymptotic form when N and K are large. A result of combinatorics says that the base-2 log sum of binomial coefficients on the left-hand side

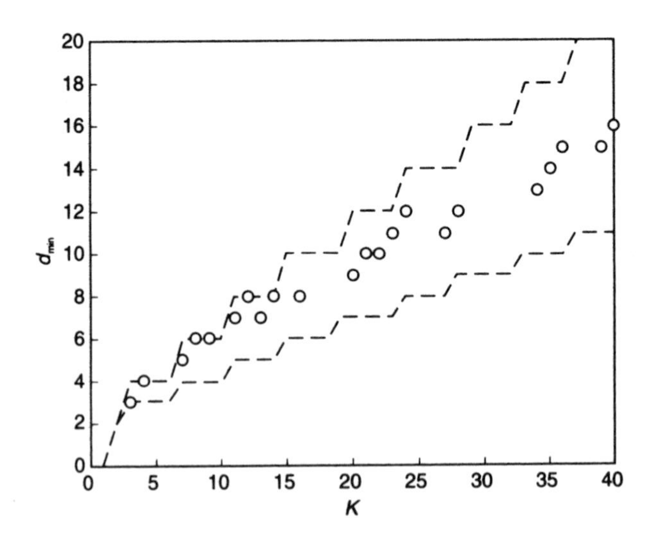

Figure 6.8 Minimum distance versus K for best known (N, K) binary codes of rate $\frac{1}{2}$. Also shown: Hamming upper limit and Gilbert–Varshamov lower limit to distance of best codes. [Courtesy of P. Ståhl, Lund University. Used with permission.]

of (6.3-9) is approximately $Nh_B\,(d^h_{\min}/N)$ when $d = d^h_{\min}$, plus terms that grow more slowly than N; h_B is the familiar binary entropy function. We can conclude that for large $d^h_{\min}$ and N, codes exist at least when $h_B(d^h_{\min}/N) < 1 - R$. Since $h_B(\cdot)$ is monotone increasing over the range of interest, it must be that

$$\frac{d^h_{\min}}{N} < h_B^{-1}(1-R), \qquad R = \frac{K}{N} \tag{6.3-11}$$

For instance, at rate $\frac{1}{2}$, $h_B^{-1}(1-\frac{1}{2}) = 0.110$, and the asymptotic Gilbert–Varshamov relation between $d^h_{\min}$ and K in Fig. 6.8 is $d^h_{\min} \approx 2K(0.110) = 0.220\,K$.

6.4 TRELLIS CODING

We turn now to a kind of codes whose words can be mapped onto a kind of regularly repeating structure called a *trellis*. Trellises have paths and states and branches, and these ideas provide a language with which to discuss encoders and decoders. The notion of a trellis closely relates to the most efficient way to build an ML decoder, but this topic will be delayed until Section 6.5. We begin here with convolutional codes, a class of parity-check codes that have the trellis structure. In the last 20 years, important trellis codes have been devised that are not parity-check codes and do not work with the BSC. A brief introduction to these will be given at the end.

6.4.1 Convolutional Codes

A convolutional code may be viewed as a parity-check code with the usual generation $\boldsymbol{x} = \boldsymbol{uG}$, and we will start with that. The generator matrix $\boldsymbol{G}$ has a special side-slipped form

$$\boldsymbol{G} = \begin{bmatrix} \boldsymbol{G}_0 & \boldsymbol{G}_1 & \boldsymbol{G}_2 & \cdots & \boldsymbol{G}_m & & & \\ & \boldsymbol{G}_0 & \boldsymbol{G}_1 & \boldsymbol{G}_2 & \cdots & \boldsymbol{G}_m & & \\ & & \boldsymbol{G}_0 & \boldsymbol{G}_1 & \boldsymbol{G}_2 & \cdots & \boldsymbol{G}_m & \\ & & & \cdots & & & & \\ & & & & \cdots & & & \\ & & & & & & \boldsymbol{G}_0 & \boldsymbol{G}_1 \\ & & & & & & & \boldsymbol{G}_0 \end{bmatrix} \tag{6.4-1}$$

Each row is made up of the same $m + 1$ G-blocks of binary symbols, and successive rows are shifted right by one block. (Elements not specifically shown are zeros.) The shifting can go on as long as desired. The individual blocks $\boldsymbol{G}_k$ have a size that depends on the rate of the code. The rate can take the form $R = (\log_2 \beta)/a$, where a and β are integers; in all practical codes, β is 2^b, a power of two, and so we will take the rate as the ratio $b/a < 1$.

To start, we take the easier case of b=1, which leads to convolutional codes with the rates $\frac{1}{2}, \frac{1}{3}, \frac{1}{4}, \ldots$. In this case, each block $\boldsymbol{G}_k$ is a length-a row with components

$[G_k(1), G_k(2), \ldots, G_k(a)]$, and there are $m + 1$ such rows, arranged end to end in one long row, as in (6.4-1). [Here $G_k(n)$ denotes the nth component of the vector $\boldsymbol{G}_k$.] The parameter m is the *memory* of the convolutional code. This is a reference to the fact that in the generation of words by $\boldsymbol{x} = \boldsymbol{uG}$, a span of $m + 1$ data bits in $\boldsymbol{u}$ affect a code word bit in $\boldsymbol{x}$. The convolutional code's complexity and approximate minimum distance are set by m.

Convolutional codes can be systematic, which means, as it did with other parity-check codes, that the data bits all appear somewhere in the code word. The following is a short systematic convolutional block code with rate $\frac{1}{3}$. Convolutional codes like this one are actually used in a different, extended form that will be explained in Section 6.4.2, but the short block form in the example is a simple way to get started.

Example 6.4-1: Rate $\frac{1}{3}$ Convolutional Code With Memory 2. The building blocks for the code's generator matrix will be $\boldsymbol{G}_0 = 111$, $\boldsymbol{G}_1 = 011$, and $\boldsymbol{G}_2 = 001$. Each of these has length 3, which is a; there are $m + 1 = 3$ blocks and the blocks have $b = 1$ row in them, so the rate is $b/a = \frac{1}{3}$. The side-slipped form of $\boldsymbol{G}$ in (6.4-1) may be continued indefinitely, but we stop $\boldsymbol{G}$ after 5 rows, and we truncate the matrix at the right after an even block to form a proper $\boldsymbol{G}$ for a block code. This gives a (15,5) block code whose generator matrix is

$$\boldsymbol{G} = \begin{bmatrix} 111 & 011 & 001 & & \\ & 111 & 011 & 001 & \\ & & 111 & 011 & 001 \\ & & & 111 & 011 \\ & & & & 111 \end{bmatrix} \tag{6.4-2}$$

The 5-bit data word 10011 will produce the code word 111 011 001 111 100 (spaces are inserted here and throughout for legibility). The first bits in each group of three are the data bits; this must be true, since columns 1, 4, 7, 10, and 13 in $\boldsymbol{G}$, taken alone, form an identity matrix. The code is thus systematic. The minimum distance of the code, it can be seen, is only 3. In Section 6.4.2 we will improve the distance by avoiding the truncation of the code generation.

Figure 6.9 is a circuit that produces the code word bits in the example. The circuit is easy to implement and easier to contemplate than the matrix multiplication. The data bits enter the shift register at the left and produce the code word bits by means of sets of taps. All arithmetic is modulo-2. The tap sets, called *generators,* are defined by the blocks in (6.4-1) according to the following scheme:

$$\begin{aligned} g_1 &= [G_0(1), G_1(1), \ldots, G_m(1)] \\ g_2 &= [G_0(2), G_1(2), \ldots, G_m(2)] \\ &\cdots \\ g_a &= [G_0(a), G_1(a), \ldots, G_m(a)] \end{aligned} \tag{6.4-3}$$

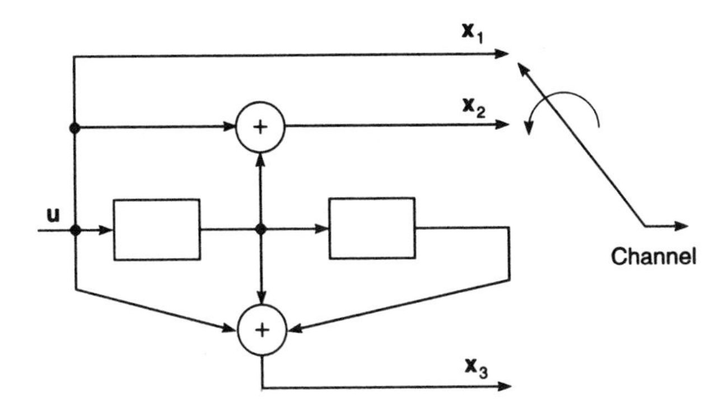

Figure 6.9 Shift register generation of (4, 6, 7) rate $\frac{1}{3}$ convolutional code.

As before, $G_k(n)$ denotes the nth component of row vector $\boldsymbol{G}_k$. The second generator, for example, consists of the second bits of each block, and the second tap set will have a tap for each 1 in the second generator.

In Example 6.4-1, the generators are $\boldsymbol{g}_1 = 100$, $\boldsymbol{g}_2 = 110$, and $\boldsymbol{g}_3 = 111$. As shown in Fig. 6.9, the top tap set, which is generator $\boldsymbol{g}_1$, is just a single tap on the data stream, without delay. The tap set produces the first of each group of three code word bits, respectively, 1 0 0 1 1. Tap sets at the middle and bottom produce the second bits in each group (which are 1 1 0 1 0) and the third bits (1 1 1 1 0). These sets are denoted $\boldsymbol{x}_1$, $\boldsymbol{x}_2$, and $\boldsymbol{x}_3$. The three sets need to be interleaved to produce the complete code word, 111 011 001 111 100. The shift register method here produces code word bits naturally in groups of a at a time. It is easy to build and easy to understand, and, as things turn out, it produces excellent code words.

The shift register in Fig. 6.9 performs a formal convolution, which is the source of the term convolutional code. Another way to generate the code words is by the mathematical convolution formula

$$x_k(n) = \sum_{j=0}^{m} g_k(j)u(n-j), \qquad k = 1, \ldots, a \tag{6.4-4}$$

Here $u(1)$ denotes the first bit to enter the shift register. The formula expresses the convolution of the kth generator $\boldsymbol{g}_k$ and the data sequence $\boldsymbol{u}$, to form the sequence $\boldsymbol{x}_k$. All arithmetic is modulo-2.

High-Rate Convolutional Codes (b > 1). It remains to look at the case of $b > 1$, which yields a rate of the form b/a. The chief motivation is to obtain a high rate, such as $\frac{2}{3}$ or $\frac{3}{4}$. It will be easiest to think of the data as arriving in blocks of b bits, which are deinterleaved into the b streams $\boldsymbol{u}, \ldots, \boldsymbol{u}_b$, as demonstrated for the $b = 2$ case in Fig. 6.10. Each stream feeds its own shift register, and taps on these b registers lead to a group of a summing junctions, which produce the a output streams $\boldsymbol{x}_1, \ldots, \boldsymbol{x}_a$. Any input register may connect to any junction, meaning that any of the (m

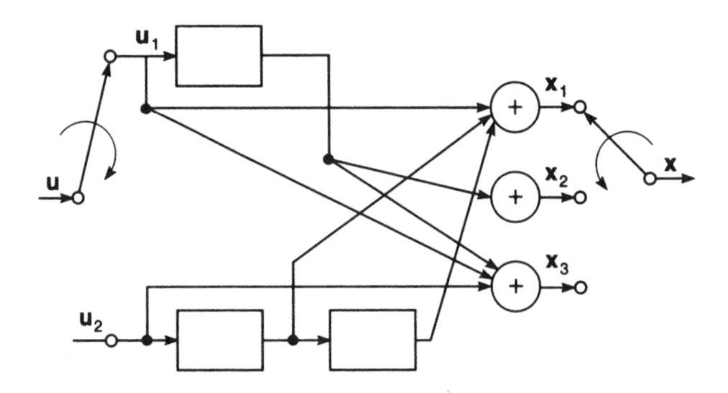

Figure 6.10 Shift register generation of convolutional code in Example 6.4-2.

$+ 1)b$ most recent data bits can affect the present group of a output bits. There are, in general, ba length-$(m + 1)$ generator sequences, which we will denote as $g_{k,\ell}(j)$, $j = 0, \ldots, m$, for each $k = 1, \ldots, a$ and each $\ell = 1, \ldots, b$. The convolution in (6.4-4) becomes

$$x_k(n) = \sum_{\ell=1}^{b} \sum_{j=0}^{m} g_{k,\ell}(j) u_\ell (n - j), \qquad k = 1, \ldots, a \tag{6.4-5}$$

The $\boldsymbol{x}_k$ are interleaved to form $\boldsymbol{x}$ in the same way as in the $b = 1$ case. Instead of being length-a rows, the $\boldsymbol{G}$–matrix blocks are now $b \times a$ matrices, which are obtained from the ba generators by the relation

$$\boldsymbol{G}_j = \begin{bmatrix} g_{1,1}(j) g_{2,1}(j) & \cdots & g_{a,1}(j) \\ g_{1,2}(j) g_{2,2}(j) & \cdots & g_{a,2}(j) \\ & \cdots & \\ & \cdots & \\ g_{1,b}(j) g_{2,b}(j) & \cdots & g_{a,b}(j) \end{bmatrix} \tag{6.4-6}$$

The notation here is admittedly confusing to the eye, but it is easy to implement in software or digital hardware. Convolutional codes like these have excellent performance, and they account for much of the practical application of coding. There are special notations for the generator bits that make them easier to handle by human engineers. The most common one is to let each group of three bits in a $\boldsymbol{g}_k$ or a $\boldsymbol{g}_{k,\ell}$ form an octal integer, starting with the initial group of three. For example, the rate 1/2 code with $\boldsymbol{g}_1 = 1101$ and $\boldsymbol{g}_2 = 1111$ is said to have generators 64 and 74. It is called the (64, 74) rate $\frac{1}{2}$ code; note the slight conflict with block code notation, where (N, K) means a rate K/N code. The last bit in 1101 and in 1111 form the initial bit in each second octal. The code in Ex. 6.4-1 is a (4, 6, 7) rate $\frac{1}{3}$ code.

Example 6.4-2: Rate $\frac{2}{3}$ Convolutional Code With Memory 2. The shift register generation circuit in Fig. 6.10 corresponds to the six generators

$$g_{1,1} = 100, \quad g_{1,2} = 010, \quad g_{1,3} = 110$$

$$g_{2,1} = 011, \quad g_{2,2} = 000, \quad g_{2,3} = 100$$

A commutator at the input deinterleaves $b = 2$ input bits, and one at the output interleaves $a = 3$ output bits. The code is not systematic, because neither input appears alone at any summing junction. Note that the second bit of each input pair does not affect the second bit of each output triple; this is what is meant by $\boldsymbol{g}_{2,2} = 000$. If a length-12 block code is desired from this generation, its $\boldsymbol{G}$ matrix form (6.4-6) would be

$$\boldsymbol{G} = \begin{bmatrix} 101 & 011 & 000 & \\ 001 & 100 & 100 & \\ & 101 & 011 & 000 \\ & 001 & 100 & 100 \\ & & 101 & 011 \\ & & 001 & 100 \\ & & & 101 \\ & & & 001 \end{bmatrix} \tag{6.4-7}$$

Note that the side-slipped block here has total dimension 2×9. As an example of the generation, we create the length-12 word that results from the input $\boldsymbol{u} = 10\ 01\ 11\ 00$ (the input has been broken into pairs only for clarity). Deinterleaved, $\boldsymbol{u}$ becomes $\boldsymbol{u}_1 = 1010$ and $\boldsymbol{u}_2 = 0110$. These flow through the shift registers and produce the three output streams $\boldsymbol{x}_1 = 1000$, $\boldsymbol{x}_2 = 0101$, and $\boldsymbol{x}_3 = 1001$. Interleaving these gives $\boldsymbol{x} = 101\ 010\ 000\ 011$ (spaces again for clarity), which is $\boldsymbol{uG}$. In all these words, the first bit to enter the encoder circuit is the leftmost one.

A huge literature exists about convolutional codes. Summaries of it appear in Refs. 6–9 and Ref. 10 is devoted entirely to convolutional codes. A major preoccupation, of course, is the search for good codes. A selection of the best ones at certain rates and memories appears in Table 6.3. The performance measure there is a special kind of minimum distance called the *free distance, d_f*. It will be easier to give a precise definition for this after we introduce the trellis idea. The error rate of a convolutional code in a good channel stems from its minimum distance, just as it does for other parity-check codes. Hard and soft coding gain follow as in Section 6.3.2. In particular, the error probability in a binary-input soft-output AWGN channel is asymptotically

$$P_e \sim Q(\sqrt{2d_f^h R E_b / N_0})$$

in which d_f^h is the Hamming free distance in Table 6.3, $R = b/a$ is the code rate, and E_b is the energy per data bit.

We have discussed only feedforward shift registers, but codes may be generated by feedback tapped shift registers as well, and the arithmetic need not be binary.

TABLE 6.3 Selected Binary Feedforward Convolutional Codes with Best Free Distance

Memory	Free Distance	Generators
	Rate $\frac{1}{3}$	
2	8	(5, 7, 7)
3	10	(54, 64, 74)
4	12	(52, 66, 76)
5	13	(47, 53, 75)
6	15	(564, 624, 764)
7	16	(454, 664, 754)
8	18	(557, 663, 711)
	Rate $\frac{1}{2}$	
2	5	(7, 5)
3	6	(54, 74)
4	7	(46, 72)
5	8	(75, 55)
6	10	(634, 564)
7	10	(626, 572)
8	12	(751, 557)
9	12	(7664, 5714)
10	14	(7512, 5562)
11	14	(6643, 5175)
	Rate $\frac{2}{3}$	
1	3	(0, 4, 6)
		(6, 6, 2)
2	5	(2, 7, 7)
		(5, 3, 4)
3	7	(10, 54, 54)
		(64, 30, 64)

Note: Memory m Octal generator notation, with, for example, octal 6 denoting binary 110. The six rate $\frac{2}{3}$ generators are grouped as in Example 6.4-2. Various sources.

One interesting point is that systematic encoders have as good a distance at each m as nonsystematic ones if the generation allows feedback taps. Another point is that codes that perform well in a good channel may not in a poor one, and they can in fact have negative coding gain there. Codes that perform well in both types of channels are mostly feedback systematic ones and are not the same as the ones in the table. A list of these appears in Ref. 11.

6.4.2 Code Trellises

A trellis is a generalization of a finite-state machine diagram. Such diagrams are a well-known tool for describing sequential digital circuits. The state trellis, an idea suggested by G. D. Forney around 1970, shows how the states of such a machine evolve in time. Any realizable code generation machine is a finite-state machine, and therefore has a trellis diagram, but the word *trellis* usually refers to a certain type of regularly repeating diagram, and we will confine ourselves to those. The trellis concept leads directly to an optimal decoder called the Viterbi algorithm, which we take up in Section 6.5.

We begin with the finite-state machine itself—and, in particular, the one for a convolutional encoder. The machine is the tapped shift register. The machine has a *state,* which is conveniently defined at time n by the data bits in the register, namely $u(n-1), u(n-2), \ldots, u(n-m)$. The output of the machine, which is a code word a-tuple, depends on these m bits plus the present one $u(n)$. Figure 6.11 is the standard state transition diagram, drawn in the computer science style, for the rate $\frac{1}{3}$ encoder in Fig. 6.9. It has four nodes, one for each state, labeled with the state bits $[u(n-1)u(n-2)]$. A transition from one state to another produces the output triples shown; depending on the present input bit, there are two possible transitions, and the respective input bit $u(n)$ is shown in parentheses. The bits $[u(n)u(n-1)]$ will become the new state at time $n+1$.

The state transition diagram for a general rate b/a code has bm state bits and 2^{bm} states, and the transitions depend on these plus the present set of b data bits. For example, the diagram for Fig. 6.10 has, in principle, four state bits (actually three because of a missing top delay), and there are four transitions out of each node.

In a trellis diagram, states are plotted vertically down the paper and time advances to the right. The diagram shows all possible evolutions of the machine state in time. Figure 6.12 is a trellis for the rate $\frac{1}{3}$ machine in the previous figure. The assumption has been made that the machine begins in state 00. This removes some of the early structure, but the trellis develops and eventually repeats a standard pattern shown in the last two sections. The name "trellis" is taken from the resemblance of the structure to a garden trellis, and one of the trellises in the next section (Fig. 6.14 later) can indeed be found in garden stores.

The nodes in the trellis represent states of the finite-state machine. The trellis breaks down into successive vertical columns, or *stages,* one for each time unit. A transition from one node to another is a *branch.* A succession of branches is a *path*

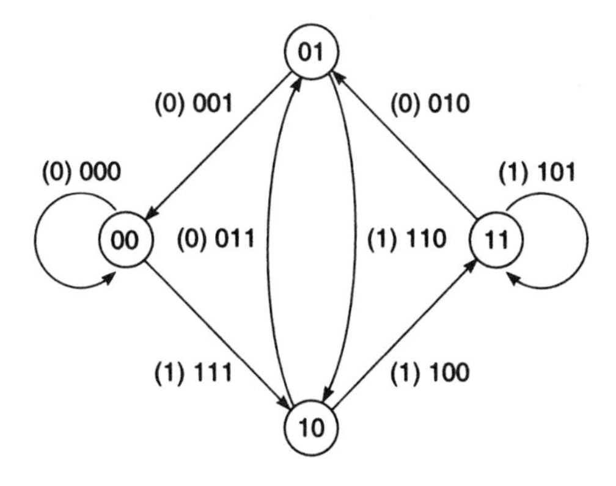

Figure 6.11 State transition diagram for rate $\frac{1}{3}$ encoder in Fig. 6.9.

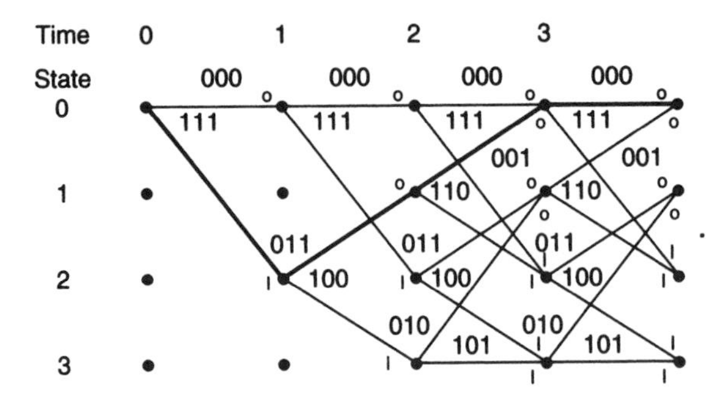

Figure 6.12 State trellis for rate $\frac{1}{3}$ code in Figs. 6.9 and 6.11. Encoder is assumed to start from state 0.

through the trellis. At each stage, two branches leave each state, one for each present data symbol, and once the trellis develops fully, two branches enter each node. The branches in the figure are labeled with the encoder output that corresponds to the transition, called a *branch label*, and in smaller figures the branches are labeled with the data bit that drove the transition. The states are named by the decimal representation of the state, with the oldest state bit $u(n-m)$ being the least significant binary in the representation.

Any path through the trellis spells out a code word, and the set of all possible paths is the set of all code words. For example, the data bits $u(0)u(1)u(2)\ldots = 100\ldots$ drive the encoder down the heavily shaded path in Fig. 6.12, with the bit $u(0)$ the first one to enter the register. The ideas here, that incoming data drive the encoder down paths and that the trellis graphs all such paths, are easy to grasp and are the reason that the trellis has become that main way to talk about a convolutional code. The heavy path rejoins the all-zero path (from $\boldsymbol{u} = 000\ldots$) after three stages. Such a joining at a node is called a *merge*. At this node the data symbols driving the two paths have been the same just long enough so that the register contents, and hence the two encoder states, are identical. It is said that the two code word paths, labeled [000 000 000] and [111 011 001], merge at state 0 after the third stage. The path structure makes it clear that the set of future paths in front of the merge node is identical for both 9-bit precursors; this must be so because the states of the encoder shift registers are the same after stage 3 for both paths.

In reality, convolutional codes are seldom truncated as they were in Examples 6.4-1 and 6.4-2; almost always, the shift-register generation of the words is allowed to run on for a long time. The code word created is still a block code word, but its length N is indefinite. The "finiteness" in the word creation is the generator memory m. The concept of indefinite length is part of a different attitude toward decoding that we can call the *trellis coding view*. In this way of thinking, encoding and decoding are done sequentially, in a series of similar, small operations, and these continue indefinitely. As each group of b data bits arrives at the encoder, the corresponding group of a code word bits is generated. We have seen how easy this is with a shift register. Decoders also can work in such an incremental way; that is, after each arrival of a code word bits, the decoder performs similar, small operations on b groups of data. This kind of decoder, a trellis decoder, will be studied in Section 6.5. It contrasts with a block decoder, which decodes an entire block at once, in one large calculation.

One of the advantages of a trellis is that it makes it easy to visualize and compute the minimum distance. First we need to adapt to trellis coding the idea of block code minimum distance. If a convolutional code is truncated and treated as a straight linear block code, then its minimum distance is, as usual, the least weight of any code word $\boldsymbol{uG}$. For example, some study of generator matrices (6.4-2) and (6.4-7) will show that these generate a set of words of minimum weight 3 and 1, respectively. But these weights stem from the place where the matrices are truncated, and in a trellis code this occurs far in the future.

What kind of distance measure is the natural one for trellis decoding? To attack this question, suppose that the decoder is to decide the first b-group of data bits, on the basis of the first L_D trellis stages (aL_D code word bits) and the first aL_D received channel bits. The quantity L_D is called the observation window, or sometimes the decision depth of the decoder. There are 2^b possible b groups to decide among, and each corresponds in the trellis to a branch out of the first node. We may as well assume that the transmitted trellis path is the all-zero one—that is, the one at the top that passes through a succession of 0-states and corresponds to data $\boldsymbol{u} = \boldsymbol{0}$; this can be assumed because even as indefinitely long words, the convolutional words retain the group code property. The first transmitted branch is then the first one on the all-zero path. The problem of deciding the first branch then comes down to the following: What is the closest path of L_D stages that does *not* start with the all-zero branch? This one is the least Hamming distance away from the all-zero path, and it is most likely to be confused with that path.

Some study of Fig. 6.12 shows that the least-weight path that splits from the all-zero path at the leftmost node has weight 6, no matter how long the trellis runs. This path is the shaded one. It splits away at time 0 and merges again to the all-zero path at time 3, accumulating weight 6 on the way. After the merge, the path has many possible descendants, but no elongation of the part before the merge can give a lesser distance than 6. The least distance of any such merge to the all-zero path is the *free distance* of the convolutional code. An organized way to find the free distance is based on the Viterbi algorithm, which will be discussed in Section 6.5. For now, we can say that when the trellis observation width is long enough, the free distance is the least distance between the all-zero path and any length-L_D path, merged or not, that splits from it at the initial node. Clearly, the decoder needs to observe enough width so that this free distance is the one that dominates the decision.

Free distances for good codes appear along with their generators in Table 6.3.

6.4.3 Some Other Trellis Codes

Much recent coding research has been devoted to trellis codes that do not work with binary words and the BSC. This subject is outside the scope of an introductory chapter, but we can briefly summarize three of the new classes. All of these work directly with the AWGN channel, and two are touched on in other places in the book. All make use of convolution in some way.

Partial Response Signaling (PRS) Codes. Consider the following convolution, which is similar to (6.4-4) except that the arithmetic is over the real numbers:

$$x(n) = \sum_{j=0}^{m} g(j)u(n-j) \tag{6.4-8}$$

As before, $x(n)$ refers to component n of a row vector $\boldsymbol{u}$ and $\boldsymbol{g}$ is a length-$(m + 1)$ generator sequence, which this time is made up of real numbers; $\boldsymbol{u}$ is composed of real-number transmission symbols such as $\{\pm 1\}$ or the QAM transmission symbols in Section 3.5. Equation (6.4-8) models the intersymbol interference and the filtering that were discussed in Section 2.4 and the partial response modulation scheme in Section 3.5. In the first case one can argue that the effect is unintentional, and in the second that it is intentional, but the end result is the same, a convolution of some $\boldsymbol{g}$ with the data sequence. Viewed in a more general way, Eq. (6.4-8) can be thought of as specifiying a class of coded modulation called partial response signaling (PRS) codes.

The code words generated by (6.4-8) have a trellis structure. As a matter of fact, the structure stems from the convolution form and is identical to that of a binary convolutional code having the same memory. Figure 6.13 shows an example code created by the generator sequence (0.4, 0.8, 0.4) with the binary data inputs +1 and –1.[4] The figure shows both the encoder shift register and a typical section of the four-state trellis. The state bits are $u(n-1)$ and $u(n-2)$, and these are the node labels; the branch labels are the real-valued outputs of the shift register that has the corresponding state history.

Continuous Phase Modulation (CPM) Codes. Another type of trellis code is the CPM class which was introduced at the end of Section 3.6. The signals here are T-length pieces of RF sinusoids, joined together in such a way that their phase is continuous at the symbol boundaries. The phase of the overall signal obeys (3.6-10), which is a convolution (yet again) of the real-valued transmission symbols with a certain generator. This time the generator is $q(t)$, a continuous function of time, which satisfies certain properties as described in Section 3.6; Fig. 3.27 shows some examples of $q(t)$.

This time, the trellis does not have shift register form.[5] A concrete example is given by CPFSK with modulation index $\frac{1}{2}$, which uses the $q(t)$ in Fig. 3.27b. This simple CPM code is just the MSK modulation in Example 3.6-3. The first four sections of this code's trellis are shown in Fig. 6.14. The value of the excess RF phase

[4]In Section 3.5 it was more convenient to represent this generator by the z-transform $0.4 + 0.8z^{-1} + 0.4z^{-2}$; the transform notation will be used again in Chapter 7.

[5]Although CPM codes are based on convolution, the operation applies to the signal phase offset, not to the signal itself. The trellis thus does not have the standard convolutional form.

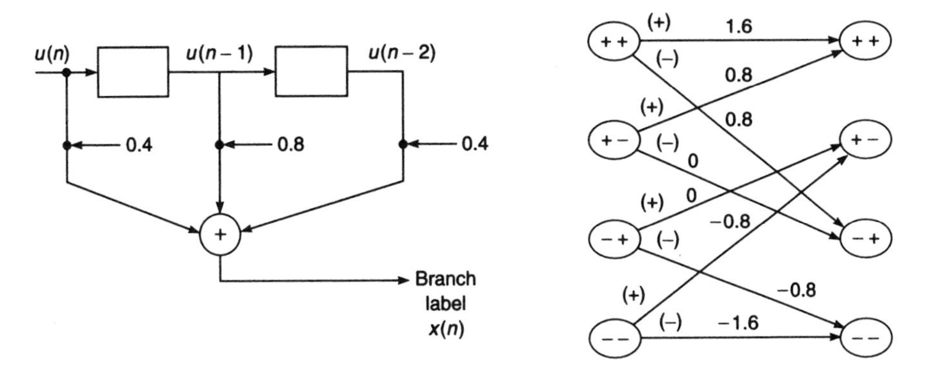

Figure 6.13 Encoder and trellis stage for PRS code with taps (0.4, 0.8, 0.4). All arithmetic is real.

at time nT provides a sufficient state variable for this code. States are shown on the trellis as multiples of $\pi/2$ radians, either 0, 1, 2, or 3, that the phase $\psi(t)$ relative to the carrier occupies at the start of the symbol interval; the whole signal is $s(t) = \cos[\omega_0 t + \psi(t)]$. If the transmission symbols in $\boldsymbol{u}$ take values ±1, then the excess phase in the nth interval is given by

$$\psi(t) = \frac{(t - nT)\pi u(n)}{2T} + \frac{\pi}{2}\sum_{i<n} u(i), \qquad nT < t < (n+1)T \qquad (6.4\text{-}9)$$

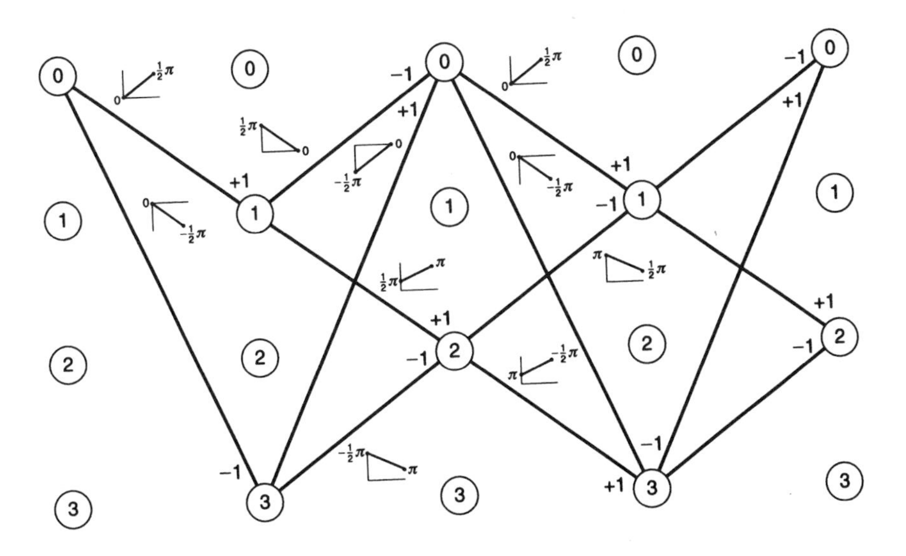

Figure 6.14 Four stages of trellis for CPFSK with index $\frac{1}{3}$. Branch labels are transitions of carrier phase.

The branch labels, as shown on the figure, are these small pieces of phase. Observe that with this particular state description, the trellis consists of two standard sections that alternate. More sophisticated CPM trellises are shown in advanced texts, such as Ref. 12.

The codes in Figs. 6.13 and 6.14 are examples of *coded modulations,* meaning that the coding scheme directly manipulates part of the actual transmitted signal, such as its amplitude or phase. This is in constrast to parity-check codes, which manipulate only symbols. When the parity-check code is binary, the encoder must add extra check symbols, but a coded modulation need not do this, because it ordinarily works with an alphabet of phases and amplitudes that is larger than binary and can even be infinite. With coded modulations, it makes better sense to think of coding as the imposition of one of a set of *patterns* onto the transmitted signal. At the receiver, the decoder finds a legal signal pattern that is close to the received signal, and the data symbols that correspond to this pattern are the output. With the PRS encoder of (6.4-8), code words are patterns of signal amplitude; with the CPM encoder of (6.4-9), code words are patterns of phase. Parity-check words are patterns, too: They are the patterns of symbols that satisfy the parity-check equation (6.2-6). It is important to realize that coding deals fundamentally with patterns, not with added symbols.

Set-Partition Codes (TCM). An important class of trellis codes is based on the idea of *set partitioning,* published by Ungerboeck in 1982. These are coded modulations that work with QAM-modulated signals (see Section 3.5), and they find particular use in telephone line modems, which normally use QAM modulation. The most common name for these is TCM codes, and so we will use this name.[6] The design of a TCM code starts with a large QAM constellation in the *I–Q* plane, such as the 16QAM one in Fig. 3.16. First the constellation is partitioned into a group of disjoint subsets. The data symbol stream is divided into two streams. Bits from one stream drive a finite-state machine that selects one of the subsets of the QAM constellation; bits from the other stream select a point from the subset, and this *I–Q* point is transmitted. The data are carried both in the choice of the subset and in the choice of a point within it. The subset choices can follow only certain coded patterns.

Figure 6.15 describes a TCM code based on the 8PSK constellation and a finite-state subset selector that is just a standard rate $\frac{1}{2}$ convolutional encoder circuit. A section of the fully developed trellis appears in Fig. 6.15a. The labels on its branches are one of four *subsets* C_0, C_1, C_2, C_3, not the individual symbols, amplitudes, and phases that we have seen before. The four subsets of the 8PSK master constellation are shown at the right of the section. The correspondence between subset and trellis branch is set by the selector circuit in Fig. 6.15b, in the following way. The incoming data are split into two equal-rate streams. The subset-determining stream

[6]TCM stands for trellis coded modulation, which is perhaps not the best choice of words. As common nouns, it simply means a signal with a trellis structure, which, obviously, is any signal in this section; as a proper noun it means a set-partition code.

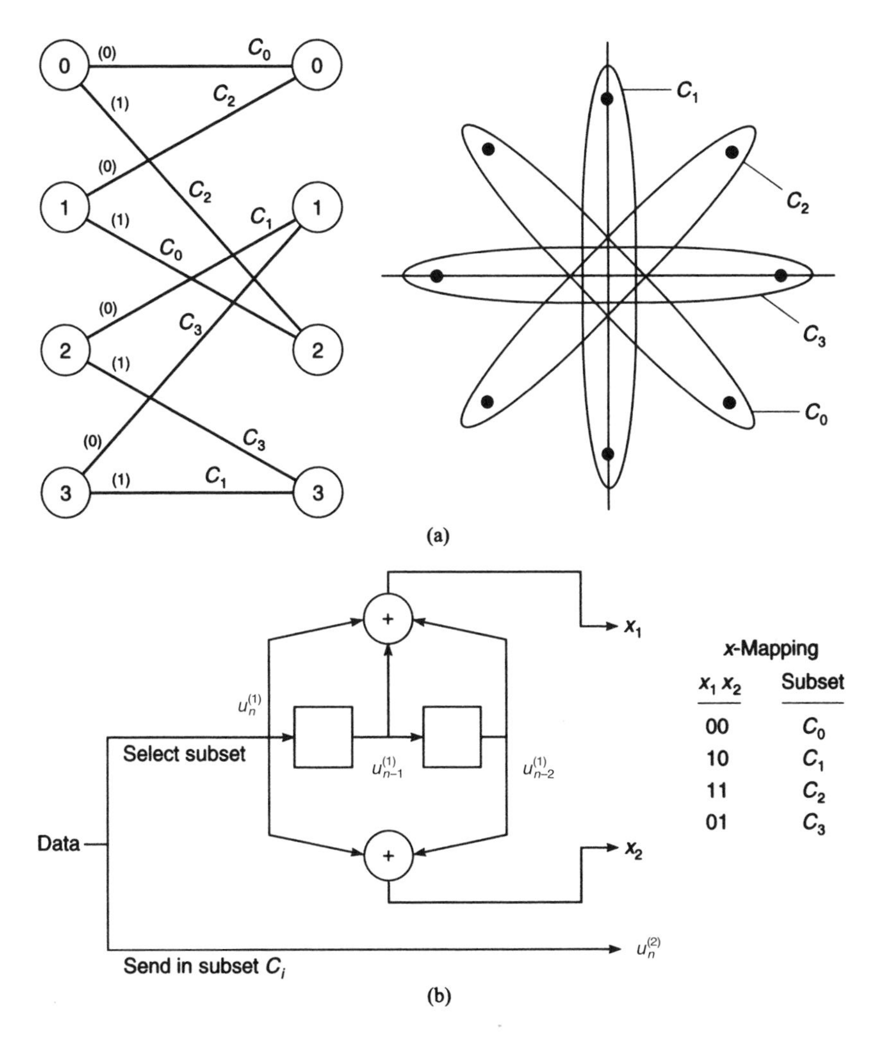

Figure 6.15 **(a)** One-stage of set-partition (TCM) code trellis. Original constellation is 8PSK; subsets are as shown. **(b)** The (7.5) rate $\frac{1}{2}$ convolutional mechanism for selecting subsets.

passes into a (7, 5) rate $\frac{1}{2}$ convolutional encoder circuit, which produces two bits x_1 and x_2 per input bit; these determine the subset according to the table at the encoder's right. The subsets each have two points in them, and the bit in the other data stream selects which point is sent. The total rate of the TCM encoder is 2 bits/symbol time, made up of one set-determining bit and one carried in the set.

A great many variations on this theme are possible by varying the split between subset-determining bits and bits in the sets. In addition, the size and type of the QAM constellation can be varied, as well as the mapping from bits to QAM points. Finally, the breakdown of the constellation into subsets can be optimized. Just as with other trellis codes, the major design object is to maximize the distance between the closest pair of trellis paths. A glimpse of this variety, as well as a more complete explanation, can be seen in Refs. 13 and 14 and in Ungerboeck's later survey article [15].

6.5 TRELLIS DECODERS

The obvious design for a decoder is the ML principle. We have introduced this principle in Section 2.6 and applied it to binary symbol decoding in Eq. (6.2-3). For the AWGN channel, which is the soft-output case, the ML decoder seeks the closest code word to the received word in the sense of Euclidean distance; for the hard-output BSC case, the ML decoder seeks the closest word in Hamming distance. For this second case, we introduced the syndrome decoder as an ML decoder for linear codes. Unfortunately, the number of syndromes grows as $2^{N(1-R)}$ for an (N, K) block code, which for a fixed rate becomes impractical as the block length N grows. A major aim of this section is to introduce the Viterbi algorithm, a scheme based on the trellis structure, which is the most efficient full ML decoder for a code with a given state description. Unlike the syndrome decoder, the Viterbi algorithm does not grow exponentially in size with N, but depends only on the code state space size. In convolutional codes we have seen an important example of a kind of code whose length can grow independently of its state size.

There are many other decoding algorithms, and some are much more efficient than the Viterbi algorithm at obtaining a set error probability for a certain code and SNR. For linear block codes, a large number of schemes have been devised, which depend on algebraic properties of the code. For trellis codes, decoders called variously *sequential* or *reduced-search* decoders have been invented. These search only a small part of the trellis for a decoding, presumably the part where the transmitted word is most likely to lie. Still, the most important single decoder is the Viterbi algorithm, and this gets most of the attention in this introductory section. Details about the other methods may be found in the starred references at the end of the chapter.

We will start with an overall view of trellis decoding. It will be useful to refer ahead to the trellises in Fig. 6.16. For a general trellis decoder, two basic measures of cost can be defined: the size of the path storage and the work needed for a path manipulation. As for path storage, this is made up of two factors, the number of paths in storage and their length. It turns out that the Viterbi algorithm must store at a given stage one path leading to each trellis state; this is 2^m paths for a convolutional code with memory m. Reduced-search decoders store many fewer trellis paths than this. In either case the decoder error probability or the number of errors corrected can be related to the number of paths stored, and more details of that appear in Refs. 6, 8, 9, and 14.

The length of path retained, as mentioned in the last section, is called the decision depth, denoted L_D. In earlier times, L_D was taken to be the entire block length, but today the length is more often shortened to the minimum that does not degrade decoder

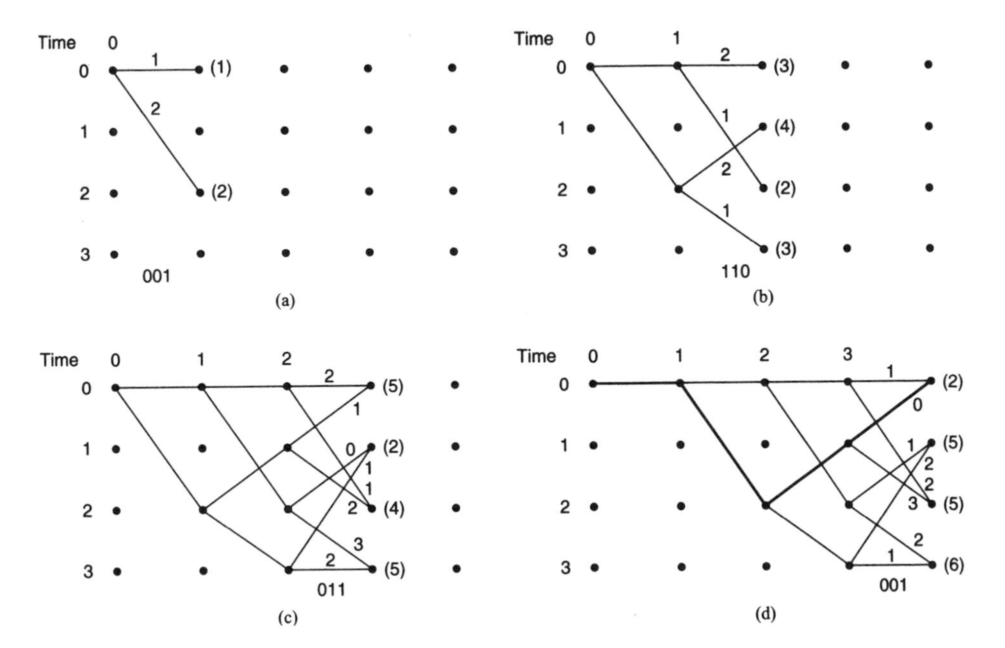

Figure 6.16 Progress of four-state Viterbi decoder when received sequence is 001 110 011 001 · · ·. Trellis is rate $\frac{1}{3}$ one in Fig. 6.12. Numbers on branches are Hamming distance increments; numbers in parentheses are accumulated distances.

performance. A way to view a trellis decoder is to think of it as a length-L_D processor that chews its way across a trellis, at each stage ingesting one new trellis stage at the front and putting out a decided b group of data symbols at the back. Parts of paths that lie farther back can be released, since all these data symbols have been decided. Recall from Figs. 6.12–6.14 that every encoded data b group corresponds to a trellis branch, and thus a decided symbol implies that the retained parts of the trellis must begin with this branch. Hopefully, all remaining paths to the right share this same single ancestor on the left. If not, in theory the paths beginning with any other branch should be dropped, but this step is often ignored in practice, with few side effects.

The delay in the decoder is L_D. It is known that a convolutional code with rate $R = b/a$ and Hamming free distance d generally requires a BSC decoder decision depth of about

$$L_D \approx \frac{d/a}{h_B^{-1}(1-R)} \qquad \text{(stages)} \tag{6.5-1}$$

Here $h_B^{-1}(\cdot)$ is the inverse of the binary entropy function, and for small d (6.5-1) is a mild overestimate.[7] As an example, a rate $\frac{1}{2}$ code with free distance d corrects about

[7]The law applies to block codes as well, with $N = d/h_B^{-1}(1-R)$. Decision depths can also be derived for the trellis codes in Section 6.4.3. See Refs. 8 and 14.

$\frac{1}{2}d$ errors and needs $L_D = (d/2)/h_B^{-1}(0.5) \approx 4.5d$. At this rate it is known that the free distance is about equal the code memory m, and so L_D in terms of m is about $4.5m$. Note that L_D is much longer than the code memory itself.

As for the work performed by a trellis decoder, this occurs on a per-path basis. A path is retrieved from storage, and its 2^b extensions are made, which comprises computing the branch labels, the distance increments, and the accumulated distance of each new path. Then the new paths are dropped or returned to storage. Here is an example of the piece-by-piece computation idea in trellis decoders in Section 6.4.2. We will illustrate now these ideas of work and storage with the Viterbi algorithm.

6.5.1 The Viterbi Algorithm

The Viterbi algorithm (VA) searches the active part of the trellis in the most efficient possible way. The central concept that makes this kind of decoder work, called the *optimality principle,* was noted by A. Viterbi in 1967 [16]. It was pointed out somewhat later by J. Omura that the scheme amounts to a dynamic program to find a least-distant code word; the concept of the trellis came still later. Today we understand the algorithm in connection with the trellis. The algorithm finds wide use despite its exhaustive searching character, because it repeats a similar step at each trellis node and stage, because it adapts easily to soft channel outputs, and because it happens that convolutional and TCM codes have attractive gains even with small trellises.

We will focus on the BSC channel case and explain the VA with reference to the rate $\frac{1}{3}$ code in Fig. 6.12. The algorithm's progress is shown in Fig. 6.16. Assume that the encoder began in state 0. In Fig. 6.16a, the bit triple 001 has been received and the available trellis branches, given starting state 0, are generated. Data 0 out of state 0 causes branch label 000 and data 1 causes label 111, which is a branch leading to state 2. The figure shows the Hamming distances to these branches, which are 1 and 2, respectively. In Fig. 6.16b, channel output 110 has arrived after time 1 and the four possible path extensions, given the starting state, have been made. The distances between 110 and each of the four new branches are shown in the figure. In addition, the total distance of each two-branch path is shown in parentheses. In Fig. 6.16c, the eight extensions out of stage 2 are made and 011 arrives from the channel. Branch distance increments are shown. Stage 3 represents the first fully developed trellis section, and now it appears that two paths through the trellis merge at each node at time 3.

The critical step in the VA now occurs. From the two paths entering each node, the one having the least total distance, called the *survivor,* is kept, and the other entering path is deleted back to the point where it diverges from another path. If both paths have the same total distance, either may be deleted without affecting the optimality of the decoder. After one such deletion at each node, the best path distances at the nodes after stage 3 are, respectively, (5), (2), (4), and (5), and the surviving trellis looks like the stages 1–4 of Fig. 6.16d. In optimization theory, this kind of procedure is called a dynamic program. The reason that the deletions can be made, and still have an ML decoder, is the optimality principle of dynamic programming:

When paths through a graph merge at a node, only one least-distant (i.e., most likely) alternative need be kept, since a least-distant path through the whole graph exists which does not contain the dropped sections.

A classic dynamic programing optimization is finding the least road distance between two cities, such as New York and Los Angeles. A code trellis is more regular than a road map, but otherwise the problems are the same. Nodes in the trellis correspond to the intermediate cities, paths are roads, and survivor sections are the shortest routes between intermediate cities.

Returning to Fig. 6.16d, we see that 001 has arrived during stage 4 and the path extensions from time 3 lead to total time-4 survivor distances of (2), (5), (5), and (6). A least-distant path with four branches is shaded; it runs through states 0, 0, 2, 1, 0 and it has total distance 2. No other path of four branches lies at distance 2, although in general there can be several least-distant paths. The data symbols that drive this path are $\{u(0) \ldots u(3)\} = 0100$ and the first 12 bits of the code word are 000 111 011 001. The optimality principle says that these are the first 12 bits of any optimal code word that passes through state 0 at time 4. It does not, however, guarantee that the optimal word passes through this particular node. Finding the survivor into a node is the basic VA step. By repeating it at each node, stage by stage, until the end of the trellis, a least-distant code word will be found.

It can happen, as occurs in the figure, that choosing survivors eliminates all branches except one at an earlier trellis stage. This has happened in stage 1. Effectively, this decides the first data symbol (it is 0). The second symbol has not been decided with certainty, although if we had to decide on the basis of stages 1–4, the most likely data symbol would be 1, because it lies on the shaded path. Aside from these outcomes, it can happen that two paths out of a node stay in contention for a long period. They may have identical distances, in which case it will be necessary to select a path at random.

To sum up the VA for a convolutional code of rate b/a, the trellis will have 2^b branches out of and into each node, once the trellis is fully developed, which occurs after m branches, m being the memory. There are a bits on each branch, $b2^m$ branches per stage, and 2^m nodes per stage.

In a software implementation, the path histories and distances into each node at stage n are stored in two arrays. Paths are identified by the data symbols that drive them. During the extension of the trellis to stage $n + 1$, two new arrays are created as the paths are extended one by one, and survivors, having a lower distance, are allowed to overwrite deletions. At the end of the cycle, the new arrays contain the survivors and their distances and are written onto the old arrays. Both the total work and the path storage depend linearly on the state space size. Many VA hardware chips are also available, and these work by similar principles.

The encoder here began from a known state, and most trellis encoders do so, even if the state is not 0. Very often, the transmission has a defined end state too, in which case it is said that the trellis is *terminated*. It can take as many as m trellis stages to bring the encoding to a specified end state, but in return the decoder looks at a proper trellis section even for the last data symbols, and the error rate for these symbols is considerably improved.

Although we have focused on binary convolutional trellis codes, the VA differs little when applied to other codes or to a soft-output channel. In every case, a channel output is compared to a stage's worth of branch symbols, whether these symbols are binary, real, or whatever. Hamming or Euclidean distances are computed and survivors are chosen. After some L_D, a data decision is made. The VA's easy handling of real-valued channel outputs can be a major advantage: As we saw in Section 6.3, soft outputs in themselves lead to a potential 3-dB coding gain.

In addition to being an effective decoder, the Viterbi algorithm provides a simple means of finding the free distance of a linear code. Because of the group property, it is sufficient to find the closest merged trellis path to the all-zero code word path. By setting the received channel sequence to all zeros, we obtain the closest path automatically from the VA, by checking the merging survivors at each stage. The check can stop when all node distances at a stage equal or exceed the closest merge so far observed. Since distances are nondecreasing, no closer merge can ever occur.

6.5.2 Trellis Error Events

Just as the mechanism of error in block coding was word error rather than data bit error, error in trellis decoding involves a whole path event, not a single branch. A trellis *error event* is defined to begin when the decoder output path splits from the transmitted path and to end when it merges again. Whatever data bit errors occur stem from such events. A data error generally occurs at the split, but in some codes (e.g., feedforward convolutional) one does not occur at the merge.

It is useful to characterize error events as long or short. No error event can be shorter than $m + 1$, where m is the memory of the encoder; this is because the symbol difference at the path split must work its way through the shift register (or other) memory. Events classified as "short" are those that confuse the transmitted path with one that lies at the free distance away, or a little more; these occur typically in good channels, because events lying at greater distances are much less likely. The data bit error rate is 1–2 times the event error rate except for TCM codes, where it can be many times larger.

The signal space theory in Chapter 2 can be used to show that if a trellis decoding is correct up to some node, and if κ future path alternatives split away there and lie at distance d, the probability that one of these alternatives will be chosen in AWGN is about

$$\kappa Q\left(\sqrt{\frac{d^2 E_b}{N_0}}\right) \tag{6.5-2}$$

In a good channel, alternatives farther away than the minimum distance have little effect. If the decoder decision depth L_D is long enough, only merged alternatives need to be considered. The free distance is the smallest such merged d that can be encountered. If L_D is too small, then unmerged neighbor paths must be considered, some will lie closer than the free distance, and error performance at larger E_b/N_0 will not be as good as (6.5-2).

In poorer channels, "long" error events can occur. In these, the decoder wanders lost for some time, and a 50% data bit error rate typically occurs while the event lasts. Long events can happen with reduced-search decoders even when the channel is good. Whether or not this sort of event is more serious than just a few bit errors depends on the application. Often, data frames are viewed as either corrupted or not and all failures are equally serious, and then only the probability of an event is of interest, not its length. In other applications, such as speech transmission, short events are preferred.

In many practical data systems, codes act as much to signal failures as to correct errors. If a frame appears to be corrupted, the receiver can ask for a retransmission, a method called *acknowledge/request repeat* (ARQ). It is important in these systems that the decoder not falsely think that it has corrected all the errors.

Some actual error rates for convolutional codes are shown in Fig. 6.17. It shows BER for two convolutional feedforward rate $\frac{1}{2}$ codes, the memory $m = 2$ code with generators (7, 5), labeled '2', and the $m = 4$ code (72, 62), labeled '4'. These have among the lowest BERs for codes with their memories (see Ref. 11); their Hamming minimum distances are 5 and 7. Transmission is over the two binary-input antipodal modulation channels in Section 6.3.2, the AWGN channel with ± 1 hard-decision outputs (labeled 'BSC'), and the same with the real-value matched filter outputs (labeled "AWGN" in Fig. 6.17 and modeled in Fig. 6.2). The soft-output bound (6.5-2), with the substitution $d^2 = 2Rd^h_{\min}$ from (6.3-6), is plotted for both

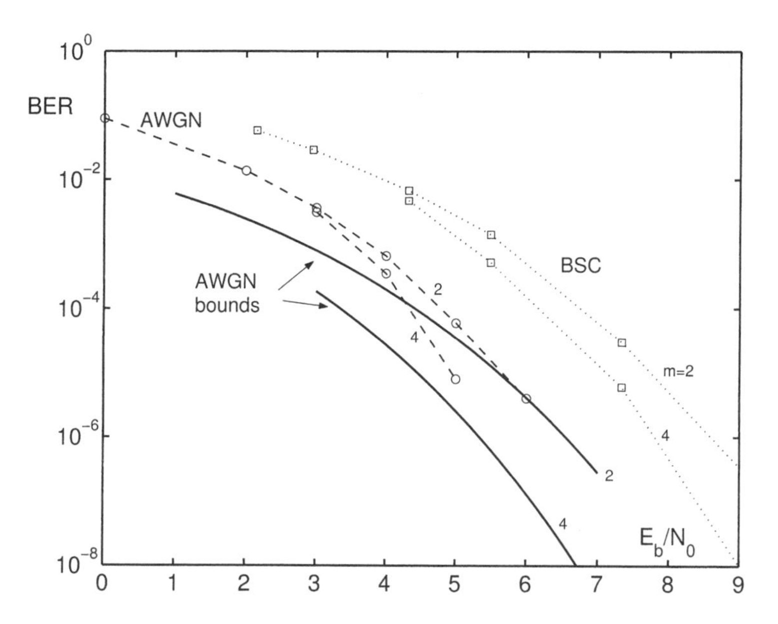

Figure 6.17 Measured bit error rate (BER) versus E_b/N_0 in dB for good memory 2 and 4 rate $\frac{1}{2}$ convolutional codes over the hard ("BSC") and soft ("AWGN") output AWGN channel, with comparison to distance-based bound (6.5-2).

codes. The decoder is the BCJR algorithm, a trellis decoding method similar to the Viterbi algorithm but with better BER in poor channels.

The predicted coding gains for the $m = 2$ code, taken from Eqs. (6.3-5) and (6.3-7), are $G_h = 10 \log \frac{1}{2}(2 + 1) = 1.8$ and $G_s = 10 \log 2(\frac{1}{2})(2 + \frac{1}{2}) = 4.0$ dB for the hard and soft channel, respectively. At BER 10^{-5} the actual gains are 1.9 and 4.0 dB (9.6 dB for antipodal signaling minus about 7.7 and 5.6). For the longer $m = 4$ code, G_h and G_s are 3.0 and 5.4 dB, and the actual gains are about 2.5 and 4.7 dB. The outcomes here are typical of well-selected convolutional codes: The match is poorer with more complex coding and the gain of the soft over the hard output channel is 2–2.5 dB. One can also compare Fig. 6.17 to the uncoded antipodal signaling error rate in Fig. 2.28 and see that only the AWGN curves are better than uncoded transmission all the way down to an E_b/N_0 of 0 dB. Codes can generally be found that improve the error rate at all E_b/N_0 (see Ref. 11), but not all do, and soft channel outputs are required.

Figure 6.18, reproduced from Ref. 14, shows the BER for TCM codes of the type in Fig. 6.15. The data are adapted from Ref. 17. The codes are (a) the one in the figure with 4 states and (b) two with the same 8PSK subsets but 8 and 128 states. Bounds of type (6.5-2) are given. All these codes have rate 2 data bits per use of the QAM channel, or 1 bit/AWGN channel use, and the decoder is the VA. Compared to QPSK at BER 10^{-5}, these codes have 2.5- to 4.0-dB gain *in the same bandwidth*. By comparison, the convolutional codes in Fig. 6.17 double the transmission bandwidth, and we thus see a reason why TCM codes are attractive.

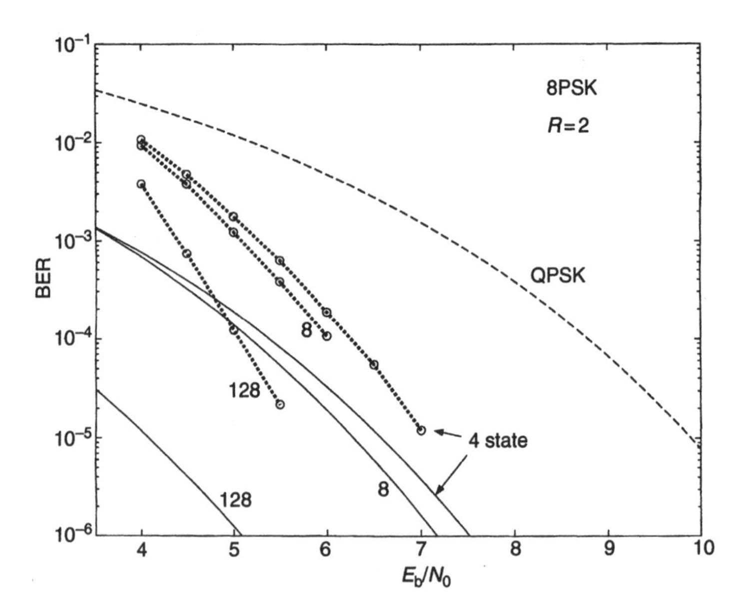

Figure 6.18 Measured BER versus E_b/N_0 in dB for three TCM codes based on subsets of 8PSK, with comparison to bound of type (6.5-2) and BER of QPSK. Codes have 4,8,128 states and rate 2 data bits/QAM channel use. (From Anderson and Svensson [14]. Copyright Plenum Publishing Co., 2003. Reproduced with permission).

PROBLEMS

Entropy and Capacity

6-1. Surprisingly, it has been discovered that the initial 1 Euro coin is not balanced, but in at least one version (Belgium), heads come up with probability of 0.56.[8]

(a) Find the entropy of the 1 Euro coin.

(b) Nonetheless, successive tosses are independent. Find the entropy in 0 tosses.

(c) An enterprising gambler finds a way to toss the Euro so that the second of two tosses matches the first with probability 0.9. Find the entropy in the dual two-toss outcome. There are four such outcomes.

6-2. Two independent and fair dice are tossed.

(a) What is the entropy in their joint outcome?

(b) Let z by the *sum* of the two outcomes. What is $H(z)$?

6-3. In an analogy to (6.2-3), derive the MAP receiver instead of the ML receiver. This takes into account the a priori probability of each word.

6-4. A binary channel has transition matrix

$$p(y|x) = \begin{bmatrix} .9 & .1 \\ .4 & .6 \end{bmatrix}$$

(a) Find $p(y)$ if the inputs x are equiprobable.

(b) Find the entropy of Y.

(c) Find the mutual information $I(X; Y)$.

6-5. Find the capacity of the channel in Problem 6.4. What BSC has the same capacity?

6-6. A BSC with crossover probability 0.001 is a very good channel.

(a) Find its capacity.

(b) Consider the single parity-check (6, 5) code. Over this channel, what is its probability of correct decoding, error detection, and undetected error?

(c) At what crossover probability does the rate of this code equal the capacity of the BSC?

6-7. Suppose two BSCs with crossover p are arranged so that one feeds the other, in tandem. Find the transition matrix of the two in tandem. Find an expression for its capacity.

6-8. Suppose that three BSCs work in parallel. The same symbol is sent over all three, and at the receive end a majority logic decision is made from the three received bits (i.e., receiving 1, 1, and 0 means the decision is 1).

(a) If each BSC has crossover p, give the transition matrix of a binary-input binary-output channel that models the whole transmission system.

[8]See *The Guardian,* Manchester, UK, 4 January 2002.

(b) Compare the error probability of this system to that of one using a single BSC with a (3, 1) repetition code.

6-9. Consider a BSC that is in reality a BPSK modulator/demodulator over an AWGN channel with $E_b/N_0 = 10$ dB.

(a) Find the capacity of this channel.

(b) Find the word error probability of the (7, 4) Hamming code with this BSC.

(c) Repeat both calculations when E_b/N_0 is 0 dB.

Block Codes and Minimum Distance

6-10. Show that when a parity-check code is used over an AWGN channel with binary antipodal modulation, two code words that are $d_{\min}$ apart in Hamming distance are $4E_s d_{\min}$ apart in Euclidean signal space distance, where E_s is the modulator symbol energy.

6-11. Consider a (N, K) binary parity-check code with odd minimum distance. Suppose that a bit equal to the mod-2 sum of the code word bits is added to the end of every code word, so that a $(N+1, K)$ code is formed. This is called an *extended* code.

(a) Form the (8, 4) extension of the (7, 4) Hamming code and list its words.

(b) Find the minimum distance.

(c) Show that the minimum distance of an extended code is always one plus the original code's distance.

(d) Show that the (8, 4) code corrects all single errors and detects all double errors.

6-12. Consider the (15, 11) Hamming code over the BSC.

(a) Give an expression in terms of the crossover p for the probability of word error.

(b) Give the coding gain for this code over the BSC and the AWGN (soft output) channels.

(c) Give the $\boldsymbol{H}$ and $\boldsymbol{G}$ matrices.

6-13. Consider the (6, 3) code in the standard array example of Fig. 6.7.

(a) Give an expression in terms of p for the probability of word error over the BSC.

(b) Compute the probability for $p = 0.01$.

(c) Give the asymptotic coding gain for this code over the BSC and the AWGN (soft output) channels, and compare the result to the gains for the (7, 4) Hamming code.

6-14. Show that all Hamming codes have minimum distance 3.

6-15. Prove that $(1/N)\log \binom{N}{d}$ tends to $h_B(d/N)$ as $N \to \infty$. This is the basis of the result (6.3-11). (*Hint:* Use Stirling's estimate for factorials.)

6-16. According to the Gilbert–Varshamov bound, binary codes of rate $R = K/N$ bits/channel use exist, whose distance grows with block length N at least as fast as Eq. (6.3-11).

(a) Assume antipodal signaling, the AWGN channel, and ML decoding. Give an expression for the soft gain G_s of such codes in terms of K, N, and R.

(b) Consider a given probability of decoding error, P_e. As N grows at a fixed R, is this gain available at P_e? Argue why or why not.

(c) Make of plot of G_s against K for the best known codes in Fig. 6.8. According to (a), what slope should this line have?

Convolutional Codes

6-17. Table 6.3 lists convolutional codes that are among the best known in terms of minimum distance.

(a) For memories 2–6 (1–3 for rate $\frac{2}{3}$), compute the asymptotic soft and hard coding gains, G_s and G_h, and add them to the table.

(b) The lower rate $\frac{1}{3}$ codes have more parity checks and we might suspect that they have more gain than the rate $\frac{1}{2}$ codes. Is this necessarily true? Considering the fact that rate $\frac{1}{3}$ has twice the bandwidth of rate $\frac{2}{3}$ and 50% more than rate $\frac{1}{2}$, which codes are most effective in Table 6.3?

6-18. Consider the rate $\frac{1}{2}$ convolutional code whose generators are (40, 64) in octal notation.

(a) Draw an encoder based on a shift register.

(b) Give the generator matrix $\boldsymbol{G}$ when the code is to be used in block form with block length $N = 12$.

(c) From your $\boldsymbol{G}$, find $\boldsymbol{H}$.

(d) What is the code's minimum distance?

6-19. The rate $\frac{1}{2}$ memory 2 convolutional code with the best free distance is the one with octal generators (7, 5).

(a) Draw a shift register encoder circuit.

(b) Draw a few sections of the trellis.

(c) Using the trellis, verify that the free distance of this code is 5.

6-20. The rate $\frac{1}{3}$ memory 2 convolutional code with the best free distance has octal generators (7, 7, 5). Repeat Problem 6-19 for this code.

6-21. Consider again the (7, 7, 5) rate $\frac{1}{3}$ code in Problem 6-20. The sequence 000 110 110 110 001 is received from the BSC channel (the first triple is the first to arrive). Carry out the Viterbi algorithm (VA) and find the closest 5-stage trellis path to this sequence.

6-22. Using the VA setup in Problem 6-21, find the free distance of the (7, 7, 5) rate $\frac{1}{3}$ code.

REFERENCES

1. *R. G. Gallager, *Information Theory and Reliable Communication,* McGraw-Hill, New York, 1968.

[9]References marked with an asterisk are recommended as supplementary reading.

2. *T. M. Cover and J. A. Thomas, *Elements of Information Theory,* Wiley, New York, 1991.

3. *C. E. Shannon, A mathematical theory of communication, *Bell Syst. Tech. J.,* vol. 27, July 1948, pp. 379–429 and 623–656; all of Shannon's papers are reprinted in *Claude Elwood Shannon: Collected Papers,* Sloane and Wyner, eds., IEEE Press, New York, 1993.

4. C. E. Shannon, Communication in the presence of noise, *Proc. IRE,* vol. 37, 1949, pp. 10–21.

5. F. J. MacWilliams and N. J. A. Sloane, *The Theory of Error-Correcting Codes,* North-Holland, New York, 1977.

6. *S. Lin and D. J. Costello, Jr., *Error Control Coding,* Prentice-Hall, Englewood Cliffs, NJ, 1983, 2004.

7. *R. E. Blahut, *Theory and Practice of Error Control Codes,* Addison-Wesley, Reading, MA, 1984.

8. *J. B. Anderson and S. Mohan, *Source and Channel Coding,* Kluwer, Boston, 1991.

9. *G. C. Clark and J. B. Cain, *Error-Correction Coding for Digital Communications,* Plenum, New York, 1981.

10. R. Johannesson and K. Zigangirov, *Fundamentals of Convolutional Coding,* IEEE Press, New York, 1999.

11. J.B. Anderson, Best short rate 1/2 tailbiting codes for the bit-error criterion, *IEEE Trans. Commun.,* vol. COM-48, April 2000, pp. 597–610.

12. J. B. Anderson, T. Aulin, and C.-E. Sundberg, *Digital Phase Modulation,* Plenum, New York, 1986.

13. J. Proakis, *Digital Communications,* 3rd ed., McGraw-Hill, New York, 1995.

14. J. B. Anderson and A. Svensson, *Coded Modulation Systems,* Plenum, New York, 2003.

15. G. Ungerboeck, Trellis-coded modulation with redundant signal sets—Parts I and II, *IEEE Commun. Mag.,* vol. 25, February 1987, pp. 5–21.

16. A. J. Viterbi, Error bounds for convolutional codes and an asymtotically optimum decoding algorithm, *IEEE Trans. Information Theory,* vol. IT-13, April 1967, pp. 260–269.

17. W. Zhang, Finite-state machines in communications, Ph.D. Thesis, Digital Communications Group, University of South Australia, Adelaide, 1995.

CHAPTER 7

ADVANCED TOPICS

We live in an imperfect world. The theme of this final chapter is adapting the ideal receiver designs that come earlier to the imperfect channels in the practical world.

The chapter falls into three divisions. We found in Section 2.4 and later in 3.8 that filtering of signals produces intersymbol interference, but did not show how to modify the linear or maximum likelihood receivers in order to relieve the effects. A processor that does this is called an *equalizer,* and these are the subject of Sections 7.2–7.3. An important and troublesome topic here and in all receiver design is converting signals to discrete time, and this comes before in Section 7.1. The development in Sections 7.1–7.3 focuses on the Gaussian channel, but the techniques are applied much more widely. The second part of the chapter is motivated by Chapters 4 and 5. Section 7.5 presents some special receivers for fading channels. A different set of maladies appears when carrier phase or symbol timing is poorly known. This is a failure of the equipment in Chapter 4, and some receivers for these cases appear in Section 7.4.

A final topic, spread spectrum, appears in Section 7.6. This has been around many years but has gained importance as the basis of third-generation mobile telephony. Spread spectrum combats fading, but it also provides an alternative method of multiple access.

7.1 DISCRETE-TIME CHANNEL MODELS

Every transmission link somewhere contains information bearing signals that are continuous functions of time. Yet we would like to work with signal samples in discrete time in a digital communication system if possible, especially if the processing has any sophistication. Thus two interrelated design problems present them-

Digital Transmission Engineering, Second Edition. By John B. Anderson

selves: How to convert efficiently from continuous to discrete time and how to model continuous data signals and channels in discrete time.

As digital processors become faster, they are finding their way more and more toward the front of radios. Even IF amplifiers sometimes work in discrete time, and digitally implemented detectors and matched filters are common. Such a radio may begin with an analog RF amplifier and channel tuning circuit, which performs a basic bandpass filtering. The conversion to discrete time that might come before the IF amplifier can be a simple analog-to-digital converter. The processing that follows can even be lines of software in a microprocessor; in this case the receiver is called a *software radio*. Such a radio can shift from analog FM to 16QAM simply by changing its program.

For their part, equalizers are almost always discrete-time processors. As well, they are applied mostly to PAM and QAM modulations when the channel distorts but is linear. Therefore, we aim to develop discrete-time models for distorted linear channels that carry standard orthogonal-pulse linear modulation. The basics of this scenario were explained in Section 2.2 for baseband and 3.5 for carrier modulation. In the classic scenario, the receiver chain is: IF ampifier–matched filter–sampler–detector. If the first two are analog, a rate-$1/T$ discrete-time conversion occurs at the sampler. The classic modeling problem is to model the sequence after the sampler. Alternately, A-to-D conversion can come before the matched filter (or earlier), and then the matched filter–sampler are digitally implemented at a higher rate. But the fundamental modeling problem remains the same: How do the discrete-time sampler outputs depend on the data.

By means of the quadrature receiver (Section 3.3), the detection of QAM is reduced to the detection at baseband of two independent PAMs. In order to simplify the discussion, we focus on just the baseband PAM case, both for modeling and for the equalizer designs in Sections 7.2–7.3.[1]

To start, then, here is a standard PAM model, taken from Chapters 2 and 3. The transmitted signal is

$$\sqrt{E_s} \sum a_n v(t - nT) \tag{7.1-1}$$

in which

$$\mathcal{E}[a_n] = 0, \qquad \mathcal{E}[|a_n|^2] = 1$$

and $\{a_n\}$ are IID transmission symbols. When we last saw the channel modeling subject, we proved the modeling theorem, Theorem 2.6-2, which shows that a pulse train (7.1-1) over an AWGN channel can be precisely modeled by the discrete-time model

$$r_n = \sqrt{E_s} a_n + \eta_n \tag{7.1-2}$$

[1]What follows can be organized for QAM into equivalent carrier modulation notation. One way is to employ complex variables in the equations that follow. See Refs. 1 and 3.

in which $\{r_n\}$ are received values and $\{\eta_n\}$ are zero-mean IID white-noise variates with variance $\frac{1}{2}N_0$. The theorem holds if and only if the pulses $v(t - nT)$ are orthogonal and the receive filter is matched to $v(t)$ (see Fig. 2.33). Now the channel distorts $v(t)$ and we need to replace model (7.1-2).

7.1.1 Receiver Models

Sometimes a discrete-time model is simply given, perhaps as a result of measurements, and we know not what lies behind it. Other times the model derives from the channel and the chosen receiver structure. An AWGN intersymbol interference channel model is shown in Fig. 7.1; the ISI is modeled by the filter $G(f)$ with impulse response $g(t)$, and the total response of the channel is $v * g$. This response is denoted $h(t)$. As for the receiver, many structures have been proposed, but we shall focus on three fundamental ones that are shown at the right in Fig. 7.1. These are summarized now.

1. The *MLSE receiver* is equivalent to Fig. 2.30 or 2.32. As explained in Section 2.6, it computes the ML estimate of the symbols embodied in Eq. (2.5-4). It detects the entire sequence $\{a_n\}$ at once by finding the whole signal $s_i(t)$ that lies closest in the square Euclidean distance

$$\|\mathbf{s}_i - \mathbf{r}\|^2 = \int |s_i(t) - r(t)|^2 \, dt$$

to the received signal $r(t)$. All signals $s_i(t)$ are convolutions of a PAM signal with $g(t)$, the impulse response of $G(f)$, with the form

$$\sqrt{E_s}\Big[\sum a_n\, v(t-nT)\Big] * g(t) = \sqrt{E_s}\sum a_n\, v(t-nT) * g(t) = \sqrt{E_s}\sum a_n h(t-nT)$$

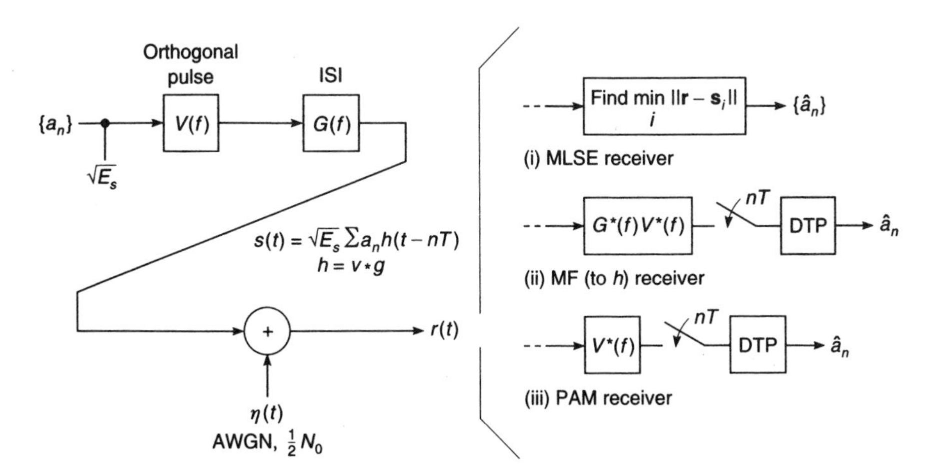

Figure 7.1 Continuous-time model for linear ISI with three standard receiver models (h normalized to unit energy; DTP, discrete-time processor).

The receiver here is called the MLSE receiver because it performs a maximum likelihood sequence estimation. Its importance is that it provides a performance benchmark, since no other receiver that detects the whole sequence has lower error probability.

2. The *matched filter (MF) receiver* begins with a filter matched to the total response $v * g$, after which comes a symbol-time sampler. The sampler sequence consists of a transmitted component, denoted y_n, plus a noise component ξ_n; the $\{\xi_n\}$ are in general correlated Gaussians. Following the sampler is some some kind of discrete-time processor. It is possible to design a complicated processor that makes the whole receiver equivalent to the MLSE receiver; in this case we will call it an MF receiver. Other times, the processor is simpler, and we will then call the processor an MF equalizer.
3. The *PAM receiver* also begins with a matched filter followed by a symbol-time sampler, this time matched only to the orthogonal PAM or QAM pulse $v(t)$. The receiver is not necessarily ML. Because there is ISI, the samples, even if noise-free, are not the transmission symbols, as they were in (7.1-2). The noise samples $\{\eta_n\}$, however, are still IID Gaussian with variance $\frac{1}{2}N_0$, because of the orthogonal filter response and Theorem 2.6-2. (For this reason we keep the notation η from Chapter 2.) The transmitted signal component y_n will differ from case 2, because the model differs. As in 2, a processor box follows. If the box is a simple equalizer, it will be called a PAM equalizer.

These, then, are the three major channel model/receivers. The first serves as a performance benchmark for a variety of equalizers that can be based on the other two.

The MLSE Benchmark. Many receivers have been proposed that attain the MLSE benchmark. Of these we can mention variants of the MF receiver 2 in Fig. 7.1 by Messerschmitt [2] and Ungerboeck [5], among others, that add to 2 a special distance measure that deals with the correlated noise. The whitened-matched filter of Forney [4] contains the $V(f)G(f)$-matched filter in the figure, but adds to it a filter that creates a new sample set with white noise, so that the usual Euclidean distance measure may be used. It is shown in Ref. 7 that all these receivers perform the same calculations but in a different order. There are other MLSE receivers as well, but we will work only with the simplest receiver to understand, which is the basic distance-minimizing receiver at the top in the figure.

If there are L M-ary transmission symbols, the basic MLSE receiver must compare M^L signals $s_i(t)$. The probability that it decides in error is computed by the standard argument in Section 2.6, applied to M^L signals, and the error estimate there, (2.6-17), becomes

$$P \leq (2K/M^L)\, Q(\sqrt{d^2_{\min} E_b/N_0}) \tag{7.1-3}$$

This estimate becomes tight at large E_b/N_0. For the benchmark we will use the normalized minimum distance $d_{\min}$ that appears here.

Formally, $d^2_{\min}$ is $D^2_{\min}/2E_b$, in which E_b is the average energy per data bit over all the signals and

$$D^2_{\min} \triangleq \min_{i,j} \|\boldsymbol{s}_i - \boldsymbol{s}_j\|^2 \tag{7.1-4}$$

Although the minimization is simple in principle, it grows overwhelmingly in size, at least at first glance. Fortunately, techniques have been invented that simplify the calculation. The generation of signals in Fig. 7.1 is really a form of convolutional encoding, as in Chapter 6, but with numbers in the field of reals. With some modification, distance-finding techniques and insights devised for ordinary convolutional codes can be applied to these new "convolutional codes." For our purposes we can summarize the findings as follows:

- For finite ISI, Forney in Ref. 4 pointed out that a trellis structure exists, just like the ones in Chapter 6. Thus a dynamic program finds $d_{\min}$.
- For infinite ISI, such as arises from filters with poles, the complexity of (7.1-4) depends on the effective length of the response. Efficient methods have been devised to find $d_{\min}$ for the infinite-response case in Refs. 8–10.
- For linear modulation with linear ISI, $d^2_{\min}$ often equals and is never greater than the *antipodal signal distance* d_0^2. The distance d_0 is also called the *matched filter bound* and is the minimum distance of unfiltered PAM with orthogonal pulses.

The antipodal distance d_0 is at the heart of a practical procedure to calculate the MLSE error expression, and so we explore it now further. Before proceeding, we need to calculate the average signal energy E_s. The filtered PAM signal is

$$s(t) = \sqrt{E_s} \sum a_n h(t - nT), \qquad E_s = E_b \log_2 M \tag{7.1-5}$$

where $h = v * g$. E_s is the average symbol energy when h is normalized to unit energy. This is proven as follows. Let there be N pulses. The average energy is

$$\frac{1}{N} \mathcal{E}\left[\int |\sqrt{E_s} \sum a_n h(t - nT)|^2 \, dt\right]$$

$$= \frac{E_s}{N} \mathcal{E}\left[\int \sum_n \sum_{m \neq n} a_n a_m h(t - nT) h(t - mT) \, dt + \int \sum_{n=1}^{N} |a_n h(t - nT)|^2 \, dt\right]$$

$$= \frac{E_s}{N} \sum_n \sum_{m \neq n} \mathcal{E}[a_n a_m] \int h(t - nT) h(t - mT) \, dt + \frac{E_s}{N} \sum_{n=1}^{N} \mathcal{E}[|a_n|^2] \int |h(t - nT)|^2 \, dt$$

The first term here is zero from the model (7.1-1) conditions; the second is E_s, since both the energy in $h(t)$ and $\mathcal{E}[|a_n|^2]$ are unity. Thus the average energy is E_s.

A binary antipodal signal pair consists of two equal and opposite signals, such as would be generated by the two single-symbol sequences +1 and –1. This signal pair is $\{+h(t), -h(t)\}$. The normalized Euclidean distance between these is easy to find; it is

$$d_0^2 = \frac{\int |h(t) - (-h(t))|^2 \, dt}{2\int |h(t)|^2 \, dt} = 2 \tag{7.1-6}$$

The same calculation gives the minimum distance of straight binary PAM [see Eq. (2.6-5)]. The upper bound idea mentioned above means that at large E_b/N_0, p_e for any filtered binary transmission approximately obeys the universal bound[2]

$$p_e \leq Q(\sqrt{2E_b/N_0}) \tag{7.1-7}$$

The extension to quaternary one-pulse signaling is not much harder. The four signals are $\{\pm\sqrt{2/10}h(t), \pm 3\sqrt{2/10}h(t)\}$, as explained further in Example 3.5-1. The antipodal signal distance is the worst-case normalized distance between any two of them, which works out to be $d_0^2 = 0.8$. Thus the p_e bound for uniformly spaced quaternary transmission is $Q(\sqrt{0.8E_b/N_0})$.

By normalizing the energy of $h(t)$, we place the energy of the channel in E_s and the ISI property of the channel separately in $h(t)$. The energy is measured at the receiver. There are several cases to distinguish. In a radio channel, the ISI probably stems from the transmitter filter, and the energy is measured when it arrives at the receiver; a given energy at the filter output yields the same E_s at the receiver regardless of the filter, if all filters are energy-normalized. The ISI loss is contained in $d_{\min}$. In a wire or cable channel, the filtering is in the medium itself. The energy entering the medium is set by its energy-handling capability; E_s is what survives at the receive end. The additional loss from ISI is again contained in $d_{\min}$.

Now we can put all these results together in a procedure, the object of which is to find the benchmark MLSE error estimate for a linear modulation with ISI. The procedure finds $d_{\min}$ from a given $h(t)$. It is based on finding the Euclidean distance between two signals, whose difference is

$$\sum a_n^{(1)} h(t - nT) - \sum a_n^{(2)} h(t - nT) = \sum \Delta a_n h(t - nT)$$

The sequence of *symbol differences,* $\{\Delta a_n\}$, is nonzero over a short range where the symbols differ. In the binary case, Δa takes the three values 0, ±2. The procedure tries out a number of short difference sequences in order to find the one leading to the least distance. While this is brute force, with modern computing equipment there is usually no reason to employ one of the more efficient methods that have been devised [8–10].

Here is the procedure:

1. Given the pulse $v(t)$ and the ISI as $g(t)$, convolve these to obtain $h(t)$. Normalize $\mathcal{E}[|a_n|^2]$ and the energy of $h(t)$ to unity.

[2] To be more precise: $\log p_e / \log Q(\sqrt{2E_b/N_0})$ tends to a limit ≤ 1.

2. Compute square distances via

$$\frac{1}{2}\log_2 M \int \left|\sum \Delta a_n h(t-nT)\right|^2 dt \tag{7.1-8}$$

for a number of symbol difference sequences $\{\Delta a_n\}$. For the binary case, begin with $\{\Delta a_1, \Delta a_2\} = \{2, -2\}$. Note that sequence 2, 0, 0, . . . gives d_0^2, the antipodal distance.

3. Longer sequences tend to lead to larger distances. When it appears that the minimum over all sequences has been found, stop. The estimate to $d_{\min}$ is the minimum of d_0 and the least distance found. The error estimate is $p_e \leq Q(\sqrt{d_{\min}^2 E_b/N_0})$.

In case the model presented is an $H(z)$, already discrete time, the same procedure estimates $d_{\min}$, except that the signals are samples and the distance integral in step 2 is replaced by the discrete convolution

$$\frac{1}{2}\log_2 M \sum_m \left|\sum_n \Delta a_n h_{m-n}\right|^2 \tag{7.1-9}$$

Next come three infinite response examples that illustrate the MLSE benchmark and provide the basis for the equalizer examples in later sections. Afterward, Program 7.1-1 is given, which is an implementation of the above distance procedure. The first example is a standard case of radio channel filtering that we have seen before in Chapter 2.

Example 7.1-1: Six-Pole Butterworth Binary Channel. Example 2.4-1 describes a standard binary transmission system consisting of 30% root RC pulses followed by a six-pole Butterworth channel filter. Figures 2.16 and 2.17 show the frequency and impulse responses of this filter in its baseband version. Figure 2.18 shows the binary eye patterns as they would appear after a filter matched to the root RC filter only. Were there no equalizer, the PAM receiver would have to deal with the ISI in these pictures. This example focuses on the $0.5/T$ bandwidth filter, a filter that interferes with just the outermost spectrum of the root RC pulse. Following the $d_{\min}$ procedure via Program 7.1-1, we compute the total response $h(t) = v * g$ and normalize it. Here $v(t)$ is the pulse and $g(t)$, the six-pole Butterworth impulse response, is computed as in the example calculation in Program 7.1-1. The response $h(t)$ is compared to $v(t)$ in Fig. 7.2. The antipodal signal square distance d_0^2 stems from the symbol difference sequence $\{2\}$, and, as always with binary signaling, it is 2. The actual antipodal signals are shown in Fig. 7.3. Next, we try out a few more symbol difference sequences, such as $\{2, -2\}$; this one yields

$$\int |\{2, -2\} * h(t)|^2\, dt = 3.7$$

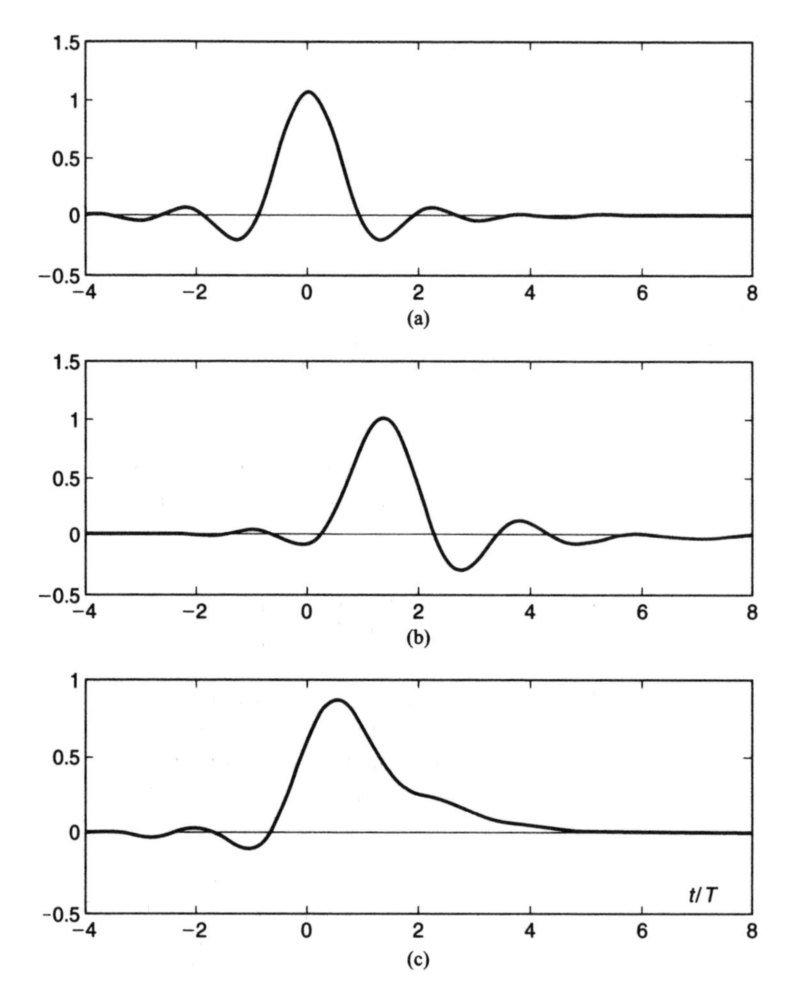

Figure 7.2 Plots of continuous-time responses from Examples 7.1-1 and 7.1-2: **(a)** original 30% root RC orthogonal pulse $v(t)$; **(b)** total $h(t) = v * g$ after six-pole Butterworth ISI; **(c)** total $h(t)$ after exponential ISI in Eq. (7.1-10).

Since this value is much larger than 2 and since the integral grows rapidly with longer sequences, we can be quite sure that $d^2_{\min} = d^2_0 = 2$; that is, the MLSE receiver suffers no loss compared to the interference-free PAM error probability, which is $Q(\sqrt{2E_b/N_0})$.

Example 7.1-2: Analog One-Pole Exponential Binary Channel. In this channel, the impulse response is $e^{-at}u(t)$, $a > 0$, corresponding to an s–plane pole at $-a$. For a numerical value, take $a = \ln 2$. As in the previous example, use Program 7.1-1 to compute

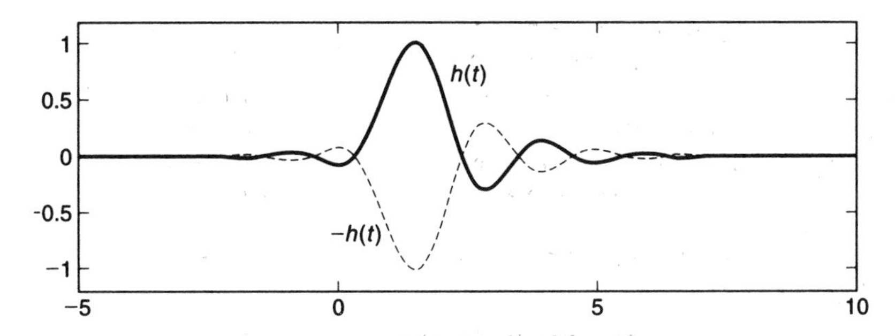

Figure 7.3 Pair of antipodal signals in Example 7.1-1; these stem from concatenation of 30% root RC pulse with six-pole Butterworth filter.

$$h(t) = v(t) * e^{-at}u(t) = \int_{-\infty}^{t} v(\tau)2^{-(t-\tau)}\, d\tau \qquad (7.1\text{-}10)$$

Then normalize to unit energy. The normalized $h(t)$ appears in Fig. 7.2c. Next evaluate the distance in some symbol difference sequences:

$\{2, -2\}$	gives 1.60
$\{2, -2, 2\}$	gives 2.31
$\{2, 0, 2\}$	gives 2.88
$\{2, -2, 2, -2\}$	gives 2.46
$\{2, -2, -2\}$	gives 4.88

Longer sequences give large distances. Apparently, the square minimum distance is 1.60, and the MLSE error probability estimate is $Q(\sqrt{1.60\, E_b/N_0})$. This represents a 1.0-dB loss in energy efficiency, compared to binary PAM without ISI.

Example 7.1-3: Discrete-Time One-Pole Exponential Binary Channel. Here a discrete-time model has been given to us, whose z-transform $H_D(z) = 1/(1 - \alpha z^{-1})$ has a pole at $z = \alpha$. The situation here needs to be distinguished carefully from the previous example, where the pole was an analog one. The present example might derive, for example, from a multipath channel with nT-delayed components whose analog response was

$$h(t) = \sum_{n=0}^{\infty} \alpha^n v(t - nT)$$

After a filter matched to $v(t)$ and a symbol rate sampler, the noise-free received sequence would be $\{1, \alpha, \alpha^2, \ldots\}$, which has z-transform $H_D(z)$. The sequence has energy $1 + \alpha^2 + \alpha^4 + \ldots = 1/(1 - \alpha^2)$, so that the normalized $H_D(z)$ is

$$\frac{\sqrt{1-\alpha^2}}{1 - \alpha z^{-1}} \qquad (7.1\text{-}11)$$

We could compute square distances by means of (7.1-9) for some symbol difference sequences, but the minimum distance of response (7.1-11) is available in the literature [8]. For $\alpha < 1/2$, $d^2_{\min} = d^2_0 = 2$, and the MLSE receiver suffers no distance loss from ISI; for $\alpha = 1/\sqrt{2}$, $d^2_{\min} = 1.17$ and so $p_e \approx Q(\sqrt{1.17\, E_b/N_0})$. Compared to an antipodal system, this has energy loss $10 \log_{10} (2/1.17) = 2.3$ dB.

It remains to give the program for calculating the distance between two transmitted signals.

Program 7.1-1: Distance Between Signals in ISI. The following MATLAB program computes the Euclidean square distance (7.1-8) between the two baseband signals:

$$s_1(t) = \sum a_n^{(1)} h(t - nT)$$

$$s_2(t) = \sum a_n^{(2)} h(t - nT)$$

With no loss of generality, set $T = 1$. In what follows, the vector of nonzero differences $\Delta a = a_n^{(1)} - a_n^{(2)}$ is the variable `delta`; `delta[1]` is the first nonzero difference. The variable `m` is the alphabet size of the PAM; `hh` is the convolution of the ISI response `gg` and the basic pulse `vv` (the last need not be orthogonal). All three of these are vectors of samples taken `fs` times per second. The pulse `vv` has total width `wid` symbol intervals.

The formal input variables are:

`gg`—samples of $g(t)$
`delta`—symbol difference sequence
`fs`—sample rate per symbol
`m`—alphabet size
`wid`—total pulse width (even)
`a`—pulse parameter

The following functions must be supplied:

`y=pulse(a, t)`—a function that returns samples `y` of a pulse at times in the vector `t`, where the pulse has unit symbol interval and pulse parameter `a` [Programs 2.2-1 and 2.2-2 are examples].

`y=pulsetr(pulse,a,fs,wid,delta)`—a function that returns samples `y` of the pulse train caused by symbols `delta` [This is given in Program 2.2-3].

The output is `dist`, the normalized square minimum distance; $h(t)$ appears as `hh`, $v(t)$ as `vv`, and the difference signal is `difsig`.

The program is then:

```
        % Find and normalize the total response
vv = pulse(a,[-wid/2:1/fs:wid/2]);        %Find total response
hh = conv(vv,gg)/fs;                      %Approx. to v*g
norm = sqrt(fs/sum(hh.^2));               %Normalization factor
hh = norm*hh;                             %h now has unit energy
        % Find the square distance
difsig = pulsetr(pulse,a,fs,wid,delta);   %Difference signal before ISI
difsig = norm*conv(gg,difsig)/fs;         %Difference signal after ISI.
dist = sum(difsig.^2)/fs);                %Energy in difference signal
dist = dist*log10(m)/(2*log10(2))         %Norm to 2Eb
```

In case the ISI model is discrete time, `hh` can be taken as $[h_0 0 \ldots 0 h_1\ 0 \ldots 0 h_2 0 \ldots 0 \ldots\ldots]$, in which h_0, h_1, . . . are the model coefficients and 0 . . . 0 represents `hh` – 1 zeros; this simulates the effect of a sequence of isolated impulses.

Example: For Example 7.1-1, the six-pole Butterworth response with cutoff at 0.5 Hz and sampling rate `fs = 20` per symbol has z-transform numerator and denominator

```
[num,den] = butter(6,.5*2/fs)
```

Then 201 samples of the impulse response `gg` are given by

```
gg = filter(num,den,[1,zeros(1,200)])
```

The remaining inputs are `m = 2` (binary transmission), `a = .3` (30% pulse) and `wid = 10` (v pulse width 10). Take the root-RC pulse of Program 2.2-2. With these inputs and `delta = [2]`, the program gives `dist = 2`, which is the antipodal signal distance. For `delta = [2 -2]` , it gives 3.70.

7.1.2 Models Derived from MF and PAM Receivers

The last two receivers in Fig. 7.1 create discrete-time channel models, because of the symbol-time sampler that follows the matched filters. Now we will show how to compute these models and give examples. Throughout, $h(t)$ has unit energy and the channel is AWGN.

First the matched-filter receiver model. The input to the receive filter $V^*(f)G^*(f)$ is the signal $s(t) = \Sigma a_n h(t - nT)$ as usual, with $h = g * v$. Accordingly, the response after the sampler to a single symbol a at time zero without noise is a sequence $\{y_n\}$, where y_n is the value of the convolution

$$[ah(t)] * [v(-t) * g(-t)] = ah(t) * h(-t)$$

at time nT, $n = \ldots, -1, 0, 1, 2, \ldots$. We define a special notation for the sequence of samples of $h(t) * h(-t)$; let

$$\rho_k \triangleq h(t) * h(-t)|_{kT} = \int h(\tau)h(\tau - kT)\, d\tau, \qquad \text{each } k \qquad (7.1\text{-}12)$$

Then the sample sequence $\{r_n\}$ produced by a sequence of symbols $\{a_n\}$ is, by the linearity of the channel and receiver,

$$r_n = \sum_j \rho_j\, a_{n-j} + \xi_n = \xi_n + \{a_n\} * \{\rho_n\}\Big|_k, \qquad n = 0, \pm 1, \pm 2, \ldots \qquad (7.1\text{-}13)$$

The noise sample ξ_n here is in general correlated with previous noise samples. From stochastic process theory, the variance of ξ_n is given by the integral of the power spectral density $\frac{1}{2}N_0$ times $|H(f)|^2$, which is Eq. (2.6-22). By Parseval's identity this becomes

$$\sigma^2 = \int \frac{1}{2}N_0\, |H(f)|^2\, df = \int \frac{1}{2}N_0\, |h(t)|^2\, dt = \frac{1}{2}N_0 \qquad (7.1\text{-}14)$$

That is, the sample variance in the channel is unchanged by our unit energy $h(t)$. The autocorrelation of successive samples is given by the inverse transform of $|H(f)|^2$ at times $0, \pm T, \pm 2T, \ldots$, which is equivalently the value of $h(t) * h(-t)$ at these times. These values are precisely the $\{\rho_k\}$ given in (7.1-12). The sequence $\{\rho_k\}$ here is the *autocorrelation* of the total analog response $h(t)$. The form of (7.1-12) shows that $\rho_k = \rho_{-k}$, and $\rho_0 = \int |h(t)|^2\, dt = 1$ since h has unit energy.

Since the convolution of the symbols and $\{\rho_k\}$ gives the sampler output, $\{\rho_k\}$ is also the discrete channel model we are looking for. In z-transforms,

$$Y(z) = H_D(z)A(z) \qquad (7.1\text{-}15)$$

plus noise, where

$$H_D(z) = \sum \rho_k\, z^{-k}, \qquad k = 0, \pm 1, \pm 2, \ldots$$

By the notation $H_D(z)$ we mean the discrete model of the analog channel $h(t)$. One interesting property of the model $H_D(z)$ is that it ordinarily contains both positive and negative powers of z; this is a statement that the peak response to a symbol a_n appears only after a delay. The frequency response $H_D(e^{j\omega T})$ has important properties too, but these are held off until Section 7.2.

Next we derive the PAM receiver model, the bottom one in Fig. 7.1. Now the receive filter is $V^*(f)$. The response to a single transmitted unit symbol this time is the convolution[3] $h(t) * v(-t)$. Denote the outcome of this convolution at time nT as

$$\gamma_n \triangleq h(t) * v(-t)|_{nT} = \int h(\tau)v(\tau - nT)\, d\tau, \qquad \text{each } n \qquad (7.1\text{-}16)$$

[3]The PAM convolution may also be written as $g(t) * p(t)$, where $p(t) = v(t) * v(-t)$. If $v(t)$ is orthogonal, then Theorem 2.2-3 states that $p(t)$ satisfies the Nyquist pulse criterion. If $v(t)$ is root RC, then $p(t)$ is an RC pulse.

Then the response to the whole symbol sequence $\{a_n\}$ is

$$r_n = \sum_j \gamma_j a_{n-j} + \eta_n = \eta_n + \{a_i\} * \{\gamma_i\}\Big|_n \tag{7.1-17}$$

Since $v(t)$ is orthogonal, Theorem 2.6-2 shows that the noise samples $\{\eta_n\}$ are IID Gaussians with variance $\frac{1}{2}N_0$.

In z-transforms the model is

$$Y(z) = H_D(z)A(z) \tag{7.1-18}$$

plus noise, where

$$H_D(z) = \sum \gamma_k z^{-k}$$

Once again, $H_D(z)$ can have both positive and negative powers of z. When there is danger of confusion, we will write this model as $H_D^{\text{PAM}}(z)$ and the model (7.1-15) as $H_D^{\text{MF}}(z)$.

As examples of the modeling process, we turn to the channels in Examples 7.1-1 and 7.1-2. The calculations are performed by Program 7.1-2, which follows the examples. We will design equalizers for both models in Section 7.2.

Example 7.1-4: Discrete Models for Six-Pole Butterworth ISI. Figure 7.4 shows the matched filter (MF) model $\{\rho_k\}$ and PAM model $\{\gamma_k\}$ for the combination of 30% root RC pulses and the baseband six-pole filter with cutoff $0.5/T$ in Examples 7.1-1 and 2.4-1. The continuous curves are the unit normalized receiver filter output (in Fig. 7.1), and the circles are the symbol-time samples at $0, \pm T, \pm 2T, \ldots$ which make up the actual models $H_D^{\text{MF}}(z)$ and $H_D^{\text{PAM}}(z)$. The zero of the time axis marks the appearance of the time-zero symbol. Note that the root RC pulse is noncausal, as are both receiver filters; noncausality is particularly apparent in the MF case. This is not so frightful as it looks. In a practical system, the transmit pulse should be limited to $[-\tau_0, \tau_0]$, delayed τ_0, and then sent as a physical pulse over time $[nT, nT + 2\tau_0)$ for symbol n; in the figure $\tau_0 = 5T$. Similarly, the theoretical receive filter would be limited to interval $[-\tau_L, \tau_R]$, with $\tau_L > 0$, and delayed τ_L to create a physical filter response on the interval $[0, \tau_L + \tau_R]$. The total system delay for a symbol is then $\tau_L + \tau_0$. In Fig. 7.4 a total delay $3T$ might be needed in the MF case and perhaps less in the PAM case. There are three to six terms in $H_D(z)$, depending on the desired model accuracy.

Example 7.1-5: Discrete Models for One-Pole Exponential ISI. Figure 7.5 shows the matched-filter model $\{\rho_k\}$ and the PAM model $\{\gamma_k\}$ for 30% root RC pulses combined with the channel response in Example 7.1-2, which is $g(t) = (0.5)^{t/T}\, u(t)$. Continuous curves show receive filter output and circles show symbol-time samples. Because the significant response of the ISI is much longer, the ISI

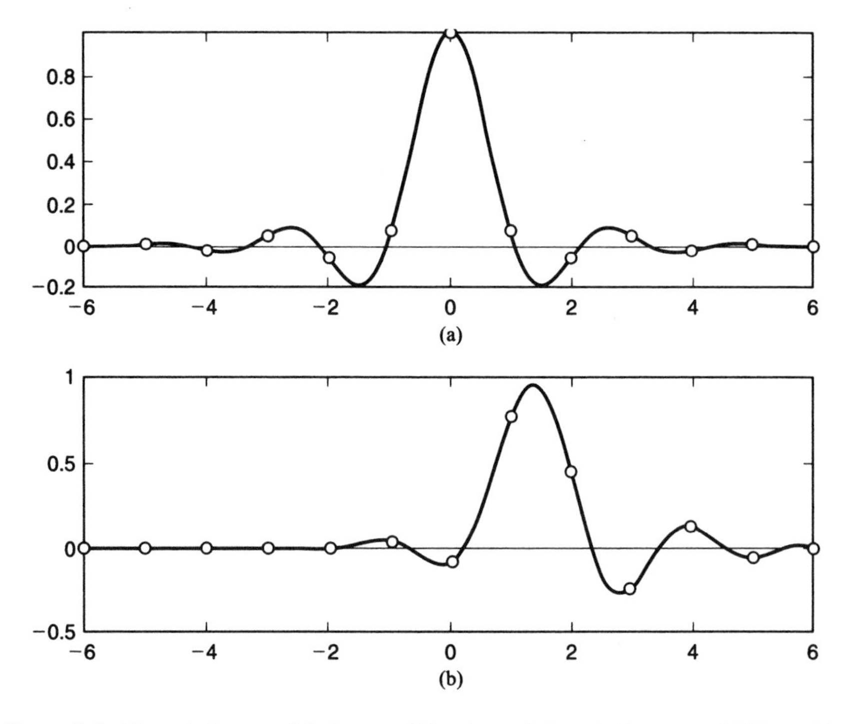

Figure 7.4 Discrete-time models for root RC pulse and six-pole Butterworth ISI combination in Example 7.1-4; **(a)** MF receiver model; **(b)** PAM receiver model. Continuous curve is MF output, circles mark discrete model values. Both are plotted versus symbol time, with 0 marking appearance of unit symbol impulse.

models need more terms. Recall from Example 7.1-2 that no equalizer can achieve the antipodal signal bound with this ISI. The models here are quite different from the discrete channel in Example 7.1-3, which has $H(z)$ given by (7.1-11).

The models in the last two examples were found with the following program.

Program 7.1-2: Matched Filter and PAM ISI Models. The next MATLAB program finds the coefficients $\{\rho_k\}$ and $\{\gamma_k\}$ in the models of (7.1-15) and (7.1-18) for a given analog channel ISI impulse response $g(t)$ and m-ary PAM pulse $v(t)$. Without loss of generality, the symbol interval is $T = 1$. In the program, the vector `vv` contains samples of $v(t)$, `gg` contains samples of $g(t)$ and `hh` contains $h(t) = v * g$, all taken at rate `fs` per symbol interval. `rho` is the MF model and `gam` is the PAM model. The program works by evaluating (7.1-12) and (7.1-16).

The formal input variables are:

`gg`—samples of $g(t)$

`fs`—sample rate per symbol

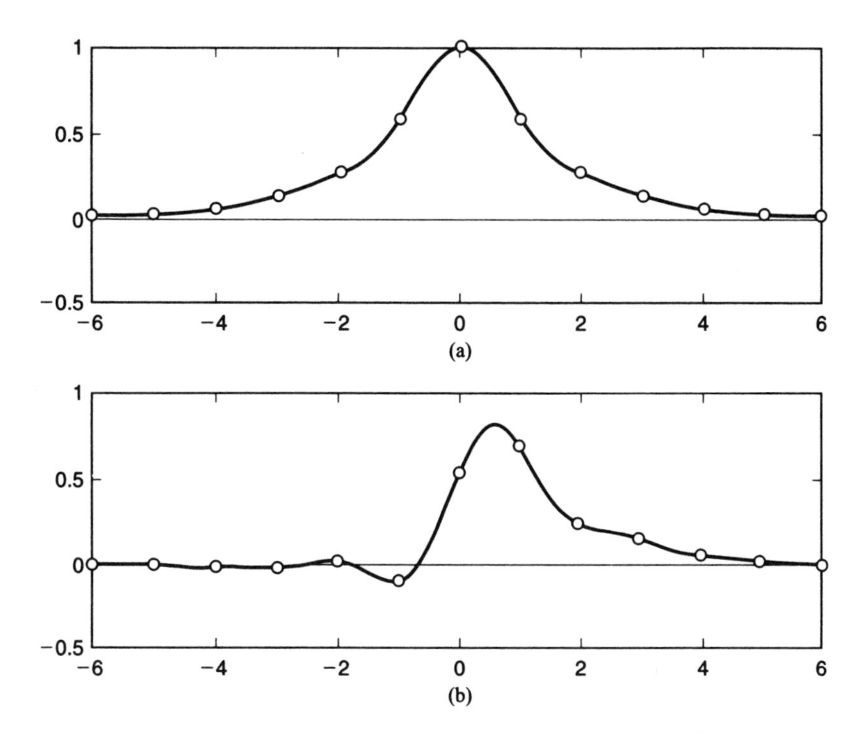

Figure 7.5 Discrete-time models for root RC pulse and exponential ISI combination in Example 7.1-5; **(a)** MF receiver and **(b)** PAM receiver models and matched filter outputs, as Fig. 7.4.

`m`—PAM alphabet size

`wid`—total pulse width (even)

`a`—pulse parameter

The following function must be supplied:

`y=pulse(a, t)`—a function that returns samples `y` of a pulse at times in the vector `t`, where the pulse has unit symbol interval and pulse parameter `a` (Programs 2.2-1 and 2.2-2 are examples).

The output is `rho` and `gam`. The continuous receive filter outputs appear as `crho` and `cgam`; the continuous channel response is `hh`.

The program is then:

```
          % Find and normalize the total response
vv = pulse(a,[-wid/2:1/fs:wid/2]); %Find total response
vv = vv*sqrt(fs/sum(vv.^2));       %Be sure v is normed to 1
hh = conv(vv,gg)/fs;               %Approx. to v * g
norm = sqrt(fs/sum(hh.^2));        %Normalization factor
hh = norm*hh;                      %h now has unit energy
```

```
        % Find the MF model
lh = length(hh); hlim = 2*lh-1;
crho = conv(hh,hh(lh:-1:1))/fs;      %Conv h and reversed h
rho = crho(1:fs:hlim)                %Rho is every fs-th point
        % Find PAM model
lv = length(vv); vlim=lh+lv-1;
cgam = conv(hh,vv(lv:-1:1))/fs;      %Conv h and reversed v
gam = cgam(1:fs:vlim)                %Gamma is every fs-th point
```

A sample calculation is shown in Fig. 7.4 for the inputs `fs = 20, m = 2, wid = 10, a = .3`, a root-RC pulse, and `gg` equal the outcome of

```
        [num,den] = butter(6,.5*2/fs);
    gg = filter(num,den,[1,zeros(1,200)]);
```

7.2 EQUALIZERS I: THE LINEAR EQUALIZER

With the receiver models of Section 7.1, we can turn to a study of the more common equalizers. In practice, the PAM receiver is most often encountered, because it uses a simple, standard receive filter, and the noise samples are IID Gaussians. However, the PAM receiver cannot always achieve ML detector performance. The MF receiver structure can do this, with enough discrete-time processing, although the processor may not be one of the simple equalizers in this section.

The many equalizers in the present day fall into two broad classes, the *linear* and the *feedback* equalizers. Linear equalizers are those limited to a linear filtering of the channel outputs, and they will be the subject of this section. This type of processor can indeed remove some or even most of the ISI, but sometimes the side effects are too great. In that case, a feedback equalizer can be designed that detects the symbols, regenerates the ISI, and then subtracts the ISI away from the incoming signal. The presence of the detector in this feedback loop makes the processor nonlinear. Among nonlinear processors in general, the most powerful is the MLSE receiver in Section 7.1. Nonlinear equalizers and the topics of adaptation and fractional spacing of taps will be taken up in Section 7.3.

Equalizers can also be adaptive or precoded. What approach works best with an ISI channel depends on the overall link scenario. It is sometimes not easy to see what characteristics define a transmission system. One needs to think about at least the following.

1. Is the channel fixed, or is it time-varying? If the channel varies, the equalizer needs to be *adaptive*. Most often, such an equalizer is given a known channel starting state and it then tracks variations; when an adaptive equalizer must also identify its starting state, it is called a *blind* equalizer.
2. Is the medium "owned" by the link, or is it shared with other links? For example, if the medium is a wire pair channel, a modem transmitter might pre-

filter the transmission by something similar to an inverse of the medium, so as to make the overall path wider band. This is called *precoding*. But it would make no sense in a radio channel if the neighboring spectrum were occupied by other users.

3. Do both receiver and transmitter know the channel or just the receiver? Ordinarily, the receiver learns the channel; it might inform the transmitter if a reverse channel is available. Other times, as with a known transmit filter, both ends know. If we are free to change the transmit filter, precoding is a possibility.
4. What is the channel response like? Responses can be finite or they can have poles and be infinite. Wire pairs and coaxial cables do not have rational transfer functions.

Because what follows is an introduction, we take the simplest case, a fixed, linear Gaussian channel, without precoding.

7.2.1 Zero-Forcing Linear Equalizers

A linear equalizer (LE) is a filter, the output of which is hopefully the interference-free symbol stream. Only a thresholding is required to remove the channel noise from the symbols. The basic linear equalizer is shown in Fig. 7.6 together with the discrete channel models. *Zero forcing* (ZF) refers to forcing the ISI effect on a symbol to zero by a filter or other processor. We denote a linear equalizer based on this idea as an LE-ZF equalizer.

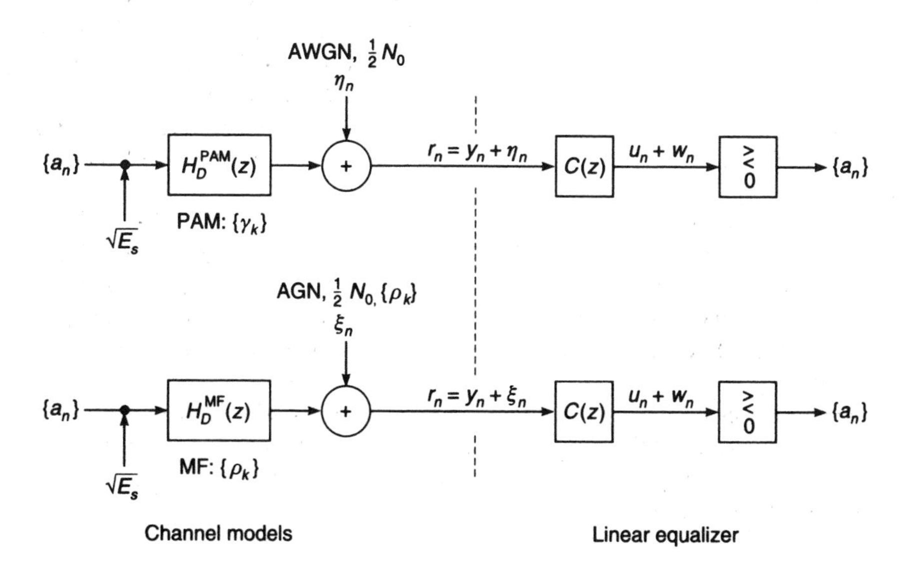

Figure 7.6 Discrete-time basic linear equalizer systems for PAM and MF channel models. Thresholding shown is for binary transmission case.

To see how zero forcing works, remove the noise temporarily from Fig. 7.6, and consider the noise-free sequence $\{u_n\}$ at the output of the equalizer $C(z)$. We can write the entire effect of the channel and $C(z)$ on the symbols $\{a_n\}$ as

$$u_n = \sum_{k\geq 0} c_k y_{n-k} = \sum_{k\geq 0} c_k \left(\sum_j h_j \sqrt{E_s}\, a_{n-k-j} \right)$$

With some rearranging of the sums, this becomes

$$u_n = \sqrt{E_s} \sum_{\ell} \left(\sum_j c_j h_{\ell-j} \right) a_{n-\ell} \tag{7.2-1}$$

If we define $q_\ell \triangleq \Sigma_j c_j h_{\ell-j}$, then (7.2-1) can be written as

$$u_n = \sqrt{E_s}\, q_0 a_n + \sqrt{E_s} \sum_{\ell\neq 0} q_\ell a_{n-\ell} \tag{7.2-2}$$

where the first term is the scaled nth symbol and the remaining sum is ISI. To force the ISI to zero requires a $C(z) = \Sigma c_k z^{-k}$ such that $q_\ell = 0$, $\ell \neq 0$; that is, the set of equations

$$\sum_j c_j h_{\ell-j} = \begin{cases} 1, & \ell = 0 \\ 0, & \ell = \pm 1, \pm 2, \ldots \end{cases} \tag{7.2-3}$$

must hold. These are the zero-forcing equations: Whenever $H(z)$ has an inverse, $C(z)$ is in fact $1/H(z)$; when $H(z)$ has no inverse, $c_0, c_1, c_2, \ldots$ simply satisfy (7.2-3).

That (7.2-3) must have a solution is easy to see, but a number of difficulties nonetheless are lurking. We have ignored the channel noise in the derivation. There is no guarantee that $C(z)$ is stable. If $C(z)$ is a stable but infinite response, the best finite equalizer will not necessarily be a finite part of the sequence that arises from (7.2-3). We will take these difficulties one at a time, starting with the stability problem.

Consider first the simple ISI model $H(z) = 0.873[1 + 0.5z^{-1} + 0.25z^{-2}]$. This model has the stable inverse $1/H(z) = 1.15 - 0.57z^{-1} + 0.14z^{-3} - 0.07z^{-4} + 0.02z^{-6} + \ldots$. With this zero-forcing $C(z)$ for an equalizer in Fig. 7.6, we can expect the circuits there in the absence of noise to deliver an ISI-free symbol output, no matter how extended the input to the model. It is easy to demonstrate that this in fact happens.

The transform $1/H(z)$ is unstable whenever the zeros of $H(z)$ lie outside the unit circle in the z-plane. Typically, this happens when the sequence $\{h_k\}$ has a *precursor*—that is, some early activity that arrives before the main peak. As an example of an unstable inverse, take the PAM model of the six-pole filter in Fig. 7.4. The symbol causing the response arrives at time zero. Although some elements in the model come earlier, the significant ones are at zero or later. We can take as the model the six elements

$$\{h_0, \ldots, h_5\} = \{-0.08, 0.79, 0.46, -0.24, 0.12, -0.06\}$$
$$h_k = 0 \qquad \text{otherwise}$$

With these coefficients, $H_D(z)$ has several roots outside the unit circle. The solution of (7.2-3) gives

$$c_0, c_1, \ldots = -12.3, -119, -1225, -12526, \ldots \tag{7.2-4}$$

This rapidly diverging solution confirms what we already know, that $H(z)$ has no stable inverse. Yet $C(z)$ made from these values will remove the ISI from any transmission $A(z)H_D(z)$; that is, if $A(z)$ represents a length-L symbol sequence and $C(z)$ represents the first L elements in (7.2-4), the product $[A(z)H_D(z)]C(z)$ will give back $A(z)$.

Of course, as L grows, the elements in (7.2-4) rapidly become unusable. For zero forcing to work well, the model should really have a stable inverse. Unfortunately, it is hard to find short models for the ISI in Example 7.1-4 that are also invertible. A likely cause is the set of six poles near the filter band edge. An invertible model for Example 7.1-5 is problematic as well. The same is true for the MF receiver models.

Another strategy for zero forcing is to allow an infinite model for the ISI, in the hope that its inverse will be finite order or, in any case, be stable. This works especially well for the model (7.1-11), which contains a single pole at $z = \alpha$. The inverse of (7.1-11) is clearly $C(z) = (1 - \alpha z^{-1})/(1 - \alpha^2)$, which is a two-tap filter with zero at $z = \alpha$. This $C(z)$ is stable and completely removes the model ISI.

The same technique works passably well for the MF model of the exponential response in Fig. 7.5. The elements in the model follow closely the law

$$0.58e^{-0.73|k|}, \qquad k = 0, \pm 1, \pm 2, \ldots$$

It can be shown that the sequence $0.58\beta^{|k|}$, $k = 0, \pm 1, \pm 2, \ldots$ has the transform

$$B(z) = \frac{0.58(1 - \beta^2)}{(1 - \beta z)(1 - \beta z^{-1})} \tag{7.2-5}$$

The inverse of $B(z)$ is evidently the three-tap filter

$$\left[\frac{1}{0.58(1 - \beta^2)}\right][-\beta z + (1 + \beta^2) - \beta z^{-1}] \tag{7.2-6}$$

Therefore, an approximate zero-forcing equalizer for the MF model in Fig. 7.5 might be (7.2-6) with $\beta = e^{-0.73}$, which is

$$C(z) = -1.08z + 2.77 - 1.08z^{-1} \tag{7.2-7}$$

The seven largest taps in the Fig. 7.5 model are $\{0.07, 0.14, 0.28, 0.58, 0.28, 0.14, 0.07\}$; taking these times $C(z)$ gives the overall z-transform up to the threshold device

$$H_D^{\text{MF}}(z)C(z) = 0.08z^4 - 0.04z^3 \cdots + 1.00z^0 \cdots -0.04z^{-3} + 0.08z^{-4}$$

(dots indicate insignificant terms). The Fig. 7.5 model assumes unit symbol energy after the ISI in the original continuous channel. What this means is that a binary PAM transmission with energy E_s will create in the absence of noise the value $\pm\sqrt{E_s}$ in the receiver after the equalizer, with a worst-case reduction by $\approx$ 24% due to ISI [i.e., 2(0.08 + 0.04)], if the nearby symbols occur in just the wrong pattern. This is equivalent to an energy loss of $10 \log_{10}(1 - 0.24)^2 = 2.4$ dB against whatever noise appears at the threshold device. That may be surprisingly large, as we will soon see.

The discussion here illustrates some facets of the zero-forcing idea. Another important practical question is the following: For a given ISI, what is the best equalizer $C(z)$ of a given finite length. Lucky [11] showed that the residual ISI is a convex function of the coefficients of $C(z)$. Furthermore, if $\Sigma_{\ell>1}|q_\ell| < 1$ in (7.2-2), which means an open eye at the receiver filter output, then it can be shown that the optimal $(2K + 1)$-tap equalizer is Eq. (7.2-3) for $|\ell| < K$.

Noise Enhancement. The zero-forcing argument completely ignores channel noise. In fact, noise can seriously affect a linear equalizer, so much so that another kind must be employed.

The heart of the LE noise problem is that the same $C(z)$ that lessens the ISI can amplify the noise. Consider first the PAM receiver case. In Fig. 7.6, IID Gaussian noise with variance $\frac{1}{2}N_0$ passes through the equalizer $C(z)$, but since the rest of the LE receiver makes a single-interval decision, only the variance of the output matters here. The decision is in fact the standard Gaussian decision problem defined at Eq. (2.6-4) in Chapter 2: Given a noise-free equalizer output u_n with the right sign and zero-mean Gaussian noise w_n with variance σ_w, the probability of decision error is $P[w_n > |u_n|] = Q(|u_n|/\sigma_w)$. For any $C(z)$, σ_w is given by[4]

$$\sigma_w^2 = \text{var}[w_n] = \sum |c_i|^2 \text{var}[\eta_n] = \tfrac{1}{2}N_0 \sum |c_i|^2 \tag{7.2-8}$$

For example, the one-pole discrete ISI model in Example 7.1-3 with $\alpha = 1/\sqrt{2}$ has unit energy model

$$H_D(z) = \frac{1}{\sqrt{2} - z^{-1}}$$

but the ZF equalizer $C(z) = \sqrt{2} - z^{-1}$ has $\Sigma|c_i|^2 = 3$. Thus the growth in the noise power σ_w is this factor 3. The error calculation at the thresholding in Fig. 7.6 is the usual Gaussian one but with 3 times the noise variance; the error probability is thus approximately

[4]This follows from the fact that w_n is the sum $\Sigma c_i \eta_{n-i}$ of IID variates; $\text{var}(w_n)$ is then $\Sigma\, \text{var}(c_i\eta_{n-i}) = \Sigma|c_i|^2 \text{var}(\eta_i)$. Note also that the argument here holds for any data symbol alphabet, since only the noise variance expansion is at issue.

$$p_e = Q\left(\sqrt{\frac{2E_b}{3N_0}}\right) = Q\left(\sqrt{\frac{\frac{2}{3}E_b}{N_0}}\right)$$

This is equivalent to an energy loss of $|10 \log_{10} 3| = 4.8$ dB. The MLSE benchmark loss, by comparison, was 2.3 dB. Thus the ZF equalizer carries with it an extra penalty of $4.8 - 2.3 = 2.5$ dB.

A different calculation is needed when an MF model is used, since the noise passes through a nonorthogonal receive filter. The details of this calculation are given in Appendix 7A.

The noise growth factor, assuming a model $H_D(z)$ derived with unit symbol energy, is called the *noise enhancement factor* F_{NE} of an equalizer. A classic cause of enhancement is poles in the equalizer transfer function, especially when they are close to the unit circle. Such poles must happen when the ISI model has zeros near the unit circle, if the ISI is to be strongly reduced.

Frequency Domain Interpretations. The noise-free outputs of the matched filters in the MF and PAM receivers have an interesting interpretation in terms of the ISI spectrum $H(f)$ and the Nyquist pulse criterion. Before looking at these, we need to review what it means to Fourier transform a sequence.

Consider the sequence of samples $\{g_n\}$. We can make it into a time function by weighting each sample with an impulse to obtain

$$g_{\text{sa}}(t) = \sum g_n \delta(t - nT) \tag{7.2-9}$$

This is called the *sample function* for $\{g_n\}$. The formal Fourier transform of $g_{\text{sa}}(t)$ is

$$G_{\text{sa}}(f) \triangleq \int g_{\text{sa}}(t) e^{-j2\pi ft}\, dt = \sum g_n e^{-j2\pi fnT} \tag{7.2-10}$$

which is a periodic function of f with period $1/T$. Observe that this *sample transform* may be obtained from the z-transform $G(z)$ via

$$G_{\text{sa}}(f) = G(z)|_{e^{j2\pi fT}} = G(e^{j2\pi fT}) \tag{7.2-11}$$

As for an inverse transform, it is easily verified by direct substitution of (7.2-10) into the following that sample g_n is given by

$$g_n = T \int_{-1/2T}^{1/2T} G_{\text{sa}}(f) e^{j2\pi fnT}\, df, \qquad \text{all } n \tag{7.2-12}$$

This then is the formal inverse transform.

With this equipment, we can prove the following lemma, which gives the sample spectra of the PAM and MF receiver models $\{\gamma_n\}$ and $\{\rho_n\}$.

Lemma 7.2-1. Let the total channel response to a symbol $h(t)$ have Fourier transform $H(f)$. Then the sample Fourier transform of the discrete model $\{\gamma_n\}$ for the PAM receiver is the *folded* spectrum

$$H_{\text{fold}}(f)V_{\text{fold}}(f) = \frac{1}{T}\sum_k H\left(f+\frac{k}{T}\right)V^*\left(f+\frac{k}{T}\right), \qquad \text{all } f \tag{7.2-13}$$

The sample transform of the MF model $\{\rho_n\}$ is the folded power spectrum

$$|H(f)|^2_{\text{fold}} = \frac{1}{T}\sum_k \left|H\left(f+\frac{k}{T}\right)\right|^2, \qquad \text{all } f \tag{7.2-14}$$

Proof. To prove (7.2-13), start with (7.2-12), which states that if $X(f)$ satisfies

$$\gamma_n = T\int_{-1/2T}^{1/2T} X(f)\, e^{j2\pi fnT}\, df, \qquad \text{all } n$$

then $X(f)$ is the sample transform of $\{\gamma_n\}$. Now take definition (7.1-16), which says that $\gamma_n = \int h(\tau)v(\tau - nT)d\tau$. By Parseval's theorem, this may also be written

$$\gamma_n = \int_{-\infty}^{\infty} H(f)\,[V(f)\, e^{-j2\pi fnT}]^*\, df$$

This can be broken into a sum of separate integrals

$$\gamma_n = \sum_k \int_{(k-1/2)/T}^{(k+1/2)/T} H(f)\,[V(f)e^{-j2\pi fnT}]^*\, df$$

which by the variable change $f \to f + k/T$ becomes

$$\gamma_n = \int_{-1/2T}^{1/2T} \sum_k H\left(f+\frac{k}{T}\right)V^*\left(f+\frac{k}{T}\right)e^{j2\pi fnT}\, df$$

Thus the sample transform $X(f)$ must be (7.2-13). The proof of (7.2-14) works the same way, except that ρ_n is $\int h(\tau)h(\tau - nT)\, d\tau$ and Parseval's theorem gives

$$\rho_n = \int_{-\infty}^{\infty} H(f)H^*(f)e^{j2\pi fnT}\, df = \int |H(f)|^2 e^{j2\pi fnT}\, df$$

Breaking into separate integrals leads to (7.2-14).

A transform of the form $X_{\text{fold}}(f) = \Sigma X(f + k/T)$ is called the folded (or aliased) spectrum of $x(t)$ because of the way frequency-shifted replicas of $X(f)$ add up near the frequencies $\pm 1/2T, \pm 3/2T, \ldots$ hertz. Similarly, the quantity $|H(f)|^2_{\text{fold}}$ in (7.2-14) is a folded power spectrum. An example of the PAM folded spectrum (7.2-13) is shown in Fig. 7.7 for the case of 30% root RC pulses $v(t)$ and the six-pole Butterworth total response $h(t)$ in Example 7.1-4; this is the spectrum of the sample response of the whole channel taken as one discrete-time object. Because the filter cuts off at a frequency close to half the sample rate, there is only a little aliasing. But there will in general be aliasing with any excess bandwidth pulse and any realizable filtering.

If it should happen that $|H(f)|^2_{\text{fold}}$ is a constant, then the Nyquist pulse condition is satisfied. This happens if $h(t)$ is just an orthogonal pulse, without, for example, the filtering in Fig. 7.7. Figure 2.4 shows such a constant summing up for a 100% excess bandwidth root RC pulse. Whatever the case with $h(t)$, if the equalizer following has the inverse of (7.2-13) or (7.2-14), then the total system response is a constant, the samples yield the symbols, and there is no ISI. This is the principle of zero forcing in the frequency domain.

7.2.2 Least-Square Error Linear Equalizers

Is is usually easier to design a workable equalizer if the strict zero-forcing requirement is relaxed somewhat. The many cases where the equalizer $1/H_D(z)$ is unstable will not be so troublesome, and with luck a performance near the MLSE benchmark may be obtained. The most common such design is the least-square linear equalizer, denoted LE-LS for short.

In principle, the new equalizer $C(z)$ in Fig. 7.6 can have both poles and zeros, but in practice the poles seldom improve performance, and so we will consider a finite-response $C(z)$ with zeros only. The idea is to *minimize the expected square error* between a transmission symbol a_{n-D} lying D before the present and the equalized version of it u_{n-D}: That is, find $C(z)$ such that we minimize

$$J = \mathcal{E}[|a_{n-D} - u_{n-D}|^2] = \mathcal{E}\left[\left|a_{n-D} - \sum_{k\geq 0} c_k\, r_{n-k}\right|^2\right] \qquad (7.2\text{-}15)$$

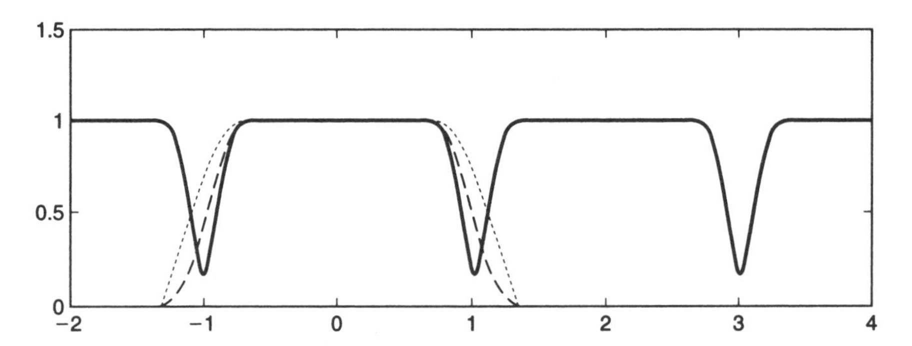

Figure 7.7 Folded spectrum (7.2-13) for 30% root RC pulses filtered by six-pole Butterworth filter with cutoff close to $0.5/T$ hertz, plotted versus frequency in multiples of $1/2T$ hertz. Dotted curve is pulse spectrum $V(f)$; dashed curve is filtered pulse spectrum $V(f)H(f)$. All spectra shown as absolute values.

The variable D allows the processor to work at some extra delay in hopes of better performance. The expectation is over both the IID symbol outcomes and the channel noise variates. Both contribute to the square error.

The optimal set $\{c_k\}$ depends on both the channel model H_D and the noise variance. This is a change from zero forcing, where only H_D affects the design. While the optimal $\{c_k\}$ minimize the error variance in u_{n-D} for an H_D and a particular noise, the solution can certainly be used at other channel noises and it ordinarily provides an attractive, stable equalizer. The details of an optimal solution are put off to Appendix 7A, along with Programs 7.2-1 and 7.2-2, which perform the minimization and design $C(z)$.

The variation in a LE-LS output u_n is not Gaussian, since it is in part residual ISI from other symbols. In the limit of a long equalizer length, Berger and Tufts [12] showed that the variance is

$$J_{\min} = \int_{-1/2}^{1/2} \frac{N_0/2}{|H_D(e^{j2\pi f})|^2 + N_0/2}\, df \tag{7.2-16}$$

Here, $H_D(e^{j2\pi f})$ is the sample Fourier transform of the channel impulse response, assuming unit-interval sampling. For the finite equalizer, (7.2-16) is only an estimate, and Appendix 7A gives a value for the true mean-square error.

For an estimate of error probability, we need to separate out the ISI and Gaussian channel noise components in the square error. Suppose the channel noise is AWGN with variance $\frac{1}{2}N_0$, where N_0 is not necessarily the value used to design the equalizer. A threshold decision follows $C(z)$ (see Fig. 7.6). As usual, the decision there is the standard Gaussian detection problem of Eq. (2.6-4), but both the symbol value and the noise variance have been modified by $C(z)$. For simplicity, consider binary transmission symbols, so that the threshold comparison in Fig. 7.6 will be to zero. The worst-case symbol modification may be computed as follows. Suppose the impulse response of the total communication chain is $H_D(z)C(z)$ is $\{\iota_k\}$. We can presume that the largest element in the sequence, ι_μ, represents the estimate of the sent symbol a_{k-D} and that μ is its index; presume further that $a_{k-D} > 0$. The largest possible residual ISI for unit energy $H_D(z)$ will be $\Sigma_{k\neq\mu} |\iota_k|$, which occurs when all the other symbols have just the wrong sign. Thus the lowest that output u_n can be with unit symbols is

$$\kappa_{\text{wc}} \triangleq \iota_\mu - \sum_{k\neq\mu} |\iota_k| \tag{7.2-17}$$

It remains to compute the noise enhancement factor F_{NE} . In the PAM model case, the component $\{w_n\}$ in the equalizer output is the AWGN sequence $\{\eta_n\}$ filtered by $C(z)$; the result is a zero-mean Gaussian sequence with variance (7.2-8). In the MF equalizer case, the AWGN gets filtered by both $H_D^{\text{MF}}(z)$ and $C(z)$, and the variance, from Appendix 7A, is (7A-12). Now scale the symbol energy to $\sqrt{E_s}$. The threshold decision, which in the ISI-free case was between $\pm\sqrt{E_s}$ in noise $\frac{1}{2}N_0$, has degenerated to a decision between $\pm\kappa_{\text{wc}}\sqrt{E_s}$ in Gaussian noise of variance $\frac{1}{2}N_0 F_{\text{NE}}$.

The standard Gaussian detection problem that led in Chapter 2 to (2.6-4) and then to $Q(\sqrt{2E_b/N_0})$ now leads to $Q(\kappa_{wc} 2\sqrt{E_b}/\sqrt{2F_{NE}N_0})$ instead. The estimate for low error probability, based on the worst-case ISI, is thus

$$p_e \approx Q\left(\sqrt{\frac{2\kappa_{wc}^2 E_b}{N_0 F_{NE}}}\right) \tag{7.2-18}$$

The performance here at energy E_b has degraded to that of an isolated pulse at energy $(\kappa_{wc}^2/F_{NE})E_b$.

We conclude the section with a return visit to some of the earlier examples. Most of the difficulties with the LE-ZF equalizer will be resolved.

Example 7.2-1: LE-LS Equalizer for Six-Pole Butterworth ISI. This example designs several LE-LS binary equalizers. The PAM and MF channel models are given for this example in Example 7.1-4; the pulse there was 30% root RC and the channel was a six-pole Butterworth filter with cutoff at $0.5/T$ hertz. For the PAM model in Fig. 7.4b,

$$\{h_k\} = \{-0.082, 0.792, 0.460, -0.240, 0.124, -0.059\}$$

Program 7.2-1 with $N_0 = 0.1$ (true variance 0.05) and delay 13 yields a 31-tap equalizer with mean-square error about 10.2 dB less than the symbol value, which is unity. Shorter equalizers or ones with the large center tap displaced left or right give a somewhat larger mean-square error. The noise power enhancement (7.2-8) via $C(z)$ is 1.07. The residual ISI factor κ_{wc} from (7.2-17) is actually less than zero, which means that the output of this equalizer does not have an open eye: A few symbol streams drive the equalizer output negative at a time when it should be positive, and vice versa. Since the mean-square error is small, we can assume that these streams seldom occur and that the combination of ISI and channel noise creates an error typically 10 dB below the symbol value. A charitable assessment of this equalizer might say that it has a 3-dB loss compared to antipodal signaling, since the assumed noise variance (0.05) at the start is 13 dB below the symbol value but the total noise after $C(z)$ is 10 dB below.

A better equalizer works as follows. By delaying the sampling in the receiver by an additional $0.4T$, the model becomes

$$\{h_k\} = \{0.144, 0.986, -0.050, -0.039, 0.042\}$$

With N_0 set to 0.1 and delay 4, Program 7.2-1 gives the seven-tap equalizer

$$\{c_0, \ldots, c_6\} = \{0.016, -0.133, 0.952, 0.041, 0.047, -0.034, -0.002\}$$

which has mean-square error 0.05; $F_{NE} = 0.929$ for this $C(z)$ and $\kappa_{wc} = 0.925$. This degrades E_b/N_0 by $(0.925)^2/0.929 = 0.921$, or 0.36 dB, according to Eq. (7.2-18). It means that the error probability at reasonably high SNR is close to that of an isolated pulse, which is $Q(\sqrt{2E_b/N_0})$. The ISI has thus been virtually removed.

Turning to the MF model, recall from Example 7.1-4 that it was the autocorrelation at . . . , $-2T$, $-T$, 0, T, $2T$, . . . of the Butterworth analog response. Seven actual model values are

$$\{h_k\} = \{\rho_k\} = \{0.0435, -0.061, 0.075, 1, 0.075, -0.061, 0.0435\}$$

Program 7.2-2 with $N_0 = 0.1$ and delay 8 gives the nine-tap equalizer

$$\{c_0, \ldots, c_8\} = \{0.011, -0.051, 0.070, -0.086, 0.977, -0.086, 0.070, -0.051, 0.011\}$$

This has mean-square error 0.05, noise power enhancement 0.93, and residual ISI factor 0.91, which leads to $Q(\sqrt{1.77E_b/N_0})$ in (7.2-18); this is a loss of 10 $\log_{10}(2/1.77) = 0.5$ dB. Figure 7.8 shows the channel output before and after equalization.

To conclude, we have investigated three LS equalizers. Both the MF receiver and the optimized delay strategy with the PAM receiver can lead to performance near the MLSE benchmark error $Q(\sqrt{2E_b/N_0})$. Without modification, the PAM receiver equalizer loses at least 3 dB.

Example 7.2-2: LE-LS Equalizer for Exponential ISI. The PAM and MF channel models for this example are given in Example 7.1-5; the channel is analog-exponential with response $(0.5)^{t/T}$ and the pulse is again 30% root RC. The PAM model values in Fig. 7.5b are

$$\{h_k\} = \{-0.088, 0.568, 0.733, 0.268, 0.174, 0.073\}$$

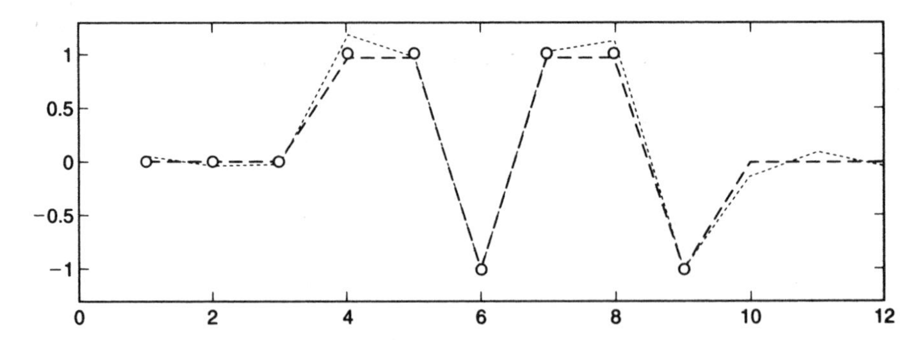

Figure 7.8 Input to (dotted) and output from (dashed) MF model LE-LS equalizer in Example 7.2-1. Channel is six-pole Butterworth filter with cutoff $0.5/T$. Circles indicate original symbol values. All signals are discrete time; dots and dashes are interpolated only to clarify picture.

Program 7.2-1 with N_0 set to 0.01 and delay parameter $D = 3$ gives the 19-tap binary equalizer plotted in Fig. 7.9a. The mean-square error is 0.040 (14 dB below the symbols) and the SNR loss (7.2-18) compared to antipodal signaling is more than 11 dB, due mostly to large noise enhancement. Figure 7.9b shows the equalizer in action in a no-noise example, for which it performs quite well. However, this is not a good equalizer, since a 1 dB loss was predicted by the MLSE benchmark (Example 7.1-2). Lengthening the taps or changing the delay does not improve this equalizer. The MF-model equalizer, on the other hand, performs much better. Program 7.2-2 with $N_0 = 0.01$ and delay 16 gives a symmetric 21-tap equalizer having mean-square error 0.012 (19 dB below symbols) and an asymptotic SNR loss of only 4.1 dB compared to antipodal signaling; that is, Eq. (7.2-18) is $Q(\sqrt{0.78E_b/N_0})$ in a low noise channel. The SNR loss is again mostly due to noise enhancement. Since the MLSE performance is $Q(\sqrt{1.60E_b/N_0})$, the MF model design runs at a 3.1-dB loss compared to the ultimate MLSE equalizer.

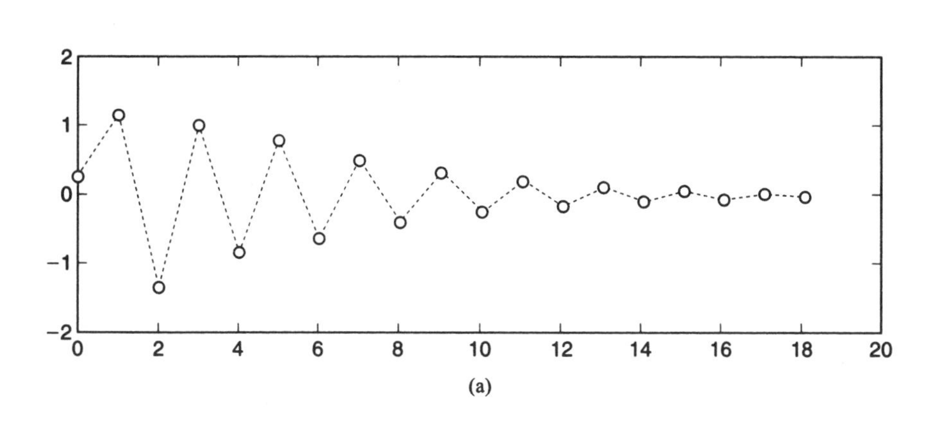

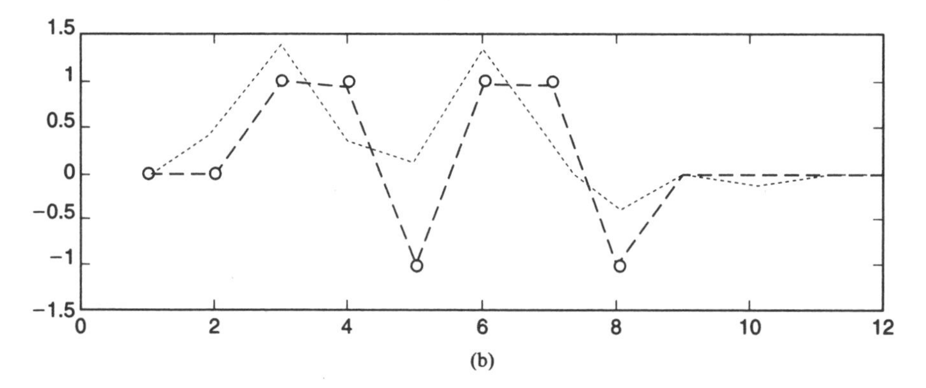

Figure 7.9 **(a)** Nineteen-tap PAM model LE-LS equalizer in Example 7.2-2 for exponential ISI channel. **(b)** Input to (dotted) and output from (dashed) equalizer; circles show original symbols.

7.3 EQUALIZERS II

7.3.1 Nonlinear Equalizers

The foregoing pages describe the simplest and most widely used equalizer, the linear equalizer. When the channel model has zeros on or near the unit circle or when no linear scheme can come near the MLSE benchmark, it is time to consider a nonlinear approach. The next pages concentrate on the decision feedback equalizer (DFE), which works by regenerating the ISI from the detected symbols and then precisely subtracting it away. The feedback equalizer idea traces back to work by a number of researchers [14–17]. Providing that the fed-back symbols are correct, the equalizer will achieve the antipodal signaling error performance. But the feedback is sometimes wrong and can then lead to a long error event that feeds upon itself. When the MLSE benchmark falls short of the antipodal signaling distance, it is clear that the DFE must *consistently* go wrong.

We will conclude with a brief discussion of practical MLSE equalizers.

Simple Zero-Forcing DFE. Consider the basic feedback equalizer shown in Fig. 7.10. This circuit will not work well in many situations, but it provides a good starting point for discussion. We consider only the PAM model case. A threshold device converts the summing junction outputs to the nearest legal transmission symbol $\hat{a}_n$. The sequence of these feeds the filter $1 - H(z)$, the output of which adds immediately to the present channel output r_n. If all the previous $P - 1$ detected symbols are correct, where P is the length of the channel model, the z-transform of the signal at the summing junction output settles to

$$N(z) + A(z)H(z) + A(z)[1 - H(z)] = A(z) + N(z)$$

Here $A(z)$ is the z-transform up to the present of the symbols and $N(z)$ of the noise.

For example, the PAM model for the six-pole Butterworth channel in Examples 7.1-4 and 7.2-1 was $H(z) = -0.082 + 0.792z^{-1} + 0.460z^{-2} - 0.240z^{-3} + 0.124z^{-4} - 0.059z^{-5}$. Consider the fate of the present time n symbol a_n. The channel output at n

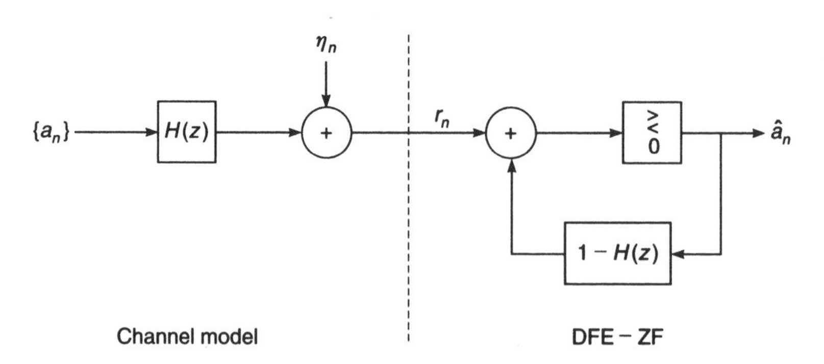

Figure 7.10 Basic zero-forcing decision feedback equalizer.

is $\eta_n - 0.082a_n + 0.792a_{n-1} + 0.460a_{n-2} \ldots$, to which the junction adds $(1 + 0.082)\hat{a}_n - 0.792\hat{a}_{n-1} - 0.460\hat{a}_{n-2} \ldots$. If somehow $a_i = \hat{a}_i$, $i \le n$, the junction output is $1 \cdot a_n + \eta_n$, which the thresholder converts to a_n if η_n is small enough.

The crucial assumption in the above is that all $P - 1$ previous symbol estimates are correct. In actuality, it is easy for the feedback loop to settle into the wrong decision about a_n when even a small noise exists. Letting $\hat{a}_{n-1}, \ldots, \hat{a}_{n-P+1}$ be correct, we can see that the small component $-0.082a_n$ of the channel output is easily overwhelmed by a noise $\eta_n > 0$, which leads to the opposite threshold output, which adds a feedback term of 1.082, which reinforces the decision. The demon here is the small value $h_0 = -0.082$, the presence of a precursor. Figure 7.10 is problematic whenever there are precursors or the symbols are nonbinary and therefore often small.

The MLSE benchmark minimum distance for the six-pole channel is the antipodal signaling distance d_0, which tells us that it is at least conceivable that Fig. 7.10 can cancel the ISI over the long term. With the exponential channel of Example 7.1-5, the MLSE benchmark falls short of antipodal signaling, and we conclude that the zero-forcing DFE cannot consistently work as advertised.

Least-Square Error Feedback Equalizer. These equalizers trace back to Salz [13] in 1973. There are several ways to modify the least-square linear equalizer in Section 7.2 to work as a feedback equalizer. The result is a DFE-LS scheme, more stable than the DFE-ZF with less error feedback. Error feedback, however, is a problem with all DFEs; as well, stability can be hard to achieve when the DFE is adaptive. There is thus a tendency to avoid the DFE-LS in practice.

A standard DFE-LS design appears in Fig. 7.11 and consists of two sections. A standard LE-LS equalizer $C(z)$ from Section 7.2 makes up a feedforward section through which flows the channel sequence $\{r_n\}$. This length $D + 1$ equalizer produces a least-square error estimate of symbol value a_{n-D}, which is u_{n-D}. A feedback section $B(z)$, through which symbol values flow, computes an ISI component in u_{n-D}. This component is subtracted from u_{n-D}, and what remains is mapped to the nearest symbol value by the threshold device. This value $\hat{a}_{n-D}$ feeds back through $B(z)$, and it also forms the output of the DFE equalizer.

The object of the feedback section is to remove all the ISI from u_{n-D} that stems from symbols already decided—that is, from symbols $\ldots, \hat{a}_{n-D-2}, \hat{a}_{n-D-1}$. The estimate becomes

$$u_{n-D} = \sum_{k=0}^{D} c_k \left(\sum_{j=0}^{P-1} h_j a_{n-k-j} \right) + \text{noise} \tag{7.3-1}$$

A careful look at these terms shows that those coming from symbols already decided are

$$\sum_{k=1}^{M} b_k a_{n-D-k} \tag{7.3-2}$$

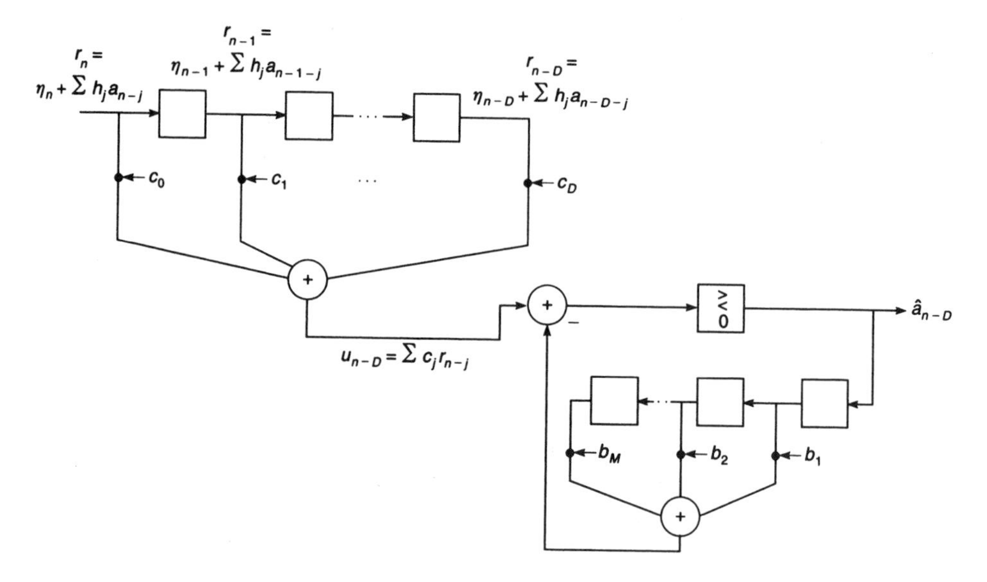

Figure 7.11 Standard least-square error feedback equalizer, denoted DFE-LS.

where

$$b_k = \sum_{j=0}^{D} c_j h_{k+D-j}, \qquad k = 1, \ldots, M$$

The $\{b_k\}$ are the coefficients of feedback $B(z)$. In order to squeeze out every ISI contribution, M, the order of $B(z)$, needs to satisfy

$$M = P - 1 \tag{7.3-3}$$

where P is the number of taps in the ISI model. The last feedback tap will be just $b_{P-1} = c_D h_{P-1}$.

In choosing the equalizer size L, it is a good strategy to insist that $c_0, \ldots, c_{L-1}$ at least reach past the precursor of the ISI response. Then the LS estimate u_{n-D} will be a function of the main response to symbol a_{n-D}, and the decision on a symbol is not forced before its main influence is felt. The LE-LS section can be designed with Programs 7.2-1 or 7.2-2 given in Appendix 7A, with equalizer length $D + 1$ and delay D. This done, the taps of $B(z)$ come from (7.3-2). A routine for finding $B(z)$ is built into the two programs.

The success of this DFE at improving upon the feedforward section alone depends of course on error-free decision feedback. In reality, error occurs. Estimating DFE error probability without assuming error-free detection up to the present is a challenging problem, summarized to some extent in Refs. 1 and 3. With the assumption of no errors, Salz [13] showed that the smallest mean-square error achievable with a long DFE-LS is

$$J_{\min} = \exp\left\{\int_{-1/2}^{1/2} \ln\left[\frac{N_0/2}{|H_D(e^{j2\pi f})|^2 + N_0/2}\right] df\right\} \tag{7.3-4}$$

This formula gives a mean-square estimation error, not a probability of decision error. Perhaps the best way to estimate the latter is by testing the complete design with Gaussian noise.

Proakis [1] shows that for the discrete-exponential model in Example 7.1-3, which is $H(z) = \sqrt{1-\alpha^2}/(1-\alpha z^{-1})$, estimate (7.3-4) predicts a mean-square error of about $\frac{1}{2}N_0/(1-\alpha^2)$ in a DFE with correct feedback. Equation (7.2-16) for the linear equalizer predicts $\frac{1}{2}N_0(1+\alpha^2)/(1-\alpha^2)$. Antipodal signaling would yield simply $\frac{1}{2}N_0$. For $\alpha = 1/\sqrt{2}$, as in Example 7.1-3, these square errors are 3.0 and 4.8 dB larger than $\frac{1}{2}N_0$, respectively. The DFE thus improves the LE-LS, if we ignore error propagation. A more extensive example of a DFE is the following. It illustrates the interplay of noise and residual ISI. It also shows when the DFE may fail.

Example 7.3-1: DFE-LS on a Channel with Troublesome Zeros. A standard example of a channel with zeros near the unit circle is the binary PAM channel model $\boldsymbol{h} = (0.408, 0.817, 0.408)$ [1]. The zeros in fact lie at –1.05 and –0.95, and they make the linear equalizer all but useless. Yet the MLSE benchmark square distance for this channel, found with Program 7.1-1, is 1.33, a loss of only 1.76 dB compared to antipodal signaling.[5] So we are motivated to look for DFE-LS equalizers that might perform better than the linear one. The candidates will contain a length $D+1$ LE-LS equalizer with delay D; the feedback section can have up to $P-1=2$ taps. Program 7.2-1 with $\boldsymbol{h}$, $N_0 = 0.1$, equalizer length 4 and delay 3 yields the DFE with $C(z)$ and $B(z)$ as in Fig. 7.12. Its mean-square error is 0.30. A part of this is Gaussian noise of variance $\Sigma|c_k|^2 = 1.53(\frac{1}{2}N_0) = 0.077$, an enhancement of 1.9 dB compared to the channel $\frac{1}{2}N_0$. A longer forward section will not improve performance; the length-3 section shown (for which $D = 2$ covers the precursor and main peak) performs almost as well. It appears that the feedback section can be reduced to one tap, 0.116. Thus u_{n-3} is corrected by $-0.116\,\hat{a}_{n-4}$ before reaching the thresholder. Consider now low noise, far below the design N_0: mostly, the residual ISI remains and an rms estimation error of typically $\pm\sqrt{0.30-0.077} \approx \pm 0.45$ occurs in u_{n-D}. If we assume correct feedback, the one significant tap in the feedback circuit reduces 0.45 to $\pm(0.116)(0.45) \approx \pm 0.05$. It is thus probable that the estimate presented to the thresholder lies in the range [0.9, 1.1], compared to a nominal value of 1 (taking twice the rms variation). Between the 1.9-dB channel noise enhancement and the reduction sometimes of the noise-free threshold input from ±1 to ±0.9, we might expect the threshold compare to show perhaps 3-dB degradation in error probability compared to an ISI-free decision. Proakis [1] reports tests of a similar DFE, with and without the assumption of correct feedback; a few of his data are compared to the MLSE benchmark and the ISI-free performances in Fig. 7.13. There is indeed a 3-dB loss beyond MLSE with correct feedback, and another 2 dB if actual detected

[5]We can assume orthogonal pulses on the original continuous channel; any orthogonal pulse leads to the same loss.

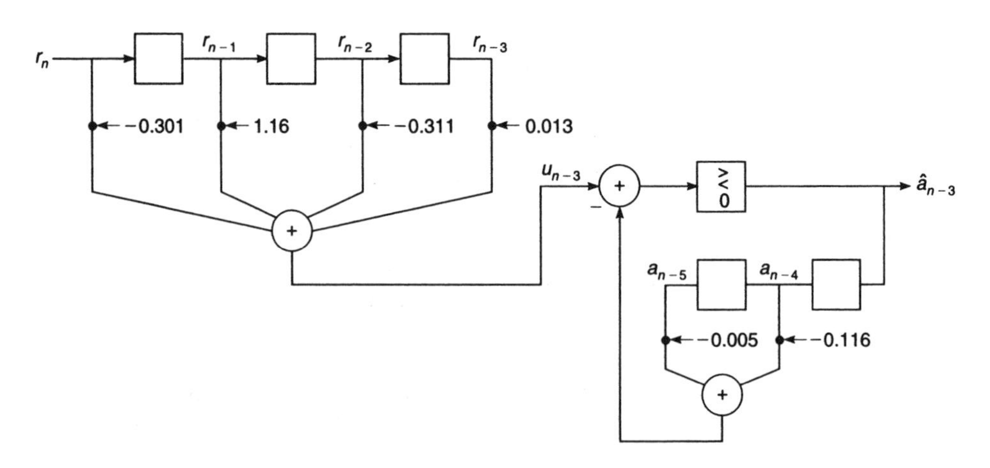

Figure 7.12 The DFE-LS, with $N_0 = 0.1$ and $D = 3$ for channel (0.407, 0.815, 0.407) in Example 7.3-1. Circuit assumes $E_s = 1$.

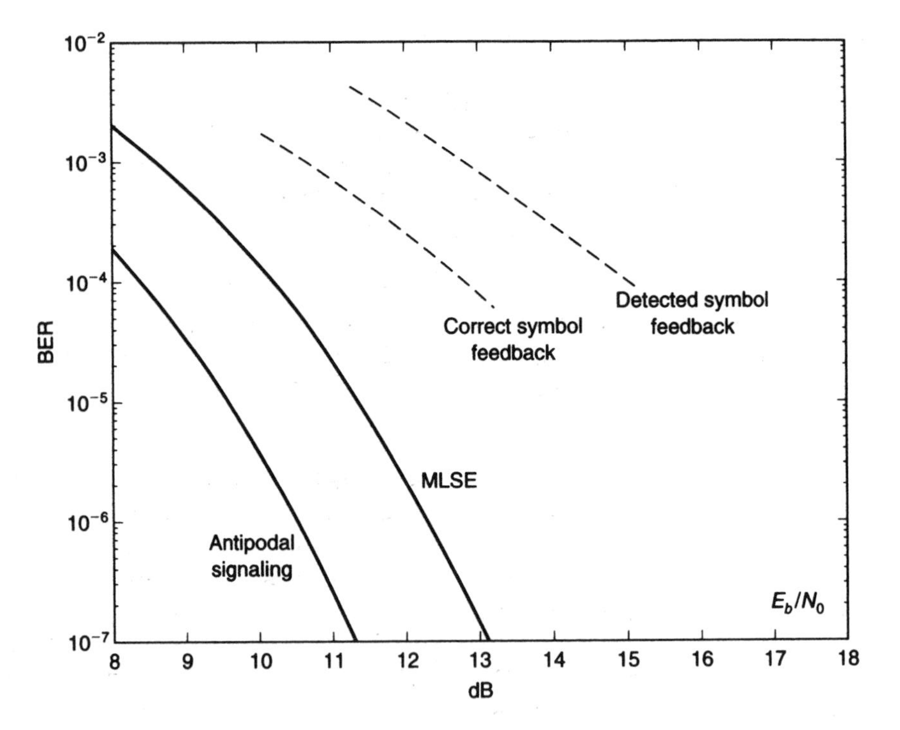

Figure 7.13 Comparison of antipodal pulse error performance to approximate MLSE performance and DFE-LS performances with and without assumption of earlier correct decisions. Data in the last case are adapted from Proakis [1, Chapter 6].

symbols are fed back. This is a total 5 dB, and it is tempting to resort to MLSE detection.

Other Nonlinear Equalizers. A number of alternate feedback equalizers are suggested in the literature, particularly in Refs. 1, 3, and 6. One that we can mention is a special scheme in Ref. 3 for designing the DFE forward and backward filters when the MF model receiver is used.

When the ISI is long and severe, has a significant precursor, or has zeros near the unit circle, we have seen that only MLSE can come near the performance theoretically attainable. In some cases, the full Viterbi decoder is small enough and can be used with the ISI trellis. The trellis of Example 7.3-1, for example, is essentially that given in the PRS example in Section 6.4.3. But infinite ISI, such as comes from high-order narrowband filters, lacks a workable finite-state trellis model, and the Viterbi decoder consequently cannot be used to remove the ISI.

An interesting alternate approach is to use low-cost simplifications of the MLSE receiver, which nonetheless achieve close-to-MLSE error performance. The ISI can be modeled as a large tree structure, and a so-called limited search decoder employed, which searches only small parts of the structure. Research has shown that these limited decoders need to explore only a very small part of the ISI tree in order to achieve near-MLSE error performance, even in very severe ISI (see Refs. 8 and 10). They are thus candidates in future practical systems.

7.3.2 Fractional Spacing and Adaptivity

Up to now, all of our equalizers have worked with channel output samples taken at the rate of one per symbol interval. A fundamental result of communication theory, which we do not prove here, states that this rate of sampling is sufficient under ideal conditions if the MF model receiver is used. This is a remarkable result when one considers that the bandwidth of the signaling usually exceeds the corresponding Nyquist bandwidth, $1/2T$ hertz. The samples at the matched filter output have an aliased spectrum, as we have seen in Lemma 7.2-1.

When there is phase or symbol timing error, or when the simpler PAM model receiver is used, the story is quite different. An inkling of this occurred in Example 7.2-1, where it was found that a judiciously chosen sampling delay greatly improved equalizer performance. What if this delay were unknown or applied in error? What if the ISI law itself were inaccurately known? Very often in these cases, it turns out that sampling times that deviate from the optimal lead to large performance loss. A practical way to reduce the risk is to sample at a rate faster than the symbol rate. An equalizer that does this is said to be *fractionally spaced*; the extra samples allow the scheme to repair timing errors later during its discrete-time processing. The most practical fractional rate is twice the symbol rate, and so we will concentrate on that.

Such a double-speed equalizer/receiver is shown in Fig. 7.14 for the MF and PAM cases. In either case, the analog filter output is sampled at rate $2/T$ instead of

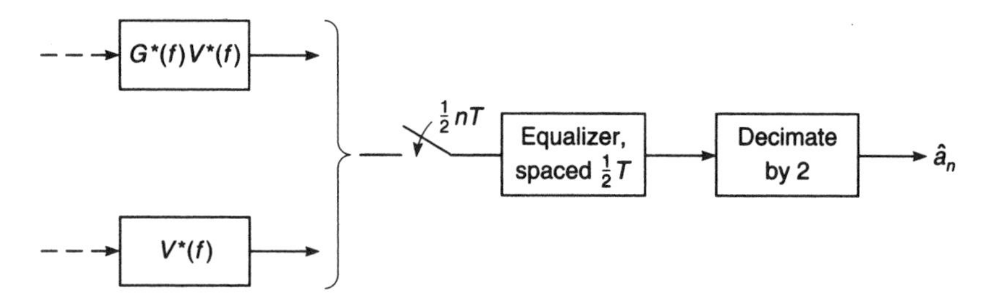

Figure 7.14 General structure of fractionally spaced equalizer showing half-symbol spacing.

$1/T$; an equalizer, either least square or whatever, linear or not, is designed for this sample train. Its output is decimated by two; that is, every other output is taken as a symbol estimate $\hat{a}_n$.

The equalizer design calculation and the evaluation of performance are beyond our scope (see, e.g., Refs. 1 and 6). We can summarize by saying that the fractional equalizer will typically compensate better for a variety of channel distortions, with less noise enhancement and less sensitivity to model and sampling errors, all at the same or less complexity than a symbol-spaced equalizer.

Some further insight into fractional spacing comes from a receiver structure that was suggested in Ref. 18. It is shown in Fig. 7.15. The system there consists of the usual linear modulation with orthogonal $V(f)$, and filtering $G(f)$, to which is added white Gaussian noise. The receiver is the PAM structure, with an orthogonal filter $U(f)$ having *twice the bandwidth,* and the sampler and processor that follow work at double speed.[6] It can be shown that if the flat part of the $U(f)$ spectrum encloses all the nonzero spectrum of $V(f)G(f)$, then the exact received signal $r(t)$ can be recovered from the double-rate samples, without aliasing. This means that in principle an MLSE detection is possible in the processor that follows, even though the receiver is only the PAM structure. A symbol-rate PAM receiver cannot always achieve MLSE performance, as we have seen repeatedly.

When the excess bandwidth factor in $V(f)$ and the double-width $U(f)$ both satisfy $\alpha \leq \frac{1}{3}$, the flat-top requirement is met. It is not necessary that $U(f)$ has the same shape otherwise as $V(f)$, nor does $U(f)$ need to change when $G(f)$ varies. Adaptation to $G(f)$ can take place in the discrete-time processor, where it is easy to do, rather than in the continuous-time filtering. All these advantages stem from fractional spacing.

Adaptive Equalizers. This large subject is discussed in many books and papers; see particularly Refs. 1, 3, and 6. We can focus on a few of the simpler approaches here, with an emphasis on those that relate to least-square equalization.

The best known of the simple methods was suggested as a general technique by Widrow and Hoff [19] and is called the *gradient search* or *stochastic gradient* algorithm. Consider the estimation error

[6]The fractional spacing is not restricted to twofold.

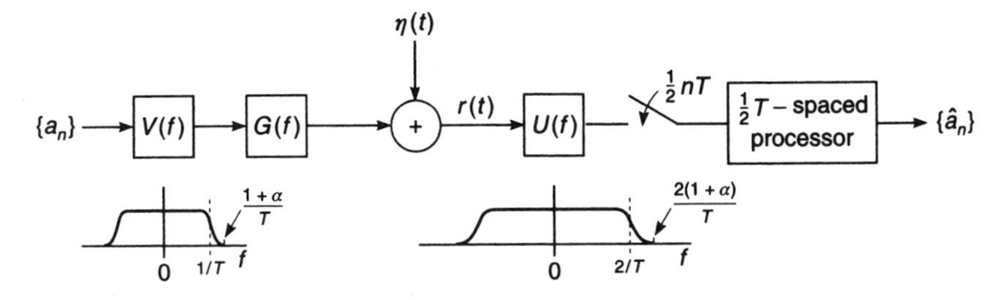

Figure 7.15 Illustration of double-bandwidth PAM receiver with double-rate sampling.

$$\varepsilon = a_{n-D} - \sum_{k=0}^{L-1} c_k r_{n-k} \tag{7.3-5}$$

of a linear equalizer; the expected square value of this was minimized in (7.2-15). Now the equalizer taps $\{c_k\}$ may not be correct for the ISI. How might the error be driven smaller by modifying the taps? One strategy is to focus on the square of (7.3-5) and find its gradient—that is, the derivative of the square error with respect to the taps. The gradient is an L-dimensional vector, and the kth component of it is

$$\frac{\partial \varepsilon^2}{\partial c_k} = \frac{\partial}{\partial c_k}\left[a_{n-D} - \sum_{j=0}^{L-1} c_j r_{n-j}\right]^2 = -2\varepsilon r_{n-k}, \qquad k = 0, \ldots, L-1 \tag{7.3-6}$$

The function being minimized here is a quadratic, which means that enough small steps in the vector $\boldsymbol{c}$ taken always in the direction of the gradient will find the minimum of ε^2 over $\boldsymbol{c}$.

The Widrow–Hoff procedure takes a step $-\delta\,\varepsilon$ in the direction of the gradient, where δ is called the step size. It takes one new step for each r_n that arrives from the matched filter. The error ε is a function of the last L r_n. As the $\{r_n\}$ and the symbols evolve, the algorithm tends to wander stochastically in the rough direction of the global square error minimum. An idea of this is given in Fig. 7.16. It can be seen

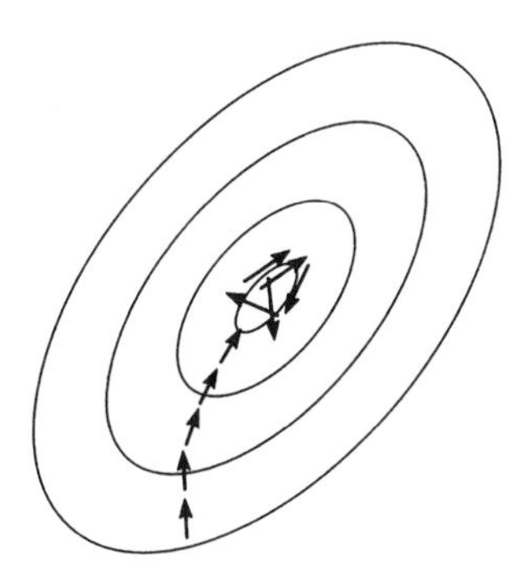

Figure 7.16 Example of gradient search.

that as the search approaches the minimum, it tends to jitter back and forth. This is called misadjustment error, and it can be shown that the scheme never gets closer in an rms sense to the minimum than a value proportional to δ. One solution to this problem is to reduce δ as the search appears to converge.

As a rule, gradient adaptive equalizers take much longer to converge as the object being identified—the best set $\{c_k\}$—becomes more complicated. Equalizers $C(z)$ with poles near the unit circle are particularly difficult. Nonetheless, the algorithm is simple and quite effective, particularly when it tracks a target that changes only slowly.

Quite a different approach is simply to solve directly for the best $\{c_k\}$, given a set of symbols and filter outputs. There are several ways of performing the calculation, some based on matrix inversions and others based on sequential calculations. A type of the last, called recursive lattice methods, is perhaps the most effective.

7.4 MODULATIONS OVER THE FADING CHANNEL

We now shift to quite a different channel, the Rayleigh fading channel. Instead of time dispersion, the medium multiplies the signal amplitude by a Rayleigh-distributed factor, as introduced in Section 5.4.1. A white Gaussian background noise has constant power density $\frac{1}{2}N_0$. Figure 5.19 shows some basic fading conditions. It has been said that channels are good, bad, or awful (Sklar [20], p. 150): Good is AWGN, bad is simple flat fading, awful is selective fading, perhaps with interference from other users. Error probability in a good channel has a Q-function form, and in a bad channel it is inversely proportional to E_b/N_0, as we will see in this section. In an awful channel there can be an *error floor,* meaning the error rate always lies above a threshold. When the channel is bad or awful, the first resort is equalization, which mitigates time dispersion. If this is not enough, one turns to *diversity*—that is, to several independent transmissions. This is the second theme of the section.

We first calculate error probabilities for the "bad" condition, slow flat fading. This occurs when the coherence bandwidth of the fading is wide and the Doppler bandwidth is narrow. This is the desirable case in the design formula (5.4-13)–(5.4-14): It means that the whole bandwidth of the transmission fades up and down as a unit, and the transmission stays in one fading mode for a considerable time.

Error Probabilities in the Flat Rayleigh Channel. For our first case, assume that the fading is so slow that there is time to obtain a good phase reference and that the coherence bandwidth exceeds the signal bandwidth. Then a phase-shift keyed signal arrives with some faded amplitude $A(t) = A$ and can be detected coherently. As explained in Section 5.4.1, the amplitude during an interval of time surrounding the present is Rayleigh distributed according to (5.4-5). Under Rayleigh conditions, A^2 is the sum of two independent squared zero-mean Gaussian variates A_I^2 and A_Q^2, where A_I represents an in-phase amplitude factor and A_Q is a quadrature amplitude factor and the whole signal is

$$s(t) = \sqrt{\frac{2E_s}{T}}[A_I I(t) \cos \omega_0 t - A_Q Q(t) \sin \omega_0 t] \tag{7.4-1}$$

If σ^2 is the variance of both variables A_I and A_Q, then an elementary calculation shows that the probability density function (PDF) of their sum $y = A_I^2 + A_Q^2$ is Gaussian, with

$$f_Y(y) = \frac{1}{2\sigma^2} e^{-y/2\sigma^2}, \qquad y \geq 0 \tag{7.4-2}$$

The mean value of y, which is the second moment of A, is $2\sigma^2$. Here, $I(t)$ and $Q(t)$ bear the data in (7.4-1), and A_I and A_Q scale them slowly up and down with the fading.

Compared to a standard, unfaded signal $\sqrt{2E_s/T}[I(t) \cos \omega_0 t - Q(t) \sin \omega_0 t]$ with SNR $E_b/N_0 = E_s/N_0 \log_2 M$, the faded signal (7.4-1) can be written as

$$s(t) = A\sqrt{\frac{2E_s}{T}}[I(t) \cos \omega_0 t - Q(t) \sin \omega_0 t]$$

which is a signal with SNR equal to $A^2 E_b/N_0$. A convention in Rayleigh signal analysis is to denote this faded SNR by the single parameter $\gamma = A^2 E_b/N_0 = A^2 E_s/N_0 \log M$. An important quantity is the average SNR over the fading process, denoted by

$$\overline{\gamma} = \frac{\mathcal{E}[A^2]E_b}{N_0} = \frac{2\sigma^2 E_b}{N_0} \tag{7.4-3}$$

The PDF of γ is obtained from (7.4-2) by the variable transformation $\gamma = yE_b/N_0$, which gives

$$f(\gamma) = \frac{1}{2\sigma^2}\frac{N_0}{E_b} e^{-\gamma/2\sigma^2} = \frac{1}{\overline{\gamma}} e^{-\gamma/\overline{\gamma}} \tag{7.4-4}$$

Turning now to the case of BPSK, we have from (2.6-5) that the error probability in the no-fading case is $p_e = Q(\sqrt{2E_b/N_0})$. In a local region of time, where SNR γ holds, this becomes in the fading case $Q(\sqrt{2\gamma})$. By averaging this $Q(\cdot)$ over the distribution of γ, we get a statement of the p_e that applies to any symbol at some point in the future. The calculation is

$$\int_0^\infty \frac{1}{\overline{\gamma}} e^{-\gamma/\overline{\gamma}}\, Q(\sqrt{2\gamma})\, d\gamma$$

This integral comes out to be

$$\begin{aligned} P_{\text{BPSK,f}} &= \frac{1}{2}\left(1 - \sqrt{\frac{\overline{\gamma}}{1+\overline{\gamma}}}\right), \qquad \overline{\gamma} \geq 0 \\ &\approx \frac{1}{4\overline{\gamma}}, \qquad \text{large } \overline{\gamma} \end{aligned} \tag{7.4-5}$$

The approximation at large $\overline{\gamma}$ shows a fundamental result for fading that generalizes far beyond BPSK: The error probability is proportional to the *inverse of the average SNR*. One way to explain this is that when a strong fade occurs, communication fails no matter what the modulation; thus the error probability is the probability of a deep fade, which, it turns out, is inversely proportional to the signal average.

Expressions like (7.4-5), and how to thwart them, are basic drivers of communication design in fading. Another fundamental case with a similar outcome is orthogonal signaling with two signals. Its constellation is Fig. 2.26b, and the error probability is (2.6-6), which becomes $Q(\sqrt{\gamma})$ at SNR γ. Binary FSK is a practical scheme with this constellation and probability (Section 3.6). By following through with similar steps to those that led to (7.4-5), we arrive at the integration

$$P_{\text{2FSK,f}} = \int_0^\infty \frac{1}{\overline{\gamma}} e^{-\gamma/\overline{\gamma}}\, Q(\sqrt{\gamma})\, d\gamma = \frac{1}{2}\left(1 - \sqrt{\frac{\overline{\gamma}}{2+\overline{\gamma}}}\right), \qquad \overline{\gamma} \geq 0$$
$$\approx \frac{1}{2\overline{\gamma}}, \qquad \text{large } \overline{\gamma} \tag{7.4-6}$$

Observe that the 3-dB loss in energy efficiency for orthogonal compared to antipodal signaling in AWGN carries over to the fading case: the large-$\overline{\gamma}$ outcome is 3 dB weaker in SNR in (7.4-6) than in (7.4-5). Both methods, of course, need much more average energy under fading than under AWGN.

One can carry out such fading channel integrations for a great many modulations. A general approach appears in Ref. 21. For M-ary PSK at large average energy, the fading symbol error probability is close to

$$P_{\text{MPSK,f}} \approx \frac{M-1}{2\overline{\gamma} M \log_2 M \sin^2(\pi/M)}, \qquad M \geq 2 \tag{7.4-7}$$

Some calculation with this shows that QPSK in the fading channel needs some 1.8 dB more average energy $\overline{\gamma}$ to achieve the same approximate error probability as BPSK; 8PSK needs 6 dB more. By comparison, it was shown in Section 3.4 for the AWGN channel that QPSK requires the same energy as BPSK and 8PSK requires just 3.6 dB more [Eq. (3.4-5)]. Thus fading affects the large-alphabet schemes relatively more.

Diversity Reception. Often the error formulas just presented are simply not good enough. Or the channel may be not merely "bad," but "awful." Then the key to fighting fading is the use of multiple, independent reception paths, a technique called *diversity*. When fading occurs, it does so disastrously, but total faded time as a fraction of the whole is usually small. If reception is arranged along several paths, one is likely to be clear when another is faded, so that an open path is available with high probability.

Diversity has been an established technique at least since the 1920s, and many mechanisms exist to produce diversity paths: spaced antennas, spaced frequencies, and differing times, reception angles, and polarizations. The only requirement is that the fading act independently on different paths. For frequency diversity, this means that diversity channels should be separated by at least the coherence bandwidth; in time diversity, successive transmissions should be delayed more than the coherence time (definitions are in Section 5.4.2). Spaced antennas can experience independent fading at spacings as short as a few wavelengths, a length that makes space diversity useful at UHF even on cars. A classic treatment of the physical basis of diversity, as well as its mathematics, was written by Stein in Ref. 22, Chapter 10.

Once L diverse transmission paths are arranged, it remains to choose the strongest among them, a process called *combining*. *Selection* combining refers to simply choosing the path out of L that has the best SNR. Choosing a nonfaded path is generally not difficult because it will be much stronger than faded paths, and radio AGC (automatic gain control) circuits easily and quickly sense SNRs. Figure 7.17 shows selective combining of two signals. The outcome, the heavy line, has much less fading.

A second method is called *equal gain* combining. Here all the diversity signals are brought into phase synchronism and added constructively. The signals can be brought into lock two at a time with a succession of PLL circuits. The simple addition of the signals tends to emphasize the stronger, more reliable diversity paths.

While these methods are effective, the ultimate linear combining technique to reduce error probability is *maximal ratio* combining (MRC), devised originally by Brennan [26]. Here signals are combined synchronously, but each is emphasized by

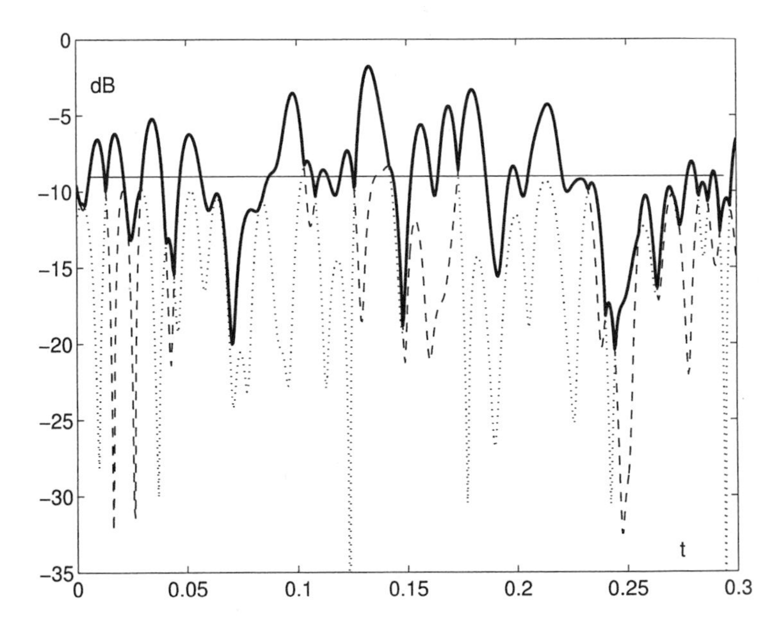

Figure 7.17 Selective combining of two signals. Horizontal line shows average power after combining.

a factor proportional to its observed SNR. To be more precise about this, let the result of a general linear combining of L signals be

$$y(t) = \sum_{\ell=1}^{L} \kappa_\ell r_\ell(t) \tag{7.4-8}$$

The short-time strengths of the signals are measured to be $A_1, \ldots, A_L$. A constant AWGN with density $\frac{1}{2}N_0$ appears at the receiver front end. It is easiest to consider both the signals and the weights in complex-number notation. It can be shown that the SNR of the combined $y(t)$ is maximized when $\kappa_\ell = A_\ell^*$, for all ℓ [1, 22]. The conjugate operation simply brings all signals into phase synchronism by subtracting the phase offset of each; the scaling by strength optimizes the SNR of y. The optimal SNR leads to the minimum error probability, at least with simple single-interval detectors.

A precise calculation of the MRC error probability can be based on the statistics of the intensity of y, a variable that has a chi-square distribution with L degrees of freedom. While the mathematics is tedious, some tight approximations may be derived, and the BPSK error probability at high average SNR $\overline{\gamma}$ with L independent diversity paths turns out to be

$$P_{\text{BPSK,MRC}} \approx \binom{2L-1}{L} \frac{1}{(4\overline{\gamma})^L} \tag{7.4-9}$$

Compared to (7.4-5), this estimate tends to zero as the Lth power of $\overline{\gamma}$ rather than just inversely with $\overline{\gamma}$. Some example cases are shown in Fig. 7.18. While (7.4-9) applies to MRC combining precisely, the critical factor in it, $1/(4\overline{\gamma})^L$, stems from the simple fact that the receiver can select among L diverse channels.

Provided that one is willing to multiply the RF receiver chain and the consumption of channels, it is clear from the figure that diversity can quickly bring the fading channel error performance down near the ordinary AWGN performance. For the MRC case, the general M-PSK probability (7.4-7) becomes

$$P_{\text{MPSK,MRC}} \approx \frac{\binom{2L}{L}(M-1)}{M[\log_2 M \sin^2(\pi/M) 4\overline{\gamma}]^L} \tag{7.4-10}$$

The Rake Receiver. Assume now that the channel is still slow fading, but that it is now frequency selective. An equivalent time-domain view is that it has multipath propagation. Assume further that the multipath impulse response consists of several response peaks and each of these is subject to its own slow fading process. Such a scenario can happen in cellular or local-area communication when there are several diffuse propagation paths. The peaks are called "fingers." One strategy is to measure the response fingers as if they were separate diversity paths and add them in phase, in a way similar to (7.4-8) and MRC combining. This is the principle of the

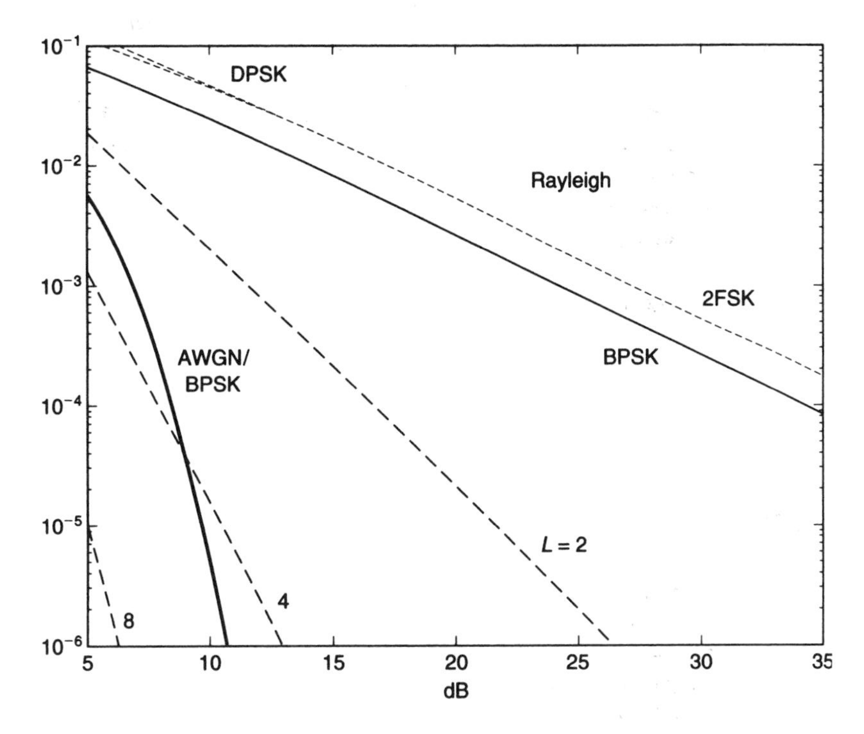

Figure 7.18 Symbol error probabilities under Rayleigh fading for BPSK, DPSK, and 2FSK. Shown for baseline is AWGN/BPSK case. Dashed lines are MRC diversity estimates from (7.4-9) for BPSK with $L = 2, 4, 8$ diversity paths.

rake receiver. The name derives from the idea of the fingers of a rake combing out a signal. The receiver has begun to play a role in wireless standards as more computation power has become available, and so we review its concept.

The classic rake receiver is a set of tapped delay lines, each like a linear equalizer with an all-zero response $C(z)$. There is one line for each possible I/Q transmitted symbol. The line taps consist of the complex-conjugate impulse response to a symbol; each line in effect performs a correlation with the received signal and the decision is made in favor of the symbol with highest correlation (complex notation is once again more convenient here). We can depict the correlation as the convolution at time zero of the received $r(t)$ sampled every T_q and the conjugate $c(t)$ sample sequence

$$\cdots, c_{i1}^*, \cdots, c_{i2}^*, \cdots, c_{i3}^* \cdots\cdots\cdots, c_{i\phi}^*, \cdots \qquad (7.4\text{-}11)$$

where $c(t)$ is the response to one of the symbols. The samples $c_{i1}, \ldots, c_{i\phi}$ represent ϕ fingers. Research has shown that tap positions should be ignored that are not clearly active fingers, since noise in a large number of inactive positions has a harmful effect. The positions in (7.4-11) denoted '$\cdots$' should therefore be thought of as zeros. Correlation (7.4-11) effectively carries out the MRC formula (7.4-8). The analysis is similar and details are available in [1].

Practical rake receivers allow 2–6 nonzero taps. Extra tap values that are in reality noise are a disadvantage, but 20–60 may be needed in order to space out the fingers. The tap sampling rate $1/T_q$ should be considerably larger than the coherence frequency B_{coh} of the selective fading, but should not be needlessly fast. The fading needs to be slow, slow enough that the responses $c(t)$ can be reliably estimated; this surely means that the symbol time T satisfies $T \ll T_{\mathrm{coh}}$. Rake receivers have a bank-of-processors structure like that of the receivers to come in Section 7.5. The rake idea can be extended to DPSK and other modulations, and simplified versions are available. One can hope that the error performance approaches that of formal diversity with ϕ paths. In a real sense, the rake receiver takes awful-channel multipath, and turns it into an advantage.

7.5 NONCOHERENT DEMODULATION

In Section 3.7 we began the topic of coherency in detection. A coherent detector is one that knows the signal phase reference and presumably makes use of it during reception. A noncoherent detector makes no use of the phase reference, other than to assume that it is constant. In between lies a family of partially coherent detectors that use some kind of partial knowledge, such as a probability distribution. We return now to the noncoherent case and explore some topics, starting with the Rayleigh fading channel in the previous section.

Detectors for the various channels and degrees of coherency are a huge subject. An exhaustive treatment on the subject is the book by Simon et al. [23].

DPSK over the Rayleigh Channel. Differential phase-shift keying was introduced in Section 3.4. It is BPSK with a certain type of differential encoding (see Tables 3.3 and 3.4) and a one-delay differential detection (as in Fig. 3.14). The probability of an initial detection error in the AWGN channel is Eq. (3.4-7), which is $p(e|\gamma) = \frac{1}{2}e^{-\gamma}$, if $\gamma = E_b/N_0$ is the SNR at the present time. Since DPSK detection depends only on the difference in phase from symbol to symbol, the detector need not know the absolute phase reference; it needs only to know that the phase is stable across the two intervals or at worst slowly varying. This is an example of noncoherent detection.

The DPSK scheme may be used to advantage over a fading channel as well as an AWGN one. Suppose the fading is the flat slow kind, which means that the phase reference varies slowly. With an ordinary modulation, time is needed for a synchronizer to obtain an absolute reference. As we saw in Chapter 4, this can be quite a long time. With DPSK, the phase need hold for only two intervals. This is a strong advantage in a fading channel.

The error probability of DPSK in Rayleigh fading can be derived as follows. As with the BPSK case, integrate the DPSK error probability conditioned on the SNR γ versus the distribution $f(\gamma)$ given in (7.4-4). The result is the simple integral

$$\int_0^\infty p(e|\gamma) f(\gamma)\, d\gamma = \int \frac{1}{2\overline{\gamma}} e^{-\gamma/\overline{\gamma}} e^{-\gamma}\, d\gamma, \qquad \overline{\gamma} \geq 0$$
$$= \frac{1}{2(1+\overline{\gamma})} \tag{7.5-1}$$

A plot of this is compared to BPSK in Fig. 7.18; from there or from Eq. (7.4-5) we see that DPSK in energy performance is about 3 dB poorer than BPSK. By contrast, they have the same energy efficiency on the AWGN channel.

Diversity reception may be applied to DPSK with a similar outcome, namely, that DPSK with the same diversity performs 3 dB worse than BPSK ([1], Section 7.4).

Noncoherent Detection over the AWGN Channel. We consider next the ML detection of a set of M signals in AWGN when the phase reference is unknown but stable. The receiver is easiest to conceptualize when it is a correlation receiver (as in Fig. 2.30) and when it works with baseband signals. This receiver is shown in Fig. 7.19. The details of its operation are explained next.

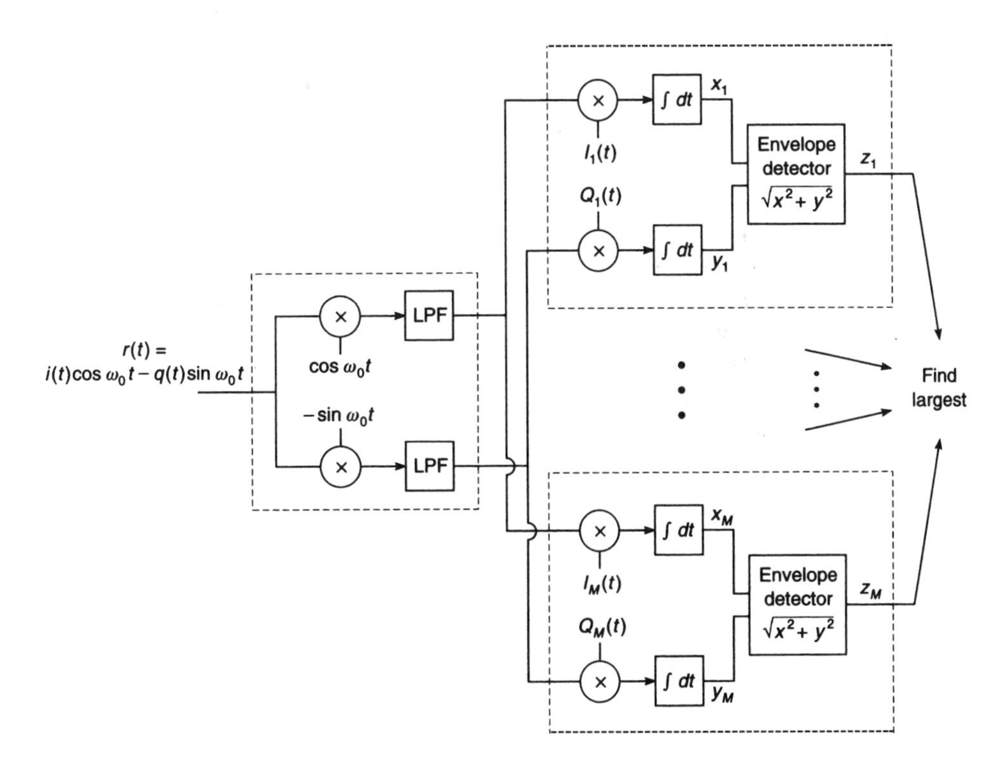

Figure 7.19 The I and Q correlation receiver for optimal noncoherent detection of M signals. A bank of envelope detectors follows a double balanced mixer.

A derivation [1, 23], which we will not repeat here, shows that the optimum demodulator compares the envelopes of the correlations of the M signals $s_1(t), \ldots, s_M(t)$ with the received signal $r(t)$. The signal with the largest correlation envelope value is the ML one under the noncoherent conditions. To define the envelope concept more precisely, consider the bandpass correlation of $r(t)$ with the kth signal $s_k(t)$. Here $I_k(t)$ and $Q_k(t)$ are the in-phase and quadrature components of $s_k(t)$, and $i(t)$ and $q(t)$ are those for $r(t)$; in addition, $r(t)$ is phase-shifted by an unknown θ compared to $s_k(t)$. All signals exist over the interval $[0, t_0]$. The bandpass correlation is given by the integral over $[0, t_0]$ of

$$\begin{aligned} r(t)s_k(t) = {} & [i(t) \cos(\omega_0 t + \theta) - q(t) \sin(\omega_0 t + \theta)] \\ & \times [I_k(t) \cos \omega_0 t - Q_k(t) \sin \omega_0 t] \end{aligned} \tag{7.5-2}$$

Some trigonometry shows that the received signal in the first brackets is

$$r(t) = [i(t) \cos \theta - q(t) \sin \theta] \cos \omega_0 t - [q(t) \cos \theta + i(t) \sin \theta] \sin \omega_0 t \tag{7.5-3}$$

The brackets in (7.5-3) are the in-phase and quadrature components of $r(t)$, respectively, from the receiver's point of view. Carrying out the integration, we get

$$\begin{aligned} \int_0^{t_0} r(t)s_k(t)\, dt = {} & \frac{1}{2} \int [i(t) \cos \theta - q(t) \sin \theta] I_k(t)\, dt \\ & + \frac{1}{2} \int [q(t) \cos \theta + i(t) \sin \theta] Q_k(t)\, dt \end{aligned} \tag{7.5-4}$$

plus a number of terms that tend to zero as $\omega_0 \to \infty$.

The outcome here consists of baseband quantities only, and it is clear that the correlation can be computed directly by correlating the baseband signals of $r(t)$ and $s_k(t)$ as well as by correlating the bandpass signals. In Fig. 7.19, the baseband signals in (7.5-3) are formed in the balanced mixer box. The dashed boxes that follow carry out the integration in (7.5-4).

It remains to define the envelope detector. The kth detector takes the value

$$z_k = \sqrt{x_k^2 + y_k^2}$$

as shown in Fig. 7.19, where

$$\begin{aligned} x_k &= \frac{1}{2} \int [i(t) \cos \theta - q(t) \sin \theta] I_k(t)\, dt \\ y_k &= \frac{1}{2} \int [q(t) \cos \theta + i(t) \sin \theta] Q_k(t)\, dt \end{aligned} \tag{7.5-5}$$

x_k and y_k being the outputs of the in-phase and quadrature branches of the correlation. Here z_k functions as the decision variable, and the rest of the receiver finds the largest of $z_1, \ldots, z_M$ and thus decides the most likely signal. An alternative to the envelope detector is the square law detector, which simply forms $x_k^2 + y_k^2$; the receiver can equally well find the maximum of these decision variables. In Section 3.1, the term *envelope* was reserved for the quantity $\sqrt{I^2(t) + Q^2(t)}$, where $I(t)$ and $Q(t)$ are the baseband signals for some bandpass signal. Although we will not take the time to do it, it can be shown that x_k and y_k are the baseband signals for the output of the bandpass matched filter; z_k is the value of the output's envelope at the sample time. The circuit in Fig. 7.19 is a convenient way to find this envelope value.

With considerable mathematics, the probability of error of this ML noncoherent receiver can be computed (Ref. 1, Chapter 4; or Ref. 23, Chapter 5). An example is the case of two orthogonal signals, which is 2FSK when the frequencies are orthogonal (Section 3.6; constellation in Fig. 2.26b). The error probability turns out to be

$$p_e = \tfrac{1}{2} e^{-E_b/2N_0} \qquad (7.5\text{-}6)$$

This expression has the same asymptotic energy efficiency as $Q(\sqrt{E_b/N_0})$, the error probability of *coherent* orthogonal 2FSK. A similar conclusion applies with M-ary orthogonal signaling. Thus noncoherent detection of orthogonal FSK in a good AWGN channel leads to no real loss in energy efficiency.

Signals that are alike over an interval except for a phase shift, such as PSK signals, cannot be distinguished by Fig. 7.19. It will put out the same $\{z_k\}$ for every such signal.

Noncoherent Detection of M-DPSK Signals. While Fig. 7.19 cannot distinguish signals that are phase shifts, noncoherent detection of M-ary PSK signals is nonetheless possible when the signals are differentially encoded. The analytical framework that leads to Fig. 7.19 in fact does not apply, because it is not possible to define a set of differentially encoded signals solely on a fixed interval $[0, t_0]$.

In Section 3.4 we saw that a combination of differential encoding, binary PSK, and differential decoding—the DPSK scheme—had approximately the energy efficiency in AWGN that ordinary BPSK has. But when the idea is extended to four or more phases, there is a significant asymptotic loss in energy, which becomes worse as the signaling alphabet grows. The receiver was actually a noncoherent one, because it did not use the phase reference.

The key to a better noncoherent reception of this M-DPSK is to observe many intervals of the signal in the detector, rather than just the two in DPSK. If many are observed, detection can be improved, and it is generally true that a long enough observation by the right receiver will achieve the coherent energy efficiency of the corresponding nondifferential MPSK. Reference 23, Chapter 7, gives many such receivers.

Noncoherent Detection of Nonorthogonal FSK. In Section 3.6, the chief examples of FSK with nonorthogonal signals were CPFSK and its generalization, the CPM class of constant-envelope coded modulations. These schemes do not fit into the framework leading to Fig. 7.19: They are properly considered to be trellis codes, and their signals cannot be thought of as limited to a short interval. In addition, the signals are not orthogonal. As in the case of M-DPSK, the key to efficient noncoherent detection is to view several intervals of the signal. Here again, a variety of such receivers exists and may be found, for example, in Ref. 23, Chapter 7 and Ref. 24, Chapters 7 and 10. As with M-DPSK, it is usually true that the right receiver with a wide enough signal observation can achieve the coherent energy efficiency.

Figure 7.20 shows a classic example, a noncoherent CPFSK receiver due to Osborne and Luntz [25]. The receiver has a rough similarity to Fig. 7.19, but differs in some details. It estimates the middle symbol out of a group of L. To look at its operation more closely, we define some special notation for the CPFSK signals.

Suppose $\sqrt{2E/T}\cos[\omega_0 t + \phi(\boldsymbol{a}, t)]$ is the CPFSK signal over LT seconds generated by transmission symbols $a_1, \ldots, a_L$ [see Eq. (3.6-1)]. Let L be odd, and for convenience let the a_ℓ be binary and set $\sqrt{2E/T}$ to 1. The receiver will estimate the *middle* symbol out of the L, having observed the signal created by the $\frac{1}{2}(L-1)$ symbols to either side. Let $\phi(+1; \boldsymbol{a}', t)$ denote the signal generated by the symbols $a_1', \ldots, a_{(L-1)/2}'$ to the left, +1 in the middle, and $a_{(L+3)/2}', \ldots, a_L'$ to the right, and analogously for $\phi(-1; \boldsymbol{a}', t)$. Let $\boldsymbol{a}_i'$ denote the ith such set of left and right symbols.

The receiver in Fig. 7.20 begins by multiplying the input $r(t)$ by the pair $\cos[\omega_0 t + \phi(+1; \boldsymbol{a}_i', t)]$ and $-\sin[\omega_0 t + \phi(+1; \boldsymbol{a}_i', t)]$ for each i and for each binary middle symbol. There is then a calculation of the envelope correlation value as in Fig. 7.19. The whole operation forms the unit at bottom of Fig. 7.20. These units feed blocks that scale the z outputs by a zero-order modified Bessel function as shown. This is a tabulated function that can be stored as some simple approximation. All the outputs stemming from the +1 middle symbol form one sum; those from –1 form another. The ML decision goes to the larger sum.

The Osborne and Luntz receiver and the earlier ones in this section are alike in several ways. Although they add a few new elements, their blocks are similar to structures in Chapters 2 and 3. There is extra complication, but as a reward the receivers achieve nearly the coherent detector error performance. Present-day receivers for moderate speed data often have extra processing power available. A conclusion we can make for this section is that if the extra receiver cost is acceptable, what is important in a communication channel is not a known phase reference, but a *stable* one.

7.6 SPREAD SPECTRUM

Spread spectrum offers a solution to range of real-life problems in digital communication. It itself is a range of technigues that have in common one attribute: They transmit in a much wider bandwidth than the value suggested in Chapters 2 and 3. A review of the pulses in these chapters shows that *any* useful, more or less orthogonal pulse that appears at rate $1/T$ has a mainlobe bandwidth of about $1/T$ positive hertz at baseband, and twice this at bandpass. Independent I and Q signals can be carried

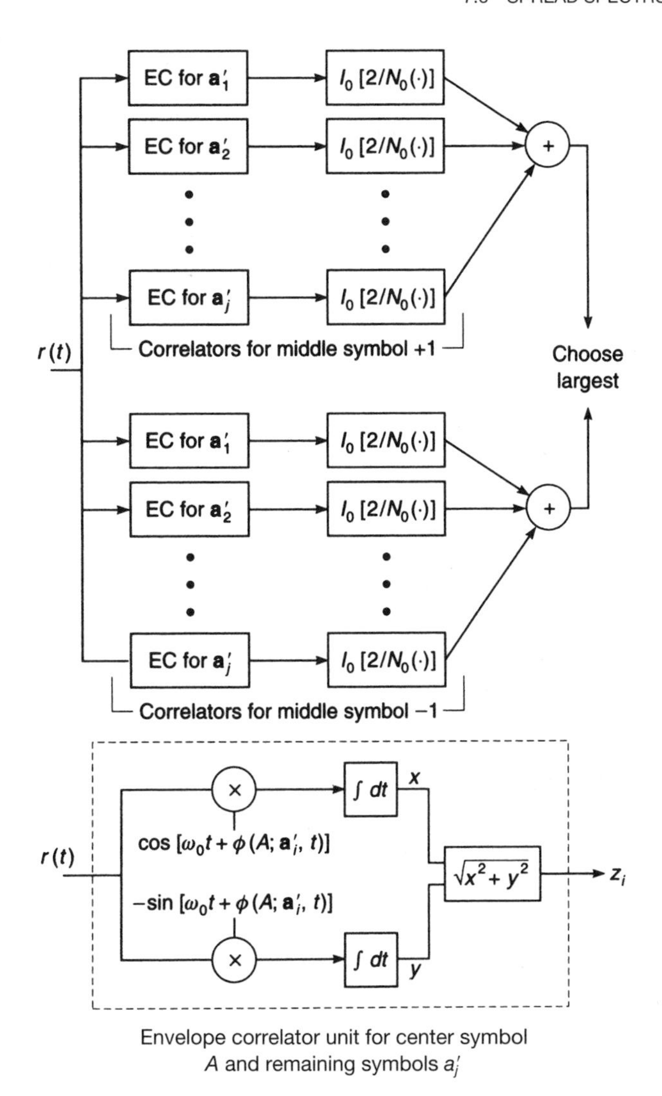

Figure 7.20 Osborne and Luntz receiver for noncoherent binary CPFSK. Receiver estimates middle symbol out of I. Top half finds decision variable for middle symbol +1, bottom half for −1. Each half contains $J = 2^L$ quadrature envelope correlators (ECs), details of which appear in lower box.

there, so the bandwidth is roughly $1/T$ in either case. When a transmission system has significantly wider bandwidth than $1/T$, we classify it as a spread spectrum system.

Hopefully, we get something in return for this expansion. Communication engineers learn that bandwidth in the practical world is expensive. A similar view is given by Shannon's theory of the Gaussian bandpass channel; it begins with Eq. (6.1-14) and arrives at the formula

$$C_W = W \log_2(1 + 2E_s/N_0), \qquad \text{bit/s} \tag{7.6-1}$$

for the channel capacity, in which W is the positive bandwidth and E_s/N_0 is the usual SNR.[7] Capacity thus grows linearly with bandwidth and only logarithmically with energy. A spread spectrum system generally spreads bandwidth while keeping the same E_s, and it thus displaces rapidly more capacity, while carrying no additional data. What can we get in return for the bandwidth?

Spread spectrum is in fact useful in a variety of situations where the AWGN model has degenerated into something less pleasant. To motivate the section, here are some of these.

- *Jamming*. A jammer can place a strong narrowband signal in a critical spot. By spreading the desired signal enough, the jammer cannot block enough of it.
- *Secrecy*. By spreading the signal enough, it becomes hard to find by one who lacks the key to inverting the spreading. This is a low probability of intercept (LPI) system.
- *Multiple Access*. Many simultaneous signals can use the same bandwidth, relatively independently of each other. This is multiple-access communication.
- *Fading*. Many channels, especially mobile ones, have frequency- and time-selective fading (as described in Section 5.4). A signal that is spread across a frequency and time range that is wider will not be destroyed by any one fade event.

Spread spectrum began to develop around 1950 in military communication, where jam resistance and secrecy were the motivators. It has come to greater prominence since 1990 as a method of wireless communication, with multiple access and fade resistance the drivers.

The plan of this section will be to discuss first *direct sequence* spread spectrum (DS-SS), a method where an ordinary modulated signal is spread via multiplication with another much wider band signal. A particular area of interest is the application of the pseudo-noise sequences from Section 4.8. Then we turn in Section 7.6.3 to *frequency hopped* spread spectrum (FH-SS), which works by shifting about the carrier of the ordinary signal. The FH and DS methods each have their own strengths and detection schemes. The discussion is necessarily a summary of a very large field. Ziemer and Peterson [29], Dixon [30], and Holmes [31] are specialized texts in spread spectrum. Further reading is available in chapters and survey articles, especially [1, 32–34].

7.6.1 Direct Sequence Spread Spectrum

Figure 7.21 shows the essential functions in a DS-SS communication system. The system begins and ends with an ordinary carrier digital modulator and demodulator

[7] His argument is that a transmission of bandwidth W and total length τ seconds makes available $2W\tau$ orthogonal signal space dimensions; each one of these is an independent AWGN channel with the capacity C in (6.1-14), which gives total capacity (7.6-1). See further in Refs. 1 and 28.

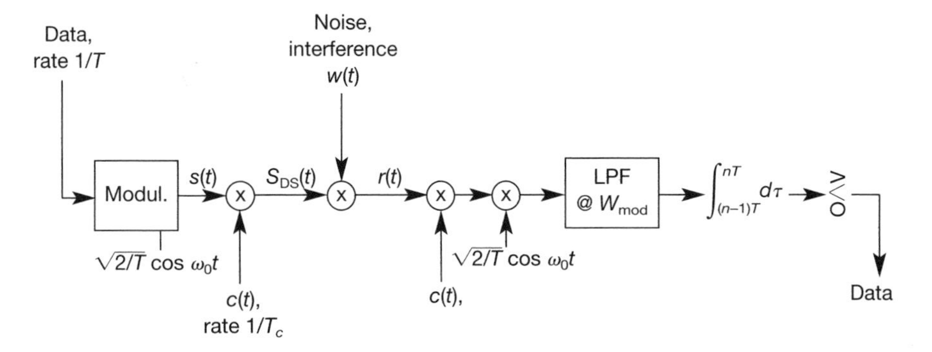

Figure 7.21 Essential functions in a DS-SS communication system. Square-pulse modulator assumed. Integrator and thresholder perform detection.

like those in Chapter 3. We can assume that the modulator signal is a straightforward rate-$1/T$ orthogonal pulse train

$$s(t) = \sqrt{2E_s} \sum_n a_n v(t - nT) \cos \omega_0 t \tag{7.6-2}$$

with nearly square pulse shape $v(t)$ and bandwidth W_{mod} according to some measure. In fact, we will most of the time assume that the modulation is binary and square pulse; that is, $s(t)$ is the NRZ BPSK modulation in Eq. (3.2-1) and can be written as simply $\sqrt{2E_s/T}\cos[\omega_0 t + \psi(t)]$, where $\psi(t)$ takes values $\{0, \pi\}$ in response to the data. Since spread spectrum will multiply the basic bandwidth $1/T$, there is little use worrying about the square-pulse sidelobes, either in this section or in practice.

Next comes the *spreading signal* $c(t)$, which multiplies $s(t)$ to form $s_{\text{DS}}(t)$, the DS-SS signal. The signal $c(t)$ is a rapidly modulated signal with a wide, continuous spectrum, which has energy

$$\int_{(n-1)T}^{nT} |c(t)|^2 \, dt = 1$$

over data symbol interval n. In most systems it is a binary ± 1 sequence with symbol rate T_c, where $T_c \ll T$. The signal can be written during interval $[nT, (n+1)T]$ as the rate-$1/T_c$ pulse train

$$c(t) = \sqrt{\frac{2}{T}} \sum_k b_k v_c(t - kT_c - nT) \tag{7.6-3}$$

If $v_c(t)$ is a width-T_c square pulse, this is effectively another NRZ BPSK modulation. Both sender and receiver know the sequence $\{b_k\}$ and are in perfect synchronism. The time T_c is called the *chip time* and $v_c(t)$ is the chip waveform. With the simple BPSK assumptions here, we can say that each binary symbol in $s(t)$ is each broken into T/T_c chips, a ratio called the DS *spreading factor*.

To $s_{DS}(t)$ is added noise, other signals, or jamming, to form the received $r(t)$. The receiver performs a multiplication by $c(t)\sqrt{2/T}\cos\omega_0 t$, followed by a low-pass filter whose width is W_{mod}, and an integration over the nth T-interval. These three perform a baseband correlation of the known spreading signal with $r_{DS}(t)$; that is, after the LPF is formed

$$\int_{(n-1)T}^{nT} a_n |c(t)|^2 \frac{2}{T}\frac{1}{2}\,dt + \text{noise} + \text{interference} \tag{7.6-4}$$

Most of the interference is wider band than W_{mod} and is blocked by the LPF.

To see that the spectrum is indeed spread, we perform some spectral calculations. The analysis can be carried through in the manner of Section 2.3.2 and the linear modulation spectrum theorem. But technically this produces an average power spectrum—the power spectrum of the signal, averaged over the data—and in the sequel we need the true power spectral density of a stochastic process. By the Wiener–Khintchine theorem, this is the Fourier transform of the autocorrelation $R(\tau)$ of the (stationary) process in question.

Continuing the NRZ BPSK assumption above, let $I(t) = (1/\sqrt{T})\Sigma_n a_n v(t - nT)$ denote the unit-energy in-phase baseband signal in (7.6-2). Here $v(t)$ is a square pulse and $\{a_n\}$ is an IID infinite sequence of equiprobable ±1s. The random process $I(t)$ is called a random telegraph signal, and analyzing it is a classic problem. Its autocorrelation may be written as

$$R_{II}(\tau) = \lim_{L\to\infty} \frac{1}{2L}\int_{-L}^{+L} [I(t)I(t-\tau)]\,dt$$

If $|\tau| > T$, then $I(t)I(t - \tau)$ is the product of two independent a_n, and its expectation is zero. Otherwise, pulses driven by the same a_n each overlap by exactly τ; the value of the integral is $2La_n^2\tau/T = \tau$. The autocorrelation of a long $I(t)$ is thus the triangle function

$$R_{II}(\tau) = \begin{cases} 1 - \dfrac{|\tau|}{T} & |\tau| < T \\ 0 & \text{otherwise} \end{cases} \tag{7.6-5}$$

The power spectral density of I is the Fourier transform, which from standard tables is

$$S_{II}(f) = T\,\text{sinc}^2(fT) \tag{7.6-6}$$

We recognize here the square of the NRZ pulse spectrum in Fig. 2.2: Either averaging the pulse train spectrum of Theorem 2.3-1 or finding a true PSD in fact yields the same outcome.

Continuing on, we compute the spectrum of $c(t)I(t)$. Let the spreading sequence $c(t)$ be another, much faster random telegraph process and assume that c is long and

c and I are *independent* processes. Then probability theory shows that the autocorrelation of the product $p(t)$ is the product

$$R_{\rm pp}(\tau) = R_{\rm cc}(\tau) R_{\rm II}(\tau) \tag{7.6-7}$$

The PSD of the baseband process $p(t) = c(t)I(t)$ is therefore the convolution of $S_{\rm II}(f)$ and $S_{\rm cc}(f)$. The PSD $S_{\rm cc}(f)$ is $T_c\ \mathrm{sinc}^2(fT_c)$, found the same way as (7.6-6). It is a factor T/T_c *wider* than $S_{\rm II}(f)$, and so also must be the total PSD $S_{\rm pp}(f)$, by the properties of convolution. We have spread the original signal.

The PSD of the process $s_{\rm DS}(t)$ is that of $p(t)$ shifted to $f_0 = \omega_0/2\pi$ hertz; from Fourier properties this is

$$S_{\rm DS}(f) = \tfrac{1}{2} S_{\rm II}(f - f_0) + \tfrac{1}{2} S_{\rm II}(f + f_0) \tag{7.6-8}$$

We can think of the Ordinary Modulator box as taking care of this shift, but the box can just as well have a baseband output and the shift then takes place after the $c(t)$ multiplication. The NRZ BPSK then works directly on a rate-$1/T_c$ high speed binary sequence. In either case, the DS-SS signal is expanded to $W_{\rm DS}$ hertz, and whether measured at baseband or at RF this exceeds $W_{\rm mod}$ by the factor T/T_c.

In arriving at (7.6-6)–(7.6-8), there have been several assumptions: NRZ pulses, and long spreading and IID data sequences that are independent of each other. Practical systems, however, come close to these assumptions: Sequences are long, symbols are nearly independent, and pulses are close to square. The spectral formulas here are quite accurate.

Pulse-Noise Jamming. What is jamming? At first glance one might think that a strong sinewave can be a jammer, but single tones are easy to defeat. A filter with a very narrow notch can be designed and the sine tuned to lie directly under the notch. Such "notch filters" are a feature of high-quality receivers. They work because the notch removes all the jammer but no significant part of the desired signal.

A more devious jammer will block a whole carefully chosen band and will concentrate its power at embarrassing times such as synchronization words. Suppose an average power P_J watts is available[8] and jamming takes place a fraction δ of the time across a bandwidth W_J that is approximately $W_{\rm mod}$. With little loss of generality, we let the "friendly" transmission be antipodal signaling in AWGN with N_0, with the familiar (2.6-5) error probability $p_e = Q(\sqrt{2E_b/N_0})$. It is difficult to say in advance what is the jammer's best signal, except that it should fill $W_{\rm mod}$. An all-purpose signal might be bandpass Gaussian noise, and this assumption leads to an easy derivation of results that hold for practical jamming. This is called *pulse-noise jamming*.

Our jammer now sets W_J equal the modulator $W_{\rm mod}$ and concentrates an average P_J during fraction δ. Define the jamming noise density $N_J = P_J/W_J$ and assume that

[8]Throughout this discussion, powers and noises are as measured at the jammed receiver. Obviously, jammers should place themselves close to receivers.

$N_J/\delta \gg N_0$ (otherwise little jamming takes place). The long-term average error probability becomes

$$\overline{p_e} \approx \delta Q\left(\sqrt{\frac{2E_b}{N_0 + N_J/\delta}}\right) + (1-\delta)Q\left(\sqrt{\frac{2E_b}{N_0}}\right) \qquad (7.6\text{-}9)$$

We can ignore the right-hand term. The optimally bad δ may be found by maximizing (7.6-9) over δ with the aid of the accurate Q-function approximation $Q(x) \leq (1/\sqrt{2\pi x})\exp(-x^2/2)$. The outcome is

$$\overline{p_e} \approx \begin{cases} \dfrac{1}{\sqrt{2\pi e}\,2E_b/N_J}, \text{ with } \delta = \dfrac{N_J}{2E_b}, & E_b/N_J \geq \dfrac{1}{2} \\ Q\left(\sqrt{\dfrac{2E_b}{N_J}}\right) > 0.16, & \text{otherwise} \end{cases} \qquad (7.6\text{-}10)$$

We would hope that $\overline{p_e}$ has at worst the form $Q(\sqrt{2E_b/N_J})$, but in fact it is only *inversely* related to E_b/N_J, as it would be over the Rayleigh fading channel in Section 7.4.

DS Processing Gain. A solution is to spread the signal over a much wider bandwidth. The jammer must then cover a wider W_J, and consequently the density N_J is less for the same jammer power. As we have seen, the spread signal bandwidth and the data signal bandwidths are around $1/T_c$ and $1/T$. The ratio T/T_c is the spread factor. We can hope that the jamming density is reduced by the same ratio, and this indeed is usually the case in practical systems. A reduction in noise like this is called a *processing gain,* denoted G_p. In what follows we give a simple derivation that the gain against a pulse-noise jammer is indeed T/T_c for the binary square-pulse DS scheme in this section.

Suppose the jammer continues with bandwidth $W_J = 1/T$, the approximate data signal bandwidth. We can ignore front-end white noise to the receiver. After despreading by $c(t)$ in Fig. 7.21, the DS signal can be written as two terms

$$[c(t)s(t) + w(t)]c(t) = c^2(t)s(t) + c(t)w(t) \qquad (7.6\text{-}11)$$

Let the first term be the despread data signal $s_d(t)$ and let the second be the despread jamming signal $s_j(t)$. Both $s(t)$ and $w(t)$ here are bandpass and $c(t)$ is low-pass. Under pulse-noise jamming, $w(t)$ is bandpass Gaussian noise (see Appendix 3B) whose PSD has width W_J centered on f_0 and density $\frac{1}{2}N_J$. The spreading sequence $c(t)$ has PSD $T_c\,\mathrm{sinc}^2(fT_c)$ as before; the processes $c(t)$, $w(t)$, and $s(t)$ are independent. Therefore $R_{jj}(\tau) = R_{cc}(\tau)R_{ww}(\tau)$ and the jamming component $s_j(t)$ has PSD given by the convolution

$$S_{jj}(f) = S_{cc}(f) * S_{ww}(f) \qquad (7.6\text{-}12)$$

This can be computed, but it is enough to argue as follows. As T/T_c grows, $S_{ww}(f)$ tends to impulses of weight $W_J N_J/2$ at frequencies $\pm f_0$, in comparison to the much wider $T_c\,\text{sinc}^2(fT_c)$; thus $S_{jj}(f)$ is approximately

$$S_{jj}(f) = \tfrac{1}{2}T_c P_J[\text{sinc}^2([f-f_0]T_c) + \text{sinc}^2([f+f_0]T_c)] \qquad (7.6\text{-}13)$$

where we replace $W_J N_J$ with the jammer power P_J. This wide PSD is shown in Fig. 7.22, together with a width-$1/T$ passband equivalent to the LPF in Fig. 7.21. The wide PSD is approximately flat in the passband and the power that passes through is about $T_c P_J W_J = T_c P_J/T$.

The baseband PSD of $s(t)$ has been shown to be (7.6-6). If this is shifted to $\pm f_0$ and the arriving power at the receiver is P, then the data component PSD is

$$S_{dd}(f) = \tfrac{1}{2}TP[\text{sinc}^2([f-f_0]T) + \text{sinc}^2([f+f_0]T)] \qquad (7.6\text{-}14)$$

This also is shown in Fig. 7.22. The total power in $s_d(t)$ after the BPF is about P, ignoring the clipped sidebands. The ratio of data to jamming power is therefore $(P)/(T_c P_J/T) = (T/T_c)(P/P_J)$. The original power ratio has been improved by the factor T/T_c, and this we take to be the processing gain; that is,

$$G_p \approx \frac{W_{\text{DS}}}{W_{\text{mod}}} \approx \frac{T}{T_c} \qquad \text{(DS processing gain, NRZ BPSK)} \qquad (7.6\text{-}15)$$

A similar calculation may be performed for the single tone jammer (see Ref. 29, Chapter 7), for interference from other users in multiple access communication, and

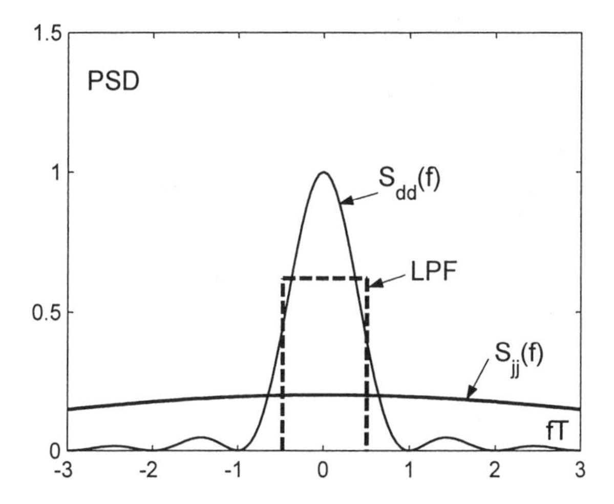

Figure 7.22 Power spectra in DS-SS communication after despreading: The despread data spectrum $S_{dd}(f)$, the jammer spectrum $S_{jj}(f)$ and the filter passband. The spreading factor here is $T/T_c = 10$ and the jamming power density is twice the signal density.

for other DS-SS situations. The result is usually the same: A performance measure improves by T/T_c; that is, it improves by approximately the spreading in bandwidth.

The ideas and calculations in this section extend easily to DS-SS systems based on QPSK and continuous phase modulations such as MSK [1, 29]. The QPSK systems are somewhat more complex but work in half the bandwidth and are more effective in LPI applications. The MSK systems have reduced sidelobes, if this is needed.

7.6.2 Pseudonoise Sequences in Spread Spectrum

In Section 4.8 we introduced pseudonoise (PN) sequences. Their near-perfect randomness and the near-zero correlation of a sequence's cyclic shifts with each other make PN sequences ideal for use in DS spread spectrum. As spreading sequences $c(t)$, PN sequences are close enough to random so that the spectral and processing gain results of the last section hold with good approximation. In multiple access communication based on DS-SS with correlation detection, each user can have a different shift or a different PN as its $c(t)$; their interference with each other will be reduced by about G_p. There is also a rich literature on sequences that are not PN but are close enough in properties to be useful in spread spectrum.

We will begin with a simple DS-SS multiple access communication system that features the length-7 PN sequence[9] in Example 4.8-1 and the discussion after. There will be three simultaneous user pairs, who are assigned the spreading sequences

$$\text{User 1:} \quad \boldsymbol{c}^{(0)} = \{+1, +1, -1, -1, -1, +1, -1\}$$

$$\text{User 2:} \quad \boldsymbol{c}^{(1)} = \{-1, +1, +1, -1, -1, -1, +1\}$$

$$\text{User 3:} \quad \boldsymbol{c}^{(3)} = \{-1, +1, -1, +1, +1, -1, -1\}$$

where $\boldsymbol{c}^{(0)}$ is the basic PN sequence in the example and $\boldsymbol{c}^{(1)}$ and $\boldsymbol{c}^{(3)}$ are its first and third right cyclic shifts, as discussed after the example. If transmitter 1 wishes to send data +1, it spreads this to the sequence $+\boldsymbol{c}^{(0)}$; otherwise, it spreads –1 to $-\boldsymbol{c}^{(0)}$. Transmitters 2 and 3 behave the same with $\boldsymbol{c}^{(1)}$ and $\boldsymbol{c}^{(3)}$. The seven-times-faster sequences are separately BPSK-modulated and add coherently in the channel. Receivers 1, 2, and 3 correlate the composite signal with their respective $\boldsymbol{c}$ in order to obtain their own data. How well does this work? What is the error mechanism in this system?

Figure 7.23 shows an example transmission in which user pairs 1, 2, and 3 communicate data +1, +1, and –1. The picture is drawn at baseband. The sum signal $\boldsymbol{s}_\Sigma = \boldsymbol{c}^{(0)} + \boldsymbol{c}^{(1)} - \boldsymbol{c}^{(3)}$ is shown. Receivers 1, 2, and 3 adapt Eq. (4.8-3) to form the correlations of their spreading sequences $\boldsymbol{c}^{(0)}$, $\boldsymbol{c}^{(1)}$, $\boldsymbol{c}^{(3)}$ against $\boldsymbol{s}_\Sigma$. The periodic correlation between $\boldsymbol{c}$ and $\boldsymbol{s}$ is defined by

$$C_{\text{per}}(\boldsymbol{c}, \boldsymbol{s}) \triangleq \frac{1}{m} \sum_{i=0}^{m-1} c_i s_i \tag{7.6-16}$$

[9]As always, the equivalences $\{1, 0\} \leftrightarrow \{-1, +1\}$ are used to convert the symbols in the example to ± 1.

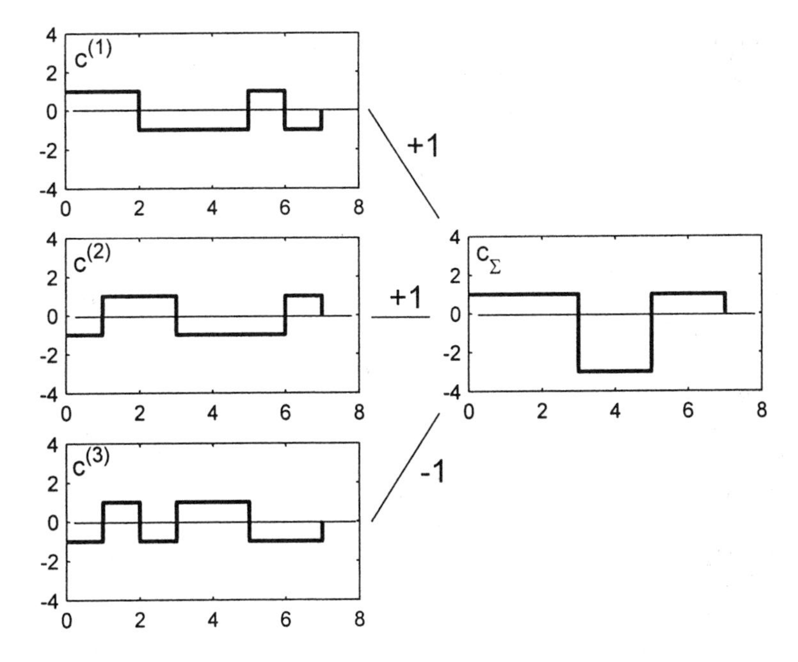

Figure 7.23 Example transmission in which user pairs 1, 2, and 3 communicate data +1, +1, and –1. Drawn at baseband.

where m is the sequence length. The outcome is

$$\begin{aligned} C_{\text{per}}(\boldsymbol{c}^{(0)}, \boldsymbol{s}_\Sigma) &= \tfrac{1}{7}[+1 + 1 - 1 + 3 + 3 + 1 - 1] = 1 \\ C_{\text{per}}(\boldsymbol{c}^{(1)}, \boldsymbol{s}_\Sigma) &= \tfrac{1}{7}[-1 + 1 + 1 + 3 + 3 - 1 + 1] = 1 \\ C_{\text{per}}(\boldsymbol{c}^{(3)}, \boldsymbol{s}_\Sigma) &= \tfrac{1}{7}[-1 + 1 - 1 - 3 - 3 - 1 - 1] = -1\tfrac{2}{7} \end{aligned} \tag{7.6-17}$$

The two positive correlations indicate that transmitters 1 and 2 sent data +1; the negative correlation indicates transmitter 3 sent –1. In the absence of noise these correlations can actually be guessed in advance. From Section 4.8.1, we know that the correlation of any shift with itself is $R_{\text{per}}(0T_c) = 1$ and the cross-correlation of any two is $R_{\text{per}}(jT_c) = -\frac{1}{7}, j \neq 0$. It must be true, for instance, that

$$\begin{aligned} &C_{\text{per}}(\boldsymbol{c}^{(3)}, \boldsymbol{c}^{(0)} + \boldsymbol{c}^{(1)} - \boldsymbol{c}^{(3)}) \\ &\qquad = C_{\text{per}}(\boldsymbol{c}^{(3)}, \boldsymbol{c}^{(0)}) + C_{\text{per}}(\boldsymbol{c}^{(3)}, \boldsymbol{c}^{(1)}) - C_{\text{per}}(\boldsymbol{c}^{(3)}, -\boldsymbol{c}^{(3)}) \\ &\qquad = -\tfrac{1}{7} - \tfrac{1}{7} - 1 = -1\tfrac{2}{7} \end{aligned}$$

A similar calculation gives the other correlations in (7.6-17) and any other correlations for spreading sequences that are shifts of $\boldsymbol{c}^{(0)}$.

An error occurs in this multiple access system if the summing of transmissions changes the sign of a correlation. If all the spreading sequences are shifts of the same length-m PN sequence, it is clear that up to $m - 1$ users can be supported since the total interference to a user data ± 1 has absolute value $\leq (m-2)/(m-1)$. This is in the absence of noise, often a good assumption because spread spectrum systems tend to be limited by other-user interference. In case of AWGN, the products $c_i s_i$ in (7.6-16) for a square-pulse system need to be replaced by a chip-time integration $\int_{Tc}[s_i(t) + \eta(t)]c_i(t)\, dt$, which will be disturbed somewhat by the noise $\eta(t)$. The presence of K interferors adds K terms of size $\pm 1/m$ that reduce the worst-case noise resistance by the energy factor $1 - K/m$. This again shows that with small noise about m users can be supported. One can also look at the effect of spreading the other way around: The interference and loss of noise resistance caused by a single interferor is reduced m-fold. It is therefore said that the processing gain of a length-m PN sequence DS-SS system is $\approx m$.

Direct sequence systems thus provide a straightforward method of multiple access, but some main weaknesses are also evident from the example. First, DS-SS depends on spreading sequence synchronization. Without knowledge of where $c(t)$ starts, users cannot correlate with their partner signal. The reader can verify that different user pairs do not need to know each other's timing, but the ends of any one pair need to be in perfect synchronism. A second weakness is that a strong interferor will dominate the correlation even if it is nearly orthogonal. An $(m - 1)$-fold more powerful interferor will contribute the same to (7.6-16) as the desired incoming signal. It is crucial in DS-SS mutliple access that incoming power levels be nearly equalized. This is not always easy. A threat by strong, local interferors like this is called a *near–far effect*.

The discussion here has featured PN sequences but near-orthogonality among user pairs is all that is required and several types of less perfect spreading sequences are in use. These include Gold, Kasami, and m-sequences [1, 29]. Some systems attempt to correlate with only a part of the spreading sequence. Generally, spreading sequences are much longer than 7. The name given to all systems of this sort is CDMA, for code-division multiple access; "code" refers to the set of spreading sequences.

Finally, we turn to a well known example of DS-SS multiple access.

Example 7.6-1: The Third-Generation WCDMA System. The basic Third-Generation cellular telephone standard now more or less accepted in the world is a DS-SS method that occupies 60 MHz in both the forward (down to the handset) and reverse directions. We will review the standards that presently apply to this system. A number of DS-SS systems can operate in parallel within the 60 MHz. The RF frequency bands are typically 2110–2170 MHz (forward) and 1920–1980 MHz (reverse). There are a number of voice and data bit rates; the low-speed ones are in the range 4.75–12.2 kbit/s. The basic DS-SS channel, to be shared by users in the same cell, has width 5 MHz (although bandwidths in the range 1.25–20 MHz are allowed in the standards). The basic chip rate applied to a low-speed bit transmission is 4.096 Mchip/s (rates 1.024–16.384 Mchips/s are allowed). At the highest of the data rates the theoretical processing gain is thus $G_p \approx 400$; values 100–200 are also typical. The DS-SS system uses QPSK modulation, but BPSK spreading. Operation in different cells can be asynchronous. Significant attention is devoted to controlling the power levels of the users; a number of pilot symbols are transmitted to make this and synchronization easier. Rake receivers are allowed by the standards, and rate $\frac{1}{3}$ convolutional coding is employed for voice, with stronger coding for data. If the links are indeed carrying voice, an interesting phenomenon called *voice activity* comes into play: A person speaking over a one-way voice link in fact uses it only 40–45% of the time. The rest is silence, and a DS-SS link can be shut off during this time. The result is less interference, by about the same factor. Spread spectrum systems are uniquely able to make practical use of voice activity. The capacity of the system depends on many other factors as well, such as cell antenna design and interference from neighboring cells. The design capacity of each 5 MHz is said to be about 390 one-way voice links.

A complete discussion of cellular capacity appears in, for example, Ref. 35. Adachi et al. [34] give a detailed picture of DS-SS third-generation transmission, and Pandya [36, Chapter 6] gives a higher-level network view.

7.6.3 Frequency-Hop Spread Spectrum

Figure 7.24 shows the essential functions of an FH-SS system. As in Fig. 7.21, the system begins and ends with an ordinary modulator and demodulator, whose symbol time is T. This time, however, the modulator output $s(t)$ is simply shifted about in carrier frequency through mixing with a sequence of sinusoids $\cos \omega_i t$, $i = 0, \ldots, L - 1$, to produce $s_{FH}(t)$.[10] Noise, jamming, and other interferors add to $s_{FH}(t)$, and the received $r(t)$ is mixed down hopefully by $\cos \omega_i t$. A lowpass filter removes out-of-band components and the signal is demodulated. The carrier is said to *hop*, and the set $\{\cos \omega_i t, i = 0, \ldots, L - 1\}$ is the hop set. The hopping at the send and receive ends must be in nearly perfect synchrony. The new bandwidth, W_{FH}, is L times the basic modulation bandwidth.

Hopping occurs at a chip rate $1/T_c$, paralleling the DS method, but now T_c can be larger or smaller than T. A slow hop system has $T_c \geq T$ and may send several symbols during each hop; otherwise the system is fast hop. A frequency synthesizer performs the hopping, driven by some sort of random signal $c(t)$. The system bandwidth W_{FH} is not necessarily expanded by the factor T/T_c as in a DS-SS system. As an example we can take $c(t)$ as the PN sequence in Example 4.8-1. The hop set size L is seven, and converting the 3-bit binary 'XYZ' in Fig. 4.35 in the natural way, we get the pseudorandom frequency sequence $\{4, 6, 7, 3, 5, 2, 1\}$. With slow hopping, $T_c \geq T$ and W_{FH} is 7-fold expanded, no matter what the chip time T_c is.[11]

[10]For simplicity, we assume that ω_i is the final RF carrier radian frequency; this implies that the modulator produces a baseband output. A practical FH system would have a number of intermediate frequency steps.

[11]With $T_c < T$ and very fast hopping, a Fourier calculation shows that the bandwidth expansion is greater than L, and in the limit it tends to the DS-SS value T/T_c.

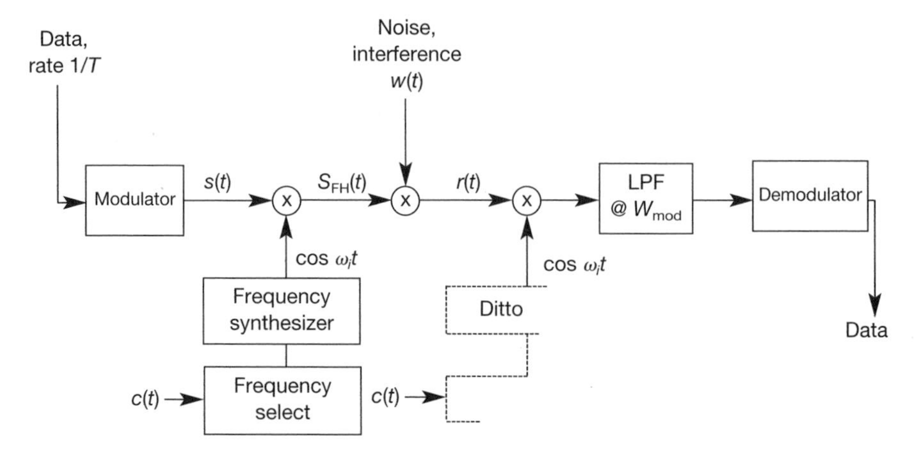

Figure 7.24 Essential functions of an FH-SS system. With square pulses, the demodulator can be that in Fig. 7.21.

With either a long sequence, a wide-ranging sequence, or rapid hopping, it is difficult for an adversary to learn and follow the sequence. Much of the technology in a FH-SS system comes down to the synthesizer; the party with the fastest one cannot be tracked by others and has a measure of security. Otherwise, the FH-SS system is just a straightforward narrowband modulator/demodulator. Synchronization demands are much less in FH-SS. With moderate frequency hopping these are implementation advantages over DS-SS.

A slow FH system can employ a coherent BPSK or other modulation, since there is time to lock to the carrier during each hop. Fast FH probably needs an incoherent basic modulation. A favorite choice is M-ary incoherent wideband FSK, case 1 in Section 3.6.1. Often the FSK is nonbinary: Since a frequency synthesizer is used in any case, it is natural to switch the FSK among M frequencies during each chip. The system thus carries data at the rate of $\log_2 M/T$ bit/s. The FSK baseband bandwidth is $\approx M/T$ (assuming an index h near 1—see Fig. 3.24a), and the entire FH-SS RF bandwidth is $\approx L(M/T)$ hertz. For example, an $R = 1$ kbit/s data transmission via 16-ary FSK with the length-7 PN hop sequence will have a symbol time $T = \log_2 M/R = 4/1000 = 0.004$ ms and a total RF bandwidth of $M/T_c = 16/(\frac{1}{7}T) = 28$ kHz. The incoherent M-ary demodulator looks for the presence or absence of energy in one of M bands. A detector of this sort is called a *radiometer,* because it measures radio energy.

An FH-SS error probability based on on-off signaling in AWGN can be computed. Under pulse-noise jamming with power P_J spread over W_J set to the basic FSK bandwidth, noise density N_J results and jammed FSK error formulas similar in form to (7.6-9)–(7.6-10) can be derived. As with DS-SS, FH-SS forces the jammer to cover a bandwidth W_{FH} rather than the FSK bandwidth, and the noise density seen at the demodulator is reduced by the hop set size. Under additive noise jamming, then, the FH-SS processing gain is $G_p \approx 1/L$.

Error Probability Based on Collisions. The dominant error sources in FH-SS are ordinarily jamming and other-user multiple access interference, not Gaussian noise. The error calculation is based on signal collisions. After the mix-down by cos $\omega_i t$, the radiometer simply looks for energy lying above the receiver noise floor. If it finds energy in exactly one of the M FSK bands, detection is successful; several interferors may contribute, but since no other band is occupied, the desired sender must have been one of them. If it finds energy in two or more bands, either can be the sender and a detection failure is declared.

If we assume random hopping sequences over L frequencies by K independent interferors who send random data, what is the probability P_f of this failure? A collision of a given interferor I and the desired signal occurs with probability

$$P[I \text{ misses FSK frequency} | I \text{ hits hop channel}]\ P[I \text{ hits hop channel}]$$
$$= \frac{M-1}{M}\ \frac{1}{L}$$

For successful transmission, this must fail to happen K independent times, so that

$$P_f = 1 - \left(1 - \frac{M-1}{ML}\right)^K \tag{7.6-18}$$

A tight approximation is given by

$$\begin{aligned} P_f &= 1 - \left(1 - \frac{M-1}{ML}\right)^{\frac{ML}{M-1}\ \frac{K(M-1)}{LM}} \\ &\approx 1 - \exp\left[-(K/L)\frac{M-1}{M}\right] \end{aligned} \tag{7.6-19}$$

The collision probability is thus set by the ratio of interferors to hop set size, K/L. This differs from DS-SS multiple-access, where error probability could be improved by increasing power, providing $K < L - 1$.

As an example, take a binary FSK system $K = 5$ interferors and hop set size $L = 20$. Equation (7.6-18) yields $P_f = 0.119$, and the approximation (7.6-19) is 0.118. To reduce P_f to 0.01 requires size 249. This performance is not good, but parity-check coding improves it rapidly with little expense. The collisions count as erasures, not errors, and parity-check decoders correct up to $d - 1$ erasures, where d is the Hamming minimum distance.

Other Spread Spectrum Methods. The frequency hop and direct sequence ideas can be combined to form a hybrid system with desirable properties. Systems with very large spreading (100–1000) will typically be formed in this way. Frequency hop systems can transmit more than one tone at once; this is called independent-tone FH.

There is a large literature on combining error-correcting codes with spread spectrum. Request-repeat coded transmission (Section 6.5) works well because many systems know when there has been interference (e.g., FH-SS) and can ask for repeats until successful transmission occurs. Low-rate parity-check coding by itself achieves an effect similar to DS spreading; rate $1/L$ achieves L-fold spreading, and the error correction ability of the code combats the interference.

PROBLEMS

MLSE Performance and Minimum Distance

7-1. Prove that the databit-normalized square minimum distance among the four signals $\{\pm\sqrt{2/10}h(t), \pm 3\sqrt{2/10}h(t)\}$ is 0.8 for any $h(t)$. (This is the quaternary antipodal signal distance.)

7-2. *d_0 for general M.* Consider ordinary PAM transmission of the form $\Sigma a_n v(t - nT)$ with T-orthogonal v and modulation symbols a_n from an M-ary equispaced alphabet. Show that the matched filter bound is

$$d_0^2 = \frac{6 \log_2 M}{M^2 - 1}$$

7-3. Consider Eq. (7.1-8). It computes the distance between two continuous signals in an ISI channel whose total response is $h = v * g$. Give some conditions under which (7.1-8) leads to the same number with different pulses $v(t)$; that is, the number depends only on $g(t)$.

7-4. *Worst-case ISI.* Consider the three-tap ISI model $h_0 + h_1 z^{-1} + h_2 z^{-2}$. Show that the worst MLSE benchmark distance for this model lies about 2.3 dB below the antipodal signaling bound. (*Suggestion:* Find the worst error difference sequence and express its distance as a function of h_0, h_1, h_2; then optimize. The result here first appeared in Ref. 27).

7-5. *One-pole model.* In Example 7.1-3 is presented the normalized one-pole discrete-time model

$$H_D(z) = \frac{\sqrt{1-\alpha^2}}{1-\alpha z^{-1}} = \sqrt{1-\alpha^2} \sum_{k \ge 0} \alpha^k z^{-k}$$

The MLSE method has error rate $\approx Q(\sqrt{d_{\min}^2 E_b/N_0})$, where $d_{\min}^2$ is the normalized square minimum distance for this model.

(a) For $0 < \alpha < 1$, find $d_{\min}^2$. In particular, show that $d_{\min}^2 = 2$ for $\alpha \le 0.5$. (Demonstrate this either by analytical proof or by trying out a few symbol difference sequences with Program 7.1-1. For further discussion, see Ref. 8.)

(b) Let $C(z)$ be the zero-forcing equalizer for $H_D(z)$. Find the noise enhancement caused by $C(z)$.

(c) For $0 < \alpha < 1$, what is the asymptotic decibel loss of zero forcing compared to MLSE?

Discrete-Time Modeling

7-6. Let the multipath in a UHF fading channel be modeled by the baseband channel response $g(t) = u(t) - u(t - 0.000001)$—that is, by a square-wave response of length 1 μs.

(a) Assuming 30% root RC pulses and symbol rate 1 Mb/s, find the PAM receiver ISI model for this multipath. In principle, can the effect of this ISI on the asymptotic error probability be removed?

(b) Repeat for a 500-kHz symbol rate.

7-7. Repeat the previous problem for the MF receiver model of the ISI.

7-8. Repeat Example 7.1-4 for ISI created by a two-pole Butterworth filter, instead of the six-pole filter in the example. That is, show how much the outcome changes when a channel filter with the same cutoff but a weak spectral rolloff is used.

7-9. A PAM Model is to be used to describe in discrete time the receiver–transmitter impulse response given by

$$h(t) = \begin{cases} 1, & -T/2 < t \leq T/2 \\ -0.5, & T/2 < t \leq 3T/2 \\ 0, & \text{otherwise} \end{cases}$$

The modulation is NRZ BPSK, so that the standard square pulse

$$v(t) = \begin{cases} 1, & -T/2 < t \leq T/2 \\ 0, & \text{otherwise} \end{cases}$$

carries data symbols a_n according to the linear modulation formula $\Sigma a_n v(t - nT)$. Due to intersymbol interference this pulse arrives as $\Sigma a_n h(t - nT)$.

(a) Find the PAM model $H(z)$. Assume that the PAM receiver samples at times nT, n an integer. Confirm that your model has a stable inverse.

(b) Give the standard zero-forcing equalizer $C(z)$.

7-10. *Timing offset.* Symbol timing offset in a receiver causes ISI. To investigate this, consider binary linear modulation with 30% root RC pulses $v(t - nT)$. The receiver has a filter matched to $v(t)$ followed by a sampler that samples late by $T/10$. That is, the samples are $y(nT + T/10)$, $n = 0, 1, \ldots$, where $y = v * v$ is an RC pulse. There is no other source of ISI.

(a) For a single transmission symbol +1, find the first three samples (use RC pulse programs in Chapter 2).

(b) Give the normalized discrete-time model $H(z)$.

(c) Design a zero-forcing equalizer.

Equalizers

7-11. Consider the two-tap ISI model $H(z) = h_0 + h_1 z$.

(a) Demonstrate that the binary MLSE benchmark $d^2_{\min}$ is 2, for any choice of h_0 and h_1.

(b) For which h_0 and h_1 does $H(z)$ have a stable inverse, so that zero forcing is a reasonable option?

(c) For the cases in (b), give the noise power enhancement factor. Be sure $H(z)$ is normalized.

7-12. Consider the three-tap model (before normalizing) $H(z) = 1 + 0.9z^{-1} + 0.9z^{-2}$.

(a) Show that a straightforward zero-forcing linear equalizer is stable for this model.

(b) Find the equalizer; find its noise enhancement.

(c) What error performance can be expected here?

(d) What is the MLSE benchmark error performance?

7-13. The discrete-time model of a transmission channel has z-transform

$$H(z) = \frac{K}{1 - z^{-2}/4}$$

in which K is a factor that normalizes the model to unit energy. The impulse response of this model, found by MATLAB or by hand, is the sequence $K \times \{1, 0, \frac{1}{4}, 0, (\frac{1}{4})^2, 0, (\frac{1}{4})^3, 0, \ldots\}$. You are to design a zero-forcing equalizer for this channel model.

(a) Find the normalizing factor K and show that $H(z)$ is stable.

(b) Find the standard zero-forcing equalizer $C(z)$.

(c) Find the noise enhancement factor for $C(z)$ in decibels.

(d) A short sequence made from the symbols ± 1 or 0 is sent through the model $H(z)$. The result is the sequence $\{-0.9682, 0, 1.2103, 0, 0.3026, 0, \ldots\}$. It is applied to the equalizer. Find the first three symbols that were transmitted.

7-14. Design a good least square-error linear equalizer for the model in the previous problem. Compare its performance to that of the zero-forcing equalizer. (For this and the following LS equalizer designs, use Appendix 7A).

7-15. Consider the three-tap model $H(z) = 1 + 0.5z^{-1} + 0.25z^{-2}$ that was featured in Section 7.2.

(a) Design a good least-squares equalizer and estimate its asymptotic error probability using (7.2-18).

(b) What error probability is expected for the zero-forcing equalizer in 7.2?

(c) Using a few test symbol difference sequences and Program 7.1-1, estimate the asymptotic probability for MLSE.

7-16. *Tandem channels.* A tandem channel is one in which a smaller channel unit repeats one or more times. Suppose a discrete-time unit with z-transform $H(z) = \sqrt{4/5}[1 + \frac{1}{2}z^{-1}]$ repeats twice to form the tandem channel $[H(z)]^2$.

(a) Give the normalized z-transform $H_T(z)$ for the tandem channel. Prove or disprove: The zero-forcing equalizer for this channel is stable.

(b) Find the first eight coefficients $c_0, \ldots, c_7$ of $C_T(z)$, the zero-forcing equalizer for $H_T(z)$.

(c) Considering these eight, what is the noise enhancement factor for this equalizer? Give a Q-function estimate of the error probability. (Note that for this part to make sense, the entire noise $\frac{1}{2}N_0$ must come between the tandem channel H_T and the equalizer C_T.)

(d) For the impulse input 1, 0, 0, . . . , demonstrate that the first eight outputs of the system $H_T(z)C_T(z)$ are 1, 0, 0, ... This demonstrates zero forcing.

(e) By trying a few simple symbol difference sequences, estimate $d^2_{\min}$ for the tandem model H_T.

7-17. *Tandem equalizers.* In the previous problem, the elements in the binary-transmission channel model came in the following order: tandem channel H_T—AWGN—equalizer C_T, where C_T is the zero-forcing equalizer for H_T. In a link with repeaters it would be more natural that the elements came in the order

$$H(z)\text{—AWGN—}H(z)\text{—AWGN—}C_T(z)\text{—[Decide symbol]}$$

or

$$H(z)\text{—AWGN—}C(z)\text{—}H(z)\text{—AWGN—}C(z)\text{—[Decide symbol]}$$

In the first, C_T equalizes two $H(z)$'s. In the second, each tandem unit has its own equalizer $C(z)$; here we say the equalizers are in tandem. Assume in what follows that H and H_T are normalized and that AWGN means the addition of white noise with variance $\frac{1}{2}N_0$.

(a) Derive the zero-forcing equalizer $C_T(z)$, if it is not already available. Then compute the noise enhancement factor for the first model above and give a Q-function expression for the error probability in terms of E_b/N_0.

(b) Derive the zero-forcing equalizer $C(z)$ for the model $H(z)$. Then compute the noise enhancement factor for the second model above and give a Q-function expression for the error probability. Which equalization approach is the preferred one?

Fading and Noncoherent Reception

7-18. Show that the Rayleigh fading error probability $P_{\text{BPSK,f}}$ in (7.4-5) tends to the large-$\overline{\gamma}$ value as the average channel energy $\overline{\gamma}$ grows.

7-19. **(a)** How much more average energy is required for BPSK to achieve error probability 0.00001 in a Rayleigh channel compared to an AWGN channel?

(b) Estimate the diversity that is required to reduce this by 10 dB.

7-20. For passband signals $r(t)$ and $s_k(t)$, show that integral (7.5-4) holds in the limit of large carrier frequency.

7-21. Show that the receiver in Fig. 7.19 cannot distinguish among signals in the signal set which are phase shifts of each other; or equivalently, show that the receiver cannot distinguish among phase shifts of the received signal $r(t)$.

Spread Spectrum

7-22. Find the power spectrum of the length-7 PN sequence in Example 4.8-1 and compare it to the theoretical power spectral density for the infinite-length random telegraph signal given in (7.6-6). *Hint:* Let the symbol time be $T = 1$. Use MATLAB or a similar engine to find the Fourier transform. Remember that the power spectrum is $1/7T$ times the energy spectrum [see the discussion after Eq. (2.3-3)].

7-23. Carry out the derivation of the pulse-noise jammed receiver error probability, Eq. (7.6-10).

7-24. A DS-SS system is to be designed to block a Gaussian pulse-noise jammer. The modulation is BPSK with spreading factor 10. The jammer can apply P_J over a before-spreading modulator bandwidth W_{mod} in such a way that the noise density in that bandwidth becomes N_J instead of the original N_0.

(a) Adapt the optimization of (7.6-9) in Section 7.6.1 to the DS-SS case, finding the optimum fraction δ and the new bit error probability. Plot the probability versus the one that applies when there is no spread spectrum protection.

(b) Compare the optimum fractions δ with and without the DS-SS system.

(c) What is the processing gain of this DS-SS system?

7-25. In the DS-SS multiple-access system based on a length-7 PN sequence in Section 7.6.2, full synchronization of all users was assumed. Assume now that only each *pair* of users has spreading sequence synchronization. In particular, assume that user pair 2's chips are delayed $\frac{1}{2}T_c$ compared to user pair 1's, and pair 3's are delayed $\frac{2}{3}T_c$. Recompute the correlations in (7.6-17). How are they affected by the lack of synchronization among user pairs?

7-26. An FH-SS system is based on incoherent binary FSK, with 100 orthogonal hop set bands, 1000 hops per second, and a user data rate of 100 kbit/s. A user pair transmits in only one direction at a time (simplex mode). Hop sequences are random and mutually independent. Take the positive-frequency RF bandwidth of the FSK as $2/T$ Hz, with T the symbol time.

(a) Making reasonable assumptions, estimate the total bandwidth of the system.

(b) Suppose three users transmit at once in multiple-access mode with perfect power control. Use the collision method in Section 7.6.3 to calculate the bit error rate that each one sees.

(c) Now suppose there is just one user, but a pulse-noise jammer tries to disrupt the transmission. Without jamming, the user has error rate $Q(\sqrt{E_b/N_0})$ with $E_b/N_0 =$ 10 dB. The jammer has available enough Gaussian jamming power P_J so that it can reduce this SNR at the receiver to 0 dB if it knows the hop position; that is, it can place $P_J = 10N_0\,(2/T)$ watts at the receiver. But the jammer does not know the hopping sequence. How should the jammer best spread its power? All in one $2/T$-wide hop position? Half of it each in two positions? More widely?

REFERENCES[12]

1. *J. G. Proakis, *Digital Communications,* 3rd ed., McGraw-Hill, New York, 1995.
2. D. G. Messerschmitt, A geometric theory of intersymbol interference—Parts I and II, *Bell Syst. Tech. J.,* vol. 52, November 1973, pp. 1483–1539.
3. *E. A. Lee and D. G. Messerschmitt, *Digital Communication,* 2nd ed., Kluwer, Boston, 1994.
4. G. D. Forney, Jr., Maximum-likelihood sequence estimation of digital sequences in the presence of intersymbol interference, *IEEE Trans. Information Theory,* vol. IT-18, May 1972, pp. 363–378.
5. G. Ungerboeck, Adaptive maximum-likelihood receiver for carrier-modulated data transmission systems, *IEEE Trans. Commun.,* vol. COM-22, May 1974, pp. 624–636.
6. *S. Qureshi, Adaptive equalization, *Proc. IEEE,* vol. 73, September 1985, pp. 1349–1387.
7. G. E. Bottomley and S. Chennakeshu, Unification of MLSE receivers and extension to time-varying channels, *IEEE Trans. Commun.,* vol. COM-46, April 1998, pp. 464–472.
8. N. Seshadri and J. B. Anderson, Decoding of severely filtered modulation codes using the (M,L) algorithm, *IEEE J. Sel. Areas Commun.,* vol. SAC-7, August 1989, pp. 1006–1016; *see also* Asymptotic error performance of modulation codes in the presence of severe intersymbol interference, *IEEE Trans. Information Theory,* vol. IT-34, September 1988, pp. 1203–1216.
9. A. Said and J. B. Anderson, Bandwidth-efficient coded modulation with optimized linear partial-response signals, *IEEE Trans. Information Theory,* vol. IT-44, March 1998, pp. 701–713.
10. A. Said, Design of optimal signals for bandwidth-efficient linear coded modulation, Ph.D. Thesis, Department of Electical, Computer and Systems Engineering, Rensselaer Polytechnic Institute, Troy, NY, February 1994.
11. R. W. Lucky, Automatic equalization for digital communication, *Bell Syst. Tech. J.,* vol. 44, April 1965, pp. 547–588; Techniques for adaptive equalization of digital communication systems, *ibid.,* vol. 45, February 1966, pp. 255–286.
12. T. Berger and D. W. Tufts, Optimum pulse amplitude modulation, Part I: Transmitter–receiver design and bounds from information theory, *IEEE Trans. Information Theory,* vol. IT-13, Janurary 1967, pp. 196–208.
13. J. Salz, Optimum, mean-square decision feedback equalization, *Bell Syst. Tech. J.,* vol. 52, October 1973, pp. 1341–1373.
14. M. E. Austin, Decision feedback equalization for digital communication over dispersive channels, Technical Report No. 437, MIT Lincoln Laboratory, Lexington, MA, August 1967.
15. D.A. George, D. R. Bowen, and J. R. Storey, An adaptive decision-feedback equalizer, *IEEE Trans. Commun. Tech.,* vol. COM-19, June 1971, pp. 281–293.
16. P. Monsen, Feedback equalization for fading dispersive channels, *IEEE Trans. Information Theory,* vol. IT-17, January 1971, pp. 56–64.
17. J. Proakis, Advances in equalization for intersymbol interference, in *Advances in Communications Systems,* Vol. 4, A. Viterbi, ed., Academic, New York, 1975.

[12]References marked with an asterisk are recommended as supplementary reading.

18. K. Balachandran, Mismatched receivers for linear coded modulation, Ph.D. Thesis, Dept. Electrical, Computer and Systems Engineering, Rensselaer Polytechnic Institute, Troy, NY, Aug. 1996.
19. B. Widrow and M. E. Hoff, Jr., Adaptive switching circuits, *Conv. Rec.,* Part 4, IRE WESCON, August 1960, pp. 96–104.
20. *B. Sklar, Rayleigh fading channels in mobile digital communication systems, Parts I and II, *IEEE Commun. Mag.,* vol. 35, September 1997, pp. 136–155.
21. S. Chennakeshu and J. B. Anderson, Error rates for Rayleigh fading multichannel reception of MPSK signals, *IEEE Trans. Commun.,* vol. COM-43, February/March/April 1995, pp. 337–346.
22. *M. Schwartz, W. R. Bennett, S. Stein, *Communications Systems and Techniques,* McGraw-Hill, New York, 1966; reissued by IEEE Press, New York, 1995.
23. M. K. Simon, S. M. Hinedi, W. C. Lindsey, *Digital Communication Techniques,* Prentice-Hall, Englewood Cliffs, NJ, 1995.
24. J. B. Anderson, T. Aulin, and C.-E. Sundberg, *Digital Phase Modulation,* Plenum, New York, 1986.
25. W. P. Osborne and M. B. Luntz, Coherent and noncoherent detection of CPFSK, *IEEE Trans. Commun.,* vol. COM-22, August 1974, pp. 1023–1036.
26. D. G. Brennan, Linear diversity combining techniques, *Proc. IRE,* vol. 47, June 1959, pp. 1075–1102.
27. F. R. Magee, Jr. and J. G. Proakis, An estimate of the upper bound on error probability for maximum-likelihood sequence estimation on channels having a finite-duration pulse, *IEEE Trans. Information Theory,* vol. IT-19, September 1973, pp. 699–702.
28. J. B. Anderson and A. Svensson, *Coded Modulation Systems,* Plenum/Kluwer, New York, 2003.
29. *R. E. Ziemer and R. L. Peterson, *Digital Communications and Spread Spectrum Systems,* Macmillan, New York, 1985.
30. R. C. Dixon, *Spread Spectrum Techniques,* IEEE Press, New York, 1976.
31. J. K. Holmes, *Coherent Spread Spectrum Systems,* Wiley-Interscience, New York, 1982.
32. S. Hara and R. Prasad, Overview of multicarrier CDMA, *IEEE Commun. Mag.,* vol. 35, December 1997, pp. 126–133.
33. E. H. Dinan and B. Jabbari, Spreadng codes for direct sequence CDMA and wideband CDMA cellular networks, *IEEE Commun. Mag.,* vol. 36, September 1998, pp. 48–54,.
34. F. Adachi, M. Sawahashi and H. Suda, Wideband DS-CDMA for next-generation mobile communications systems, *IEEE Commun. Mag.,* vol. 36, September 1998, pp. 56–96.
35. T. S. Rappaport, *Wireless Communications,* Prentice-Hall PTR, Upper Saddle River, NJ, 1996.
36. R. Pandya, *Mobile and Personal Communication Systems and Services,* IEEE Press, New York, 2000.

APPENDIX 7A: LEAST-SQUARE ERROR EQUALIZERS

Here we derive equalizers that minimize the mean-square error in the estimated transmission symbol due to both noise and ISI. Short programs are given that compute the equalizer taps and performance estimates. Two receiver types are considered, the PAM model and the matched filter (MF) model; these are described in Section 7.1. The equalizers here are all finite response. The discussion is mostly about linear equalizers, but the programs also extend to LS concept to the nonlinear feedback equalizer. Throughout, E_s is set equal to 1, and the SNR is thus $1/N_0$. The equalizer designs work for nonbinary PAM signals, although the residual ISI factor κ_{wc} and SNR loss calculations are correct only for the binary case.

Derivation of the LS Equalizer for the PAM Model

Let the discrete-time channel be represented by the length-P model $H_D(z)$ with impulse response $h_0, \ldots, h_{P-1}$ and white noise variates $\{\eta_n\}$ according to

$$r_n = \sum_{j=0}^{P-1} h_j\, a_{n-j} + \eta_n \tag{7A-1}$$

as shown in Fig. 7.6. The present time is n. The transmission symbols $\{a_n\}$ are IID with $\mathcal{E}[|a_n|^2] = 1$. The linear equalizer forms an estimate u_{n-D} for the a_{n-D} that is D symbols earlier, by means of the length-L finite-response filter

$$u_{n-D} = \sum_{k=0}^{L-1} c_k r_{n-k} \tag{7A-2}$$

We wish to choose the set of taps $\{c_k\}$ that minimizes the mean-square error

$$J = \mathcal{E}[|a_{n-D} - u_{n-D}|^2] = \mathcal{E}\left[\left|a_{n-D} - \sum_{k=0}^{L-1} c_k\, r_{n-k}\right|^2\right] \tag{7A-3}$$

where the expectation is over the symbols and the AWGN. This is done by setting each partial derivative of (7A-3) with respect to $c_0, \ldots, c_{L-1}$ to zero, thus generating L equations in these L unknowns. For each k this yields

$$\frac{\partial \mathcal{E}[\cdot]}{\partial c_k} = 2\mathcal{E}\left[\left(a_{n-D} - \sum_{j=0}^{L-1} c_j r_{n-j}\right) r_{n-k}\right] = 0 \tag{7A-4}$$

which gives the set of L equations

$$\mathcal{E}[a_{n-D} r_{n-k}] = \sum_{j=0}^{L-1} c_j \mathcal{E}[r_{n-j} r_{n-k}], \qquad k = 0, \ldots, L-1 \tag{7A-5}$$

These equations can be reduced to a compact matrix form. First take the left-hand side. We have

$$\begin{aligned} \mathcal{E}[a_{n-D} r_{n-k}] &= \mathcal{E}\left[a_{n-D}\left(\sum_{j=0}^{P-1} h_j\, a_{n-k-j} + \eta_{n-k}\right)\right] \\ &= \begin{cases} \mathcal{E}[|a_{n-D}|^2] h_{D-k} = h_{D-k}, & \text{if } 0 \le D-k \le P-1 \\ 0, & \text{otherwise} \end{cases} \end{aligned} \tag{7A-6}$$

since the symbols and noise are mutually independent and have mean zero. For $k = 0, \ldots, L-1$, we can arrange all of these in a length-L vector $\boldsymbol{q}$,

$$\boldsymbol{q} = (0, \ldots, 0, h_{P-1}, \ldots, h_0, 0, \ldots, 0) \qquad (7A\text{-}7)$$
$$|\leftarrow \quad D \quad \rightarrow|\leftarrow L\text{–}D \rightarrow|$$

If $D < P$, some h_j are truncated at the left. For the right-hand side of (7A-5), the expectation is the covariance matrix $\boldsymbol{R}$ of the stationary sequence $\{r_n\}$. The (ℓ, k) component of $\boldsymbol{R}$ depends only on $|\ell - k|$ and is

$$\begin{aligned} \boldsymbol{R}_{\ell,k} &= \mathcal{E}[r_{n-\ell} r_{n-k}] \\ &= \mathcal{E}\left[\left(\sum_{i=1}^{P-1} h_i a_{n-\ell-i} + \eta_{n-\ell}\right)\left(\sum_{j=1}^{P-1} h_j a_{n-k-j} + \eta_{n-k}\right)\right] \\ &= \sum_{i=0}^{P-1-|k-\ell|} h_i\, h_{i+|k-\ell|} + \mathrm{cov}(\eta_{n-\ell}, \eta_{n-k}) \qquad 0 \le \ell \le L-1, 0 \le k \le L-1 \end{aligned} \qquad (7A\text{-}8)$$

Since $\{\eta_n\}$ are IID, $\mathrm{cov}(\eta_{n-\ell}, \eta_{n-k}) = \frac{1}{2}N_0\boldsymbol{I}$, with $\boldsymbol{I}$ the identity matrix; note that the noise and symbols are independent as well, which leads to the separated terms in (7A-8).

With this reduction, the set of equations (7A-5) becomes

$$\boldsymbol{q} = \boldsymbol{R}\boldsymbol{c}_{\mathrm{opt}} \qquad (7A\text{-}9)$$

with $\boldsymbol{R}$ an $L \times L$ matrix and $\boldsymbol{c}_{\mathrm{opt}}$ the error minimizing equalizer solution. Some substitutions show that the least mean-square estimation error is

$$J_{\min} = 1 - \sum_{i=0}^{L-1} c_i q_i \qquad (7A\text{-}10)$$

The residual ISI can reduce the symbol estimate by as much as (7.2-17) (binary symbol case) and the noise variance $\mathrm{var}(w_n)$ is given by (7.2-8).

LS Equalizer for the MF Model

The derivation of this equalizer proceeds as above except that the noise sequence $\{\xi_n\}$ is colored Gaussian with autocorrelation sequence $\{\rho_k\}$ [see (7.1-13) and the following comments]. Note that the $\{\xi_n\}$ are still independent of $\{a_n\}$. Thus the covariance in (7A-8) becomes

$$\boldsymbol{R}_{\ell,k} = \mathrm{cov}(\eta_{n-\ell}, \eta_{n-k}) = \tfrac{1}{2}N_0\, \rho_{|\ell-k|}, \qquad 0 \le \ell \le L-1, \qquad 0 \le k \le L-1 \quad (7A\text{-}11)$$

The residual ISI of the MF model case is found from $H_D(z)C(z)$ as in the PAM model case, but the noise enhancement is now more complicated, since the white channel noise passes through both the matched filter and the equalizer. The variance of the noise output of $C(z)$ in Fig. 7.6 is

$$\begin{aligned}\text{var}(w_n) &= \mathcal{E}\left[\left(\sum_{i=0}^{L-1} c_i \xi_{n-i}\right)\left(\sum_{j=0}^{L-1} c_j \xi_{n-j}\right)\right] \\ &= \mathcal{E}[\eta_n \eta_n] \sum_{i=0}^{L-1} |c_i|^2 + 2\mathcal{E}[\eta_n\ \eta_{n-1}] \sum_{i=0}^{L-2} c_i c_{i+1} + 2\mathcal{E}[\eta_n \eta_{n-2}] \sum_{i=0}^{L-3} c_i c_{i+2} + \cdots\end{aligned}$$

From (7A-11), $\mathcal{E}[\eta_n \eta_{n-\ell}] = \frac{1}{2} N_0 \rho_\ell$. Some manipulation shows that $\text{var}(w_n)$ is the dot product

$$\text{var}(w_n) = \tfrac{1}{2} N_0\ \boldsymbol{\rho} \cdot (\boldsymbol{c}_{\text{rev}} * \boldsymbol{c}) \tag{7A-12}$$

in which $\boldsymbol{c}_{\text{rev}}$ is the equalizer coefficients $\boldsymbol{c} = (c_0, \ldots, c_{L-1})$ reversed. The vector $\boldsymbol{c}_{\text{rev}} * \boldsymbol{c}$ here is of length $2L - 1$ and $\boldsymbol{\rho}$ is the length $2L - 1$ autocorrelation vector

$$\boldsymbol{\rho} = (\rho_{L-1}, \ldots, \rho_1, 1, \rho_1, \ldots, \rho_{L-1})$$

The vector $\boldsymbol{\rho}$ is precisely the $2L - 1$ tap model, as given by (7.1-12). If the model is taken shorter than $2L - 1$, then some of the outer values here are set to zero.

Programs to Compute LS Equalizers

Program 7.2-1: PAM Model LE-LS and DFE-LS Equalizers. The following MATLAB program finds the least mean-square error linear equalizer when the PAM model is the case. A few additional lines are given that find the DFE-LS feedback taps. The program also computes the mean-square estimation error, the equalizer noise enhancement factor $F_{\text{NE}} = \text{var}(w_n)/\frac{1}{2}N_0$ via (7.2-8), the worst-case residual ISI factor κ_{wc} from (7.2-17), and the E_b/N_0 degradation factor $\kappa^2_{\text{wc}}/F_{\text{NE}}$ in (7.2-18). The last is expressed in decibels and can be thought of as the loss of E_b/N_0 in the link compared to transmission of antipodal signals. Note that the AWGN channel variates have true variance $\frac{1}{2}N_0$; if $E_s = 1$, the true SNR that applies to antipodal signaling is $2/N_0$. If the LE-LS equalizer effectively removes all the ISI, the mean-square error approaches this latter SNR. Finally, a negative residual ISI factor may occur and means that the equalizer "eye" is closed; correct detection may still occur most of the time if the mean-square error is small.

The formal inputs to the program are:

`h`—the model tap vector

`no`—the noise density factor

`leq`—the equalizer length L

`del`—the delay D of the estimate, with $D < L$

The outputs are:

c—the LS equalizer tap vector
q—$\boldsymbol{q}$ in (7A-7)
ms_error—the mean-square error $J_{\min}$
noise_en—noise enhancement factor F_{NE}
res_ISI—the residual ISI factor κ_{wc}
SNR_loss—the asymptotic E_b/N_0 loss, compared to antipodal signaling (in dB)

The program is as follows.

```
% Set up; normalize h
lh = length(h); h = h/sqrt(sum(h.^2)); hrev = h(lh:-1:1);
        % Form q vector
if del > lh,
  q = [zeros(1,del-lh) hrev zeros(1,leq-del)]
else
  q = [hrev(lh-del+1:1:lh) zeros(1,leq-del)], end
        % Form covariance matrix R and find c
for i = 1:leq,                                  %Preliminary matrix
  hmat(:,i) = [zeros(1,i-1) h zeros(1,leq-i)]'; end
r = hmat'*hmat + (no/2)*eye(leq);                  %Find R
c = inv(r)*q'                                      %Find c
        % Calculate parameters
ms_error = 1-sum(c'.*q)                         %Find ms error
noise_en = sum(c.^2)                            %Find noise enhancement
cc = conv(c,h); [mtap,idx] = max(cc); cc(idx) = 0;
res_ISI = mtap-sum(abs(cc))                     %Find resid. ISI factor
SNR_loss = 20*log10(sqrt(noise_en)/res_ISI)    %Find SNR loss (dB)
```

The following lines may be added in order to generate the DFE-LS equalizer feedback section taps FB_taps, as discussed in Section 7.3.1. The feedforward section taps for this equalizer are the ones given in c. Note that a DFE is allowed only if the delay del is one less than the equalizer length leq.

```
           % Find DFE equalizer
cc = conv(h,c); FB_taps = cc(del+2:del+lh);     %Select out taps
if del == leq-1,
  FB_taps, end                                  %Print taps if legal
```

Example: Compute the exponential ISI LE-LS equalizer in Example 7.2-2. Set

```
h = [-.088 .568 .733 .268 .174 .073];
no = .01;
leq = 19;
del = 3;
```

This yields the c shown in Fig. 7.9a. The remaining output is

```
ms_error = .040
noise_en = 6.59
 res_ISI = .694
SNR_loss = 11.37
```

Program 7.2-2: MF Model LE-LS and DFE-LS Equalizers. The next MATLAB program finds the least-square error linear equalizer when the MF receiver model is the case. Some additional lines are given that find the feedback taps for the DFE-LS equalizer. The input and output of the Program are the same as Program 7.2-1, except that the model input is the autocorrelation of $\boldsymbol{h}$, $1, \rho_1, \ldots, \rho_{Q-1}$, denoted rho, instead of h. The program operation differs in that covariance (7A-11) is used to find $\boldsymbol{R}$ and the noise enhancement factor comes from (7A-12) rather than (7.2-8); as well, certain symmetry properties are used.

The program is:

```
        % Set up
lr = length(rho); lh = 2*lr-1;
toep = [rho zeros(1,leq-lr)];                   %For autocorrelation
h = [rho(lr:-1:2) rho];                         %The actual model h
        % Form q vector
if del > lh,
 q = [zeros(1,del-lh) h zeros(1,leq-del)]
else
 q = [h(lh-del+1:1:lh) zeros(1,leq-del)], end
        % Form covariance matrix R and find c
for i = 1:leq,                                  %Preliminary matrix
 hmat(:,i) = [zeros(1,i-1) h zeros(1,leq-i)]'; end
r = hmat'*hmat + (no/2)*toeplitz(toep);         %Find R
c = inv(r)*q'                                   %Find c
        % Calculate parameters
ms_error = 1-sum(c'.*q)                         %Find ms error
cc = conv(c(leq:-1:1),c);
noise_en = sum([toep(leq:-1:2) toep].*cc)       %Find noise enhancement
cc = conv(c,h); [mtap,idx] = max(cc); cc(idx) = 0;
res_ISI = mtap-sum(abs(cc))                     %Find resid. ISI factor
SNR_loss = 20*log10(sqrt(noise_en)/res_ISI)     %Find SNR loss (dB)
```

The following lines may be added in order to generate the DFE-LS equalizer feedback section taps FB_taps, as discussed in Section 7.3.1. As before, the feedforward taps are those given in c and del must equal leq −1.

```
        % Find DFE equalizer
cc = conv(h,c); FB_taps = cc(del+2:del+lh);     %Select out taps
if del == leq-1,
  FB_taps, end                                  %Print taps if legal
```

Example: Compute the six pole Butterworth LE-LS equalizer in Example 7.2-1. The equalizer has 9 taps and the $\{\rho_k\}$ come from Example 7.1-4. Set

```
rho = [1 .075 -.061 .0435]
no = .1
leq = 9
del = 8
```

This yields

```
c = [.011 -.051 .070 -.086 .977 -.086 .070 -.051 .011]
```

and

```
ms_error = .049
noise_en = .927
res_ISI = .905
SNR_loss = .54
```

Figure 7.8 illustrates the operation of the LE-LS equalizer. The combination of delay and feedforward length are legal for a DFE-LS; the number of feedback taps will be the length of `h` less 1, and the taps are

```
FB_taps = [-.0039 .0032 -.0013 .0069 -.0028 .0005]
```

INDEX